Your Cosmic Context
An Introduction to Modern Cosmology

Your Cosmic Context

An Introduction to Modern Cosmology

Todd Duncan
Pacific University & Portland State University

Craig Tyler
Fort Lewis College

PEARSON
Addison
Wesley

San Francisco Boston New York
Cape Town Hong Kong London Madrid Mexico City
Montreal Munich Paris Singapore Sydney Tokyo Toronto

Publisher: Jim Smith
Executive Editor: Nancy Whilton
Project Editor: Katherine Brayton
Development Manager: Michael Gillespie
Executive Marketing Manager: Scott Dustan
Managing Editor: Corinne Benson
Production Supervisor: Shannon Tozier
Production Management and Composition: Laura Houston, Pre-PressPMG
Interior Design: Pre-PressPMG
Cover Design: Derek Bacchus
Illustrations: Dartmouth Publishing, Inc.
Director, Image Resource Center: Melinda Patelli
Photo Researcher: Elaine Soares
Image Permissions Coordinator: Michelina Viscusi
Image Rights and Permissions Manager: Zina Arabia
Manufacturing Buyer: Jeffrey Sargent

Cover Photo Credit: NASA Earth Observing System / Media Services
Credits continue in the back of the text.

Library of Congress Cataloging-in-Publication Data
Duncan, Todd.
   Your cosmic context : an introduction to modern cosmology/Todd Duncan,
     Craig Tyler—1st ed.
   p. cm.
Summary: "Provides a cumulative guide to the general lessons of modern scientific cosmology, as well as the historical back-
   ground that connects the nature of the universe with the reader's place in it"—Provided by publisher.
Includes bibliographical references and index.
   ISBN 978-0-13-240010-7
1. Cosmology. I. Tyler, Craig. II. Title.
QB981.D886 2007
523.1--dc22

                                        2007046370

ISBN-13: 978-0-13-240010-7
ISBN-10: 0-13240010-3

www.aw-bc.com

Contents

About This Book   *viii*

CHAPTER 1   Starting Points   1
1.1   What Do You Wonder About?   1
1.2   Science: A Way to Answer Questions   7
1.3   What Difference Does This Knowledge Make?   16
Reflective Essay 21   ▪   Quick Review 22   ▪   Further Exploration 23

CHAPTER 2   The Sky We See   25
2.1   Looking Up   26
2.2   First Impressions   28
2.3   Enhancing Our Vision   30
2.4   Light Waves   36
2.5   Putting Your Knowledge of Light to Use   48
Reflective Essay 59   ▪   Quick Review 60   ▪   Further Exploration 61

CHAPTER 3   The Universe We Discover
***through Heat and Light*   63**
3.1   Starlight in a Particle World   64
3.2   Standard Candles   73
3.3   The Scale of the Universe   85
Reflective Essay 95   ▪   Quick Review 96   ▪   Further Exploration 97

CHAPTER 4   The Universe We Discover
***through Motion and Gravity*   99**
4.1   Enter Gravity   99
4.2   Stars and Galaxies in Motion   111
4.3   Dark Matter   120
Reflective Essay 131   ▪   Quick Review 132   ▪   Further Exploration 133

CHAPTER 5   Clues about the Cosmos   135
5.1   Redshifts of Galaxies   135
5.2   The Distribution of Galaxies   140
5.3   Microwaves from Every Direction   142

5.4   Common Ingredients in the Universe   147

5.5   Comparing Astronomical Ages   153

5.6   Olbers's Paradox   155

Reflective Essay 158   ▪   Quick Review 159   ▪   Further Exploration 160

CHAPTER 6   The Fabric of Spacetime   162

6.1   The Curved-Space Concept   163

6.2   Hints of a Deeper Gravity Theory   165

6.3   General Relativity   168

6.4   Testing General Relativity with Photons and Gravitons   172

6.5   The Black Hole and the Event Horizon   177

6.6   Quasars   186

Reflective Essay 190   ▪   Quick Review 191   ▪   Further Exploration 192

CHAPTER 7   An Expanding Universe   194

7.1   Spacetime and the Cosmological Horizon   194

7.2   Cosmic Expansion   197

7.3   Recession and Redshift   204

7.4   Dark Energy   209

7.5   Consequences of the Expansion   213

Reflective Essay 215   ▪   Quick Review 216   ▪   Further Exploration 217

CHAPTER 8   Photons and Electrons   219

8.1   Blackbody Radiation   220

8.2   Photons and Bound Electrons   223

8.3   Photons and Free Electrons   230

8.4   The Cosmic Microwave Background   234

8.5   Polarization   241

Reflective Essay 248   ▪   Quick Review 249   ▪   Further Exploration 250

CHAPTER 9   The Nuclear Realm   252

9.1   Energy   253

9.2   Nuclear Interactions   256

9.3   Thermonuclear Fusion in Stars   259

9.4   Heavy Elements and Stellar Genetics   263

9.5   Exotic Particles   273

9.6   Elemental Abundances   280

Reflective Essay 285   ▪   Quick Review 286   ▪   Further Exploration 288

CHAPTER 10   The Big Bang Theory   290

10.1   Overview of the Theory   291
10.2   Expansion, Not Explosion   294
10.3   How the Big Bang Explains Our Observations   297
10.4   Evaluating the Big Bang   309

Reflective Essay 313   ■   Quick Review 314   ■   Further Exploration 316

CHAPTER 11   History, Density, and Destiny   318

11.1   Density of the Universe   318
11.2   History of the Universe   326
11.3   Destiny of the Universe   337

Reflective Essay 342   ■   Quick Review 343   ■   Further Exploration 345

CHAPTER 12   The Story of Structure   347

12.1   Primordial Harmonics   349
12.2   Precision Cosmology   357
12.3   Cold Halos, Galaxies, and the Dark Universe   362
12.4   Inflation Theory   369
12.5   Supplemental Topic: Quantum Fluctuations in Cosmology   376

Reflective Essay 380   ■   Quick Review 381   ■   Further Exploration 382

CHAPTER 13   The Emergence of Complex Life   384

13.1   Supplemental Section: Life on Earth   386
13.2   Life on Other Worlds?   402
13.3   Extrasolar Planets   406
13.4   Intelligence on Other Worlds?   409

Reflective Essay 416   ■   Quick Review 418   ■   Further Exploration 419

CHAPTER 14   What Does It Mean to You?   422

14.1   What Have You Learned?   422
14.2   Anthropic Thoughts   426
14.3   This is *Your* History   432
14.4   So What? Finding Your Cosmic Context   438

Reflective Essay 441   ■   Quick Review 442   ■   Further Exploration 444

Bibliography and Credits   446
Index   453

About This Book

The basic mystery underlying the topic of this book is perhaps the most important of any we can ponder. The simple question "What are we doing here?" is so obvious and fundamental that we rarely remember to think about it directly. What is this universe in which we find ourselves? What is it doing, and how do our individual thoughts, feelings, and actions fit within its context?

The book is intended as a guide to the key insights of modern scientific cosmology—insights that provide a framework for thinking about these sorts of questions. After all, the history of the universe is *your* history as well! Think of this book as your "resident's guide to the universe." It is suitable for independent reading and study, or as a text for a one-semester general education course in cosmology.

Over the last few decades, an avalanche of (almost surprisingly) consistent data has led cosmologists away from a field of relatively open speculation and into a single, unified view of the nature and history of the universe. As it turns out, it's a strange and unexpected universe, flooded with unfamiliar forms of matter and energy. For this reason, cosmology is one of the best available illustrations of the scientific method in action, because if we hold true to the principles of science, then we are compelled to accept a reality that our experience on Earth has not prepared us for. In short, cosmology is a novel and brainy survey spanning your grandest possible context.

In this spirit, we have constructed *Your Cosmic Context* around the following principles:

Questions precede answers and observations precede theory   As a reader, you are placed in the scientist's position. Beginning with questions about your surroundings, you will study the relevant observations to see what a successful theory would have to account for. Then you'll study the necessary physics, but not before you've seen why you need it. By the time you encounter the "answer" in the form of a scientific theory, you'll be equipped to genuinely appreciate what the theory accomplishes.

Original history integrated with scientific lessons   You'll see gray shaded history segments in every chapter, set next to a picture of a telescope that Galileo built and used (shown in the margin here) The historical information is not off to the side in a "box," but is built right into the flow of the lesson. You will read about the discovery directly from the journal article the discoverer published. This way, the history is intimately connected to the science: As you learn the science, you'll get a sense of how it was learned by scientists.

Spiral staircase structure   As you progress through the chapters, you'll frequently find yourself facing the same principles you've already studied, but at a higher level each time. In this way, you'll be able to recognize the core ideas without trying to memorize a collection of dimly related statements. For example, in this book, "dark matter" isn't just a quirky section in one of the later chapters, as it is in many astronomy textbooks. Rather, it's introduced early and then separately tied to a variety of cosmology topics—such as galactic motions, gravitational lensing, weakly interacting neutrinos, cosmic microwave acoustics, and structure formation—that recur throughout the book.

Cosmology made personal   Our aim in describing cosmology is to provide a framework for you to see yourself within the context of cosmic history. To this end we offer comments throughout the book designed to help you feel more directly connected to the broader universe and see how your awareness of those connections can filter into your daily decisions. These comments are largely concentrated in the first and last chapters, in short reflective essays at the end of each chapter, and in some of the end of chapter exercises.

We would like to acknowledge the many people who contributed to the development of this book. Parts of the book emerged from a previous collaboration with Kim Coble. At Pearson, Michael Gillespie provided editorial suggestions and Kate Brayton spearheaded our book through production. We thank them both for their contributions. We also thank Laura Houston, of Pre-Press PMG, and Brian Baker, of Write With, Inc., for their editorial services. And we appreciate the efforts of Erik Fahlgren and Erin Mulligan, both of whom helped us develop this project from its early stages.

Many informal reviewers, including Amanda Duncan, Jerry Duncan, Doug McCarty, Bob McGown, Jack Semura, Christina Tyler, and Katie Whalen, provided important feedback and suggestions. We would also like to thank the following reviewers for their feedback:

Christopher Burns, Swarthmore College
Shane Burns, Colorado College
Brendan Crill, California State University, Dominguez Hills
Marc Davis, University of California, Berkeley
Victor DeCarlo, DePauw University
Richard Gelderman, Western Kentucky University
Kevin Grazier, University of California, Los Angeles
Jim Imamura, University of Oregon
Steven Kawaler, Iowa State University
Sung Kyu Kim, Macalester College
Davide Lazzati, University of Colorado at Boulder
Ian M. Littlewood, California State University, Stanislaus
Thomas Lockhart, University of Wisconsin, Eau Claire
Robert Milligan, Muhlenberg College
Brian Nordstrom, Embry-Riddle Aeronautical University
K. Dennis Papadopoulos, University of Maryland, College Park
Panos Photinos, Southern Oregon University
Jean M. Quashnock, Carthage College
Michael Richmond, Rochester Institute of Technology
Jeff Robertson, Arkansas Technical University
Michael Rulison, Oglethorpe University
Barbara Ryden, Ohio State University
Daniel M. Smith, Jr., South Carolina State University
Michael Smutko, Northwestern University
Dan Stinebring, Oberlin College
Joseph S. Tenn, Sonoma State University
Emilio Toro, University of Tampa
Steven B. Zides, Wofford College

There are also a few manuscript reviewers who helped us draft this book, but have chosen to remain anonymous. You know who you are, and we deeply appreciate your help.

We are grateful to our students at Fort Lewis College, Pacific University, Portland State University, and Westview High School for helpful feedback and for suffering through early drafts of the text. Todd is also grateful to Bill Becker and the PSU Center for Science Education, for providing the environment for a new course to grow into this book, and to Angela Lowman, for her valuable expertise in using Adobe InDesign and Illustrator. Craig is grateful to Cindy Browder, Scott Dodelson, Angela Olinto, and Les Sommerville for useful conversations, and to Tom Norton, for the clever idea of including end-of-chapter crossword puzzles.

We hope this book enhances your experience of the universe. Enjoy!

—TD and CT

Starting Points

"Who are we? The answer to this question is not only one of the tasks, but the task of science."

—*Erwin Schrödinger*

It's helpful to know where we're starting from and what we're looking for as we begin our exploration of the universe. What do we already know about our universe? What else would we like to learn? What difference does it make to learn more about our cosmic context when so much of it is beyond our direct experience? How can science help in our search for answers? In this chapter, we consider these questions in order to establish a purpose and a framework for our investigations in the rest of the book.

1.1 What Do You Wonder About?

Exploration begins with curiosity about our surroundings and ourselves. Albert Einstein was famous for asking very simple questions that led to profound insights: What would I see if I could catch up to a beam of light? Why do all different kinds of objects fall to the ground at the same rate? Is the Moon still there when I don't look at it? Einstein captured the importance of good questions with these words:

"The mere formulation of a problem is far more essential than its solution. . . . To raise new questions, new possibilities, to regard old problems from a new angle requires creative imagination and marks real advances in science."

In this spirit, let's try to articulate some questions that occur to us as we contemplate the universe. Then we can focus those questions to guide our investigation. Each person's questions will be somewhat different, but here's an example of how your chain of thought might go:

How far into space could we travel and still find stars and planets? Are things generally the same even in the distant reaches of the universe, or are they completely unlike what we can see from Earth? Do other stars work the way our Sun does? Are there other planets made from the same elements as Earth?

What about ourselves? Where do we come from? You may remember a couple of decades of history, and others talk about historical events that happened long before you were born. We can find evidence of things that were happening hundreds, thousands, or more years ago. Historical artifacts tell us that there were entire civilizations that thrived for thousands of years and then collapsed. All of this history is out there in books and museums, or online—ours for the taking.

The same sort of evidence exists for events in the history of the natural world. The rings on a fallen tree indicate that it survived hundreds of seasonal cycles. A large asteroid that hit the Earth about 65 million years ago left behind trace elements in the layers of rocks that today give away some of its secrets. The history of how the solar system formed 4.5 billion years ago is partly expressed in what the solar system looks like today. The composition of stars like our Sun reveals something of their history. If we know how to interpret the evidence, we can put together a history of the entire universe.

This history includes our own history. So where were we when all of these ancient events were going on? Or where was the material that later became us? Was this same material once part of another living creature, or perhaps part of a star somewhere? The remainder of this section focuses on some of the specific questions that we can ask, and reasonably expect to make progress toward answering, during our exploration. But before we bias your thinking with our questions, you should brainstorm your own in Exercise 1.1.

EXERCISE 1.1 Your Questions

What do you wonder about when you have quiet moments to stop and reflect—when you can't sleep at night, or while jogging or driving? For this exercise, consider those questions which relate to the natural world, the universe, or your place in either one. Jot down all the questions you think of, and then write out the top five you would most like to answer.

What Is the Universe Made Of?

Look around you to see what the nearby universe is made of: skin, metal, wood, water, etc. But what about beyond Earth? Are stars made of the same substances we find here, perhaps glowing like a fire on Earth? What about the dark space between the stars? Is space as empty as it looks, or is there material out there that we can't see? If there is material floating between the stars, what is it made of?

On Earth, we know there are common building blocks (such as hydrogen, carbon, and oxygen) that make up everything else we see. Are there some basic building blocks common to everything in the universe? If so, how are those building blocks arranged? Do they organize into the same patterns of matter everywhere: stars and planets (perhaps with water), DNA and complex life? What other kinds of structures are the building blocks capable of producing?

What Does the Universe Look Like?

We're used to describing the shape, size, and structure of familiar objects. For example, Earth is roughly spherical, about 40,000 km around the equator, and has a layered internal structure with a thin atmosphere blanketing its surface. What would the universe as a whole look like if we could zoom out and view it from afar? Or is this even possible, if the universe includes everything?

Is there a limit to the objects we see in the night sky, or do they extend forever? If they come to an end somewhere, then what happens at that place, where the matter runs out? Is there empty space beyond the matter, extending without bound in all directions? Or does space itself reach an edge? Can the universe just stop somewhere, with nothing beyond it? What sort of barrier keeps us from going past the edge? Or is the universe infinite, so that every possible event and every possible arrangement of things might be repeated indefinitely?

What shape is the universe overall? Is it spherical like Earth, or is it something more complicated, like the shape of a donut or a spider web? One way to think of exploring the overall shape of the universe is to hop into a speedy rocket ship and head out in different directions, carefully keeping track of your flight path. According to high school geometry, if you get into a rocket and fly in a straight line away from the Earth, you'll always move farther and farther away from our planet. But could outer space hold hidden turns and corners, so that over long distances, straight paths are not possible? Maybe if you travel far enough in one direction, the shape of space brings you back to where you started. Is there a way to map out a "landscape" of this sort without actually visiting everywhere?

Aside from considering these deep questions about the size, shape, and arrangement of the universe as a whole, it would be interesting just to know more about the objects we see. How big are those stars you see twinkling in the night sky, and how far away are they? Are they all the same distance away, or are some near and some far? How wide are the spaces between them?

What Is the History of the Universe?

You know that you have a personal history—a series of events and stages you've passed through in your life that brought you to the person you are now. Similarly, we can ask about the history of the universe. What stages has it gone through to become what we see today?

We've directly experienced only a small fraction of the known history of our own species. Yet we know that our surroundings can change dramatically just within a single human lifetime (e.g., through new construction, forest fires, and earthquakes). What about much longer periods? Have the same stars always existed, or was there a time when they emerged from something else? And what is the origin of the building blocks of everyday things? Where did the hydrogen and oxygen that combine to form water (H_2O) come from? How did the Earth form and arrive at its current state? (See Figure 1.1.)

How far back does history go? If the universe has a beginning, then we may wonder what happened "before" the beginning. If the universe has always been around, we might wonder how this is possible. So we'd like to get a concrete sense of how old the universe and the various objects within it are, in terms we can relate to (generations, comparisons to stages of human civilization, etc.). There must be definite answers to these sorts of questions, and we'd like answers that mean something more to us than just abstract numbers.

How Did Life and Consciousness Arise?

Embedded within this universal history is our own history. The Earth must have formed sometime within the grand timeline of historical events. Life emerged

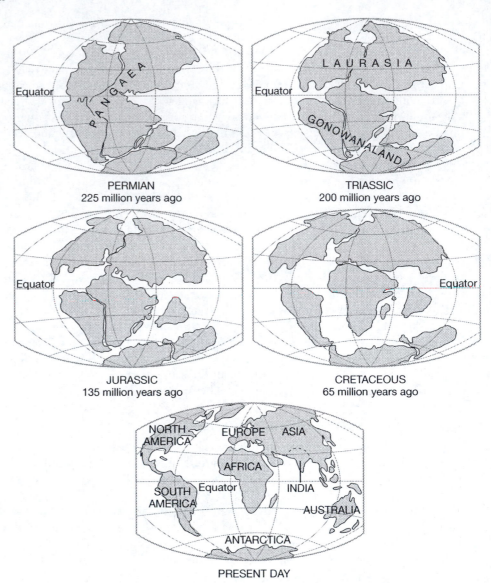

FIGURE 1.1 Geological History   The arrangement of major landmasses on Earth has changed dramatically over Earth's history. Has the rest of the universe been changing as well?

sometime within the history of Earth. What chain of events led to the emergence of life? Might similar events have happened elsewhere in the universe?

What Is the Future of the Universe?

How will the story of the universe continue to unfold in the future? Will new structures emerge that don't exist today? Will stars ever stop shining, and if so, what will the "dead stars" look like? What will be the ultimate fate of the universe? Will it eventually cease to exist, or will it last forever?

Even if the universe lasts forever, will life and consciousness last forever, too? Life as we know it depends on the balance of conditions that we experience on

Earth: not too hot or cold, the right kind of atmosphere, plenty of water. Perhaps these restrictions apply to life everywhere, or perhaps other forms of life exist under very different conditions. Can some type of life and consciousness endure within the universe of the distant future, or will conditions change so much that all life will be destroyed? Or, to view things in a different way, will some future species be able to transform part of the universe to preserve a habitat for life? After all, if our conscious species is capable of transforming our planet, then who knows what may be possible a billion years from now?

What Rules Does the Universe Follow?

Among the wide variety of things we see on Earth and throughout the universe, we notice common patterns. Just as we can watch a game (like soccer) and try to figure out the basic rules that underlie all the action (such as "only the goalie can touch the ball with his hands while on the field"), we can ask what rules the universe follows.

Scientists have uncovered many of the rules that the universe seems to follow, and some of these are familiar. You experience continual reminders of the law of gravity, for example. But we can keep asking what other rules we might find as we learn more. We can also wonder whether the same rules apply throughout the universe. For example, does the same law of gravity we observe on Earth apply as well to all of the stars we can see? Or might other parts of the universe obey rules that are entirely different from those which are familiar to us? (See Figure 1.2.)

Deeper Questions

Modern scientific cosmology (and this book on the subject) has at least partial answers to all of the questions we have just posed. But physical science often progresses by posing not the most important questions, but rather the easiest ones. The usual textbook word problem might read: "A train leaves Boston headed west at 45 mph, while another train leaves Chicago headed east at 52 mph. . . ." This type of problem never takes into account the realistic aspects of the situation, such as the stops and turns the two trains have to make. That's okay, because we can often get at the fundamental physics by ignoring trivial details. But now we're studying the whole universe, and who's to say how to separate the fundamental from the trivial?

We tend to focus on the simpler questions because we might actually be able to answer them and because they help us to formulate the harder ones better. For example, an important step taken on the road to modern physics was simply to describe in detail how far different objects move in equal intervals of time as they fall. This might seem like a trivial detail compared with more interesting questions like "Why does this water balloon fall on the neighbor's kids when I drop it from my upstairs deck?" Yet investigation of the simple-sounding "how" question provided the needed groundwork for the later understanding of deeper "why" questions about gravity. Among other things, such investigation led scientists to discover that gravity causes the same constant acceleration for falling objects of different weights. As you'll see in Chapter 6, this discovery turns out to be a very deep insight that played into Albert Einstein's profound reformulation of gravity several hundred years later. It also tells you that a massive water balloon and a light one will fall at the same rate. So you might as well drop the massive one to maximize water delivery.

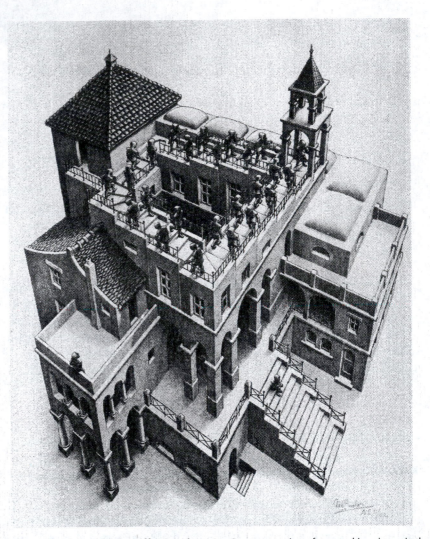

FIGURE 1.2 Different Rules in Different Places?   One assumption often used in science is that things behave according to the same rules here as everywhere else. But what if there are places in the universe where different rules apply? In this picture by artist M. C. Escher, a staircase climbs continuously uphill, but ends at the same height at which it started ("Ascending and Descending" by M.C. Escher, 1960.)

As we gain a deeper understanding of the universe from answering simple questions, we put ourselves in a position to ask harder questions. Some of these harder questions involve seeking a deep understanding of mathematical notions like "infinity" or "randomness." Others center around physical entities like "space and time" or "matter and energy." We might hope to learn about the most pervasive processes, such as the origin of complex forms (like the spiral shape of galaxies), or the real nature of life and intelligence. How do we learn to appreciate the most significant parts of the universe? Are they the common features (like stars), or the rare ones (like people)? Maybe we can even shed light on the grandest questions, such as the meaning of existence and why anything exists at all.

1.2 Science: A Way to Answer Questions

With the preceding long list of questions, we're ready to begin exploring the universe in search of answers. You already know some things about the universe from your immediate experience of living in it. But your experience is limited, and science has developed into a powerful method for extending our ability to make observations and answer questions. Throughout the book, we'll make extensive use of the scientific method and the knowledge it has collected over centuries of exploration. So it's important to understand what characterizes science as a way of answering questions. If we're going to make extraordinary claims of detailed knowledge, for example, about the first few seconds of our universe's existence, then we really need a solid understanding of how the scientific community justifies those claims.

What Is Science?

The most important distinguishing feature of science is that its claims are subject to observational or experimental testing. If you have one idea of how an airplane flies and we have another, then the scientific answer is determined when we build two planes according to our competing ideas and test which one flies better.

Formally, we can say that science is a process by which we come up with possible explanations or theories for what we observe in nature and then use observation and experiment to "filter out" theories that do not work. The deciding role of observation sets the tone for science. It means that theories can be falsified—proven wrong by observations that contradict their predictions. In fact, the best theories are ones that can be easily falsified if they are wrong. Imagine a theory predicting that a new compound, kryptonite, will have a melting point between 122 and 124 °C. The theory can be disproved if some kryptonite is heated up and found to melt at 490 °C. But how impressed would you be if the melting point turned out to be 123 °C? We'll see in Chapter 10 that the big bang theory is widely accepted because it's easily falsifiable, yet detailed observations have been unable to disprove it so far.

Before we proceed with such a grand scientific subject as a theory of the entire universe, we had better get our terminology straight. To a scientist,

- A *theory* is a proposed explanation for how some aspect of the universe behaves. Sometimes, when a theory is first proposed, it's called a hypothesis. Another term often used in place of "theory" is "model." Strictly, a theory must be testable, and a model can be just an idea that's used simply for convenience. In practice, however, the two terms are often uttered interchangeably; that is, most of the time, scientists expect models to be accurate and testable, too. We'll look at some examples of theory testing in the next subsection.

- An *experiment* is an observation made under conditions arranged so as to isolate and test a particular prediction made by a theory. For example, an experiment designed to test whether penicillin kills bacteria must isolate the effect of applying penicillin. One way to do this is to have two samples of the same type of bacteria, make sure that they both receive the same food supply, lighting, temperature, etc., and then apply penicillin to one sample. Since the penicillin is the only difference between the two samples, the observation that bacteria die in that sample provides a direct test of the theory. The other sample of bacteria provides a control in the experiment, showing what happens without the penicillin.

- *Uncertainty* is a numerical expression of the precision of a measurement, as in "the age of the universe is 13.7 ± 0.2 billion years." The uncertainty (the $\pm$ 0.2 billion years) specifies the repeatability of the method of measurement. For example, if three different people measure your height with a tape measure, they might report your height as 170.2 cm, 170.7 cm, and 171.1 cm. The range of these values indicates the uncertainty in the method used to measure your height (the use of a tape measure in this case). We'll discuss uncertainty further later in this section.

- A *scientific law* describes a pattern that is observed to be true. The law of gravity and Ohm's law for the electric current flowing in a wire are common examples.

It's important to recognize that the meanings of the preceding terms are sometimes totally different among nonscientists (including scientists outside of their professional lives). To a layperson,

- A theory is someone's unsubstantiated guess that might be totally wrong, as in "Oh, yeah? Well, that's just your theory!" (And a model is something you build with plastic and glue.)

- An experiment might not isolate a particular cause and thus might yield an incorrect conclusion. For example, someone may erroneously conclude, "Eating breakfast makes me sick," when the truth might be "Milk makes me sick at any time of day."

- Uncertainty is an admission of imprecision or incompetence!

- A law is a socially imposed rule prescribing certain behavior. Nothing physically prevents you from breaking such a law, but there are negative consequences if you do. (In one episode of *The Simpsons*, Homer Simpson sees that his daughter has somehow built a perpetual-motion machine, and he complains angrily, "In this house, we obey the laws of thermodynamics!")

In practice, science is often guided by a search for unifying ideas that are as simple as possible and that connect many different observations with as few explanations as possible. The classical theory of gravity, for example, allows a single framework (the force of gravity) to explain many seemingly unrelated observations: falling objects, ocean tides, and the motions of planets. The big bang theory offers a similar benefit, because many seemingly unrelated observations of outer space—the motions of galaxies, the intensity of microwaves, the amounts of certain elements in the universe, and more—are explained by its framework.

The rules governing scientific progress are somewhat different than those guiding other forms of knowledge. The differences result from the defining principle of science: Scientific theories must be tested by observations or experiments, so answers are determined by observing the natural world. This situation is unlike that existing in some other fields, in which expert opinion suffices to establish knowledge. In certain legal matters, for example, the decision of a judge is the source of correctness. Beyond that judge, there might be an appeals process, leading to the Supreme Court perhaps, in which greater and greater expertise (and authority) determines the correct interpretation of law. In this process, we expect the legal system to respect fairness as well, giving each party an equal chance to present its case.

Expertise and fairness, however, are not the rule in science. Although scientific expertise is respected, and you'll have a much easier time getting your ideas heard if you are recognized as an authority in the field, expertise and authority are not sufficient to establish correctness. And no oversight mechanism exists to grant each interested party an equal opportunity to voice his or her theory. Only the accumulation of observational evidence will ultimately move a scientific claim to global acceptance.

In science, progress is made through a system which requires that scientific claims be peer reviewed; that is, one or more other scientists must agree to allow the publication of a scientific paper, on the basis of its merit. These peers can make sure that they understand any experiments described in the paper, compare the results claimed with their own technical knowledge base, etc. Also, scientific accuracy can often be checked by repeating the experiments described in the paper: If you follow the same procedure, you should get the same result.

Key Elements of Science

Process

STEP 1   Construct theories to explain some aspect of the natural world.

STEP 2   Use theories to predict the outcome of a future experiment or observation.

STEP 3   Perform an experiment or observation to test predictions.

STEP 3.5   Return to step 1 and change something: Modify the theory, or use it to investigate a different prediction or perform a different test. Then repeat the other steps.

Science does not always progress neatly in this order. For example, an observation may trigger an idea for an experiment which only later leads to a theory that can be tested by more observations and experiments.

Principles

- Observation and experiment have the final word in settling disagreements. We pose the questions, but only observation and experiment can provide answers (e.g., experimental results).
- A theory is never proven beyond doubt, but more supporting evidence makes a theory stronger.
- The best theories generally explain many phenomena with a small set of rules.
- The practice of science respects the roles of repeatability and peer review, so that other scientists can check your work and verify that if they perform your experiment, they, too, will get your results.

Although the bullet points in this box highlight the defining features of science, it's important to understand that there is much more to how people do science in the real world. The rules provide guidance to weed out theories, but don't tell you how to come up with theories in the first place. And there are additional criteria that come into play. Personal bias and philosophical beliefs can sometimes influence what kinds of theories different scientists will propose. (This is usually not a problem, since a theory motivated by personal bias is subject to observational tests just like any other theory.) Aesthetic criteria such as beauty and simplicity also guide the choice among theories that all meet the requirements of falsifiability and observational testing. We'll explore some of these factors further as they come up in our study of the universe.

EXERCISE 1.2 Science and Its Relevance

(a) Define science in your own words. What characterizes it as a method of inquiry distinct from other ways of knowing about your world?

(b) Pick another field besides science, and describe the basic process for the acceptance of new knowledge in that field. Indicate some similarities and some differences between the acceptance process in the field you chose and the process in the field of science.

(c) Is science (as you've defined it) a necessary part of your life? If so, describe how you use science (or how you think you should use it) as part of your daily life. If not, describe what aspects of science make it difficult to integrate into your life.

How Science Works: Theory Testing in Action

To solidify your understanding of the scientific process, let's look at an example. Consider a theory which states that the world is flat. Early explorers challenged this theory by sailing around the world. Further falsification comes from modern satellite photographs showing the spherical Earth. In fact, some of the most convincing evidence that our world is spherical is well over 2000 years old.

Aristotle, one of the great philosophers of ancient Greece, outlined a set of clever scientific reasons for concluding that our world is shaped as a sphere and that its size can be measured. Perhaps his simplest and most convincing reason is that the shadow of the Earth cast onto the Moon during a lunar eclipse is always circular. In his book on astronomy, he wrote,

> "How else would eclipses of the moon show segments shaped as we see them? . . . Also, those mathematicians who try to calculate the size of the earth's circumference arrive at the figure 400,000 stades. [This is large by about a factor of two.] This indicates not only that the earth's mass is spherical in shape, but also that as compared to the stars it is not of great size."
>
> —Aristotle, *De Caelo (On the Heavens)*

He also argued against a flat Earth on the grounds that (1) it ought to have formed spherically if it formed from some sort of infall of smaller bits and (2) stars rise and set at different times for people living sufficiently east and west of one another.

But now try to prove that the world is spherical. Aristotle's reasoning, modern satellite photos, and circumnavigation of the globe by sailboat constitute evidence for the spherical world, but are they proof? Or could some other explanation satisfy those tests?

One of the first popular video games was "Asteroids" by Atari (see Figure 1.3). In this game, your triangle-shaped spaceship can drift off the left side of the screen and re-emerge on the right side. Could the Earth work that way, so that it's actually flat, but when you cross the International Date Line, you are somehow instantly transported to the other end of the map? Could space be warped in such a way as to make a flat world appear spherical from orbit?

Aristotle (384–322 B.C.) was one of the founders of Western philosophy. He studied under Plato and taught Alexander the Great. In his own academic studies, he pursued a wide variety of disciplines, including ethics, politics, logic, theology, and many sciences, such as anatomy, geology, physics, and astronomy. Among his numerous accomplishments, he made a compelling scientific case that the Earth is spherical rather than flat, and he set out a description of the universe with the Earth at the center that stood for nearly 2000 years.

Or perhaps it's even more extreme than that. In the 1999 movie *The Matrix*, a different reality was offered. You perceive the world on the basis of sight, sound, and other senses—senses that operate by electrical activity in the brain. The movie asserts that your perceptions of the world could be faked by a series of electrodes and impulses deliberately inserted into your brain. The DVD case reads as follows:

> "Perception: Our day-in, day-out world is real.
> Reality: That world is a hoax, an elaborate deception spun by all-powerful machines of artificial intelligence that control us. Whoa."

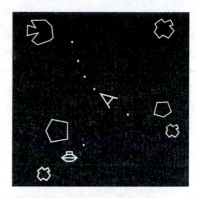

FIGURE 1.3 "Asteroids" Video Game, Atari, 1979 Your spaceship (triangle at center) can drift off the left side of the screen and automatically reemerge on the right side. Can you prove that your world works differently?

Thus we learn an important lesson from the movie: Your life could be a cleverly written dream. So is the Earth really spherical? Is there even an "Earth"?

Before you run off and complain to your teachers that everything they taught you was wrong, consider that any alternative explanations to "the Earth is spherical" and "the Earth is real" must match with the rest of our experience as well as those explanations do. The point of this example is not to convince you that your life is a dream, but rather to illustrate the nature of science.

We can't replace the spherical-Earth theory with just a vague idea that space is warped so that a flat Earth appears round from orbit and we are transported to the other side when we cross the International Date Line. To be credible, the theory must explain detailed observations about how our movements get us from one place to another on Earth, how much the Earth is curved, etc. In fact, the spherical-Earth theory itself is wrong in detail: Earth is not a perfect sphere; for instance, it bulges slightly near the equator. But the spherical Earth matches more closely with observations than a flat Earth does. A successful warped-space theory would have to incorporate the general features expected of a sphere (such as the correct geometric surface area) and also the slight equatorial bulge.

Similarly, if our experience is actually a computer-programmed illusion, it must be a very carefully written program. It would have to match what we expect of a real Earth, down to the finest details we've been able to measure so far. The *Matrix* theory of reality becomes scientifically interesting only where it makes testable predictions that distinguish it from the "real Earth" explanation. For example, we might find a way to "unplug" ourselves from the program and experience another level of reality (as Keanu Reeves's character Neo does in the movie). But if two theories predict absolutely identical observations in all possible situations, then it doesn't really matter which one you believe; you'll never be able to tell the difference.

Science admits and quantifies the ways it could be mistaken. That's why it demands such convincing evidence. A scientific theory can be disproved by experiment, but it can never be proved: There's always the possibility that some new observation will find an exception to the theory. But although scientific knowledge is always tentative at any given moment, core ideas emerge over time that are well tested and are very unlikely to be wrong within the domain in which they have been tested. The predictions of any new theory will have to match very closely with the old one within that domain of experience.

Uncertainty in measurement plays an essential part in distinguishing among possible theories. Since a core element of science involves looking for observations that surprise us and disprove our current theory, it is important to know when an observation accomplishes that. This means we can't just give numbers to describe our observations, because a single number doesn't tell you how reliable an observation is. Just as important as the number obtained from a measurement is the uncertainty in the measurement: How far off might our measurement be from the actual number? If you measure your height and read off 5 feet, 11 inches, is it possible that you really are 6 feet tall? A person reading your measurement doesn't know unless you also report the uncertainty. The simplest way to express that uncertainty is to say that a measurement is some number, plus or minus (±) some other number.

For example, if you are 6 feet tall and you're installing a new doorway, you'd like to make sure that there is plenty of room and you won't bump your head. So you might want the doorway to be 7 feet high. If the contractor tells you it will be "about 7 feet," you'll also want to know the actual uncertainty. If it's based on his tape measure, accurate to the nearest inch, you're pretty safe. The height of the doorway will then be 7 feet ± 1 inch. However, if the contractor measures feet by stacking up shoes, a method that's accurate only to within 2 feet, the same measurement ("7 feet") would make you very nervous. The measurement of the door in this case is 7 feet ± 2 feet, which places it somewhere between 5 and 9 feet—potentially a serious problem.

Similarly, if a theory predicts that the melting point of kryptonite is between 122 and 124 °C, you need to know the uncertainty in order to decide whether a new measurement rules out your theory or not. A measured value of 120 ± 1 °C would rule out the theory, while a measurement of 80 ± 50 °C would not contradict your theory, even though the actual measurement is much farther off from your prediction in the second case.

EXERCISE 1.3 Evidence

It's important to think carefully about the evidence for any statements we make about the universe and to understand the procedures for obtaining that evidence. Relate this to your own experience by writing about something you strongly believe (i.e., something you feel fairly certain is true, such as "the Sun will rise tomorrow morning"). What is the evidence that makes you so confident of that belief? Try to think of as many different pieces of evidence as you can. Also consider what kind of evidence you would need to see in order to change your belief.

Do you have any nonscientific beliefs about which you're so certain that no amount of contrary evidence would change your mind? What about the belief that "it's impolite to stare," even though staring could be construed as flattering.

Incorporating Science into Your Thinking

Although science emerged from ordinary observations as a way to answer questions about the world, it has become a specialized subject with its own terminology and methods that can be daunting to the uninitiated. For example, the way you describe your life experiences is probably different than the way physics describes the scientific universe. Physics gives a technical description that is unfamiliar in everyday life. Physics has to do with mathematical equations, exotic particles, nuclear reactions, electromagnetic fields, curved spacetime, and galaxies billions of years old and billions of light-years apart. These concepts bear little obvious resemblance to our common experience: a world of thoughts and emotions, warm sunlight and cool grass, life spans measured in decades, and distances measured in meters or miles. But as you become more familiar with scientific language and methods throughout this book, you'll be able to extend your awareness to encompass parts of the universe that were once hidden.

EXERCISE 1.4 Expanding Your Awareness

Describe as much as you can about what's going on in the room where you are reading, at as many different size scales as possible. Try to consider aspects of the room you are not normally aware of, such as the radio waves streaming past (you know they exist because you can tune into them with a radio), the tiny spider hiding in the corner of the ceiling, or the atoms in your fingers. Much of what goes on is outside of our awareness most of the time, and science can help us expand our awareness to include more of what's going on all around us.

For many people, the mathematical language in which much of science is expressed is the steepest barrier to incorporating science into their thinking. If the language is not familiar, then it's more difficult to access the insights science can provide (see Figure 1.4). The rest of this subsection covers some mathematical background that we will need throughout the book.

Units of Measurement

Much of science is about measurement: How big is a star or planet? How far away is it? In order to measure anything, we need units. Using units amounts to comparing the quantity you want to measure with some standard quantity. For example, if your stride is the standard of measurement for length, you can measure other lengths as the number of your strides required to cover distances equal to those lengths.

We'll generally use SI units in this book (SI stands for Système International, or International System, of units). SI uses meters, kilograms, and seconds as the fundamental units of distance, mass, and time, respectively. Whenever you do

NON SEQUITUR   *BY WILEY MILLER*

FIGURE 1.4 Using Math   If this cartoon expresses your sentiments about math, try to remember that mathematical tools emerged to help answer the kinds of questions we discussed in Section 1.1. If the cartoon does not express your sentiments about math, then you can just chuckle and move on.

calculations in this book, if you convert everything to units based on meters, kilograms, and seconds, and then do whatever calculations are required, you'll be assured of getting an answer in SI units also.

Large and Small Numbers: Scientific Notation

In astronomy, we work with very large numbers because the distances to most of the objects are—well, astronomical. In order to explore the building blocks inside of matter, we'll also have to deal with some very small numbers.

If we had to write out 40,000,000,000,000,000 meters every time we wanted to mention the distance to the nearest star, and 0.00000000005 meter to discuss the size of a hydrogen atom, these pages would be filled with zeros. It's much more convenient to use scientific notation to express large and small numbers in more compact form. The shorthand of scientific notation works like this: We move the decimal place from its original location until the number we are left with is between 1 and 10. Then we multiply this simpler number by 10 raised to the number of decimal places we moved over. If the actual number is 10 or larger, then the power gets a positive sign. If the actual number is less than 1, then the power gets a negative sign. (If the number is between 1 and 10, we probably wouldn't bother to use scientific notation, but in that case the power is 0.)

This sounds more complicated than it actually is; the procedure is easiest to understand with a few examples. The number 100 is 1.0×10^2 in scientific notation, because we have to move the decimal point 2 places to the left to get 1.0. The number 0.00042 is 4.2×10^{-4}, because we have to move the decimal point 4 places to the right in order to get 4.2. The advantage of this notation becomes more apparent for numbers like those we mentioned earlier. The number 40,000,000,000,000,000 m can be written compactly as 4×10^{16} m, and 0.00000000005 m can be written as 5×10^{-11} m.

We're also far less likely to make mistakes when we use scientific notation. When we write out a number or type it into the calculator, it's very easy to skip a zero or add an extra zero. ("Was that 15 zeros or 16? Now I have to start all over . . . .") This becomes a nightmare when we try to multiply or divide big or small numbers. But with scientific notation, it's easy. To multiply numbers, add the powers. To divide them, subtract the power of the number in the denominator from the power of the number in the numerator. For example, $(1 \times 10^2) \times (4 \times 10^3) = (1 \times 4) \times 10^{(2 + 3)} = 4 \times 10^5$, and to divide 4000 by 200, we have $(4 \times 10^3)/(2 \times 10^2) = (4/2) \times 10^{(3 - 2)} = 2 \times 10^1 = 20$.

We might still be working on the problem of how many hydrogen atoms will fit between our Sun and the nearest star if we had to divide the number out longhand. But in scientific notation, the number is $(4 \times 10^{16}$ m$)/(5 \times 10^{-11}$ m$) = (4/5) \times 10^{[16-(-11)]} = 0.8 \times 10^{27} = 8 \times 10^{26}$. The number of atoms is about 8 followed by 26 zeros. Table 1.1 lists a few examples of scientific notation.

TABLE 1.1   Examples of Scientific Notation

Number	In Scientific Notation
2500	2.5×10^3
0.096	9.6×10^{-2}
10×100	10^3
0.001×0.1	10^{-4}

EXERCISE 1.5 Using Scientific Notation

(a) Multiply the numbers 100,000 × 32,000, writing them out or entering them into your calculator the long way.

(b) Multiply 100,000 × 32,000 again, this time using scientific notation and without a calculator. Check that your answer is the same as in (a).

(c) At the time of this writing, Bill Gates's fortune amounted to over $46 billion ($4.6 \times 10^{10}$). Use scientific notation to calculate how long it would take him to run out of money if he spent $10,000 per day. Then express your answer in years.

One other advantage of scientific notation is that it focuses our attention on the "order of magnitude" (power of 10) of a number. When we're trying to gain perspective from the scale of the universe, exact numbers are often unnecessary. What matters is whether an object is about 10^{13} km away, rather than, say, about 10^4 km. Thinking about orders of magnitude also gives you a rough check as to whether you're in the right ballpark and gives you a feel for the answer. For instance, when you're typing numbers into your calculator, you can check its answer with a quick calculation by hand, using only the orders of magnitude involved.

Reading Graphs

In order to understand what we observe in the universe, we will often need to display a relationship between two of its attributes. Some common examples are the relationships between the time and the location of an object, between an amount of light and the type of light, and between the speed of an object and its distance from us. It's often convenient to show this relationship visually with a two-dimensional graph.

A graph provides a condensed way to display lots of information. For example, suppose we want to know how the outside temperature changed throughout a particular day. We could display this information in a table, listing each time and its corresponding temperature (Table 1.2). But we can more easily see what is going on if we plot this information as a graph (Figure 1.5).

Each pair of numbers in Table 1.2 becomes a point on the graph in Figure 1.5. Temperature is represented by the distance along the vertical (y) axis; time of day is represented by distance along the horizontal (x) axis. To find the temperature associated with each point on the graph, follow a perfectly horizontal line to

TABLE 1.2   Temperature Variation During a Day

Time	Temperature (°C)	Time	Temperature (°C)
6:00 A.M.	10	4:00 P.M.	31
7:00 A.M.	12	5:00 P.M.	30
8:00 A.M.	15	6:00 P.M.	29
9:00 A.M.	18	7:00 P.M.	27
10:00 A.M.	20	8:00 P.M.	25
11:00 A.M.	21	9:00 P.M.	21
12:00 P.M.	25	10:00 P.M.	19
1:00 P.M.	27	11:00 P.M.	16
2:00 P.M.	29	12:00 A.M.	14
3:00 P.M.	30		

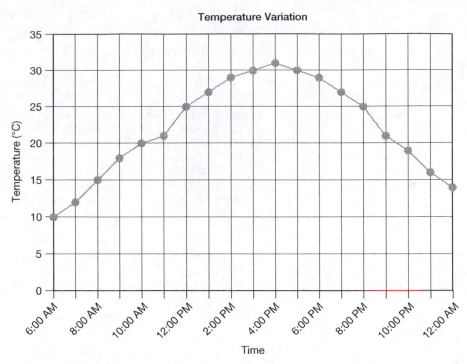

FIGURE 1.5 Representing Information Graphically   The relationship between time and temperature in a particular location on a particular day.

the left and read the temperature from the *y*-axis. Similarly, to find the time corresponding to when a given measurement was taken, go straight down from the point and read the value on the *x*-axis. Unlike the table, the graph gives you an immediate visual sense of how the temperature changed throughout the day, including times when no observations were collected. From the graph, you can quickly estimate what was happening at, say, 7:30 A.M., when no measurement was recorded.

1.3 What Difference Does This Knowledge Make?

Okay, so maybe it's fun to think about big questions and learn that science might help answer them. The science of **cosmology**—the study of the universe as a whole—might teach you some amazing facts about the distant universe. But what relevance does this have to your life here on Earth? We'll give a short answer to this question now and continue exploring the implications as we learn more about the universe throughout the book.

While scientific cosmology studies the universe with the tools of science, each of us also lives within an overarching *personal* cosmology—a cosmic picture or story that outlines our view of reality. The word "cosmology" itself derives from the Greek *cosmos*, meaning "ordered whole" (think cosmetics—making your appearance an ordered whole). Often, we don't really think much about our individual view of cosmology; we just absorb beliefs from the culture we live in. So each culture's cosmology brings order to its experience and describes how things all fit together in the big picture. In turn, the perceived order of the universe affects how people live and what they see as their role within that culture. Exercise 1.6 illustrates, in general terms, the way your cosmology affects your life.

EXERCISE 1.6 Your Self-Image

Different personal cosmologies give different answers to how an individual human fits into the big picture of the universe. We've listed a few examples of what your place in the cosmos might be. Add your own to the list if you can think of other examples. Then choose any two examples (yours or ours), and write a few sentences describing how a person with one point of view might live differently than a person holding the second point of view.

- I am one cog in a giant machine.
- I am the "eyes of the Milky Way[1]"—a way for the universe to notice and appreciate itself.
- I am a cosmic artist—a contributor to a universal creative process.
- I am a participant in a definite cosmic plan or purpose, with some assigned role to play in carrying out that purpose.
- I am a random speck in a vast uncaring universe.

Perhaps the deepest mystery of the universe is that it produces creatures like ourselves who can invent cosmologies. We are parts of the universe, trying to grasp the whole through our personal cosmologies. Our limited experience prevents us from completely capturing the whole, but it's easy to forget this limitation. The cosmology we live in takes on the appearance of the complete reality. Human history provides a wide variety of examples; we'll consider a few here.

Animist Cosmology. Many cultures have subscribed to animist cosmologies, which viewed every object in the environment as alive, with a personality and motivations similar to what humans experience. If this sounds absurd to you, then ask yourself this: Have you ever described your computer as being in a bad mood? Do you complain to the stoplight when it seems to take forever to turn green? Try to imagine looking at all of the world from within this cosmology. You watch lightning and hear thunder, and you know that the sky is angry. The Sun, Moon, trees, clouds, mountains, and rocks—each is an individual being with humanlike feelings, thoughts, and motives. Your daily interactions with the world are shaped by the belief that you are always negotiating with other living beings, persuading them to act as you wish.

Babylonian Cosmology. As another example, the Babylonians living in Mesopotamia more than 2500 years ago left records of their cosmology in the form of clay tablets describing their observations and interpretations of events in the sky. (See Figure 1.6.) In the Babylonian universe, the gods communicated with earthly kings through *omens*—occurrences in the sky such as weather conditions, phases of the Moon, or eclipses of the Sun and Moon. Omens were signs from the gods, indicating their pleasure or displeasure and warning of bad consequences if not properly responded to.

Since the omens were signals from the gods and not absolute predictions, people could follow proper courses of action (ranging from performing rituals to

FIGURE 1.6 Clay Tablet with Cuneiform Record Luckily for us, the Babylonians did not use paper for their records. They kept notes on clay tablets, which survived for thousands of years (clay tablet document shredders have yet to be invented), so we can still read their notes and records today.

---

[1] Thanks to Brian Swimme for this descriptive phrase.

simply staying indoors) to avoid undesirable consequences. Thus, events in the heavens were closely linked to events in the daily lives of the people, including such details as when to weave fabric, care for the sick, and wash one's feet. The cosmology of the Babylonians placed a high value on careful observations of the skies. Their diaries of the weather and astronomical phenomena spanning at least 600 years from about the eighth century to the first century B.C. are the longest-known continuous astronomical records.

Perhaps these first two cultural cosmologies seem easy to dismiss as ancient history, not relevant to you. But such remote examples are needed as reminders that you, too, are embedded in a personal cosmology that could seem just as irrelevant to someone outside of it. Your personal cosmology is like water to a fish. Water surrounds the fish and plays a role in everything it does. But it's so familiar that it disappears into the background. A fish might not notice that there is such a thing as water at all, unless the water changed significantly. Imagine a fish suddenly finding itself immersed in lime JELL-O®! Similarly, it's difficult to see the pervasiveness of your personal cosmology unless you imagine how your life would be under a different cosmology. So before you laugh off the influence of cosmologies as ancient history, consider a third example: the cultural cosmology rooted in Western European history that probably influences your perspective right now.

Mechanistic Cosmology. A core aspect of Western culture is the mechanistic cosmology that emerged from 17th-century physics and was further developed by scientists and philosophers in the 18th and 19th centuries. In this cosmology, reality is understood by breaking things down into their simplest, most fundamental parts and looking at how these parts interact to produce what we observe. Within this reductionist perspective, we focus on understanding the simple building blocks (e.g., atoms) out of which everything else is made and the universal forces through which these building blocks interact. Then we build our understanding of more complicated things out of these simpler parts.

Seventeenth-century physics provided an unprecedented unification of the cosmos, bringing together Earth and sky, large and small, living and nonliving, under the control of a few simple laws. In contrast to the animated cosmology we discussed earlier, this unification produced a view of the universe that was more like a machine than a living being.

The consequences of this approach are far reaching. It emphasizes universal laws that operate in the same way on everything, whether you are a rock, a person, or a planet. There is a loss of emphasis on the fundamental uniqueness of each object—a uniqueness that was present in the animated cosmology. There is also an apparent loss of free choice at the human level. After all, if we are made of simple particles, obeying laws that control their interactions and how they move, then where is there any room for our choices to influence what happens to us or to the world? To some people, the premise of the mechanistic cosmology—that cause-and-effect physical laws govern everything—just sounds like simple truth. But to accept it, you have to admit that you might not have free will. This topic has generated centuries of philosophical discussions that continue today.

The mechanistic cosmology embedded in Western culture also leads to an analytical approach to solving problems. Consider, for example, a mechanical clock as an analogy to the mechanistic universe. If a clock is broken, you look for the individual pieces that are responsible: gears, springs, etc. The clock is repaired by fixing these pieces. A mechanistic cosmology leads its adherents to approach problems in medicine, the environment, government, business, etc., as if they were fixing a mechanical clock. This implies that if some system (say, a public education system) becomes flawed, then the remedy is to fix the broken "part" (say, by giving schools more money or issuing standardized tests), rather than inventing a whole new system from scratch (say, rethinking the general philosophy of public education). Often, such an analytical approach is successful; it's the thinking behind our modern technological world. Still, it's worth being aware that this perspective has its limits.

Each perspective filters some properties out and highlights other aspects of the universe. The mechanistic cosmology tends to leave aside feelings, consciousness, etc., in favor of simple building blocks and forces. But feelings and consciousness are also core aspects of our experience, so we know that this cosmology does not completely capture the whole of reality. The animated universe brings consciousness to the forefront, but leaves out the reality that common building blocks underlie all objects in nature and they do interact with the same basic forces. Where the cosmologies intersect is interesting, teaching us lessons that can expand our awareness of reality.

With this background, we can suggest an answer to the question posed in the title of this section. The reason learning about the universe matters is that *you* also live within a cosmology that affects how you see the world. How is your own cosmology shaping your life and the ways you interact with your surroundings? How might new information change your perspective and influence how you live? For example, in your description of the universe, spiders might be unimportant and serve only as an annoyance. By that rationale, if a simple means of exterminating them became available, you might take it. But with a broader perspective, you'd be aware of the impacts on the food chain and how insects would no longer be eaten by spiders and thus would overpopulate the world and become an even bigger annoyance, forcing us to stay indoors more often, ruining flowers, trees, and crops, etc. Your idea of the most desirable course of action changes with your perspective.

Exercise 1.7 will help you articulate your own personal cosmology as it exists in your mind right now.

EXERCISE 1.7 Your Universe

Imagine traveling back in time to visit different cultures with widely varying descriptions of the universe. You listen to the sages of each society outline their view of reality—the nature and arrangement of the universe and the principles on which it operates. What each of them tells you contains some truth about the real universe. Even beliefs you know are inaccurate in some way still contain insights that accurately capture part of the universe as we experience it.

Write an essay that expresses your own personal description of the universe, as if you were explaining what the universe is really like to a visitor with a vastly different perspective. Try to make this an honest snapshot of how you actually view the universe, explained in a way that makes it understandable to someone who does not see the world as you do. Don't restrict yourself to scientific information or what you think your instructor wants to hear. Just describe anything that seems important about your universe.

Writing this down will make you more conscious of your own beliefs about what the universe is like. This will help you focus your thinking as you read the rest of the book, to be on the lookout for the connections that are most meaningful to you.

Although cosmology as a general subject is very old, cosmology as a modern science, in which the methods of physics are applied to the universe on the largest scales, is relatively new. This application of physics has given us access to information about our surroundings that once seemed beyond our reach.

Many of the findings we will report about the large-scale nature of our universe are recent. Solid evidence for even some of the most general features of our current scientific picture of the universe dates only back to the middle of the 20th century, and evidence about important details of the picture continues to pour in as you read. You therefore have a unique opportunity: You are among the first people with the chance to make use of these insights in developing your perspective.

The insights from scientific cosmology that matter most to you are those which affect your personal cosmology in some way. In the rest of this book, we'll create a framework to help you access these insights that are brought to light by the scientific approach to the universe. We'll point out ways to expand your web of experience and feel more directly connected to the broader universe. Figure 1.7 illustrates the organization of this process.

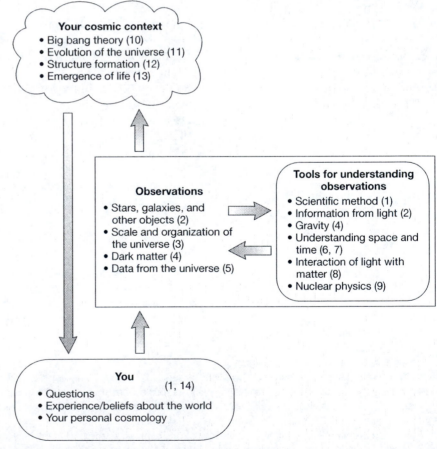

FIGURE 1.7 Exploring Your Cosmic Context.   Your exploration of the universe begins with your experience with the world so far and with the questions you would like to answer. We use science to help answer some of these questions by observing the world and formulating theories, to build up an overall picture of your cosmic context. This context in turn affects your beliefs and future observations you make. Numbers indicate chapters in which each topic plays a significant role.

Science and Your Personal Cosmology

Science has a very obvious impact on our lives through the technology it creates. The Internet, cell phones, microwave ovens, iPods, satellite TV . . . we are immersed in a technological world that is largely a product of science. But what about the view of the world that underlies these seemingly magical inventions we use so often without a second thought? Physicist Richard Feynman highlighted this more subtle influence of science during a lecture to the National Academy of Science, emphasizing, "I would like not to underestimate the value of the world view which is the result of scientific effort." He went on to illustrate this type of impact with an example:

> "For instance, the scientific article may say, 'The radioactive phosphorus content of the cerebrum of the rat decreases to one-half in a period of two weeks.' Now what does that mean?
>
> It means that the phosphorus that is in the brain of a rat—and also in mine, and yours—is not the same phosphorus as it was two weeks ago. It means the atoms that are in the brain are being replaced: the ones that were there before have gone away.
>
> So what is this mind of ours: what are these atoms with consciousness? Last week's potatoes! They can now remember what was going on in my mind a year ago—a mind which has long ago been replaced."

We could add that the atoms whose pattern helped store memories in our brains and were once part of a potato were also once inside a star that died long ago and helped make our existence possible. Imagine the story these atoms could tell if they could convey their history to us! (We'll tell you some of their story in this book if you're willing to accept it secondhand.)

Perhaps there is value in being aware of our connections to the rest of the universe, connections revealed in greater and more surprising depth as we learn more. Consider this result revealed by an understanding of the atomic model of gases (such as air): Almost every time you take a breath, the air you inhale contains one atom exhaled in the last breath of any person who has ever lived and died on Earth. So take a deep breath and hold it for a few seconds. Within your lungs is an atom of air that was in Gandhi's last breath, and Joan of Arc's, and Einstein's, and Aristotle's![2]

All of human society is still just a tiny part of the universe. We're among the first people ever to have access to some key information about the cosmic backdrop for our lives—and the information hasn't really become part of our culture yet. Therefore, you have a unique opportunity to build a personal connection to a universe that has been inaccessible for most of human history. It's your job to figure out what you believe and put yourself into the story. We invite you to continually ask how we know that each piece of cosmological information is true and how it fits into or modifies the personal cosmology you described in Exercise 1.7. How can you make this knowledge real and alive for yourself? How can you fit your current thoughts and choices, difficulties and successes, into that universe? What new questions and new mysteries will this knowledge raise for you?

---

[2] This insight is from Art Hobson's *Physics: Concepts and Connections*, 4e. New York: Prentice Hall, 2007: 40.

Quick Review

In Chapter 1, you looked at a few core questions about the universe that will guide our investigation throughout this book. Scientific cosmology applies the tools and knowledge of science to expand our awareness and try to answer some of these questions. Therefore, it's important to be familiar with the methods of science in order to understand the foundation for what you will learn in later chapters.

Try the following crossword puzzle to test your knowledge of key terms and concepts from Chapter 1:

ACROSS

5 Observation designed to test some aspect of a scientific theory
7 Body of knowledge in which correctness is decided by observations
8 Personal cosmology in which objects are considered to be alive
9 Visual representation of numerical data
10 Basic unit of knowledge in science
12 Testable claim made by a theory
13 Scientific principle that is always validated by observations of nature

DOWN

1 Quantitative expression of the precision of a measurement
2 Philosopher who presented observational evidence that the world is a sphere
3 Personal cosmology in which all things behave like clockwork
4 Meters, kilograms, and seconds, for example
6 Scientific papers are subject to _____ by other scientists
11 Communication from the gods in Babylonian cosmology

Further Exploration

(M = mathematical content; P = project idea; R = reflection; D = suitable for class discussion)

1. (R) What key characteristic distinguishes science as a way of learning about the world?

2. (M) Express the following numbers in scientific notation:
   a. 0.000352
   b. 96,000,000
   c. 4.5

3. (RD) Ludwig Wittgenstein expressed the sentiment that "Even when all scientific questions have been answered, the problems of life remain completely untouched." How does this statement relate to the different perspectives we've been considering in this chapter?

4. (MPD) Find a newspaper or magazine article that requires numerical data to support the author's point. Summarize the article and explain how numbers were used to make an argument.

5. (RDP) Read "Without earth there is no heaven: The cosmos is not a physicist's equation" (Edwin Dobb, *Harper's*, Feb. 1995, pp. 33–41), which describes a way to approach the interface between science and your personal cosmology. Write a short reflection describing your reaction to the article.

6. (RD) What aspects of the universe as you see it really matter to you and have an influence on how you live your daily life? What things form such a crucial part of the perspective you use to interact with the world that you might see things differently and live differently if those parts of your understanding of the universe changed?

7. (M) A typical atom is approximately 10^{-10} meters in length across any dimension. Estimate the total number of atoms in your body.

8. (RD) *What's in the Box?* This exercise is a microcosm for the process of science and is best done in a group, using the following procedure:

   - One person chooses objects to place inside a shoebox.
   - Each member of the group takes a turn at holding the box, shaking it, listening, etc., in order to gain information about the object(s) inside, without opening the box.
   - In a three-column format, a record keeper for each group records observations made as the box is handed around. Column 1: properties they think describe the object (metallic, plastic, size, number of objects, etc.). Column 2: guesses—what's in the box? Column 3: confidence in the guess, on a scale from 1 to 10 (1 = low confidence; 10 = high confidence).

   - Also, record notes on the following questions, for later discussion:
     1. What methods are you using to try to figure out what the unknown object is?
     2. How certain are you that you know what the object is? How does your level of certainty change over the course of the exercise? This activity is a model for how your beliefs about what's true can change on the basis of evidence, so it's useful to become aware of the process of changing beliefs.
     3. What is the most convincing evidence for you?
     4. How do you convince someone else that your guess is correct?
   - Try to reach a consensus within the group about the contents of the box. When everyone agrees or you've decided that you cannot agree, write down the best guess(es).
   - Repeat the process with several different objects.

   This activity is a model for the general scientific process. Answers in nature are not in the back of the book; we have no direct link to "truth," but have only our observations to formulate the best answer we can. This exercise gets at many of the skills and processes that are important in science: problem solving, testing knowledge claims, and persuading others that your theory is the best one, given the observations so far. There is, however, a distinction to be made: In this exercise, you can touch, but not look; in cosmology, you can look, but not touch.

9. (RD) Comment on what this passage expresses about the difference between Western European and Shawnee cosmologies:

   "When the Shawnee chief Tecumseh learned that an expedition of scientists from Harvard was traveling to Iowa during the summer of 1806 to observe an eclipse, he was curious. He asked his friend, Galloway, to explain what a scientist is. Galloway described a scientist as someone who 'studies the things of earth and heaven. . . . Scientists watch plants and animals, they watch the stars, they watch clouds and rain, and the earth. . . .' 'Does not all white men do this?' Tecumseh asked. 'All Shawnee do this.'"

   —James M. Patchett and Gerould S. Wilhelm, *Designing Sustainable Systems: Fact or Fancy*

10. (RD) What three experiences or observations about the world have most shaped your personal cosmology? Pick one choice you made this week, and trace its foundation, in detail, back to a core set of beliefs about how the world works. For example, do you try to recycle? Why or why

not? What do you believe about the universe that makes this experience seem valuable or not valuable? Or consider the following example: Imagine that scientists somehow discover that free will is an illusion of the brain and that the future is already written, even though we don't yet know what it is. How would that knowledge affect your life?

11. (R) Do stars rise in the east and set in the west each night, as the Sun does each day? How do you know?

12. (RD) Imagine that scientific investigation discovers that human beings are purely biomechanical machines. Each human–machine is slightly different, but each obeys its own "programming" at all times. Write a paragraph or two explaining how this knowledge would affect your day-to-day behavior. For example, would you experience betrayal in the same way that you do now?

13. (RD) Pick a few "big questions" from Section 1.1 (yours and/or ours). For each, explain what sort of observation, if it were made, would help answer that question. For example, if you choose "How old is the universe?" then what sort of measurement would help answer it—measuring the ages of stars, measuring some subatomic particle's half-life, etc.?

14. (RD) (a) Find examples of the "animated cosmology" in your own life. For example, some people describe their computers as being "in a bad mood."

    (b) If we know that a description is not accurate, then why hang onto it?

15. (RD) Address any three of the following issues:

    • Find examples of the mechanistic cosmology in your own life. Cite a few that are questionable and explain why.
    • As a thought experiment, try applying the mechanistic cosmology to a complex problem, like the valuation of abstract art or the improvement of physiological health in the world's population.
    • Think of a few simple-sounding cause-and-effect societal changes that probably should be enacted (on the basis of a mechanistic perspective), but haven't been.
    • Use the mechanistic cosmology as a reference point to evaluate how a company's profits will change when it chooses to initiate a policy of mandatory overtime with no extra pay for all employees. Is it possible that a different guiding cosmology would work better to increase profits?

16. (RD) Babylonian-like omens might hold a place in your life, too. Try to identify one. For example, have you ever quit something because of "signs," such as a series of relevant negative events all in one day, giving you a sense that you're "trying too hard," or that "it wasn't meant to be"?

17. (RD) Do you hold any values that aren't your own, in that you never actually thought the issue through, but rather adopted someone else's wisdom instead? It might help to consider how you feel about hard work, about looking your best, or about school. Pick one example, and write a few sentences about where the value came from and whether, upon reflection, you still believe that it's true.

18. (RD) Are you merely a character in someone else's dream? Discuss how scientific experimentation would or would not help you answer this question. Is the "dream character" idea a scientific theory? Why or why not?

19. (R) Consider the following two descriptions of sunlight: (1) the electromagnetic waves described by the mathematical equations of physicists; (2) the feeling of warmth you get when you lie on the beach on a sunny day. Comment thoughtfully on which one is correct.

20. (RM) Which of the following is the strongest evidence against a theory predicting sunrise at 6:07 A.M.?
    (a) An observation of sunrise at 6:20 ± 20 minutes.
    (b) An observation of sunrise at 6:03 ± 5 minutes.
    (c) An observation of sunrise at 6:09 ± 1 minute.

    Explain your reasoning.

21. (RD) What are the major distinctions between scientific cosmology, cultural cosmology, and personal cosmology? Are there any commonalities?

22. (M) In science, measurements are often quoted with uncertainties and with an indication of how confident the scientist is in them. For example, we might claim with 95% confidence that a certain car gets 33 miles per gallon (mpg) ± 3 mpg. This means that, for 95 out of 100 measurements, you'll get between 30 and 36 mpg. Figure out how the uncertainty and the confidence are related. If one goes up, what happens to the other? Suppose I choose to quote the mileage results at 99% confidence instead; how would the uncertainty change? Explain your answer. (*Hint:* You can choose to quote a measurement with whatever confidence you like, without remeasuring anything, as long as you adjust the uncertainty accordingly. For example, suppose you make 100 separate automobile mpg measurements as just described and 95 of them come out within the ± 3 mpg range. Then what "±" value might make sense if you included, say, 3 more of the remaining 5 measurements in addition to the original 95?)

The Sky We See

"We had the sky up there, all speckled with stars, and we used to lay on our backs and look up at them, and discuss about whether they was made, or only just happened."

—*Huckleberry Finn*

In Chapter 1, we identified some questions to investigate about the universe, learned how science can help us pursue answers, and gained a glimpse of why it matters to know as much as we can about our cosmic context. Now we're prepared to start investigating the universe in search of answers to our questions. In this chapter, we focus as much as possible on direct observations of objects in the night sky. In the next chapter, we'll discuss additional information that tells us more about what these objects actually are, and our focus will shift to explaining the observations with a model of the universe that fits the facts together into a consistent picture. For now, what we're doing is more like botany: looking for patterns—organizing and classifying our direct observations. The distinction between direct observations and the theories we construct is important to emphasize because the observations tell us facts about our universe, regardless of how we interpret them. If we choose to discard one theory in favor of an alternative one, our alternative theory will still have to reasonably account for those same observations. Thus, we want to give you a solid grounding in the observations you have to work with, before suggesting a theory to explain them. This order of presentation (discussing observations before theories) will be a recurring theme throughout the book.

The Italian scientist Galileo Galilei (1564–1642) was one of the earliest and strongest proponents of the power of direct and often simple observations in understanding our universe. His observations of the night sky through an early telescope settled questions about the nature of the universe that had been the subject of philosophical debate for centuries. We will turn to Galileo's work throughout this chapter for examples of the impact observations can and should have on our thinking.

2.1 Looking Up

If you are fortunate enough to observe the sky from a very dark and clear place, you might see something similar to the photograph in Figure 2.1. A typical view of the night sky contains a wide variety of different spots of light—some faint, some bright; some small points, some spread out like smudges; some pure white, others red or blue.

In this section, we start making sense of these splotches of light that are visible in the night sky, as the first step in our journey outward to learn more about our cosmic context. We begin with observations using our unaided eyes. This isn't just looking at the objects in the sky for the sake of looking; it's also learning to interpret the meaning of their light. To do this, we'll need to study the nature of the light which carries that information to our eyes. Since we don't have direct access to most of the things we see in the sky (we can't touch them or do experiments on them), we rely on the properties of light (such as color) to tell us more about them.

FIGURE 2.1 The Night Sky   The well-known constellation of Orion the hunter is featured in this field of view. Betelgeuse, the star at the top left of the constellation, is visibly red to the naked eye; Rigel, at the bottom right, is visibly blue. (◆ Also in Color Gallery 1, page 3.)

In later sections, we'll expand our vision and our understanding by using technology to enhance our eyesight and by using some basic knowledge about the physics of light to improve our ability to interpret the information the light brings.

What can we conclude from our initial observations? Besides your emotional reaction to looking up at the vast canopy of stars, and any neck cramping that may occur, the first thing to notice is that it's not at all obvious what exactly you're looking at. We cannot with any confidence say how far away anything is. The small points of light we call **stars** have no discernible features—nothing we can use to give us a sense of scale indicating how far away they are. But the total absence of recognizable features is not enough to halt the march of science! For example, we might at least estimate that they must be very far away, in order to make any features they do have too small to make out.

In order to navigate our way around the sky without knowing real distances, we work in terms of **angular size** instead of physical size. (By "physical size," or sometimes "linear size" or "actual size," we just mean "size" as you normally think of it; a meterstick's physical size is 1 meter.) Angles simply tell you what fraction of a total circle you have to turn in order to go from looking in one direction to looking in another direction. (See Figure 2.2.) We divide the circle into 360 parts, called degrees (°), so that if you're looking in one direction and turn around in a complete circle back to look at the same spot, that's 360°. If you have to change the direction you are facing by only one-sixth of this amount, that's 60°. The advantage of this system of angular measurement is that it makes no difference how far away things really are. Whether you're looking at the real sky or the inside of a planetarium dome, you can still navigate your way around from one star to another.

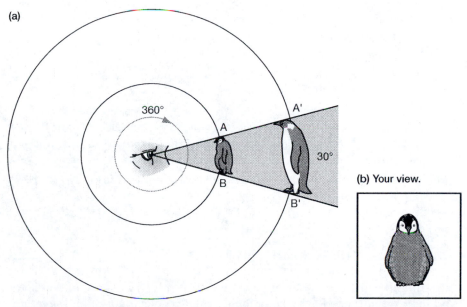

(b) Your view.

FIGURE 2.2 The Meaning of Angles   (a) The physical distance along the arc of the bigger circle from A' to B' is longer than that from A to B on the smaller circle. For example, a tall adult penguin just fits into the space A'B', while a short baby penguin just fits into the space AB. But the two arcs AB and A'B' each represent the same fraction (1/12) of the total circle. Thus, the angular distance in both cases is 1/12 of 360°, or 30°. (b) shows how this arrangement translates to what you see: The baby penguin blocks out the adult from your view, even though the baby is physically smaller.

FIGURE 2.3 The Moon and the Eiffel Tower The distant Moon (to the right of the tower, near the top) appears much smaller than the nearby Eiffel Tower, even though the Moon's physical size is much bigger than that of the tower.

Let's investigate angular distance navigation by comparing it with giving someone directions by using a map and physical distances, such as kilometers. We might tell a friend how to get from the airport to our house by saying something like "Go 5 miles east and then 2 miles north." To get from one star to another, we would say instead, "Go 5° east and 2° north." This tells the person what fraction of a complete circle to turn in each direction in order to move his or her line of sight from one star to the other.

We can also describe the size of an object as an angle, based on what fraction of the total dome of the sky the object covers in each direction. For example, the full Moon has an angular size of about half a degree. Even though the Moon is huge, it is also far away, so it covers about the same angle as the eraser on your pencil held at arm's length from your eyes. The apparent size of an object in the sky depends on a combination of its physical size and its distance away from you. In Figure 2.2, you can see an example of this relationship: Two objects of different physical sizes can have the same angular size if they're placed at the right distances from your eyes. Figure 2.3 shows the case of an object with smaller physical size having a bigger angular size because it is much closer to you.

EXERCISE 2.1   Observing the Sky

The purpose of this exercise is to encourage you to look at the universe and make your own direct observations. Do not look anything up in a book to get the "right" answers, and try to forget anything you have learned previously about stars. Rely only on what you are observing and on your own imagination to invent explanations. Much of what you read about astronomy started with people observing carefully what they saw, just as you are doing now. On a night when the sky is at least somewhat clear, step outside, look up at the stars, and answer the following questions:

(a) About how many stars can you see in the sky? A rough estimate is fine; you don't have to count them all! But try to make your estimate reasonable.

(b) Do all the stars appear to be the same color? If so, what color do you think best describes them? If not, list all of the different colors of stars you can see.

(c) What do you think the stars actually are? What do you think is making them shine? For instance, are they fires, pinholes in a tapestry, living things, light bulbs, or what? Use your imagination, but try to be realistic about what would truly be possible as an explanation if all you knew about them is what you can see with your unaided eyes.

(d) What else do you want to know about the stars? What observations do you think you could make personally to answer those questions?

2.2 First Impressions

Continuing with our observations, we see plainly that some of the stars are brighter than others. On a photograph, brighter objects appear as bigger circles, even though they are not actually bigger; rather, it's a result of how light interacts with the camera and the film or electronic detector. From our experience on Earth, we're familiar with lights of different brightness, either because they are inherently brighter (a lighthouse beacon is much brighter than a firefly) or because they are closer to us (a lighthouse seen when you stick your head in front of the lamp is much brighter than one seen from a distant ship). So the differences in brightness among various stars can tell us something about their nature and their distances from us.

At least one of the objects in Figure 2.1 (toward the bottom-left corner of the photo) is not a point or a circle like the others. It appears fuzzy and irregular, like a cloud. Some objects like this one are gas clouds called **nebulae** (singular: "nebula"). As with the stars, we cannot tell the distance. The fuzzy nebula could be as close as a spot on the camera lens, or it could be a huge cloud at the far reaches of the universe. Without something to set the scale for us, we simply don't know.

Most fuzzy irregular objects appear fairly small on the sky. But you may also have seen a very large cloudy patch looking like a faint stream of milk poured across the night sky. If you've seen it, you won't be surprised to learn that it's called **the Milky Way,** or **the Galaxy** (from the Greek word for milk; think "galactic" and "lactic," as in lactose). As you can see in Figure 2.4, the Milky Way is a strip that runs around the whole sky.

Even without knowing the distances or the true nature of the stars, we begin to notice patterns in their distribution across the sky. Obviously, the stars are not uniformly distributed. They can be organized into *constellations*—collections of stars that lie relatively close to one another on the sky and that form a recognizable pattern. For example, Figure 2.1 shows the constellation Orion the hunter. But many more patterns in the stars have been given the status of "constellation," and different cultures have organized the patterns into different constellations.

There are clumps with many stars bunched together, and regions of sky that appear dark, with no visible stars at all. On a night with a full Moon, you can see very few stars and most of the sky seems empty. On a very dark night, far from city lights, the sky seems almost filled with stars. But even then, there are gaps of darkness. Are those spaces really empty? If you had more sensitive eyes, might you see stars even in the gaps?

If you keep watching, you will notice that the stars move across the sky. The patterns of the constellations stay the same, but they move together as the night

FIGURE 2.4 The Milky Way   Panoramic view of the cloudlike Milky Way stretching over the whole sky.

goes on. This is a result of the Earth turning on its axis (the imaginary line running straight through the Earth from North Pole to South Pole). It's just as if you stood in one place and turned around, watching the scenery as you turned. All of the scenery turns together, as if it is painted on a surface spinning around you.

The Moon moves through the sky as well, but not at the same rate as the stars. If you watch it for several nights in succession, you'll see that it changes location relative to the background stars. Apparently, it has an additional motion of its own, not due to the turning of the Earth.

If you watch very closely for many nights in row, you may notice that some of the starlike points behave like the Moon, wandering among the other stars rather than maintaining the fixed pattern. These objects are the **planets,** a name derived from the Greek word for wanderers.

You've probably discovered by now that, although there is much you can notice with unaided eyes, trying to observe with *only* your eyes is pretty frustrating. In the next section, we'll introduce some tools to help us see better.

2.3 Enhancing Our Vision

This is a good place to pause and recognize that what we call "seeing" is actually receiving bits of light that have traveled through space and that tell us something about the object they last traveled from. In some cases, such as a star or the flame from a candle, the light travels directly from its source to our eyes. In other cases, such as the Moon or the book you're reading, the light is reflected to your eyes from another source (e.g., the Sun or a lamp). In either case, what we receive through our eyes is just a certain amount of light, of certain kinds, coming from certain directions. We can use technology to help our eyes extract more information from the amount and kinds of light we receive. For example, binoculars magnify distant objects like birds or football players (or birds painted on the helmets of football players), and magnifying glasses or microscopes magnify close-up objects like leaves or insects.

In astronomy, telescopes capture a greater *quantity* of light from an object than our unaided eyes do, enabling us to see fainter things and to magnify them so we can see more details. Most of the figures in the rest of this section are images of objects you can see with your unaided eye, viewed in much greater detail with the aid of telescopes. The idea here is to give you a brief tour of the kinds of objects that are visible in the sky, expanding your view of what is out there.

Using a telescope enables us to find out more about fuzzy objects such as the one located in the constellation Orion that we pointed out earlier. Some fuzzy objects turn out to be clouds of gas and dust (nebulae); the one in Orion is a prime example. (See Figure 2.5.) Others turn out to be dense collections of individual stars, which blend together to our unaided eyes and *look like* clouds (Figure 2.6). The Milky Way is a good example: If you point your binoculars or a small telescope toward it, this cloudy region is revealed as a dense grouping of stars (as well as some much smaller nebulae that remain cloudy even through a telescope).

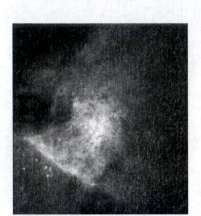

FIGURE 2.5 HST Image of Orion Nebula   Remember the tiny smudge you could see toward the bottom left of the constellation Orion (Figure 2.1) with your unaided eyes? Here is a zoomed-in image of that same region, taken with the aid of the Hubble Space Telescope. (◆ Also in Color Gallery 1, page 3.)

FIGURE 2.6 Star Cluster M80   What looks like a nebula is revealed by the Hubble Space Telescope as a dense region of individual stars. (◆ Also in Color Gallery 1, page 4.)

Galileo was among the first to point a telescope at the Milky Way to reveal the individual stars, part of a perspective-changing series of discoveries he made by looking at the sky through an early telescope. He described his discovery of these individual stars in a book called *The Starry Messenger*, published in 1610:

> "I have observed the nature and the material of the Milky Way. With the aid of the telescope this has been scrutinized so directly and with such ocular certainty that all the disputes which have vexed philosophers through so many ages have been resolved, and we are at last freed from wordy debates about it. The Galaxy is, in fact, nothing but a congeries of innumerable stars grouped in clusters. Upon whatever part of it the telescope is directed, a vast crowd of stars is immediately presented to view. Many of them are rather large and quite bright, while the number of smaller ones is quite beyond calculation. . . . And what is even more remarkable, the stars which have been called 'nebulous' by every astronomer up to this time turn out to be groups of very small stars arranged in a wonderful manner."

Galileo's discovery is an example of the power of simple, direct observation in revealing deep insights about the nature of our cosmic context.

Galileo Galilei (1564–1642) built his own astronomical telescopes and made a series of discoveries that upended nearly two thousand years of prevailing wisdom from the time of Aristotle. Galileo discovered that, inconsistent with Aristotle's universe of perfect, heavenly spheres circling around the Earth, the Moon is marred with craters, the Sun is speckled with sunspots, Venus has phases like our Moon (e.g., a crescent phase), Jupiter has moons in orbit around it, and the cloudlike Milky Way is actually made of countless individual stars.

Technology has obviously come a long way in the 400 years since Galileo's first use of the telescope for astronomy. The Hubble Space Telescope (HST) is one good example of this progress, greatly improving the power of our eyes. (See Figure 2.7.) Not only has the instrument itself improved dramatically, but by viewing from above Earth's atmosphere, its light-gathering power and the clarity of its images are further improved. Many of the figures in this section, including Figures 2.5 and 2.6, as well as many of the images in Color Gallery 1, were obtained by the Hubble Space Telescope.

FIGURE 2.7 The Hubble Space Telescope (HST)   This telescope is located beyond most of our atmosphere, about 600 km above Earth. It is about 13 m long and collects light with a mirror that is 2.4 m in diameter.

Galaxies

Many of the diffuse, fuzzy objects in the sky have similarities in structure that led astronomers to group them under the name **galaxies** (e.g., Figure 2.8). They are called galaxies because of similarities to the Milky Way Galaxy, but this connection won't be apparent until we learn more about the nature of galaxies in Chapters 3 and 4. For now, we will just try to classify them by direct observational properties, and later we'll return to understand more about what they are and the reasons for their appearance.

Astronomers have recorded images of close to a million galaxies. Like stars, they are clearly very common objects in our universe. An idea of the vast number of galaxies that exist in the universe can be gained from looking at Figure 2.9. Each of the splotches of light on the image is a galaxy, and what is seen here is typical of what you see when you look for very faint objects in any region of the sky. But the angular size of this view is tiny—about one-twentieth of a degree on a side. Remember our discussion of angles: That's only about a tenth of the size of the full Moon. So if you imagine covering the entire sky with the tiny squares represented by this field of view, and if you then multiply by the thousands of galaxies in this one field of view, you get an idea of how many galaxies must be out there.

Figures 2.10 through 2.14 show magnified images of some of the different types of galaxies that have been observed. Following a classification scheme introduced by Edwin Hubble in the 1920s, we group them into four types based on their visual appearance: **spirals**, **barred spirals**, **ellipticals**, and **irregulars**. For

FIGURE 2.8 The Andromeda Galaxy   Another fuzzy object you can see in the night sky. On closer inspection, it reveals a spiral structure. (◆ Also in Color Gallery 1, page 4.)

FIGURE 2.9 The Hubble Ultra Deep Field This is a picture of a region of the sky only about one-tenth the diameter of the full Moon that looks completely empty to our unaided eyes (and even to most telescopes!). Notice the variety of sizes and shapes of images. The telescope is aimed at a region in the constellation Fornax, which is south of Orion. Total exposure time is about 280 hours. (◆ Also in Color Gallery 1, page 5.)

now, let's use this classification scheme to help us wrap our minds around the various types of galaxies we see in the sky.

Spiral Galaxies As the name implies, these galaxies exhibit a distinct spiral structure. (See Figure 2.10.) The spiral galaxies are labeled with the letter S in Hubble's classification. They are further subdivided on the basis of the relative size of the bright spot ("bulge") that appears in their center, compared with the rest of the galaxy. Those with a large central bulge are designated Type Sa galaxies, those with smaller bulges are Sb, and those with the smallest central bulges are Sc. Other features, such as how clumpy and how tightly wound the spiral arms are, are also closely related to the size of the central bulge, with smoother and more tightly wound spiral arms typical of galaxies with a large central bulge (Sa).

It's also noteworthy that some galaxies appear to be tilted versions of others, which makes sense because we expect galaxies to be randomly scattered around, presenting themselves to us at random angles. For example, Figure 2.8 is similar to what Figure 2.10 might look like if we were seeing it from the side, not face on. Imagine looking at a dinner plate from various perspectives. Figure 2.11 supports

FIGURE 2.10 A Spiral Galaxy
Notice the distinct spiral shape of the arms spreading out from a central bulge.

FIGURE 2.11 Spiral Galaxy Viewed Edge-On What appears to be a galaxy seen from an edge-on perspective (◆ Also in Color Gallery 2, page 18.)

FIGURE 2.12 An Example of a Barred-Spiral Galaxy Spiral galaxies sometimes have a bar across their centers.

this suspicion, looking like a spiral galaxy similar to Figure 2.10, but seen from an almost exactly edge-on view.

Barred Spirals   These are spiral galaxies whose central bulges have a distinct bar shape (see Figure 2.12).

Elliptical Galaxies   These objects, as can be seen from Figure 2.13, have no spiral arms and no particularly distinguishing features. They are classified according to how circular or elongated they appear.

Irregular Galaxies   Consistent with this creative naming system, those galaxies which do not fit into any of the other categories are referred to as "irregulars." An example is shown in Figure 2.14.

Figure 2.15 provides a summary of Hubble's classification scheme for the different types of galaxies. As you can see, there are labels for various amounts of ellipticity, and for various barred spirals. These details are useful for galaxy classification, but hold no particular importance for our purpose here.

FIGURE 2.13 An Example of an Elliptical Galaxy Some elliptical galaxies are more elliptical or more spherical than others.

Planets

Planets are spherical bodies that move around stars in particular paths (which we'll discuss in Chapter 4). Images of the planets in our solar system provide dramatic examples of the power of telescopes to enhance our vision. Unlike the stars, planets can be magnified by a telescope to reveal not just a brighter image, but something that looks completely different than what we saw with our eyes alone. As you look at the images that follow, remember that the planets look like small points of light to the unaided eye. Through telescopes, they are revealed as worlds in their own right, some smaller than Earth, some dwarfing the Earth.

FIGURE 2.14 Irregular Galaxy An example of a galaxy classified as "irregular."

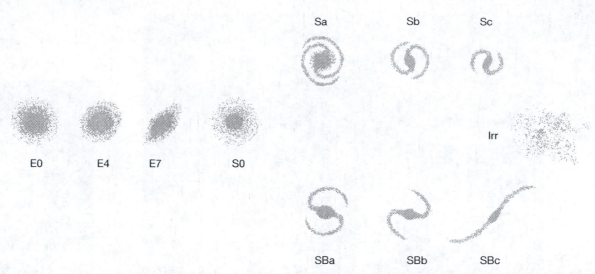

FIGURE 2.15 Hubble's Galaxy Classification Scheme Because of its shape, this diagram for classifying galaxies is often called the Hubble "tuning fork" diagram.

We return to Galileo for another example of the worldview-changing impact of careful observations. One of the targets toward which Galileo pointed his small telescope in 1610 was Jupiter. Figure 2.16 shows a sequence of sketches Galileo made of what he saw on subsequent nights. To describe his experience, he wrote in his 1610 book, *The Starry Messenger:*

"There remains the matter which in my opinion deserves to be considered the most important of all—the disclosure of four planets never seen from the creation of the world up to our own time . . . . I invite all astronomers to apply themselves to examine them and determine their periodic times . . . . Once more, however, warning is given that it will be necessary to have a very accurate telescope such as we have described at the beginning of this discourse.

"On 7 January, 1610, at the first hour of night, when I was viewing the heavenly bodies with a telescope, Jupiter presented itself to me; and because I had prepared a very excellent instrument for myself, I perceived (as I had not before, on account of the weakness of my previous instrument) that beside the planet there were three starlets, small indeed, but very bright. Though I believed them to be among the host of fixed stars, they aroused my curiosity somewhat by appearing to lie in an exact straight line . . . and by their being more splendid than others of their size. Their arrangement with respect to Jupiter and each other was the following:

East   * * O *   West

that is, there were two stars on the eastern side and one to the west . . . . But returning to the same investigation on January 8th—led by what, I do not know—I found a very different arrangement. The three starlets were now all to the west of Jupiter, closer together, and at equal intervals from one to another as shown in the following sketch:

East   O * * *   West

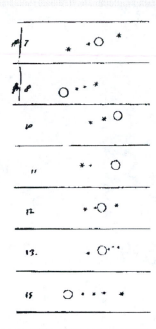

FIGURE 2.16 Galileo's Sketches of Jupiter and Its Moons Note the changing positions of the moons.

"At this time, though I did not yet turn my attention to the way the stars had come together, I began to concern myself with the question how Jupiter could be east of all these stars when on the previous day it had been west of two of them . . . .

"On the 10th of January, however, the stars appeared in this position with respect to Jupiter:

> East   * * O   West

". . . I had now decided beyond all question that there existed in the heavens three stars wandering about Jupiter as do Venus and Mercury about the sun, and this became plainer than daylight from observations on similar occasions which followed."

Imagine what it must have felt like to be in Galileo's position, seeing such clear evidence that at least three objects in the universe did not orbit the Earth, but orbited Jupiter instead. He later discovered a fourth object around Jupiter (see Figure 2.16), and these four are known today as the Galilean satellites. (In astronomy, "satellite" is a generic term: *Artificial* satellites are put in motion around a planet by NASA, for example, but *natural* satellites such as moons are already there.)

A more up-close image of Jupiter and its satellites (moons) is shown in Figure 2.17.

Modern technology—both Earth-based telescopes and cameras on board planetary exploration spacecraft—provides some remarkable views of the planets, greatly increasing the power of our eyes. More examples of these modern views are shown in Color Gallery 1.

FIGURE 2.17 Jupiter and Its Moons
This image, taken by the *Voyager 1* space-craft in 1979, shows Jupiter and two of the moons (Io and Europa) discovered by Galileo. (◆ Also in Color Gallery 1, page 2.)

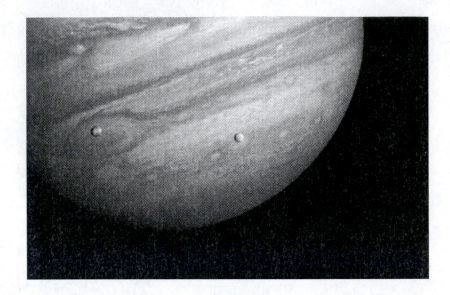

2.4 Light Waves

We've already learned that light comes in different colors. In order to learn more about the objects we see in the sky, we need a more systematic understanding of color and some other properties of light. Then we'll see how we can use these properties to find out about the objects we receive the light from.

The first thing to understand is that all the various kinds of light can be described as **waves**. So we need to learn something about the properties of waves in general, and then we'll return to the specific kind of wave light is. It's easiest if we start out thinking about familiar waves such as those on the surface of a pond or along a rope. These are referred to as *mechanical waves*. If you take hold of one end of a rope while a friend stands across the room holding the other end, you can transmit energy to her without any of the particles that make up the rope actually moving from one end to the other. Just jiggle your end of the rope in an abrupt motion, and watch the pulse travel along. This pulse is a *wave*. Even though bits of the rope themselves don't go sailing across the room, you *know* energy is traveling to your friend, because her hand is jostled by the rope a short time after you shake your end. Even if your so-called friends aren't willing to perform this experiment with you, you don't have to give up on understanding waves. A doorknob is equally capable of holding a rope and receiving the energy you send it.

To generate a more periodic pattern—one with a regular, repeating structure—that we can use to describe some standard properties of waves, move your hand up and down at a steady rate. A situation like that is shown in Figure 2.18.

The **frequency**, f, is the rate at which the cycles of the wave motion are repeated (i.e., the rate at which you move your hand up and down) and is measured in cycles per second. One cycle is completed when you move your hand up, down, and back up again.

The **wave speed**, v, is the speed at which a pulse or part of a wave travels along the rope and is measured in, for example, meters per second. We use the word **velocity** instead of speed when we want to specify both the speed *and* the direction of motion.

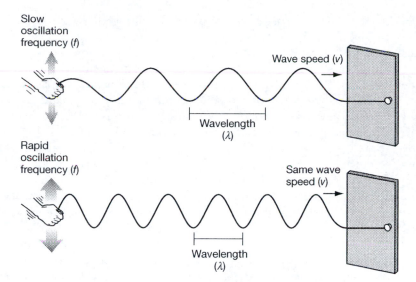

Slow oscillation frequency (*f*)

Wave speed (*v*)

Wavelength (λ)

Rapid oscillation frequency (*f*)

Same wave speed (*v*)

Wavelength (λ)

FIGURE 2.18 A Rope Wave   You hold one end and tie the other end to a doorknob (since not all of us have friends who will help with physics experiments by holding the end of a rope). Because the disturbance you cause by shaking the end of the rope moves to the right at a fixed speed determined by the nature of the rope, the waves will be more bunched up (have shorter wavelengths) if the oscillation frequency is increased. The faster you move the end of the rope up and down, the shorter is the wavelength of the resulting waves.

The **wavelength**, λ, is the distance between repeated features in the wave pattern—the distance from one peak (the highest point on each wave) to the next or from one trough (the lowest point on each wave) to the next, for example.

These three properties of waves are not independent. If you *increase the frequency* of the wave (i.e., if you generate pulses in more rapid succession by moving your hand up and down more quickly), *the wavelength decreases*. This is because the wave's speed doesn't change. If you flip the rope up and down twice as fast as before, the rope's wave will cycle up and down *twice* by the time it travels the same distance as it did with one up-and-down flip at the previous frequency. That results in *halving* the wavelength. (See Figure 2.18.)

All of this is because the speed at which any feature of the wave travels along the rope is determined by the material and tension of the rope—*not by the cause of the wave*. This is true in general for waves; their travel speed depends on the nature of the *medium* (rope, water, etc.), and not the source of the motion. Once out in the rope, the pulses or features all travel at the same rate, so if you generate them more rapidly (at a higher frequency), the first pulse doesn't get as far ahead before the next pulse sets out, causing them to be bunched closer together (producing a shorter wavelength). In other words, to increase the frequency, you have to decrease the wavelength, and vice versa. This relationship can be summarized with the equation

$$\lambda = \frac{v}{f}.$$ (eq. 2.1)

Electromagnetic Waves

To understand *light* as a wave, we need to ask about what provides the medium—the equivalent of the rope or water—in which the wave travels. This is where light differs fundamentally from mechanical waves such as waves in water or on a rope, because light travels through empty space. What sort of disturbance can travel without having a tangible medium to disturb?

You've glimpsed an answer to this question if you've ever played with refrigerator magnets and noticed that they can push and pull on each other without actually touching. Some influence, called *magnetism*, acts through empty space.

Similarly, the electric force causes two socks to pull together when you take them out of the dryer. You can create an *electromagnetic* disturbance by shaking an electrically charged object, like a sock fresh out of the dryer with (electro)static cling, back and forth. This is like shaking the end of a rope. Scientists have found by experiment that the disturbance can be "felt" some distance away through empty space. Shake a charged object in one place, and it can cause a charged object in another location to move around. This phenomenon is similar to what we saw with the rope: Shake one end of the rope, and the disturbance causes the rope to move at the other end. And in fact, an electromagnetic disturbance travels as a wave, so the amount of disturbance felt at various locations rises and falls while the wave travels, just as the mechanical disturbance in the rope does.

The real hint arrives when we look at the speed at which the electromagnetic wave travels. So if you're picturing a scientist wearing a white lab coat and shaking a sock at one end of a room, you need to put another scientist at the other end of the room with a stopwatch, to see how long it takes for the electromagnetic effects to be felt there. It turns out that there is a tiny time delay, corresponding to an electromagnetic wave speed of about 300,000 km/s. The realization that this is also the **speed of light** (which is usually referred to by the symbol *c*) led to a remarkable conclusion: *Light is an electromagnetic wave*. Therefore, once the light begins traveling, it has a life of its own; it keeps moving even if the electrified sock stops shaking. Again, this action is similar to that of a pulse on a rope, which keeps moving along the rope for a while even if your hand stops moving up and down.

At this point, it's worth noting a few things. First, stars and galaxies rarely (if ever) produce light by shaking charged socks. It's the charged particles (usually *electrons*) in the socks that produce the light, and these particles are also found in stars and galaxies. We'll return to that aspect of the story in Chapter 3, in order to start building a sense of size and distance in the universe, and again in Chapter 8, in order to understand some of the evidence at the heart of cosmology. Second, the speed of light is worth thinking about. For example, driving your car at a speed of 50 miles per hour means that you will cover 50 miles of distance in a time of 1 hour. (Please stay in the slow lane.) But at the speed of light, you could travel completely around the world more than seven times in 1 second!

For a light wave, the frequency is the rate at which the charged sock (or whatever) was shaken up and down. Evidently, for light, we can summarize the relationship we described earlier for waves in general as

$$\lambda = \frac{c}{f} \,.$$

(eq. 2.2)

Wavelengths of light are often measured in units of **nanometers** (10^{-9} m) or **angstroms** (10^{-10} m). These are very small units of distance. It takes a billion nanometers to make up 1 meter, and it takes 10 billion angstroms to make a meter. Frequency is commonly measured in cycles per second—also known as *hertz* (as in "25 Hz")—i.e., the number of times the wave completes its up–down–up cycle each second.

Frequency is also directly related to energy. You can imagine that more rapid shaking, which means higher frequency, requires higher energy. This gives us a simple, but powerful, idea of what sorts of events emit different kinds of light. Very high energy explosions, for example, produce high-frequency, short-wavelength light. A low energy glow gives off low-frequency, long-wavelength light.

The Electromagnetic Spectrum

We know that there are different kinds of light, which our eyes distinguish as different colors. Color Figures 11 and 12 in Color Gallery 1 show more vividly the variety of colors of stars, for instance.

You may also know that there are other types of light that our eyes don't detect. For example, some insects can see ultraviolet light, and you may have seen images taken at night with an infrared camera, which detects infrared light. It turns out that our eyes are sensitive only to a very narrow range of the kinds of light emitted by objects, so we can learn much more about the objects by using technology to extend the types of light we can see. Color Figure 14 in Color Gallery 1 shows the same region of the night sky, around Orion, that we looked at with our unaided eyes. But now we see that region of the sky in *x-ray*, *ultraviolet*, *visible light*, *infrared*, and *radio*. Notice how different the images look, depending on which kind of light we detect.

As an example of the utility of the different kinds of light, consider Figure 2.19, an infrared view of our Galaxy. The shape revealed in this view cannot be seen with visible light alone (see Figure 2.4), because there is dust laced throughout the galaxy, blocking the view. However, the dust actually glows in infrared light (as do stars), so we can use infrared light to make out the shape of our galaxy, including the bulge in the center.

With our new background knowledge of electromagnetic waves, we can gain a deeper understanding of these different kinds of light we observe. It turns out that each color of light—including "colors" such as radio waves that are invisible to our unaided eyes—corresponds to a different *frequency* of electromagnetic wave. For example, if you shake an electrically charged particle at a frequency of 4.3×10^{14} Hz, red light is generated. At 6.7×10^{14} Hz, we get blue light. And because of the relationship expressed by Equation 2.2, you can just as easily describe colors by the *wavelength* of light. Red corresponds to a wavelength of about 700 nm, while blue corresponds to a shorter wavelength of about 450 nm.

All of the names used for these different types of light, such as blue, red, infrared, or x-ray, are just *artificial categories*. Each was invented because at the time

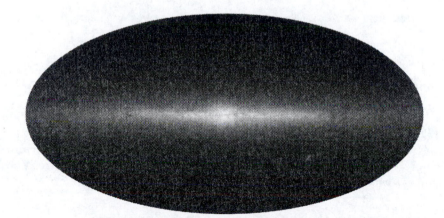

FIGURE 2.19 The Milky Way Viewed in Infrared Light   This is a wide-angle view, which means the Milky Way is much bigger on the sky than any of the galaxies we looked at previously. But note the structural similarity to the spiral galaxy shown in Figure 2.11. Stars and dust in the Milky Way Galaxy glow in infrared light, so we can see its familiar form without visible light.

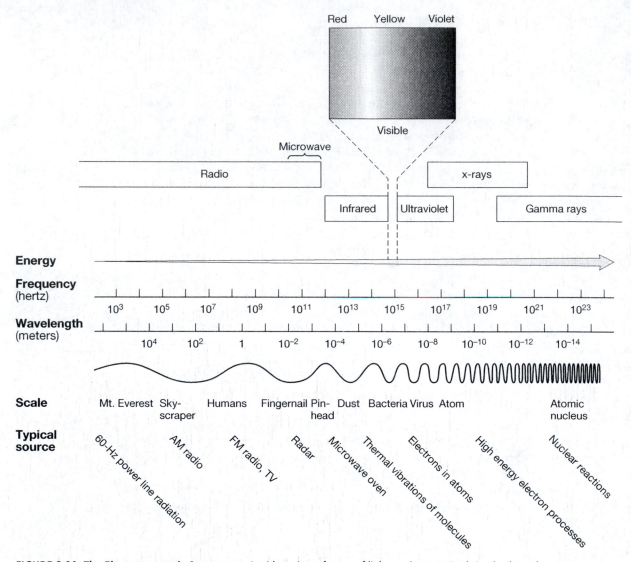

FIGURE 2.20 The Electromagnetic Spectrum   A wide variety of types of light can be organized simply along the spectrum of frequency or wavelength. (◆ Also in Color Gallery 1, page 7.)

of its discovery we didn't know the true nature of **electromagnetic radiation**: a unified entity in which the differences we observe are caused only by changing frequencies (or wavelengths). Red, blue, ultraviolet, etc—they're really all just electromagnetic waves, but we still use the category names for convenience. Taken together, from low-energy radio waves all the way to high-energy gamma rays, the range of forms of light is called the **electromagnetic spectrum.** (See Figure 2.20 for an overview of the electromagnetic spectrum and the relationships among frequency, wavelength, and energy.)

Spectroscopy: How Much of Each Color?

The light we see from a lamp or from the Sun and other stars is a mix of many different frequencies (and hence many different wavelengths and colors). We can analyze the light from a source to see how much of each color it contains. To do this, we need to separate out the colors so that we can see them individually. That is exactly what happens in a rainbow: The water droplets in the air spread the light out, sending different colors in different directions. This is because light gets deflected when it passes through the water drop, but different colors are deflected by different amounts. The same effect can be accomplished with other materials, such as a glass prism, instead of a drop of water. What matters is that light from some source is separated out, so we can study each color separately. This type of analysis is called **spectroscopy**: the process by which light of different frequencies is separated in order to measure *how much of each kind of light* is present. Look at Figure 2.21 to see what's going on, as we describe how this is done in astronomy.

A device for performing spectroscopy is known as a **spectroscope**. A spectroscope has two main components: a narrow slit for light to enter and something to disperse the light, such as a prism or a diffraction grating (a piece of glass or plastic with many closely spaced grooves cut in it). We can see the result of passing light through a spectroscope by just looking at the spread of colors (essentially a sequence of images of the slit, with each image a different color) or by taking a picture to save it for later study. This resulting spread of colors is called a **spectrum**—for example, "the spectrum of the Sun" or "the solar spectrum." (The plural of "spectrum" is "spectra.")

The brightness of each image of the slit represents the amount of that frequency which was in the original mix of light coming from the source. This idea is important: The purpose of a spectroscope is to measure how much light of each kind is hidden in the blend of light coming from a star, a galaxy, etc. Is there an unusual abundance of ultraviolet light? If so, it might be due to a recent burst of new stars

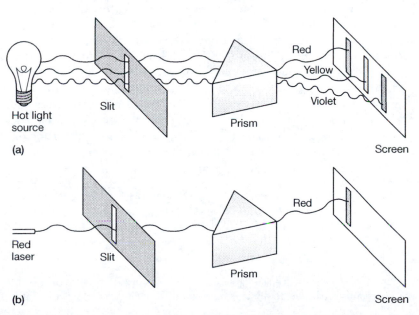

FIGURE 2.21 Schematic of a Spectroscope   (a) Light of different wavelengths passes through a narrow slit and then through a prism or grating that bends the light of each wavelength by a different amount. The resulting image of the slit is in different locations for different wavelengths, so we can separate out each wavelength and see how much of it is contained in the mix of incoming light from the source. Frame (b) shows the result when red laser light passes through the spectroscope.

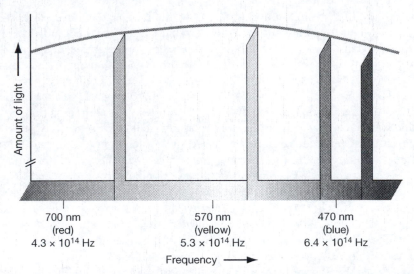

FIGURE 2.22 The Spectrum of Light From the Sun Displayed are both the image you would see on a detector at the back of a spectroscope (along the bottom) and a graph of how much light is hitting each region of the detector (and hence how much light of each kind is present in the original mix of light). (◆ Also in Color Gallery 1, page 8.)

being born. Is there a lot of red? It might indicate the presence of a certain type of nebula. If aliens from another planet shine a red laser at us, and we analyze it with a spectroscope, what would that image look like? (See Figure 2.21(b).) The point is, we can use this kind of color breakdown information to tell us about the source.

Using a spectroscope, we can look at the distribution of light from different objects, such as the Sun. A photograph of sunlight reaching the back of a spectroscope, labeled by color, wavelength, and frequency, looks something like the bottom part of Figure 2.22. One difficulty we face in displaying the information this way results from the fact that our eyes have a hard time comparing brightnesses. Does sunlight contain more red, blue, or yellow light, for example? For this reason, it is useful to make a graph of the intensity (amount) of light of each kind. The graph, as well as the colorful rainbow, are each called a "spectrum." Graphing the light just amounts to counting up how much light is hitting the detector at each region and plotting the amount of light as you move along the detector. This is also shown in Figure 2.22. The solid curve above the spectrum image graphs the amount of light received at each wavelength.

Joseph von Fraunhofer (1787–1826) observed the spectrum of light from the Sun and discovered thin, dark bands in the spectrum—now sometimes called "Fraunhofer lines"—in 1814. His catalog of these dark lines forms the basis for much of observational astronomy.

An important discovery in the history of science was that many things emit light that is not a continuous band of colors. In 1814, the German physicist Joseph von Fraunhofer observed and recorded over 600 dark bands in the spectrum of the Sun—narrow regions where the smooth spectrum of color was interrupted. He published the following account:

"In the window-shutter of a darkened room I made a narrow opening—about 15 seconds broad and 36 minutes high—and through this I allowed sunlight to fall on a prism of flint-glass . . . . I wished to see if in the color-image from the sunlight there was a bright band similar to that observed in the color-image of lamplight. But instead of this I saw with the telescope an almost countless number of strong and weak vertical lines, which are, however, darker than the rest of the color-image; some appeared to be almost perfectly black . . . . I have convinced myself by many experiments and by varying the methods that these lines and bands are due to the nature of sunlight, and do not arise from diffraction, illusion, etc."

—Joseph Fraunhofer

In 1864, English amateur astronomer William Huggins identified these lines in the Sun with some common *elements*. In fact, it turns out that each of the basic elements—the building blocks that make up ordinary matter, such as hydrogen, oxygen, aluminum, etc.—has a characteristic pattern of colors that can be used to uniquely identify that element (much as a fingerprint uniquely identifies a person). Figure 2.23 gives examples of the pattern of light for some familiar elements. In other words, Fraunhofer's statement that the lines are features of sunlight, rather than unwanted artifacts of his measurement apparatus, means that the Sun must contain all the elements associated with the "countless number" of lines. The intricate solar spectrum is shown in Figure 2.24.

It turns out that if we shine light made up of a continuous spectrum of colors—that is, light containing every color, so that its spectrum is a smooth rainbow with no lines—through a gas made up of a particular element, the element will *absorb* light at its characteristic frequencies. Since that light is absorbed, the spectrum now reveals an *absence of light* at that color—Fraunhofer's dark line. A

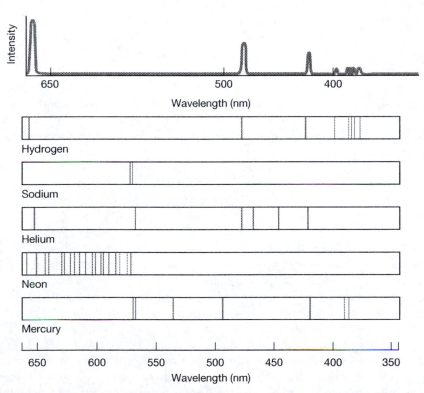

FIGURE 2.23 Emission Spectra of Some Familiar Elements   These are images at the back of a spectroscope, just as we showed in Figure 2.22. Only in this case the image is dark, except at specific wavelengths. The reason is that these elements have been energized so that they will emit only their "true colors." (◆ Also in Color Gallery 1, page 8.)

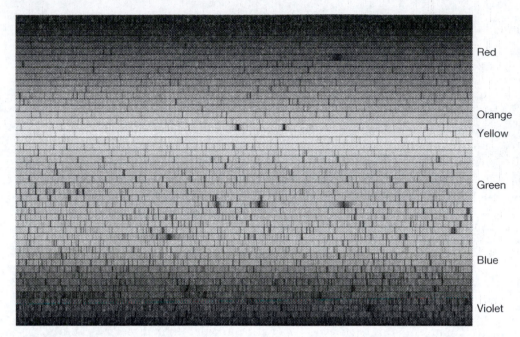

Red

Orange
Yellow

Green

Blue

Violet

FIGURE 2.24 The Spectrum of the Sun   Overlaying the continuous rainbow of colors from the Sun is a richly detailed collection of dark lines. (◆ Also in Color Gallery 1, page 9.) (These are not easy to find with an ordinary prism; this image was produced by a modern, high-resolution spectroscope system. Fraunhofer saw many of these lines with his equipment.)

spectrum that looks like this—a rainbow with dark lines where certain colors have been absorbed out of the light—is called an **absorption spectrum**.

Some sources are just the opposite; they are all dark everywhere, except for a few specific bright lines of certain colors. This is called an **emission spectrum,** caused when a gas of some particular elements is energized somehow, by, for example, electricity or an explosion. The glowing gas then contains only the colors associated with its elements, and its spectrum is not a continuous rainbow of colors. Figure 2.25 shows a comparison of emission and absorption spectra for the element sodium, a component of table salt. Figure 2.26 shows the arrangements that generate each type of spectrum. Both the emission and absorption spectra are *line spectra*, as contrasted with *continuous spectra*, which are smooth rainbows.

At this point, it's most important for you to understand that the line spectra we're discussing here (like the one in Figure 2.24) are always present in starlight and serve as "fingerprints" by which we can identify the elements that can be found in stars. This is true for emission and absorption spectra. If you're wondering what actually *causes* the lines, don't worry: We'll return to that in Chapter 8, when we need to know for the sake of understanding cosmological observations. For the time being, we're just listing *what* we can observe, not *why* we can observe it.

The line patterns have tremendous importance because they enable us to identify elements from far away just by looking at the spectrum of light they emit or absorb. This identification is crucial in cosmology for several reasons. To name just two, first it tells us that objects in the distant universe are made of the same stuff that we see locally. Second, it gives us a reference for what stars' and galaxies' spectra are *supposed* to look like. (For example, from Figure 2.23, they should have recognizable hydrogen lines.) This situation becomes cosmologically interesting

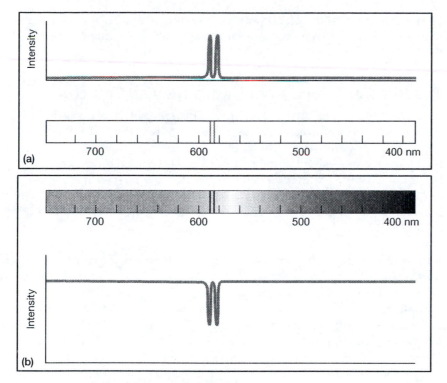

FIGURE 2.25 Emission and Absorption Spectra of Sodium   (a) emission spectrum; (b) absorption spectrum. (◆ Also in Color Gallery 1, page 8.)

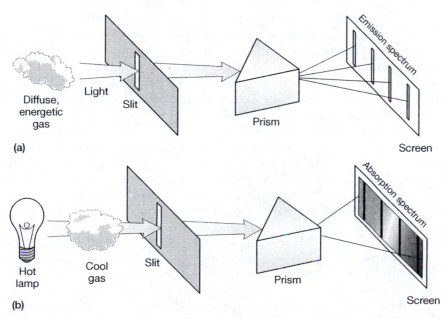

FIGURE 2.26 What Causes Each Type of Spectrum   The configuration for (a) emission, involving the glow of an energized gas, and (b) absorption, involving a nonenergized gas placed in the path of a continuous source.

when the universe starts messing with our reference lines. (Await Section 5.1.) So, because of their importance, we will use spectra like these many times throughout the book.

The Doppler Effect: Light from a Moving Source

Another consequence of the wave nature of light will prove very useful in our investigation of the universe. Known as the **Doppler effect,** the phenomenon was first explained by Christian Doppler in 1842. You have almost certainly experienced the Doppler effect on sound waves. If you watch a train go by and listen to its whistle, the pitch of the whistle is higher when the train is coming toward you and switches to a lower pitch after the train passes you and is moving away. This effect is distinct from volume: The volume is louder when the train is nearer, but the pitch is higher—a higher musical note—when the train is approaching, regardless of how close it is. As the train moves away, it gets quieter and quieter with distance, but it starts and remains at the same low pitch the whole time—lower than what you'd hear if you were riding on the train, moving along with it. The change in pitch is due to the Doppler effect. Musical pitch corresponds to a sound wave's *frequency*, so, for light, the effect causes a change in *color*.

To understand why the Doppler effect occurs, refer to Figure 2.27 as you follow the explanation. In Figure 2.27(a), a source of light is at rest relative to the observers. Waves generated with a certain frequency propagate out into space with a certain wavelength that is the same in all directions. (Think of ripples from a rock dropped into a pond.) In Figure 2.72(b), the situation is changed because the source of waves is moving to the right. Each ripple travels out through space, *centered on the point from which it was emitted*. But now this source is moving to the right, so the center of each ripple is farther to the right the more recently it was emitted. Notice the effect this has on the wavelength of light received by the observers: The observer on the right (with the source moving toward her) sees the ripples closer together (i.e., the wavelength of the light is *shorter*). The observer to the left (with the source moving away from him) sees the ripples spread apart (i.e., the wavelength of the light is *longer*). The increase in wavelength is referred to as a **redshift,** because red is at the long-wavelength end of the visible spectrum; the decrease in wavelength is referred to as a **blueshift,** because blue is at the short-wavelength end of the visible spectrum.

The Doppler effect is the principle behind radar guns used by police to catch speeders and by baseball scouts to measure the speed of a pitcher's fastball. The radar gun sends out radio waves of a known wavelength, which bounce off the moving object. By measuring how much the wavelength has changed for the return signal, the speed of the object can be calculated. We will make use of the Doppler effect in a similar way to measure the speeds of stars and galaxies.

Mathematically, the procedure works as follows: We take some known wavelength of light, such as $\lambda = 656.3$ nanometers (nm, billionths of a meter), for a particular shade of red in the spectrum of hydrogen gas. (Revisit Figure 2.23.) This line is generally present in the spectra of stars, but the star we are interested in happens to be moving toward us. As a result, we see every feature, including this red line, Doppler shifted by some amount. The shift of some particular line (e.g., this red one), in nanometers is labeled $\Delta\lambda$. For example, if $\Delta\lambda$ is 1 nm, then we observe the 656.3 nm line at 655.3 nm instead (a slightly more orange color). The formula for the Doppler shift is

$$\frac{\Delta\lambda}{\lambda} = \frac{v}{c},$$ (eq. 2.3)

which means that

> the shift in wavelength, as a fraction of   the speed toward or
> the original wavelength = away from us, as a
>   fraction of the
>   speed of light.

With this formula and the given information, we calculate that the star is approaching us at speed v = (1 nm/656.3 nm) × (300,000 km/s) = 457 km/s. If the star were moving *away* at that speed instead, then we'd observe the hydrogen line at 657.3 nm. Researchers can "measure" speeds toward or away from us with this

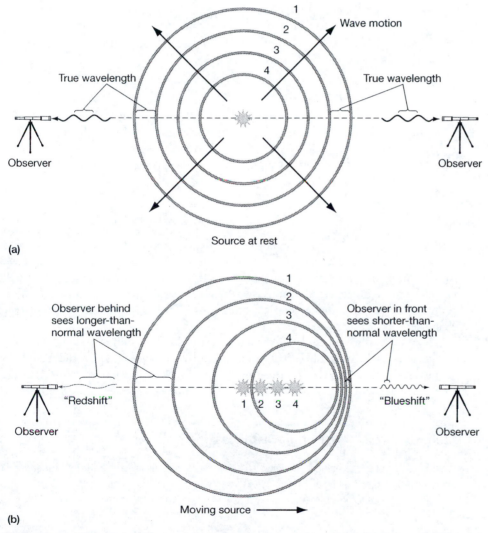

(a)

(b)

FIGURE 2.27 The Doppler Effect   Light from a moving source has its wavelength shortened if it is moving toward you, lengthened if it is moving away from you.

technique anytime they find a spectral line that they recognize, so that the "normal" λ is known. In practice, recognizable lines are almost always available.

2.5 Putting Your Knowledge of Light to Use

Let's pause for a moment to take stock of where we are. We set out to learn about our cosmic context and start answering some of the questions we came up with in Chapter 1. We began a tour of what we could see by looking at the night sky with only our eyes. We then expanded our vision, using modern telescopes as a tool. Now, armed with the deeper understanding of light gained from Section 2.4, we can learn more about the objects whose light we see with our eyes and through telescopes. For starters, we'd like to know their locations in space, identify the kinds of light they emit, and understand their motions.

Here's a list of the key pieces of knowledge we will use:

- Light travels at 300,000 km/s, so there is a delay between when light leaves its source and when it reaches an observer.
- Light comes in different frequencies (wavelengths), because it is generated by events happening with different amounts of energy.
- We can use a spectroscope to show us how much light of each frequency (or, equivalently, of each energy, color, or wavelength) we receive from a particular source.
- The Doppler shift applied to light tells us about the speed of an object moving toward or away from us. (That this is so is a consequence of light's wave behavior).

If we are going to make any progress in mapping our surroundings, we need a starting point to provide a foundation. Ideally, we would like a way to measure the distances to some of the objects we've been describing that does not rely on knowing anything about their size or what they actually are. If we can measure the distances to some objects by methods that do not depend on the details of the objects, then we can set up some benchmarks to use as references to compare with objects that are even farther away.

So, how can we measure the distances to the kinds of things we have seen so far on our tour of the night sky? First, think about some ways you know that enable you to measure distances to more familiar objects or places on Earth (Exercise 2.2). Maybe we can adapt them to measure the distances to the objects we see in the night sky.

EXERCISE 2.2 Measuring Distances

What are some of the techniques you know about for measuring distance on Earth? Which ones might work for objects in the night sky? Try to stretch your thinking to imagine as many ways of measuring distance as possible. Don't limit yourself to rulers or tape measures.

The fact that light travels at the finite speed of 300,000 km/s (rather than moving from place to place instantaneously) gives us one way to measure distances. We can simply time how long it takes light to get somewhere. You apply the same principle whenever you use time as a way to describe how far you've gone, based on the speed you're traveling. For example, if you drive at a steady highway speed of 60 miles per hour and it takes you an hour to get from one town to the next, you know that the distance between the towns is about 60 miles. This is a simple application of the relationship distance = speed × time.

In astronomy, this technique is sometimes a good way to measure distances, using light, of course, instead of a car. Every second of light travel time corresponds to 300,000 km (about 186,000 miles) of distance. Light travel time has even developed into a system of units for describing distance. One *light-second* is the distance a pulse of light travels in 1 second, a *light-minute* is the distance it travels in 1 minute, and a **light-year** is how far it gets in a year. We could invent units like this for distance traveled on the highway as well: In the driving example from before, we could just as well have said that our distance was "1 car-hour." But since cars don't always travel at the same speed, such units wouldn't be nearly as useful as those based on the constant speed of light.

The distance to the Moon can be measured with this technique by bouncing a pulse of laser light off of mirrors left at various locations on the Moon by the Apollo missions. (Of course, we also have a pretty good measure of the Moon's distance from the travel times of the rockets that made the trip there, but timing the laser light is more precise.) A tiny fraction of the original light makes it back to Earth and is picked up by a telescope. The distance to the Moon can be calculated very accurately (to within a few *centimeters*) by measuring the time it takes for the light to complete the round trip (about 2.6 seconds). For a one-way trip to the Moon, the average distance comes out to 384,401 km, or about 1.3 light-seconds. That number would be substantially bigger in "car-hours," even the way you drive. The light used for timing need not be visible light. **Radar** (an acronym for "RAdio Detecting And Ranging") is a familiar method for calculating the positions of airplanes and other objects on Earth by bouncing radio waves off the objects and timing the delay for the return signal. This method is also used in astronomy to determine distances to planets such as Mercury.

A significant difficulty with using the timing of light to measure distance is that you have to either send the light out yourself (recording the time when you sent it) and wait for it to bounce back or else have a way to know the time when the light you see now actually left the source. Since we don't have an independent way to know when the light really left an object (we only know when it arrives at our eyes), we're limited to using this method on things like the Moon and planets that are close enough to send our own signal to and wait for it to bounce back.

For anything very far away, we'd have to wait too long for the return signal, and after traveling such a long distance, it would probably be too faint to detect anyway. Even if it were possible, for example, to measure the distance to the center of our galaxy by radar, it would take over 50,000 years! So we need to continue looking for distance-measuring methods that can be used for a wider range of objects. We can always hope that there might just happen to be one in an upcoming subsection to get us started, and then maybe some more in the next chapter to keep building up our view of the universe.

Light as a Time Machine

It may have occurred to you that since light takes some time to get to our eyes after it leaves an object we see, *looking out in space also means looking back in time*. We don't see things as they are right now. Rather, we see them as they *were* when the light left them. This means that we effectively have a time machine which can peer into the past. Even though in astronomy we are limited to waiting passively for the light, we also have light from billions of objects, at different distances, to draw from. So we have the opportunity to see samples of the universe at different times in its history. We're actually seeing into the past when we look at things on Earth, too, but the speed of light is so fast that this makes little difference. If we look ahead 1 km down the road, for example, we see an object as it was 1/300,000 of a second ago. In astronomy, however, we will discover that some of the objects we can see are so far away that light takes billions of years to get to us. We literally see those objects as they were billions of years in the past. The distance to the Moon has been our first step out into the vast ocean of the universe, and we have learned that it is about 1.3 light-seconds away. So every time you look at the Moon, you're really seeing it as it was 1.3 seconds ago.

In today's fast-paced, competitive, global marketplace, 1.3 seconds is *way too long* to wait for information about current conditions on the Moon. We demand instant updates about vital information—such as stock prices. You can't be expected to wait 1.3 seconds to see a valuable new moon rock glinting in the sunlight through your backyard telescope, because you need to buy shares of Moonrock International right away! Maybe there should be an Internet relay up there, so that discoveries on the Moon can be radioed down to us the instant they happen there. That'll fix things, right? Sadly, no, it won't: The radio wave would have to travel here, too, at the speed of light, so either way, you're going to wait at least 1.3 seconds to learn about the discovery. The only good news is this: Everyone else on Earth has to wait, too.

Triangulation (Parallax)

Understanding Parallax

Another technique for measuring distance is *triangulation*. It's based on a very familiar experience: If you view a nearby object from two different vantage points, it appears to shift position relative to objects in the background. This apparent shift in the position of a nearby object relative to faraway objects when you change the location of your observation point is called **parallax.** A common example of parallax is illustrated in Figure 2.28, which shows a stop sign viewed from two different locations against a row of distant trees.

What measure should we use to describe the amount of parallax? Can we describe the shift in position of the stop sign between the two pictures in Figure 2.28 as so many meters? This would seem odd, since the stop sign itself hasn't really moved at all; only we have moved to an observation point at a different location. The situation brings to mind our earlier discussion of *angles* for measurements of sizes on the sky (Section 2.1, Figure 2.2). The view from position A in Figure 2.28(b) tells us only that the sign falls along the same line of sight as the right-hand tree. We can't determine from this view alone whether the sign is almost touching the tree or whether they are miles apart along that line of sight. Similarly, the view from position B tells us only that the sign is in the same direction as the left-hand tree, not how far apart they are along that new line of sight. So, just as we measured distances between stars on the sky as angles, parallax is properly described as an angle. In this case, the angle is how much the line of sight from the stop sign to the observation point must be rotated when we switch our viewpoint from position A to position B.

Figure 2.28(a) shows an external view of the arrangement that produces the pictures in Figure 2.28(b). The lines of sight from the two observation points to the stop sign intersect with different trees, and the parallax is the angle formed be-

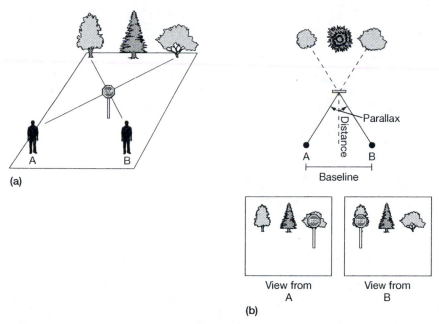

FIGURE 2.28 Parallax and Lines of Sight (a) A sketch of the arrangement and lines of sight producing the parallax of the stop sign. (b) View from above the scene, defining the geometry involved in parallax (baseline length, unknown distance to object from observer, parallax angle). The view as seen from each observer's location is inset below the observation point.

tween the two lines of sight where they intersect at the stop sign. A parallax angle of 10°, for instance, means that someone standing *at the stop sign* would have to rotate his or her line of sight 1/36 of the way around the complete circle centered on the sign in order to move from looking at observation point A to looking at observation point B.

Next, we want to figure out how to turn these observations into a method for calculating the distance from the observer to an object we see (such as the stop sign). To accomplish this, we need to look at how the amount of parallax changes as the location of our observation point changes. Figure 2.28(b) diagrams the layout as seen from above, with key features of the diagram labeled.

- The **baseline** measures how far apart the two observation points are from each other. Usually the baseline is known, because we can easily measure how far apart the two points are as we move between them.

- The **distance** refers to the distance from the observer to the nearby object (the stop sign in this case). Since the observer actually has two locations, we define the distance to mean how far it is from the nearby object to the midpoint of the baseline, because that is the average position of the observer. In practice, it won't matter much whether we define it this way or as the distance from A to the object (or from B to the object), since interstellar distances in astronomy are always much bigger than the baseline, which has to be small enough for us to physically cross.

- The **parallax** is the angle formed by the two lines of sight where they intersect at the stop sign.

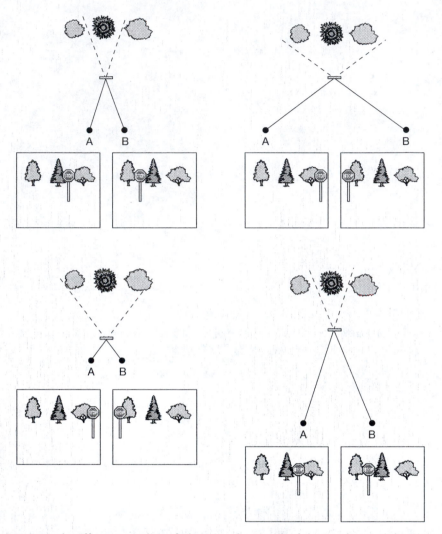

FIGURE 2.29 The Effect on Parallax of Changing the Length of the Baseline and the Distance to the Object   As in Figure 2.28, the view from each vantage point is inset below that point on each diagram.

There are two very important conclusions to take away from the diagrams in Figure 2.29:

- If the distance to an object remains constant, the parallax increases as the baseline gets longer.
- If the baseline length remains constant, the parallax increases as the object gets closer.

Since the amount of parallax is related to the distance to the object, we should be able to *measure* distance by using this effect. We can measure the parallax angle directly from the pictures we take from the two points of view. We can measure the baseline length because we traveled along the baseline to get from one observation point to another or we have a friend at the other observation point and we can communicate to figure out how far away she is. So if we can figure out the numerical relationship between the parallax and the distance we just inferred from Figure 2.29, then the parallax and the baseline length we measure directly will tell us the unknown distance. The technique is called **triangulation** because it uses the geometry of a triangle.

The basic phenomenon of parallax is something you see and use every day, although you may not have noticed it or realized how you could use it as a way to measure distance. This simple phenomenon is behind the fact that the Moon appears to follow you when you drive along a road at night. Without you even thinking about it, your eyes and brain solve the geometric problem of triangulation all the time, enabling you to perceive depth in the three-dimensional world around you. Our eyes are located on the front of our head, so we can look at an object with both eyes at the same time (unlike a horse, for example, which sees different things with each eye). The baseline for triangulation becomes the distance between your eyes. Your brain combines the images and processes the parallax (as in Figure 2.28(b)), and this is how you see the world in three dimensions.

EXERCISE 2.3   Parallax in Your Experience

Test your understanding of parallax by applying it to your experience with depth perception:

- Hold your finger a couple of inches in front of your nose, looking at your finger against a background of distant objects. Notice that if you let your eyes relax, you actually see two images of your finger, in different locations against the background. Note the similarity to the two pictures from the two points of view in Figure 2.28(b).

- Close one eye and then the other eye instead, and notice what happens. Does your finger appear to move *with respect to the background*?

- Repeat, this time with your finger farther from your face. When you alternate eyes again, does your finger appear to shift more or to shift less with respect to the background than it did before, when your finger was closer?

- Now close one eye, move your head from side to side, and notice what happens. When you move your head more, does your finger appear to shift more or to shift less with respect to the background?

- On the basis of what you observed, can you describe a qualitative relationship between the amount your finger shifts and the length of the baseline along which you move your perspective? Write down this relationship in your own words. Sketch a simple diagram showing the geometry of what is happening.

- Predict what would happen if you moved one finger even farther from your eye (say, a full arm's length) and then moved your perspective the same amount as before by switching eyes. *Then* try it and see. Was your prediction right? If not, make sure you understand what happened.

Triangulation in Astronomy

Triangulation is used by astronomers to measure distances to some of the nearest stars. In fact, it forms the foundation for the framework of astronomical distances we'll be building up throughout this book, because it's really one of the most direct methods for measuring distances to stars (i.e., it does not rely on models or assumptions). Our ability to notice parallax depends only on being able to see the object, not on the details of its properties, such as its composition or luminosity. So, it's a good match for what we need in order to measure distances to stars, which appear only as points of light, with no visible structure.

Because the distances involved are so great, the parallax angles we generally encounter in astronomy are much smaller than the angles we are used to dealing with in everyday life. Even when we use the longest baselines we can, the parallax of the closest stars seen from our two observation points is still only a tiny fraction of a degree. This is the reason we need sophisticated equipment in order to measure distances in astronomy based on parallax. It's not that the *concept* is so strange or unfamiliar; it's just that measurement of the angle requires extreme *precision*.

The limitation on measuring distances by parallax is thus a limit on how precisely we can measure angles and how long a baseline we can use. As we learned earlier, the parallax will be larger (and thus easier to measure) the bigger the baseline length is. So, when measuring distances to objects in the sky, astronomers try to use the biggest baseline they can.

FIGURE 2.30 Parallax Using Earth's Path Around the Sun as a Baseline; Defining the Parsec
(a) The geometry involved: Observations made six months apart provide a baseline that is twice the distance from the Earth to the Sun (i.e., 2 AU; see text).
(b) The view we see from the two different positions of the Earth. Note that the nearby star has shifted its apparent position relative to the background stars.

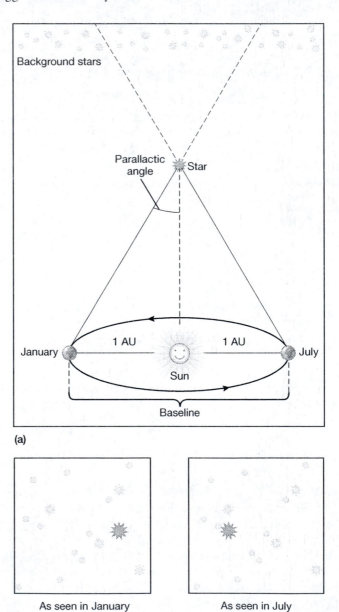

(a)

(b)

The distance to the Moon can be measured by observers using parallax in different locations on Earth. Unfortunately, nothing beyond our solar system can be measured that way. Even from opposite sides of the Earth, there is no noticeable parallax associated with *any* stars relative to each other. They must be so far away that the change in angle is too small to notice. In order to find a noticeable parallax for even the nearest stars, we must use *the Earth's motion around the Sun as the baseline* for our observations. How can we do this? Simple: Since the Earth takes 12 months to move around the Sun, we just wait half that time, or 6 months, from one measurement to the next. (See Figure 2.30).

As with our other illustrations of parallax, Figure 2.30 shows both a diagram of the situation and pictures of what we see from each of the two observation points. The average distance from the Earth to the Sun (about 150 million km) is called one **astronomical unit, or AU.** (The distance to the Sun is measured by bouncing radar signals, as mentioned in the previous section, off of the planet Venus at various points in its motion around the Sun and then using the distances measured by radar to infer a central position for the Sun. Astronomers must use this indirect method because the Sun has no solid surface capable of reflecting a radar signal.) By convention, the parallactic angle for this configuration is defined with a baseline length of 1 AU (which is confusing, since it's a 2 AU baseline that's actually used to measure parallax). Thus, as shown in Figure 2.30, *the parallactic angle for the configuration depicted is only half of the total angle.*

Parallax Distances to Stars

The *Hipparcos* space-based telescope measured parallax angles that are tiny fractions of a degree, and the planned *GAIA* telescope will be able to measure angles equivalent to the width of a human hair in San Diego, viewed from Philadelphia! To talk about angles this small, we need units much smaller than degrees. The standard convention is to divide degrees into **minutes of arc** (or **arc minutes**) and **seconds of arc** (or **arc seconds**). These have nothing to do with time; they are measures of angle, but the number of divisions works as it does with time: There are 60 arc minutes in 1 degree and 60 arc seconds in 1 arc minute.

With the convention for defining parallax as the half-angle measured with the Earth's 2 A.U. baseline, we can define a new unit of distance: the parsec. One **parsec** (short for "PARallax SECond") is defined as the distance at which an object will have a parallax of 1 arc second when viewed with a 1 AU baseline. That is, if the parallactic angle shown in Figure 2.30 is 1 arc second, then the object in question is 1 parsec from the Earth. It's also 1 parsec from the Sun, because the distance to the object in this case is much greater than the 1 AU distance between the Earth and Sun—similar to the way your nose and your left ear are both the same distance from the North Pole when you're standing near the equator.

For reference,

$$1 \text{ parsec} = 3.26 \text{ light-years.}$$

That is, it takes light 3.26 years to travel a distance of 1 parsec. As we'll see when we study astronomical distances in Chapter 3, this is wildly different from solar system distances such as the AU: parsecs and light-years are around 100,000 (that's 10^5) times larger; interstellar distances simply turn out to be much bigger than interplanetary distances. Now, with these definitions, and again referring to Figure 2.30, we see that the relationship between a star's parallax and its distance from us becomes very simple:

$$\text{distance (measured in parsecs)} = \frac{1}{\text{parallax (measured in arc seconds)}} . \text{(eq. 2.4)}$$

Professional astronomers usually quote distances to stars in parsecs rather than light-years. But the term "light-year" is somewhat more descriptive, and we will use it throughout this book.

An early measurement of the parallax of a star, with Earth's orbit as a baseline, was made by Friedrich Bessel in 1838. He found a parallax of about one eleven-thousandth of a degree for one of the closest stars, called 61 Cygni. This means that the star is about 10 light-years from Earth. (The modern measurement is 11.4 light-years.) Just imagine the radio-based conversation we could have with an alien civilization orbiting around 61 Cygni. We say, "Hello? Is anyone there?" and 11.4 years later they hear us and they reply. It's been nearly 23 years by the time we *receive* their reply: "Yes, we hear you; what's up?" By then, whoever originally sent the message from Earth could be dead! (Besides which, this is a pretty dull conversation; the real excitement would be in wondering why the 61 Cygnians speak English.) The bizarre nature of this conversation results from the vast distance to even a nearby star; a distance we can discover using the geometry of triangles.

We have gone into great detail describing parallax partly because it is such an important distance-measuring tool in astronomy, but also because it illustrates a general theme we want you to remember throughout this book: *Complex technical methods and tools used to draw conclusions in science are built up from simple, familiar ideas.* In the case of parallax, you have seen this connection evolve from your own observations of distances, to astronomical distances.

EXERCISE 2.4   Distance to an Unknown Object

A new and very bright object has appeared in the night sky. There is a debate going on about whether the object is within Earth's atmosphere or whether it is far away, among the distant stars. Use these two photographs and your knowledge of parallax to see if you can figure out the answer. Both pictures were taken at the same time, one from Oregon and the other from New York. Is the object closer than the Moon? Can you tell from these images?

Measuring Motion

In addition to knowing the locations and types of stars in our vicinity of the universe, we would like to know how those stars are moving. We would like to know the **velocities** (speeds and directions) of the stars. How much distance do they travel in a certain amount of time (e.g., how many km are they covering per second?), and in what direction are they traveling? Are they just hovering there stationary? Are they moving mostly toward or away from here (or from some other point in space)? Are they darting about randomly in all directions?

On the basis of the measurement techniques we're about to describe, it makes sense to separate the motion of a star (or any other object we see in the sky) into components of *transverse velocity* (motion across, or perpendicular to, our line of sight) and *radial velocity* (motion along our line of sight, directly toward or away from us). Figure 2.31 illustrates the meaning of the transverse and radial components of velocity.

Transverse velocity is related to the kind of motion (called *proper motion*) that we can observe directly on the sky. Proper motion is movement from one spot to another on the sky, relative to other stars (like what we see with planets or the Moon). As such, proper motion is measured with angles, while transverse velocity is physical distance divided by time. If we know the distance to an object, which enables us to convert angles into physical distances, then we can measure the transverse velocity of an object by watching its proper motion.

Radial velocity, by contrast, is exactly the type of motion we *cannot* see directly as movement, because such motion along a line of sight produces no change in position on the sky. (Refer back to Figure 2.2 to see this.) Fortunately, the Doppler effect is perfectly suited as a tool for measuring radial velocity, because it can *only* detect motion directly toward or away from us. Thus, by combining observations of proper motion with measurements of the Doppler effect, we have in hand the tools for measuring the motions of stars.

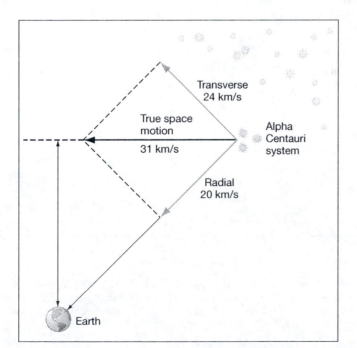

FIGURE 2.31 Components of Stellar Velocities   A star's true motion through space can be separated into the transverse and radial components of motion. The radial component is directed either toward us or away from us; the transverse component is every other kind of motion.

Measuring Radial Velocity with the Doppler Effect

How can we apply our knowledge of the Doppler effect to measure the radial velocity of a star? Recall from Section 2.4 (Figure 2.27) that the effect is based on the change in wavelength of light due to the motion of the object emitting the light. In order to apply the Doppler effect to learn the velocity of the object, we must know what the wavelength of the light *would have been* if the object were at rest. Then we compare that wavelength with the actual wavelength we receive. The more the actual wavelength has changed from the original wavelength, the faster the object must be moving.

The spectral lines we discussed in the previous subsection provide just the key we need in order to know what wavelengths we would have seen if the object had been at rest relative to us. Recall that each chemical element absorbs (or emits) light at particular frequencies (colors), so the pattern of lines is like a fingerprint for that element. If we see a fingerprint we recognize, except that all of the lines are shifted *by the same factor* from their usual frequencies, then we can use the Doppler effect to infer the speed required to produce that shift. Figure 2.32 shows how this looks.

Measuring Transverse Velocity by Proper Motion

For planets, which we now know (by parallax and radar) are much closer to us than the stars, proper motions are obvious. In fact, the proper motion of planets is exactly the "wandering" we discussed in Section 2.2, from which the name "planet" is derived. Proper motions of stars, in contrast, are very small because the stars are so far away. Even if a star physically moves a long way across our line of sight in a certain amount of time (say, a year), the angle this represents is still tiny. Think of an airplane's motion. If it flies past very close to you, its speed is obvious and you must turn your head quickly to follow it. But at 30,000 feet above you, it seems to crawl across the sky. Now imagine a plane many light-years away, and you can see that even very large velocities will not produce much proper motion.

The star with the largest observed proper motion, Barnard's star, has a motion of only about 10 arc seconds per year. Figure 2.33 shows two photographs illustrating the proper motion of Barnard's star. The images were taken 22 years apart. Given the star's distance of about 6 light-years away, this proper motion corresponds to a transverse velocity of about 90 km/s.

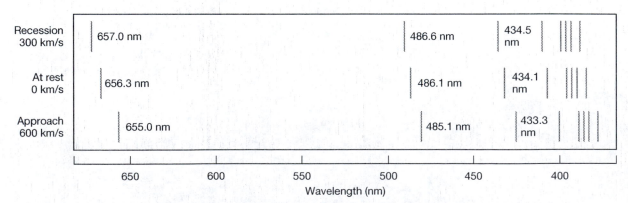

FIGURE 2.32 **Using Spectral Lines to Determine the Doppler Shift and Hence the Radial Velocity of Stars**   The middle spectrum shows the emission line pattern for hydrogen at rest relative to us (measured in the laboratory). The top spectrum shows the same pattern, but Doppler shifted because it is moving away from us at 300 km/s. The bottom spectrum shows the pattern for hydrogen moving toward us at a speed of 600 km/s. (◆ Also in Color Gallery 1, page 16.)

FIGURE 2.33 Proper Motion of Barnard's Star   Over 22 years, the position of Barnard's star moves visibly with respect to other (more distant) stars. This is not parallax, since both observations were made from the same location relative to the Sun. The shift is a proper motion, due to the star's actual motion through space.

By combining the radial and transverse components, we can make a map of the stellar motions around us. The conclusion is that stars are moving rapidly—speeds of 20 or 30 kilometers per second are common. The distances between the stars are so great, however, that even at these high speeds, a star in our vicinity would take tens of thousands of years to reach its neighbor. We find ourselves in the midst of a fleet of stars that appear to be flying around at random. But remember, we said "appear to be," because in the next chapter, when we have the machinery to measure the universe on larger scales, we'll see that the random appearance is partly deceiving.

Building on Everyday Experience

We began our exploration of the universe by using only our eyes to look up at the sky.

With the aid of technology, we've now extended our perception far beyond that direct experience under the canopy of stars. It's easy to feel disconnected from the sky we *see* when our observations are mediated by complex instruments and mathematical calculations. As we continue our investigation of the universe, we'll have to draw even more on technology to expand our awareness.

So it's worth pausing to recognize that every one of the complex instruments we have discussed so far emerged from a property of nature that we experience in daily life:

- Our eyes use natural lenses to collect and focus light, enabling us to discern properties of the objects emitting the light. Telescopes give us bigger eyes, enabling us to collect more light and see fainter things. Spectroscopes give us a greater ability to see colors, by spreading them out like a rainbow.

- Parallax is the same geometrical effect that makes depth perception possible. The only difference in astronomy is that we've refined the effect into a precise technique for measuring distance. To do so, astronomers have had to devise equipment to measure angles that are as small as a hair seen hundreds of miles (or more) away. But the principle is the same.

- The Doppler shift is just what you hear as you listen to the changing pitch of a passing train's whistle. Modern instrumentation enables us to use this effect to precisely measure the speed of a car, planet, or star from the change in color of the light emitted. But again, the underlying phenomenon is the same.

So don't let fancy equipment or math distract from what you are "seeing" with the aid of technology. It's as real as what you see with the aid of your eyes.

As you absorb the new views of the universe made possible by our enhanced vision, imagine what Galileo may have felt 400 years ago as he first turned his telescope toward craters on the Moon, the rings of Saturn, the moons of Jupiter, and the countless stars of the Milky Way.

Quick Review

In this chapter, you've begun looking out at the universe, first with your own eyes and then aided by the tools of modern technology. You've learned that we can discover many properties of the objects we see by knowing how to interpret the light we receive from them. These observations provide a first glimpse of the arrangement of objects visible in the night sky.

Try the following crossword puzzle to test your knowledge of key terms and concepts from Chapter 2:

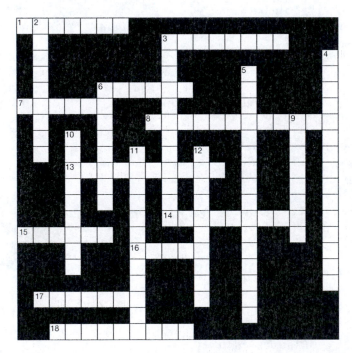

ACROSS

1 The Sun and Moon have about the same ____ size on the sky.
3 Graph of light intensity versus frequency or wavelength
6 Observable, _____ motion of stars across the sky
7 Cloudlike object in space
8 Change in the color of light from a moving source
13 A type of galaxy
14 The instrument that enabled Galileo's discoveries
15 Object that moves with respect to the stars
16 Hunter constellation, and the nebula within it
17 Discovered that the Milky Way is full of stars
18 A unit of astronomical distance, not time!

DOWN

2 Ten angstroms
3 300,000 km/s
4 Light is this type of wave
5 Distance to the Sun
6 Geometrical technique for measuring distances to nearby stars
9 A type of light with wavelength longer than visible light
10 Rate at which a wave cycles
11 Measuring how much of each color is in a light beam
12 Distance covered by one cycle of a wave

Further Exploration

(M = mathematical content; I = integrative—builds on information from more than one chapter; P = project idea; R = reflection; D = suitable for class discussion)

1. (M) How many car-hours (at 50 mph) would it take to drive to the Moon (about 240,000 miles away)?

2. (RD) A certain brand of lamp has the following claim written on its box: "This lamp emits no radiation." Do you think the claim is true? If so, would you buy the lamp? Explain.

3. (P) If Jupiter is currently visible from your location (check http://www.skyandtelescope.com/ to find out), use a pair of binoculars or a small telescope to re-create Galileo's observations of Jupiter. Look for the tiny star-like moons lined up near the planet, and sketch their positions over several nights to track their motion around Jupiter.

4. (M) A certain star has a parallax of one-tenth of an arc second when observed from a 1 AU baseline. What is its distance from us in parsecs? In light-years?

5. (M) The Moon is about 384,000 km from Earth. What would be the parallax of the Moon observed from opposite sides of Earth (assuming that you could see the Moon and stars at the same time from these locations)?

6. (IRD) Have you ever discovered or learned something about the world that really made you stop and think, and rearrange your assumptions? Write a short essay about that experience, describing the difference in how you saw the world before and after your discovery. Imagine what it must have felt like for Galileo to see that the Milky Way was made of individual stars.

7. (IRD) Consider these words by writer Edwin Dobb, and compare his sentiment with your own experience of looking up at the night sky: "What I felt was not diminishment but dilation. My whole being expanded; I felt intensely alive . . . And I think I know why. Precisely when we grasp the vastness of the universe we also glimpse an equally vast interior, the enormous geography of the soul, so to speak. Words may fail us afterward, forcing us to rely on hackneyed descriptions that emphasize our insignificance, but what we actually sense, if only for an instant, is largeness of spirit." (*Harper's*, Feb. 1995, p. 40)

8. (MR) What if you had eyes that allowed you to see with the sensitivity of the Hubble Ultra Deep Field image (Figure 2.9)? On the basis of a rough count of the number of galaxies in this image, estimate the total number of galaxies you could see by looking around the night sky, assuming that you had an unobstructed view of the entire sky. The image is about one-twentieth of a degree on each side, so you will need to estimate how many of these squares it would take to cover the whole sky and then multiply by your estimate of the number of galaxies in the image. Explain clearly the steps in your reasoning.

9. (RP) Use your knowledge of angular size to investigate the Moon illusion: the apparently bigger angular size of the Moon when it is close to the horizon than when it is high in the sky. Go outside a couple of hours after sunset on a night when the Moon is close to full. Find an object that just covers the Moon when held at arm's length. Then wait a couple of hours and repeat the experiment when the Moon is higher in the sky. Do the results match with what you expected?

10. (P) When you hold your arm fully outstretched, your fist covers about 10 degrees of angle and your first finger covers about 1 degree. Using your hand in this way, make the following angular measurements and indicate what you've discovered from each:

    - Pick a recognizable star, and measure its distance from the eastern horizon at several different times during the same night. Also, measure the angular distance from that star to another particular star several times during the night.
    - Make the same two measurements at several times during the same week, but always at the exact same clock time. If the Moon is visible when you're doing this, note the angular position of the Moon with respect to the eastern horizon each time.
    - Measure the approximate size of the Sun and the full Moon.
    - Measure the distance from the Sun to its closest horizon around noon. If time permits, do this every few days over the course of two or three weeks, always at the same time of day.

11. (P) On one clear night, sketch the location on the sky of the 10 brightest stars you can see. Draw neighboring stars so that you can identify the positions of the 10 brightest; circle each of the 10 to distinguish them from other stars. Try to make out what color those 10 stars appear, and draw that, too. Are the brightest stars always a particular color, or is the situation more complicated than that? Look up the names of your 10 stars on a star chart (you'll probably want one with you when you go outside to begin with, and one that connects the dots for constellations is easiest to use), and label your drawing with those names. (It's surprisingly pleasing to have this relatively minimal familiarity with the stars.)

12. (P) Track a planet against the background stars by sketching its location on several different nights over the course of a week. Remember that the word "planets" means "wanderers," and describe how much the wanderer wanders.

13. (P) Look at the color figure of the Hubble Space Telescope's Ultra Deep Field (See Color Gallery 1, page 5.) and count the stars. Don't worry—there aren't many. You can usually recognize stars by their "streaks"—crosses of light streaking from them, similar to what you see when you squint your eyes at a star or a streetlamp. Everything else in the photograph is a galaxy. Make a photocopy of the picture, or find it online and print it, and circle the stars.

14. (P) Go to the "Astronomy Picture of the Day" website (you can find it easily with any search engine), and search through recent pictures taken at nonvisible wavelengths like radio, infrared, or x-ray. Print a picture, and from the caption, write, in your own words, a brief explanation of what the false-color image shows.

15. (R) What is a "spectrum," for some observed light source?

16. (RI) Speculate on why there are differently shaped galaxies. (Review Figure 2.15.) To do this, invent a scientific theory, suggest what predictions that theory makes, and say what kind of observations could be made to support or disprove your theory. You don't have to be correct; you just have to be consistent in your use of the scientific method.

17. (R) Why does the speed of sound change with altitude? In very rough terms, explain how the wave nature of sound characterizes what its propagation speed should depend on. Does it matter what the actual sound is—thunder, music, etc.?

18. (MR) Visible light ranges in wavelength from about 400 nm (violet) to 700 nm (red). What frequencies are these colors? Can you see objects of this size range with your eyes?

19. (R) Imagine three clouds of hydrogen gas. One is sitting still, one is coming toward you, and the third is moving away from you. Sketch the spectrum of each, one on top of the other, with the same scale; label which is which.

20. (MP) Microwave ovens typically cook food using waves with a frequency of about 2.45 GHz (2.45 gigahertz, or 2.45 billion cycles per second). What is the wavelength of these waves? Compare your answer with the interior dimensions of your microwave oven—just measure its length, width, and height with a ruler. Note what's interesting about your result, and state the reason behind it. You can find the reason with a quick Internet search on how microwave ovens work.

21. (R) Why don't planets in our solar system visibly exhibit parallax when you observe them in the night sky with your left eye and then switch to your right?

22. (P) Sketch to scale the orbits of the four inner planets, assuming (slightly incorrectly) that their orbits are circular around the Sun. The radii are Mercury = 0.3 AU, Venus = 0.7 AU, and Mars = 1.5 AU.

23. (R) This chapter has covered observations of the night sky, often without any theory to explain them. In a few sentences, state why it might be valuable to distinguish direct observations from the theories that can explain them. Why shouldn't we just let it all run together? It's all science, after all!

24. (M) Light waves always travel at the same speed through empty space. Use that speed to find
    (a) the wavelength of your favorite FM radio station's broadcast. For example, if the station number is 99.7, then the station broadcasts at 99.7 *megahertz*, which is 99.7 million cycles per second. What kind of light is this (visible, ultraviolet, radio, etc.)?
    (b) the frequency of sunlight at a wavelength of 100 nm (100 billionths of a meter). What kind of light is this (visible, ultraviolet, radio, etc.)?
    (c) the frequency of red light that's just on the verge of becoming invisible because it's so close to infrared light.

25. (M) How much of a Doppler shift $\Delta\lambda$ would a feature with a wavelength of 500 nm exhibit if its source were moving toward or away from us at (a) 100 km/s (typical of stars within a galaxy) or (b) 1000 km/s (typical of galaxies)? How would your answers change if we were looking at a radio feature with a wavelength of 21 cm?

Color Gallery 1

COLOR FIGURE 1 Just a Dot   The *Cassini* spacecraft (NASA) snapped this photograph with the Sun positioned behind the planet Saturn. The lighting here allows us to appreciate some of the beautiful complexity of the universe: The intricate structure in Saturn's rings and the way their reflected sunlight manages to partly illuminate Saturn's night side. But there's more to see here: Just above Saturn's main rings on the left side is a small dot. That dot is our entire world, the setting for our civilization's brief history—embedded within a much larger expanse of space and time.

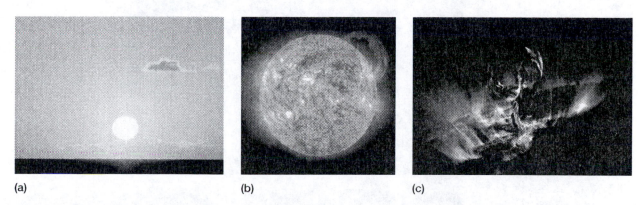

(a) (b) (c)

COLOR FIGURE 2 Zooming in on the Sun   The Sun (a) viewed from Earth, (b) viewed in ultraviolet light by a telescope on the SOHO spacecraft, and (c) even more magnified by the SOHO spacecraft telescope.

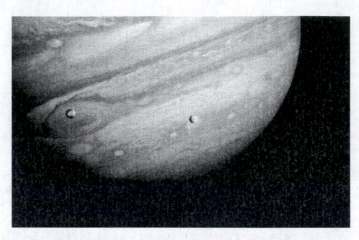

COLOR FIGURE 3 Jupiter and Its Moons.   This image, taken by the *Voyager 1* spacecraft in 1979, shows Jupiter and two of the moons (Io and Europa) discovered by Galileo. Io has active volcanoes, and Europa has a surface made of ice, probably covering an ocean of liquid water.

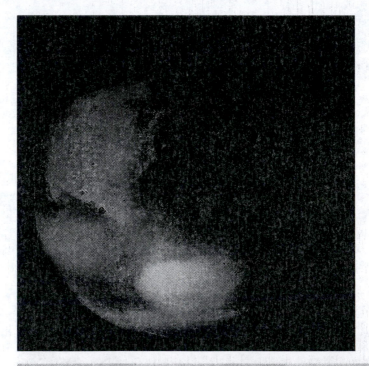

COLOR FIGURE 4 Views of the Planet Mars   What looks like a bright red star to the unaided eye is revealed as a world of its own when seen from the *Mars Global Surveyor* spacecraft (from space) and the *Spirit* rover (on the surface).

COLOR FIGURE 5 The Night Sky The well-known constellation of Orion the hunter is featured in this field of view. Betelgeuse, the star at the top left of the constellation, is visibly red to the naked eye; Rigel, at the bottom right, is visibly blue.

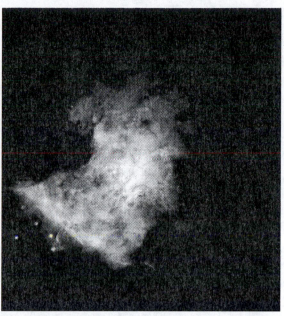

COLOR FIGURE 6 HST Image of Orion Nebula The tiny smudge in the Constellation Orion (toward the lower left in Color Figure 5) is actually an intricate nebula, revealed in this zoomed-in image taken by the Hubble Space Telescope.

COLOR FIGURE 7 Horsehead Nebula In the case of a nebula (gas cloud), a fuzzy-looking object remains fuzzy looking even when viewed through a telescope.

3

COLOR FIGURE 8 Globular Cluster M80 In the case of this object, what looks like a fuzzy cloud is revealed by a telescope to be a dense region containing hundreds of thousands of individual stars, all crowded into a region comparable in size to that typically occupied by just a few stars in our part of the Galaxy. There are about 150 known globular clusters like this one—clustered together in a cluster of clusters—associated with the Milky Way, but they do not lie in the Galaxy's flat disk.

COLOR FIGURE 9 Andromeda Galaxy This is another fuzzy object you can see in the night sky. On closer inspection, it reveals its spiral structure. The Andromeda galaxy is the closest spiral galaxy, similar in size and layout to our own Milky Way.

COLOR FIGURE 10 The Hubble Ultra Deep Field This is a picture of a region of the sky only about one-tenth the diameter of the full Moon that looks completely empty to our unaided eyes (and even to most telescopes!). Notice the apparent variety of sizes and shapes among the objects seen here. Apart from a few nearby stars, every object in this scene is a galaxy. The telescope is aimed at a region in the constellation Fornax, which is south of Orion. Total exposure time is about 280 hours.

COLOR FIGURE 11 Different Colors of Stars Stars emit different kinds of light, which are visible here as different colors. Notice that even when you enlarge stars, as in this telescope view, you don't see any more detail; they must be very far away indeed. Color Figures 5 and 8, each a few pages back, also show good color variation among stars.

COLOR FIGURE 12 Southern Cross Star Colors This image was produced by leaving a camera shutter open, so the stars make trails due to the Earth's rotation. During the exposure, the focus was changed (so the stars are increasingly out of focus) in order to make the different star colors more apparent.

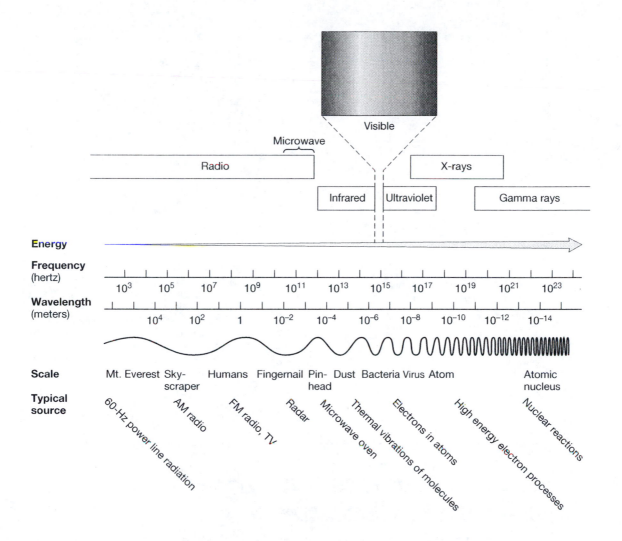

Visible

Microwave

Radio

X-rays

Infrared Ultraviolet

Gamma rays

Energy

Frequency
(hertz)

| 10^3 | 10^5 | 10^7 | 10^9 | 10^{11} | 10^{13} | 10^{15} | 10^{17} | 10^{19} | 10^{21} | 10^{23} |

Wavelength
(meters)

| 10^4 | 10^2 | 1 | 10^{-2} | 10^{-4} | 10^{-6} | 10^{-8} | 10^{-10} | 10^{-12} | 10^{-14} |

Scale   Mt. Everest Sky-   Humans   Fingernail Pin- Dust Bacteria Virus Atom   Atomic
                         scraper   head   nucleus

**Typical
source**

60-Hz power line radiation

AM radio

FM radio, TV

Radar

Microwave oven

Thermal vibrations of molecules

Electrons in atoms

High energy electron processes

Nuclear reactions

COLOR FIGURE 13 The Electromagnetic Spectrum   A wide variety of types of light can be organized simply along the spectrum of frequency or wavelength.

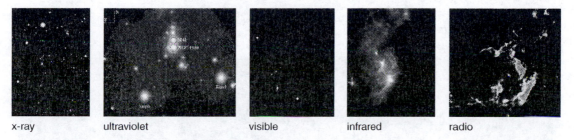

x-ray   ultraviolet   visible   infrared   radio

COLOR FIGURE 14 Orion in Different Kinds of Light   This is the constellation Orion viewed in a variety of "colors" of light. Notice how different the same region of the sky appears, depending on what kind of light we are recording.

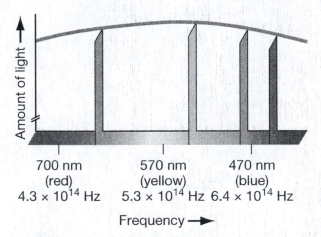

700 nm
(red)
4.3×10^{14} Hz

570 nm
(yellow)
5.3×10^{14} Hz

470 nm
(blue)
6.4×10^{14} Hz

Frequency →

COLOR FIGURE 15 The Spectrum of Light from the Sun Displayed are both the image you would see on a detector at the back of a spectroscope (rainbow along bottom) and a graph of how much light is hitting each region of the detector (and hence how much light of each kind is present in the original mix of light).

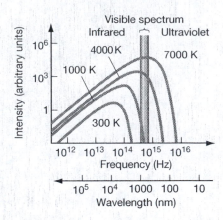

COLOR FIGURE 16 Blackbody Spectrum The blackbody spectrum reflects the pattern of photon energies for each temperature. For a cool object, the spectrum is lower in frequency (x-axis) and intensity (y-axis); that is, it's light is generally redder and dimmer. For a hot object, the opposite is true: It's bluer and brighter. This figure shows four curves representing the blackbody spectrum from each of four objects (such as stars) at different temperatures.

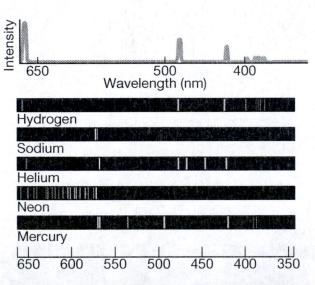

COLOR FIGURE 17 Emission Spectra of Some Familiar Elements These elements have been energized so that they will emit light, but they will emit only their "true colors." The specific set of colors for each element is distinctive, like a fingerprint at a crime scene.

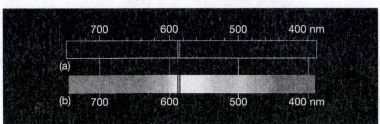

COLOR FIGURE 18 Emission and Absorption Spectra of Sodium Emission spectrum (a) and absorption spectrum (b) for sodium Atoms. The emission spectrum represents a sodium street lamp (c), for example.

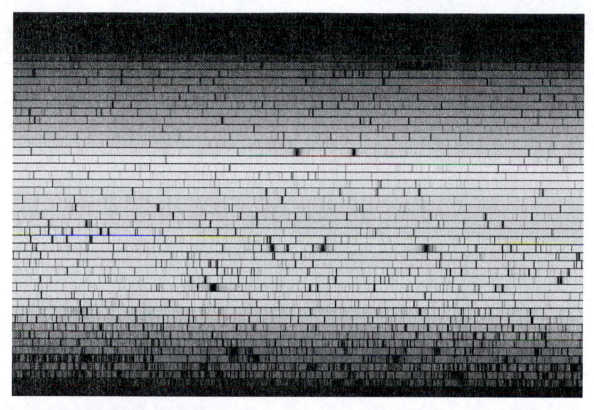

COLOR FIGURE 19 The Spectrum of the Sun   Overlaying the continuous rainbow of colors from the Sun is a richly detailed collection of dark lines due to sunlight being absorbed by atoms in the Sun's atmosphere. (These are not easy to find with an ordinary prism; this image was produced by a modern, high-resolution spectroscope system.)

COLOR FIGURE 20 Combining Doppler Measurements with Knowledge of Gravity   In this region called NGC 2467, the grouping of stars at the left appears to be a star cluster—a gravitationally bound system—because its appearance is similar to that of many other bound star clusters. In this particular case, however, the appearance is deceiving; their Doppler shifts show that these stars are moving too fast to be bound as a group.

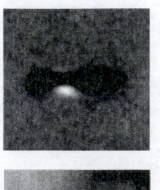
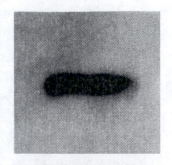
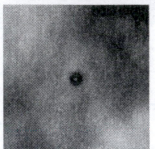
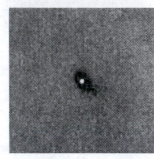

COLOR FIGURE 21 Infrared Images of Stars Forming in the Orion Region The glow observed from the centers appears to be that of small stars forming. The dark, dusty disks, seen approximately edge on in the two upper frames, are about 400 AU thick (about 7 times the diameter of our solar system). Therefore, if this system is to evolve into a planetary system like ours, the disk will need to slim down dramatically. In the two lower frames, the same situation is seen in a "top view" (we happen to view these from "above" rather than edge on).

COLOR FIGURE 22 Death of a Low-Mass Star A white dwarf star remains at the center of a planetary nebula in each of these images. These nebulae, made from material expelled by dying low-mass stars, include hydrogen, helium, carbon, nitrogen, and oxygen, some of the most abundant elements in the universe. Apart from helium, they are the most abundant ingredients in life as well.

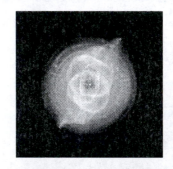

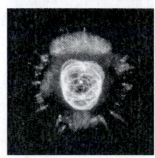

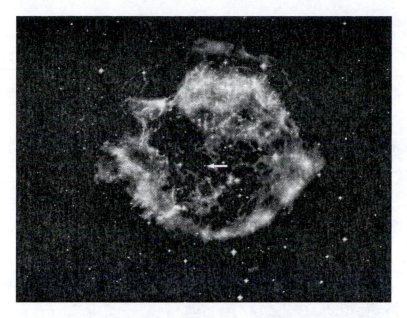

COLOR FIGURE 23 Death of a High-Mass Star   These are the remains of a nearby Type II supernova explosion, with the young neutron star it left behind visible as a small x-ray-emitting spot in the center (the blue dot pointed out by the arrow). This image combines infrared (shown as red), visible (yellow), and x-ray (blue) data. The explosion was only about 300 years ago, spawned by a high-mass star at the end of its life.

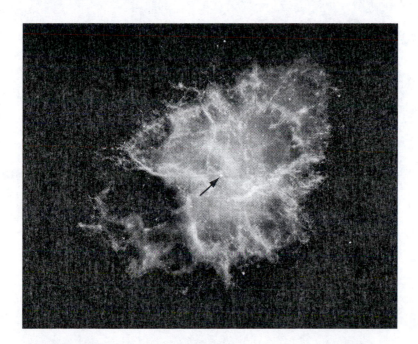

COLOR FIGURE 24 The Crab Nebula   This nebula is the remains of a relatively nearby and relatively recent supernova in our region of the Galaxy a little less than a thousand years ago. This image combines infrared (shown as red), visible (green), and x-ray (blue-purple) data. The bright x-ray spot at the center is the neutron star, called the Crab Pulsar, left over from the supernova.

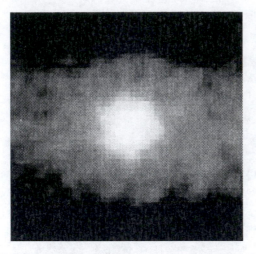

COLOR FIGURE 25 The Sun in "Neutrino Light"
Here we see what the Sun would look like if our eyes could see the neutrino particles created by the nuclear reactions in its core. False colors are assigned in such a way that white pixels appear where many neutrinos were observed, and red pixels appear where fewer were observed.

COLOR FIGURE 26 Neutrino Telescopes
Unlike normal photon telescopes for collecting light, neutrino telescopes need to be made from huge quantities of matter, in the desperate hope that a few of its many atoms will interact with at least a few of the weakly interacting incoming neutrinos. At left is the Super-Kamiokande neutrino observatory in Japan, in which neutrinos interact within a vast tank of water and are detected by photomultiplier tubes (like very sensitive cameras) lining its walls. At right is the Sudbury Neutrino Observatory in Canada, which operates similarly, but encloses a tank of heavy water (H_2O containing hydrogen-2), rather than normal "light" water (H_2O containing hydrogen-1). The heavy water makes the observatory sensitive to more neutrinos.

 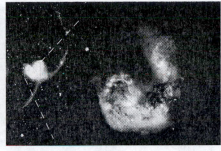 

(a)   (b)   (c)

COLOR FIGURE 27 Galaxy Mergers   (a) Two spiral galaxies, apparently likely to merge. (b) The "Antenna Galaxies," so named because their interaction has stretched them out into tails that resemble an insect's antennae. In the inset, distinct populations are being *separated* by the collision: Gas forms new stars (the brightest of which are blue, giving the blue appearance of this region), older preexisting stars are generally yellow, and dust is dark. Some stars energize the surrounding hydrogen gas, causing it to emit hydrogen lines, including the visible red color. (c) A computer simulation predicts the same antenna shape emerging.

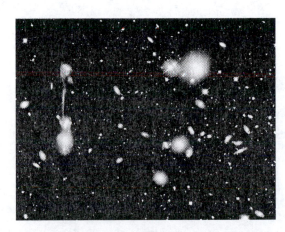

COLOR FIGURE 28 The Abell 1185 Galaxy Cluster
Here we see a galaxy cluster, rich with a variety of member galaxies: spirals, ellipticals, and even a guitar-shaped collision event in progress (on the left). The collision is visibly changing the shape of the colliding galaxies. Perhaps when it's done, the interaction will have created one or more new elliptical galaxies.

COLOR FIGURE 29 Galaxy Cluster
The densely populated "Coma galaxy cluster" contains thousands of galaxies crammed into a region so small that outside of a cluster, a region of this size would typically contain about a hundred times fewer galaxies.

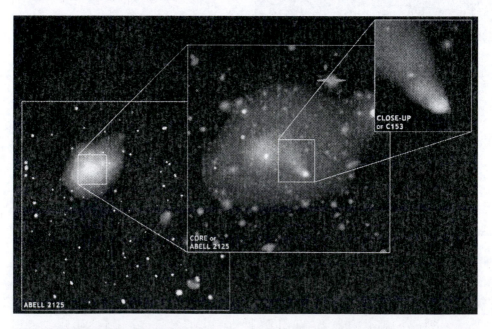

COLOR FIGURE 30 Birth of an Elliptical Galaxy?
Here in the Abell 2125 galaxy cluster, a blend of (false-color) optical and x-ray observations reveals a galaxy being reshaped and stripped of its gas (seen above as a cometlike tail) as it moves through the dense central region of the cluster. Without any gas to make new stars, short-lived stars will die, and only long-lived (red and yellow) stars will remain. If the galaxy had a spiral shape to begin with, this process is almost certainly erasing it.

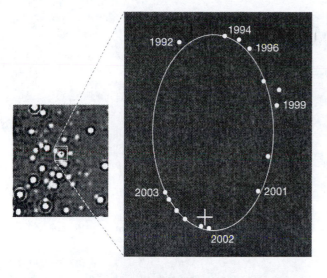

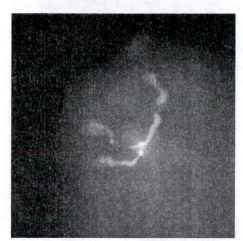

COLOR FIGURE 31 The Supermassive Black Hole in the Middle of the Milky Way Tight stellar orbits near the galactic center strongly imply that an SMBH must be their gravitational anchor. No other object would plausibly have such a strong effect on nearby orbits and yet occupy so little space; the left frame is only about 0.1 light-year across. The spots are individual stars, and the cross is the location of the SMBH. More detailed tracking of one of these stars (inset, right) reveals that it moves in a very close elliptical orbit around the central black hole, again denoted with a cross.

COLOR FIGURE 32 Binary Supermassive Black Holes X-ray (blue) and radio (pink) observations reveal two supermassive black holes at the center of a collision between two galaxies in a galaxy cluster. In this image, the two holes are orbiting one another; in the future, they will merge into one supermassive black hole.

COLOR FIGURE 33 Detection of Intermediate-Mass Black Holes The top frame is an image of the galaxy M82. The bottom frame is a false-color x-ray image of the core of the galaxy, taken by the Chandra x-ray satellite telescope. The bright sources near the center are thought to be black holes with masses 100 to 1000 times the Sun's mass.

(a)

COLOR FIGURE 34 Accretion onto a Black Hole   (a) Drawing of an accretion disk. (b) Observational evidence for an accretion disk in a galaxy's core. The evidence is based on the light spectrum from this source. We don't see an actual disk here; instead, we see that the light from one side of the source is redshifted and that from the other side is blueshifted. This implies that the accretion disk is there and is oriented in space so that at one end material is being carried away from us along our line of sight to it and the other side is moving toward us (similar to what's happening just left and right of center in frame (a)).

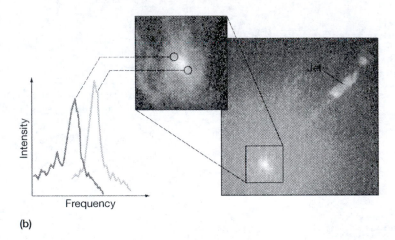

(b)

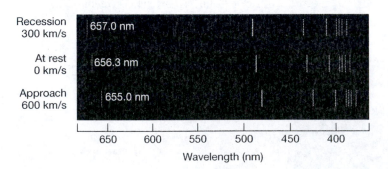

COLOR FIGURE 35 Using Spectral Lines and Doppler Shift to Determine the Radial Velocities of Stars   The middle spectrum shows the emission line pattern for hydrogen at rest relative to us (measured in the laboratory). The top spectrum shows the same pattern, but Doppler shifted toward slightly redder colors because the star emitting the light is moving *away* from us at 300 km/s. The bottom spectrum shows the same hydrogen pattern shifted toward bluer colors, from a star that's moving toward us at a speed of 600 km/s. It is this sort of redshift and blueshift that makes it possible to measure velocities toward and away from us, as in Color Figure 34 above.

The Universe We Discover *through Heat and Light*

"The universe now appeared to me as a void wherein floated rare flakes of snow, each flake a universe."

—*Olaf Stapledon*

The last chapter provided a glimpse of the challenge.

The sky is filled with points or blobs of light with varying structure and color. You looked at some of these objects with your own eyes in Exercise 2.1 and with the aid of telescopes in Section 2.3; with a web search, you can see many more. But as you can easily discover for yourself, it's difficult to figure out what (and where) the objects actually *are*. In terms of light, we see *what* they emit, but do we know *why* they emit it?

Sizes and distances present another difficulty. How would you know how big or how distant some light source in the night sky is? A little blob of light seen through a telescope might be a small cloud reflecting light in our own atmosphere, or it might be a giant glowing cloud of stars and gas, thousands of light-years away.

You now share the astronomer's challenge: You see the light from some of the objects in space, but how do you understand what you're looking at? That is, how do you figure out what something is from what it looks like?

The last chapter also provided a glimpse of the solution. To work it out, you must learn whatever you can about the nature of light itself and latch onto any little nuggets of knowledge that can help. In Chapter 2, you learned about light's speed and how to use that to measure local distances with radar. You learned about light's frequencies (and how those frequencies correspond to colors), including those beyond the range of your eyes; thus, you uncovered some properties of stars. You learned about light waves and Doppler shifts and figured out the speeds of objects moving through space. It seems that if we humans want to know about something, we can. Nature evidently gives us a series of hints, so that if we're willing to put in the effort to observe and learn (about light, about waves, etc.), then the answers are within our reach.

Nature and Human Nature

What we want to know seems to be available purely as a function of *desire* and *effort*; that is, if we desire to know something badly enough that we're willing to put in the effort to learn it, we usually can know it. Is there any obvious reason why this should be the case? We can accomplish or understand many things that we want to, but not simply by wishing them to happen. It always seems to be the case in the natural world that we must find the right pathways in order to make our wish happen. For example, a person might stand and look up at the distant Moon and wish to set foot on its surface—a seemingly ridiculous dream that was nonetheless made reality through the desire and dedicated effort of many people understanding and applying the laws of physics to build the equipment and navigate a spacecraft to land safely on the Moon and carry its passengers home again. Science often helps us make our wishes and dreams come true. As biologist E. O. Wilson writes,

> "Science offers the boldest metaphysics of the age. It is a thoroughly human construct, driven by the faith that if we dream, press to discover, explain, and dream again, thereby plunging repeatedly into new terrain, the world will somehow come clearer and we will grasp the true strangeness of the universe. And the strangeness will all prove to be connected, and make sense."

In this chapter, you'll acquire a deeper understanding of light; in the next chapter, you'll do the same with gravity. When you apply your new knowledge to the astronomical observations from Chapter 2, greater insights will follow. What are the objects in space like? How big and how distant are they? How and why do they move? All of this is necessary background for bigger questions in cosmology; by Chapter 5, you'll be equipped to appreciate our measurements of the universe itself.

We begin by learning a little bit about particles, and then we'll study a revolutionary insight into the nature of light, demonstrated by Albert Einstein in 1905. It takes some study to warm up to it, but once you do, the rest of the universe awaits. So take a slow breath, find a shady spot, and enter Section 3.1.

3.1 Starlight in a Particle World

Particles of Matter

We begin by summarizing the nature of matter—the material that makes up our physical substance:

- **Atoms** are building blocks of solid, liquid, and gaseous matter. The matter in our bodies, and in many astronomical bodies, is made up of atoms.
- Atoms come in many different varieties, called **elements**, such as hydrogen, helium, oxygen, and iron. Atoms of different elements have different masses, sizes, and chemical properties. (For example, copper conducts electricity and arsenic is poisonous to us.)
- **Molecules** are arrangements of atoms held together by chemical bonds; for instance, O_2 is molecular oxygen, with two oxygen atoms (O) chemically held together. Molecules come up less often in cosmology than do atoms or ions,

because astrophysical objects like stars are typically hot enough to dissolve chemical bonds. Molecules are important in forming a star, but they break up once things start to get hot.

Atoms are made from exactly three types of *particles*. In the current context, a particle is a distinct unit of matter. You're already familiar with the three we're talking about here:

- **Protons** are fairly massive as familiar particles go (on Earth, "massive" is the same as "heavy"), and they carry a single positive electrical charge, +1. A proton is denoted with a lowercase "p."
- **Neutrons** have approximately the same mass as protons, but they carry no electrical charge; that is, they are electrically neutral. A neutron is denoted by lowercase "n."
- **Electrons** are almost 2000 times less massive than protons or neutrons. Electrically, they carry a single negative charge, −1. They are denoted by lowercase "e^-."

As we saw above, these particles often reside in atoms, which have two distinct regions:

- The **nucleus** (plural: "nuclei") is about 100,000 times smaller than the atom itself—analogous to the size of a dime in the middle of a golf course. (The exact size ratio of atom to nucleus depends on the element.) The nucleus is made of one or more protons, zero or more neutrons, and no electrons. Therefore, its electrical charge is always positive and equal to the number of protons it contains.
- The **electron cloud** surrounds the nucleus. The cloud is a region occupied by one or more electrons moving in complicated ways around the nucleus. When people discuss the size of an atom, they refer to the electron cloud, since the nucleus is just a tiny speck within it.

Electrical interactions between particles follow the basic rule that opposite charges attract (+ and − charges attract) and like charges repel (+ and +, or − and −). That rule explains the presence of the electron cloud: The (negative) electrons in the cloud are attracted to the (positive) nucleus, so they normally don't fly away (and there are at least as many attracting plus charges in the nucleus as repelling minus charges in the electron cloud).

Still, sometimes one or more electrons do fly away from their home atom, generally because they have too much energy to stay put. This energy could be the result of a high temperature, for example. The removal of electrons from atoms is called **ionizing** or **ionization**. In a completely ionized medium, it's hot enough that the nuclei are typically stripped of all their electrons. For example, a hydrogen atom has only one electron, so a hydrogen nucleus remains behind when its atom is ionized. In a partially ionized medium, some atoms are missing one or more electrons, but other atoms are still neutral. High-temperature, ionized media come up often in astrophysics and cosmology, so it's worth distinguishing between the relevant arrangements of matter particles once more; also see Figure 3.1.

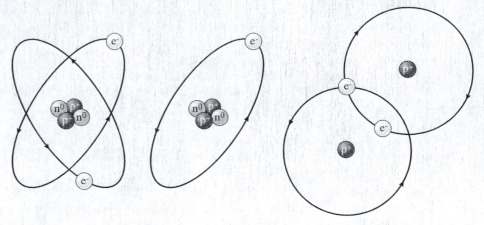

FIGURE 3.1 Atoms, Ions, Molecules   A neutral helium atom (left), a helium ion with one electron removed (center), and a molecule of hydrogen gas containing two hydrogen atoms, (H_2, right).

- **Ions** are atoms that are missing one or more electrons and are therefore positively charged, since they have more protons than electrons. A hot gas composed of ions and/or nuclei, and of electrons that have been stripped from their otherwise neutral atoms, is called a **plasma**. Plasma is a fourth phase of matter; the other three phases are solid, liquid, and gas.
- **Atoms**, or **neutral atoms**, have zero net electrical charge because they have equal numbers of protons and electrons.

Particles of Light

In the 1800s, the scientific community performed a series of important experiments on the nature of light. The big question of the time was essentially this: Is light made of particles or waves? What are the building blocks of a light beam? If the answer is particles, then a beam of light is like a spray of bullets from a machine gun: Each bullet represents a light particle. If the answer is waves, then a beam of light consists purely of a series of parallel rays, each with some sort of fluctuation back and forth, like guitar strings.

The experiments of the era were cleverly designed to pick up on wavelike behaviors—reflecting and refracting, dispersing into different wavelengths after passing through a prism, spreading in all available directions like water waves from a pebble tossed into a pond, and so forth. By 1900, every type of wave behavior had been convincingly measured in light. As a result of this evidence, the physics community overwhelmingly supported the wave model of light. A popular science author wrote this on the subject:

> *"By the end of the nineteenth century, only a genius or a fool would have suggested that light was corpuscular [made of particles]. His name was Albert Einstein."*
>
> —John Gribbin, *In Search of Schrödinger's Cat*

Einstein argued that light could also be a particle, and the experiments of the 20th century showed him to be correct. Particle-like behaviors could be measured in light, too. The best-known example is the photoelectric effect, in which a particle of light strikes an electron in a metal, sets the electron free from the nucleus it was bound to, and thereby produces an electrical current. Solar panels work by a similar effect. A significant characteristic of the photoelectric effect is that only certain colors of light produce a current. For low-energy colors of light that do not create a current, it doesn't matter what quantity (intensity) of light we shine on the metal; we still get zero current. But high-energy colors of light will produce a current even at very low intensities. It appears that what matters is the energy in each single particle of light, rather than the total energy in the beam.

None of these particle experiments, however, undoes the outcome of previous wave experiments. Interestingly, light has the properties of both a wave and a particle. When we construct experiments to find wave properties, we succeed and find them. But when we design experiments to look for particle properties, we find those instead. For astronomical purposes, we need to take full advantage of this knowledge and deduce the structure of objects in space from both the wave and the particle properties of the light we measure.

Like particles of matter (e.g., protons, neutrons, electrons), particles of light have been given a name. We call them **photons** (not the same as *protons*!). You can think of them as pennies of light—the smallest pieces of light available. Suppose you pick a particular type of light, such as red light. A red flash of light can't go any dimmer than a single photon. There are no half-photons, so if you dial the brightness down to a single photon, the only way to make it dimmer is to turn the light off entirely. As we'll see in this section and the next, characterizing light as particles can be very useful.

Now imagine the world as experienced by a photon. An individual photon wouldn't notice whole objects (trees, oceans, galaxies, department store mannequins, etc.); instead, it would sense the particles inside these objects. Knowing a little about how particles of matter interact with photons will help us learn more about stars.

EXERCISE 3.1   Energy in Light

Consider how a light wave or a light particle could deposit energy, which you feel warming your skin on a sunny day. Demonstrate that either view of light makes sense in this regard, by inventing a description of how each could cause sunburn. At this point, being vaguely realistic is important, but being fully correct is not. For example, in your wave description, remember that the wave is electromagnetic, and try to imagine how that could be harmful.

Particle Behavior and Energy

If we zoom in on our closest star, the Sun (Figure 3.2), we'll see that its surface is made up of **gas**—a collection of atoms moving around relatively freely and bumping into one another. Here's an opportunity for us to use knowledge of the behavior of particles in gases—knowledge based on theory and experimentation on Earth—in order to reveal the inner workings of distant astronomical objects. Of immediate interest to us for understanding stars is the behavior of a large collection of particles.

On the surface of the Sun, particles are bouncing around quickly, with a particular average speed. More speed means more energy. This is just as you experience things in daily life: If someone tosses a basketball gently *to* you, it doesn't transmit

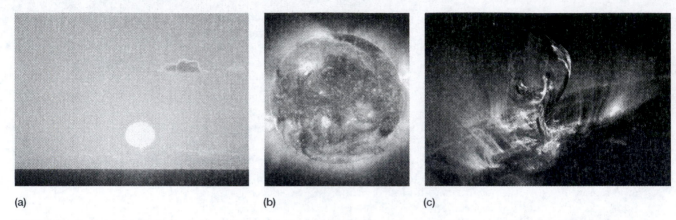

(a)   (b)   (c)

FIGURE 3.2 Zooming in on the Sun   The Sun (a) viewed from Earth, (b) viewed from a telescope in space (SOHO UV), and (c) even more magnified by the SOHO spacecraft telescope (◆ Also in Color Gallery 1, page 1.)

much energy. If someone viciously hurls the ball *at* you, you feel that energy when you try to catch it. There is a distinction to be made at this point about what kind of energy we're talking about:

- The energy associated with the *organized* motion of particles is called **kinetic energy**. For example, a speeding bullet is made of many atoms, and all of those atoms are moving *together*, from the barrel of the gun to the target.
- The energy associated with the *random* motion of particles, as in the Sun (or in air, for that matter), is called **thermal energy**. In daily conversation, this kind of energy is often referred to as "heat."

Fast, randomly moving particles in the Sun, then, correspond to its thermal energy. But it's tough to measure the speed of all the atoms in the Sun. You could go there with a high-power microscope, a tape measure, and a stopwatch, and then measure the speed of each atom, one by one. But this would take a really long time, and you would have trouble keeping track of which atoms you already measured. Also, you'd burn to death.

A better method would involve measuring some kind of *average* particle speed. We can give this average-speed concept a name; it's called **temperature**. Temperature is related to the thermal energy of a typical particle present in an object. In most cases we'll encounter, this is simply what temperature *means*: a measure of the average speed of an object's constituent particles.

The particle picture gives us an important insight into the nature of hot objects such as stars, and into the nature of thermal energy as well. Consider this: Heat can flow—for example, from a warm oven to a frozen turkey (Figure 3.3)—but can coldness also flow? The answer emerges quickly from the particle-level description of nature. Coldness must be associated with slow-moving particles. If you can cool a hot object by putting it in contact with a colder one, the fast particles will push the slow ones (imagine fast and slow billiard balls colliding), thus speeding up the slow particles and slowing down the fast ones. The hot object cools and the cold object warms. Energy flows from the hot object to the cold one; in that sense, heat flows, but coldness doesn't.

There exists an **absolute zero** temperature, corresponding to a bare minimum of atomic motion. This makes things confusing, since many temperature scales

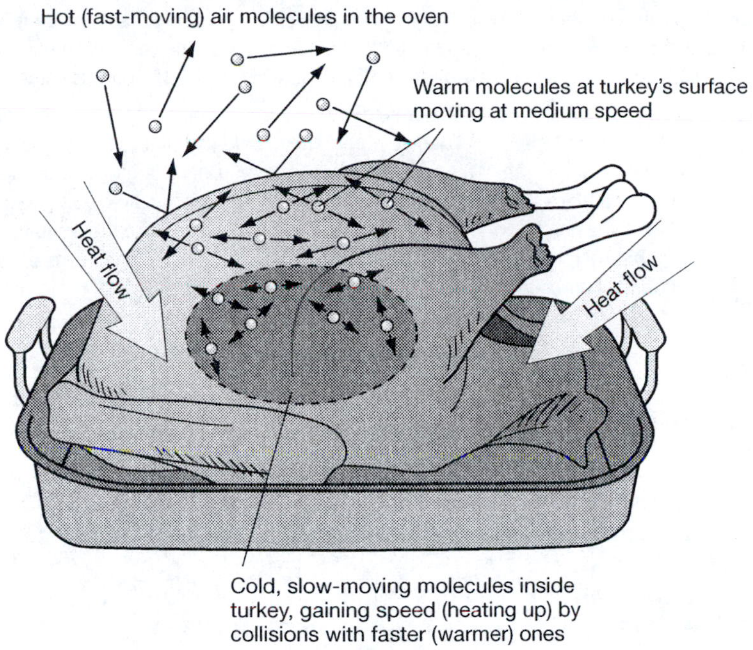

Hot (fast-moving) air molecules in the oven

Warm molecules at turkey's surface moving at medium speed

Heat flow

Heat flow

FIGURE 3.3 The Flow of Heat   The particle description of matter helps explain how heat flows. Fast-moving air molecules transfer some of their energy to the cold, slow-moving molecules in the outer layers of the turkey, which in turn transfer energy to deeper layers of turkey, and so on.

Cold, slow-moving molecules inside turkey, gaining speed (heating up) by collisions with faster (warmer) ones

assign the value zero to something *other than* absolute zero. Zero degrees Celsius, for example, is the freezing point of water, not the point at which the motion of water molecules is at a minimum. So, in science, we use a temperature scale in which zero really means absolute zero. It is called the Kelvin temperature scale, named after one of the "fathers" of thermodynamics. Absolute zero is 0 K, which is correctly read as "zero kelvins," and not "zero degrees Kelvin" (although enough people say it incorrectly that no one will make fun of you if you do, too). The Kelvin scale is by far the most useful scale in physics, astronomy, and cosmology, and it is what we will use throughout this book.

EXERCISE 3.2   Temperatures for Scientists, Americans, and Everybody Else

TABLE 3.1   Temperature Scales

	Kelvin	Celsius	Fahrenheit
surface of Sun	5800		
water boils			212
room temperature			70
water freezes			32
absolute zero	0		

(a) Most of the world uses the Celsius, or centigrade, temperature scale, which is similar to the Kelvin scale. The temperature in kelvins is just the temperature in degrees Celsius, plus 273.15°. So water freezes at 0 °C, or 273.15 K. At what temperature, in kelvins, does water boil? What temperature is absolute zero in degrees Celsius? Fill in these values in Table 3.1

(b) In the United States, the Fahrenheit temperature scale is still used. Water freezes at 32 °F and boils at 212 °F. You can subtract 32 from the Fahrenheit temperature and then multiply by 5/9 to get the Celsius equivalent. (Try it with 32 °F and 212 °F; then verify that you can go in the other direction, from °C to °F, by converting 100 °C to 212 °F.) Write the conversion equations for °C to °F and for °F to °C. What is absolute zero in degrees Fahrenheit? Add this value to your table.

(c) The surface of the Sun has a temperature of 5800 K, and its center is 15 million K. Calculate these equivalents in degrees Celsius and degrees Fahrenheit, and add them to the temperature table. Use your answers to argue that, for temperatures in the thousands or higher, it doesn't matter too much what temperature scale you use. That is, show that if someone gives you a temperature of, say, "200,000," you don't always have to ask "in what units?" because the result is the same *order of magnitude*, regardless of the units. Just consider: 200,000 K, when converted to Fahrenheit, is still hundreds of thousands of degrees, not thousands or billions, for example.

Heat and Light

We now know that, in a star, faster particles mean higher temperatures and vice versa. Let's use the particle picture to explain why stars shine. That is to say, let's put photons into the picture. To do this, we need two more nuggets of physics knowledge.

NUGGET ONE: We saw in Section 2.4 that when an electrical charge is jiggled, it *radiates* light. Now we recognize that when we say a charge "radiates," we mean that it "creates photons." So inside the Sun, whenever the charged particles change their motion (speed up, slow down, bounce, turn, etc.—all of which qualify as "jiggling"), photons appear. These particle actions are happening all the time; therefore, the composition of a star involves both *matter particles* (atoms, ions, electrons) and *photons*.

NUGGET TWO: Photons interact with charged particles, which means that the photons bounce around, too. This interaction turns out to be very important to cosmology. We'll justify everything more deeply in Chapter 8, but for now, we need to know only that the bounces of photons involve an exchange of energy. For example, when a high-energy photon (ultraviolet, perhaps) bounces around among low energy (slow-moving) matter particles, the photon surrenders some of its energy to the matter.

Now, we might speculate that the Sun gets its energy from its core and cools by shining light into space. For now, this is just a guess, albeit a reasonable one. We are assuming that the Sun is similar to any other hot object—like a turkey just removed from the oven. The turkey skin comes in contact with the relatively cool kitchen air and thus cools. Heat moves from the hot interior of the turkey through the skin and out into the room.

If the Sun behaves similarly, it must be generating high-energy photons at its center (generated by a fuel source we'll discuss in Chapter 9). These photons bounce around until they reach the surface, after which they fly off into space. The reason is this: A photon at the deep interior of the Sun will interact with many matter particles before it emerges. Each interaction bleeds some energy from the photon. This is our picture of how the Sun works (Figure 3.4). The center of the Sun is much hotter than its surface, and high-energy x-ray photons generated at the center lose energy with each bounce and emerge primarily as lower energy visible light. This simple picture explains both the Sun's light (emerging photons) and how the Sun stays hot (energy deposited by photons before they emerge).

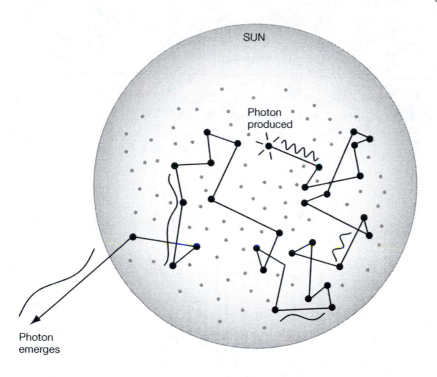

FIGURE 3.4 Life of a Photon in the Sun   High-energy photons generated near the center of the Sun must bounce their way out toward the surface, losing energy (cooling down) along the way.

Why does all of this matter? Because we need to understand the starlight we see from distant sources. To do that, we ask how a star *should* shine. Only when our answer to that question matches the starlight we actually observe can we honestly claim to understand it. Then we can use our understanding to infer, for example, the size scale of the universe, which will be our focus in this chapter.

Now, all this photon bouncing might seem as if it should be completely random, but this turns out to be far from the truth. In fact, the light produced by a hot object like a star follows a specific pattern. It is called **blackbody radiation,** which is a distribution of photons of various energies that will come up again and again in our study of cosmology. The term "blackbody" refers to a physics model of a perfect radiator—an object that reflects nothing (and hence is termed "black"), so that all of its emission is generated by its own thermal energy. Stars, although generally not black in color, turn out to be excellent blackbodies, since their light is almost entirely generated by their own thermal energy. The environment on a star's surface is determined by a long sequence of photon bounces inside; in this way, energy is shared throughout the star.

The blackbody is defined by its mix of intensities at each light frequency. This mix of frequencies, as you saw in Section 2.4, is called its spectrum. In fact, you already saw the blackbody spectrum in Figure 2.22, and we show another one here in Figure 3.5—but now you can apply the particle picture to the blackbody starlight. When photons bounce their way out of stars, they encounter particles of matter at a variety of speeds. Those speeds are characterized by a single

FIGURE 3.5 Blackbody Spectrum   The blackbody spectrum reflects the pattern of photon energies for each temperature. For a cool object, the spectrum is lower in frequency (*x*-axis) and intensity (*y*-axis); that is, its light is generally redder and dimmer. For a hot object, the opposite is true: It's bluer and brighter. This figure shows four curves representing the blackbody spectrum from each of four objects at different temperatures. (◆ Also in Color Gallery 1, page 8).

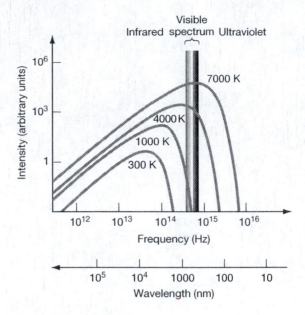

number: the temperature. So although the photons emerge at several different energies, there is still an organized pattern to how many photons emerge at each energy. That pattern is the blackbody spectrum.

Notice that the particle world gives us a deeper feel for this whole process. It shows us that the spectrum of starlight effectively boils down to a *count* of photons at each energy (color), which depends on the temperature of the star. How many red photons? How many blue, or infrared, or ultraviolet, etc.? Evidently, at the Sun's surface temperature, a collection of photons is produced that carries most of its energy in yellow and green and other visible colors, with smaller amounts in all other frequencies, such as infrared.

Understanding the physical nature of light and matter allows us to use starlight to measure properties of stars. Here we have an excellent example: Measuring the peak in a star's blackbody spectrum tells us the temperature of the star, as in Figure 3.5. Since each temperature has a slightly different blackbody spectrum—*hotter stars have brighter light, peaking at higher frequencies*—we can use the spectrum to learn the temperature of the star. Our green-peaked Sun, for example, corresponds to a 5800 K surface.

In Chapter 2, we saw that dispersing starlight through a prism separates the colors because different wavelengths bend by different amounts through such a device. That is to say, we measure stellar blackbody spectra by their *wave* properties, but we can interpret the cause in terms of photons (i.e., *particles*). This tells us that the wave and particle descriptions can be merged somehow. The intensity of a beam is a blend of the frequencies carried by that beam and the number of photons in it. A single photon, evidently, has a wavelength and a frequency. It doesn't particularly matter *how* you imagine this combination—as a particle with some feature jiggling up and down as it travels, as a particle with a certain physical length, or as a little wavelet extracted from a longer wave—any of these models will do. What's important is the set of measurable results: Light comes in discrete photons traveling at a fixed speed; each photon has a frequency and wavelength to identify its energy and color, and it exhibits wave behaviors.

3.2 Standard Candles

As noted in the previous chapter, the night sky is littered with stars and other objects that look like spots and splotches and that offer no immediate indication of how far away they are. In Chapter 2, we began the process of using our scientific understanding of light to learn how far away they are. In this chapter, we continue to apply our knowledge of light in order to measure greater distances than we can with the radar and parallax techniques from the last chapter.

As mentioned in Chapter 2, the ideal technique for indicating distance wouldn't require us to know anything about the objects whose distances we seek to measure. But that isn't always possible. For example, when you look at a tree, you always have some sense of how far away it is. That sense is based on the simple fact that you recognize trees. Because you know how big trees are, you can *interpret* a small-looking tree as *actually being* a normal-sized tree that's far away. You're so good at this that you don't even have to try; your mind does it automatically. Sadly, there is no such automatic mechanism for the points of light in the night sky.

The parallax technique from Section 2.5 depends on how our line of sight to nearby stars changes with respect to the more distant ones as we move our observation point. But it does not depend on what stars physically are. A star could actually be a giant, glowing one-eyed penguin in space, and parallax will still tell us its distance just the same. "That giant glowing one-eyed penguin is 23 light-years away," parallax would say. Parallax is great because with it we can learn an object's distance before we even know what the object is. But the effect of parallax is measurable only for astronomically local light sources, and cosmology beckons that we explore far greater distances.

In order to make any further progress, we'll have to figure out what we're looking at. The last section gets us started, because we now know how stars shine: with a blackbody spectrum determined by their surface temperatures. But we also want to know *how much* stars shine. After all, a dim nearby star looks similar to a bright distant star. It turns out that stars come in many varieties, which shine with different intensities. For those at distances too great to measure with parallax, we need to be able to discern which stars are which. Is there any way to recognize a particular type of star?

We saw in Section 2.4 that there is. When we disperse the light from a star to separate the frequencies (colors), we find dark gaps, called *lines*, in the spectrum. Many of the same lines appear in the spectra of different stars. Others are specific to particular types of stars. Still others appear in every star, but are more prominent in one star than another. Viewed with a spectroscope that reveals stellar lines, stars clearly fall into different categories (see Figure 3.6), even though the stars

FIGURE 3.6 Examples of Common Stellar Spectra To the right of each spectrum is the star's "spectral class," an identifying letter assigned to stars that share similar spectral lines. The dark lines in the spectra are places where certain colors of light have been absorbed by particular elements, some of which are labeled. The names of the spectral classes—O, B, A, F, G, K, and M—don't mean anything special, and most astronomers remember them with the mnemonic "Oh, Be A Fine Girl/Guy, Kiss Me."

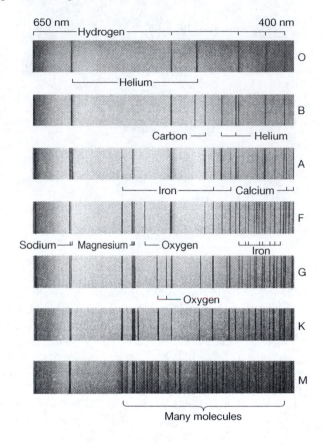

look alike to your naked eyes. The categories in the figure, based on stars' spectra, are called **spectral classes**.

The foundation for our current system of spectral classification was built up in the late 1800s and early 1900s on the basis of common characteristics observed in the brightness and placement of spectral lines—essentially, a detailed analysis of the colors of different stars. Henry Draper measured the first spectrum of the star Vega in 1872, and Annie Cannon did much of the classification of stellar spectra from 1918 to 1924. As you can see in Figure 3.6, the spectral classes are named with letters—for example, a "B-type star" has relatively few lines in its spectrum.

Even though you have not yet studied the cause of the lines (Chapter 8), you already know enough to interpret them in terms of photons. A dark line where a particular shade of green (or whatever) should appear indicates that those green photons either aren't being produced by the star or aren't escaping from the star. For this reason, dark lines are called *absorption lines*—the idea being that those photons are being absorbed somewhere and thus prevented from reaching us.

Many stars share each type of line arrangement; spectral classes are based on observations of approximately 100,000 individual stars. So suppose that many stars exhibit the same missing shade of green. Now, even without saying why the particular green photons are blocked, we can still conclude that there is some cause and that the cause is common to many stars. It seems plausible that the stars which absorb the same shade of green probably do so for the same reason—and if that's true for 20 lines or a hundred, then the stars exhibiting all these common lines must have a lot in common. Therefore, we are justified in concluding that they're all *the same type of star*. We can recognize types of stars in the sky the way we can recognize types of trees on the ground.

Why is such recognition important? Because if we knew how bright a certain type of star is—by recognizing its type—then we could figure out how distant it is, on the basis of how bright it *looks*. A star (or other source) whose actual light output we know in this manner is called a **standard candle**. (The name is intended to indicate a known, fixed light output, similar to a candle.) For example, a night-sky star of the same type as the Sun, which we recognize because it has the same G-type spectral lines as the Sun, must be much farther away than the Sun is, because it looks so much dimmer. It couldn't *actually* be that much dimmer, because we already know that it's just like the Sun.

A common example of this principle is automobile headlights. Since various cars share approximately the same brightness for their headlights, you can estimate the distance to some oncoming car, even at night, on the basis of how bright the headlights appear to you. If they're faint, then the oncoming car must still be far away. If they're bright, then the car must be very close. (If they're blindingly bright, then they must be high beams, and of course, it is your duty as a licensed driver to flash your own high beams in retaliation.)

There is a relationship that links three quantities: *apparent brightness* (what we can see from here), *luminosity* (what the source actually emits), and *distance*. We'll define these quantities formally with their mathematical relationship soon, but first it's important to appreciate the conceptual relationship.

It works like this: If you know any two of these three quantities, you can calculate the third. When determining distances with the standard-candle technique, you know the luminosity (because we've catalogued the luminosity for each type of star, and all stars of a certain type are the same) and you know the apparent brightness (because we can always measure that from Earth, usually with some sort of camera). With these two quantities known, you directly calculate the third quantity, distance. Note, however, that this technique works only if you have already learned and catalogued stellar luminosities according to their spectral classes.

Our closest star, after the Sun, is Proxima Centauri, located 4.2 light-years away. We've never sent a spacecraft there, because at the speeds our spacecraft travel, it would take about 60,000 years to get there. The point is, since we've never actually visited other stars, we've never directly measured their luminosities. (Even if we were to visit, we'd need a detector that wraps around the entire star if we're to catch every photon!)

Then how do we measure the luminosity of any star? We use the same conceptual relationship as before! This time, however, we work only with *nearby stars*. In that case, we know their distance by parallax, and we know their apparent brightness as always, so we can calculate the third quantity—luminosity. In this way, astronomers can catalog different types of stars as standard candles—like lamps with their wattage stamped on them.

Altogether, the process looks like this:

FIRST, measure spectral classes and parallax distances for many local stars. Parallax is quite reliable for relatively nearby stars.

SECOND, measure apparent brightnesses of these local stars (with a camera, say), and couple the results with distances, to create a catalog of luminosity for each spectral class.

THIRD, for more distant stars—too distant to detect their parallax—use the standard-candle technique. Measure the distant star's spectral class, look up the catalogued luminosity for that class, and combine that luminosity with the apparent brightness to calculate the distance.

Warning: When you use known star types as standard candles, this last step is sometimes called *spectroscopic parallax*, even though it has very little to do with parallax. A better name for the same method is **main-sequence fitting**. In this method, astronomers match a star's spectrum to a known star—a standard candle. That standard-candle star is typically a young or middle-aged star of any size, like our Sun, called a **main-sequence star**. (The reason for this name is not important; it relates to a particular tool used by astronomers. The name denotes a particularly long time in a star's life, like "adulthood." We'll study the lives of stars and the meaning of the main sequence in Chapter 9. The Sun will "leave the main sequence" only in the last 1% of its normal existence; that might be a good time to sell your stocks.)

A star used for main-sequence fitting is similar to a fluorescent lamp. Operating normally, all such lamps of the same wattage shine equally brightly, so they would make good standard candles. But when the bulb is about to die, it flickers less predictably, and is a standard candle no more.

The World-Famous Inverse-Square Law

Before we look at some cosmologically important uses of the standard-candle technique for measuring *extragalactic* distances (distances outside our Galaxy), let's use our understanding of light to build the precise mathematical version of the conceptual relationship we just introduced.

Day in and day out, physicists and astronomers refer to the following two quantities:

Flux—the amount of photon energy per unit time per unit area, also sometimes called *intensity*. Apparent brightness is often measured as a flux. If you double the collecting area (e.g., open both eyes instead of just one), then you double the amount of light you see, but the flux is the same, since flux is measured on a per-area basis. For example, the Sun's flux on Earth is relatively constant, so you can collect more photons per second and get more solar power by using more solar panels (more area). The same idea holds for any kind of particles streaming onto a detector, such as raindrops. The larger the area you use for your collector, the more water you will collect in a given amount of time. (See Figure 3.7)

Luminosity—the total amount of photon energy per unit time. A star's intrinsic light emission is usually quoted this way. With regard to luminosity, it makes no difference where the light goes or how much enters your eyes; the luminosity is just the total output of the star. The Sun's luminosity, for example, is fixed, and Earth intercepts only a tiny fraction of it. Your eyes intercept even less. (See Figure 3.8)

Between an astronomical source's luminosity and the flux measured on Earth is the distance the light travels. Figure 3.9 shows a source star, shining equally in

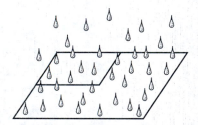

FIGURE 3.7 Understanding Flux
A steady stream of particles falls on some region, spread uniformly, but with some random scatter. (The particles could be photons, raindrops, bugs, or whatever.) For a given flux, how many particles you collect on your detector (solar panel, rain collector, windshield, or whatever) in a certain amount of time depends on how much area your detector covers. The large square drawn here will collect four times more particles than the smaller square, just because its area is four times bigger.

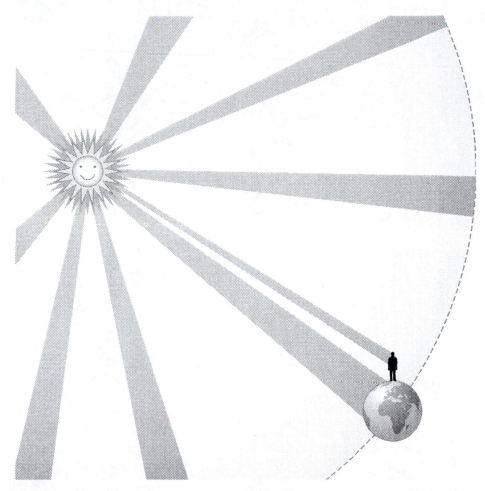

FIGURE 3.8 Intercepting Part of the Sun's Luminosity   The Earth intercepts only a small fraction of the total photons per second emitted by the Sun. A person standing on Earth intercepts an even smaller fraction.

all directions. Near that star is an undiscovered planet (seen in Figure 3.10) where a race of evil lizard people live and conduct astronomical experiments. Farther away is our own solar system. The lizard people, with their scales glinting in their sun, live on a planet that intercepts a small fraction of that sun's light. This is shown in the figure with an imaginary sphere, centered on their sun, as big as their planet's distance from their sun. The imaginary sphere is significant because *all* of the star's light goes through that sphere, but only some hits their planet.

As experienced by the more distant earthlings, an even smaller fraction of the lizard star's light intercepts the Earth. Another imaginary sphere is depicted at Earth's distance from the star, which again shows the region that all the starlight crosses (except for the little bit that has landed on the lizard planet!).

The imaginary spheres are the secret. Because the starlight spreads outward in every direction, crossing a whole (imaginary) sphere, the total amount of light observed at some location on the sphere is inversely proportional to its surface area. (See Figure 3.11.) Whenever we talk about light seen from "far away," for example, what that really means is "big imaginary sphere" with "big radius" and there-

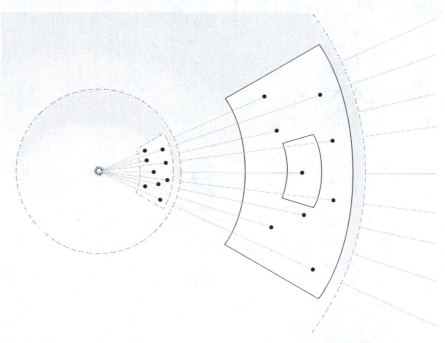

FIGURE 3.9 How Flux Varies with Distance Away from the Source   A light source such as a star sends out photons randomly in all directions, roughly evenly distributed. Here, only the paths of nine photons that happen to hit the surface area of the detector close to the star are shown. If this detector were not in the way, these photons would proceed on into space, but notice that *they continue to spread out over more area*. By the time they reach the larger imaginary sphere, three times farther away from the star, they have spread out to cover an area nine times the size of the detector (for reasons discussed in the text). As a result, only one of the photons happens to hit a detector of the same physical size as the original one. This is an illustration of how quickly flux decreases with increased distance from the source.

fore "big surface area," which in turn means "the total luminosity must be spread over a larger area, so the fraction you get in your detector is much smaller and therefore looks dim."

All of this is said quantitatively by the **inverse-square law**,

$$F = \frac{L}{4\pi r^2} \, , \qquad \text{(eq. 3.1)}$$

where r is the distance to the object, F stands for flux measured here, and L is the luminosity of the object. In words, it goes like this:

The rate at which you collect   how bright it actually *is* / something
light, or equivalently, how bright = involving how far away it is,
something *appears*   squared.

The significant thing to notice is the denominator. $4\pi r^2$ is the surface area of a sphere, and since the distance from the source to us is the radius of the imaginary spheres we've been looking at, the denominator represents the surface area of our imaginary sphere. That surface area tells us how "spread out" the light has become by the time it reaches us.

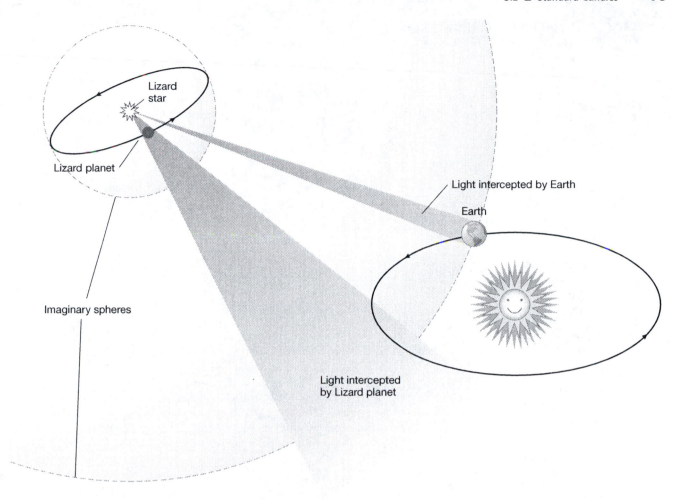

FIGURE 3.10 Starlight from Our Galactic Neighbors   The planet of the lizard people is relatively close to their star, so the imaginary sphere at their distance away from the star is much smaller than the imaginary sphere at Earth's distance from the lizard star. Since all the light from the star must pass through each imaginary sphere, the light is much more spread out by the time it reaches us. Thus, Earth intercepts a much smaller fraction of the total light than the lizard planet intercepts. This is why the star is much dimmer as seen from Earth.

Probably the easiest way to visualize this situation is with the particle view of light: The farther away you are from the source, the *fewer* photons you see, because most of the photons have gone through some other part of the imaginary sphere, far from you.

It is sometimes useful to flip the inverse-square law around. It can be algebraically rewritten as

$$L = 4\pi r^2 F \; ,$$

(eq. 3.2)

FIGURE 3.11 Surface Area of a Sphere   The surface area of a sphere grows as the square of the radius.

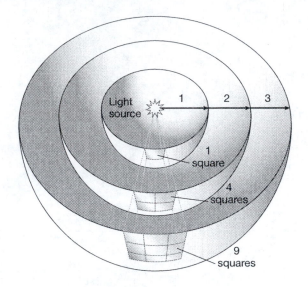

or

Total energy emitted per second = (total surface area) × (amount passing through each bit of surface area per second).

You should notice that the left side of the equation, *L*, is a constant number; it's the output of the source, like "60 watts." In terms of particles, the left side relates to the total number of photons emitted. But the right side is broken down into the intensity per unit area times the total area, which is purely a function of the distance *r* from the light source. To put it another way, *r* and *F* behave in opposite ways. For example, if you move closer to the light, *r* goes down and *F* goes up; they have to act that way, or else *L* wouldn't stay constant and the light bulb (or whatever) would have to spit out more (or fewer) photons. And there's no reason to believe that the bulb would "know" that it must do this in response to your decision to move closer to it.

The fact that distance is squared in the equation is significant. It means that if Star B is twice as distant as Star A, then Star B will look four times less bright, because $2^2 = 4$. If Star C is five times more distant, it'll appear 25 times less bright than Star A. This is the reason the formula is called the inverse-square law:

- "inverse" because *more* distance yields *less* flux.
- "square" because a *small* change in distance causes a *large* change in flux. (This is how you should read equations in general—a squared quantity is more important than one that is not squared. Double *L*, and *F* also changes by a factor of 2; but double *r*, and *F* changes by a factor of 4. The squared quantity has the greater effect.)
- "law" because, in every experiment, we find that the equation matches the measured result.

The effect is demonstrated in Table 3.2, in which we assume a nominal flux of 100 (in some units, like "orange photons per square meter per second") at a distance of 1 (in some units, like "light-years" or "inches"). These two values are listed in the first row of the table. We then see what happens as we increase the distance. For each subsequent row in the table, why is the flux the particular number that is listed?

TABLE 3.2   The inverse-square law

Distance	Flux
1	100
2	25
3	11.1

EXERCISE 3.4   Earthly Standard Candles

Try to identify the role of standard candles, and of the inverse square law, in your own life, with the following exercises:

(a) Consider the flashing lights of an airplane at night. What would it look like if you saw the airplane flying dangerously close to the ground?

(b) If the airplane were descending toward you, could you estimate how fast it was approaching? Think about what the lights will look like when the airplane was half as far away, then half again, etc.

(c) Describe some other standard candles you experience, for which you can estimate the distance to the light source.

EXERCISE 3.5   Stellar Standard Candles

(a) Two stars of the same spectral type are discovered, but one is 100 times brighter than the other. Compare their distances.

(b) Two stars of the same spectral type are discovered, but one is 100 times more distant than the other. Compare their intrinsic luminosities, and then compare their apparent brightnesses.

(c) If you could wrap mirrors around a star, so that all of its light were beamed in one direction toward you, then how would the apparent brightness change if you doubled your distance to the star?

Some Standard Standard Candles

There are several ways in which standard candles inform us about distances in the universe. The list that follows contains cosmologically important techniques for measuring extragalactic distances. All of the methods listed have the same basic standard-candle idea at their hearts. As you read about each, watch for its "trick" that tells us how bright something is intrinsically, so we can tease out the distance.

Cepheid Variable Stars   Astronomers know a great deal about stars of all sorts. When many stars reach "old age," they enter a temporary phase in which they pulsate radially. That is, they lose some of their physical stability, causing them to grow and shrink, and grow and shrink. With each oscillation, the light emission changes, too: brighter and then dimmer, brighter and then dimmer.

The most useful of these pulsating stars, or **variable stars,** is a bright one. Since brighter stars can be seen farther away, they make the best distance measurement tools. Such stars, which are significantly more massive than the Sun, are called **Cepheids** (SEF ee ids) or **Cepheid variable stars** during their period of instability. (As with many objects in space, these stars are named after the first one discovered, called Delta Cephei. Delta is the fourth letter in the Greek alphabet, and Delta Cephei is the fourth-brightest star in the constellation Cepheus.)

By itself, the pulsation of aging stars doesn't make a standard candle. It turns out that when we measure the *period* of these pulses—the time it takes to cycle from bright to dim to bright again—we discover a natural connection between that period and the star's luminosity. (Actually, it's the *average* luminosity, since the actual luminosity is continuously changing with each pulsation.) The longer the period, the brighter is the star, according to a very reliable mathematical relationship, shown in Figure 3.12. That is, Cepheid stars just happen to follow this behavior pattern: If you observe a Cepheid with a particular pulsation period, you can read its average luminosity right off the graph, because Cepheids with a particular period *always* have the same average luminosity. As you can see in the figure, the periods of variation can be days, weeks, or months.

A period–luminosity graph like Figure 3.12 was first constructed in 1912 by Henrietta Leavitt, an astronomer at the Harvard College Observatory. She accomplished this by surveying a collection of 25 variable stars, graphing their periods and brightnesses, and finding a pattern. She published the following account:

> *"A straight line can readily be drawn among each of the two series of points corresponding to maxima and minima, thus showing that there is a simple relation between the brightness of the variables and their periods . . . . Since the variables are probably at nearly the same distance from the Earth, their periods are apparently associated with their actual emission of light, as determined by their mass, density, and surface brightness."*

> Henrietta Leavitt, "Periods of 25 Variable Stars in the Small Magellanic Cloud."

As you can see from what Leavitt wrote, the stars she surveyed were all at the same distance from us. If we know that distance, we can use the inverse-square law to calculate the (intrinsic) luminosities. In the next section, we'll see that Cepheid-based standard candles enabled astronomers to establish many other large-distance measurement tools.

In Figure 3.13, we see how the brightness changes over time as the star pulses. The type of graph shown is called a **light curve**. Light curves are graphs that display measured light intensity (flux) versus time, to show how the light emission is

Henrietta Leavitt (1868–1921) discovered the period–luminosity relationship for Cepheid variable stars. Her work laid the foundation for much of the modern standard-candle method, which in turn laid the foundation for much of our knowledge of extragalactic distances.

FIGURE 3.12 Relationship between Period and Luminosity for Cepheid Variables Astronomers can use Cepheid variable stars as standard candles for distance measurement because the average luminosity of such a star is a simple function of the star's pulsation period. If one measures the period by timing how long it takes for the star's brightness to rise and fall, then the luminosity of the star can be looked up on a graph like this one.

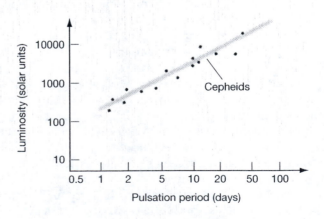

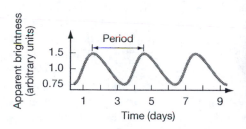

FIGURE 3.13 Light Curve for a Variable Star   Here we see variations in brightness of Cepheid variable stars, shown on the left graphically and on the right by slightly offsetting photographs taken at different times. Each pair of stars in the image on the right is the same star with the same brightness, except for the Cepheid, because its luminosity is changing.

changing. As you'll see throughout the remainder of this section, they are often useful in cosmology in general, not just for pulsating stars.

Once the luminosity is known—just measure the period and look up the luminosity from Figure 3.12—you can figure out the distance by using the inverse-square law (Equation 3.1). The period–luminosity relationship makes this distance measurement possible. And since Cepheids pulse brightly enough to be singled out in other galaxies, finding the distance to a Cepheid star enables us to learn the distance to that star's home galaxy. The maximum distance for which we can use this method is about 100,000,000 light-years—that's 10^8, or one hundred million. The closest galaxy of comparable size to our own, called the Andromeda galaxy and visible to the naked eye under dark skies, is 2.5 million light-years away, so Cepheids are very useful over a large neighborhood of galaxies around us, up to about 40 times more distant than our closest neighbor. The accuracy of the Cepheid distance technique is about $\pm 15\%$; that is, we should reasonably expect the true distance to be as much as 15% higher or lower than what we report.

The Tully–Fisher and Faber–Jackson Relations   With Cepheid variables, the standard candle is a star. But a standard candle can also be an entire galaxy. The **Tully–Fisher relation,** discovered in 1977 by two astronomers named (guess!) Tully and Fisher (first names Brent and Richard, respectively), accomplishes just that. Tully and Fisher found that the luminosity (L) of a spiral galaxy is proportional to the fourth power of its maximum rotational velocity (v^4). This relationship is the needed link—the trick for finding the intrinsic luminosity—since we can measure the velocity with the Doppler effect.

For elliptical galaxies, there is a corresponding distance tool that's also a hyphenated combination of two two-syllable last names. The **Faber–Jackson relation,** (Sandra and Robert, 1976), reveals that the luminosity of an elliptical galaxy is proportional to the fourth power of the more randomly directed stellar velocities within it. In practice, the Faber–Jackson method is more complicated, because elliptical galaxies don't have an easily defined edge, making it hard to decide how much of the outer fuzz to work with.

As applied by astronomers today, both the Tully–Fisher and the Faber–Jackson methods are accurate to $\pm 15\%$, and both work to about 700 million light-years away, a little less than a billion (10^9) light-years. (Why do galaxies as standard candles enable deeper distance measurements than stars as standard candles?)

Supernovae   When certain stars die (we'll discuss what this "death" means in Chapter 9), an interesting and complicated process ensues, resulting in an ultrapowerful explosion called a **supernova** (plural: **supernovae**). If all of these explosions happen in generally the same way (a big "if"), then the supernova becomes a standard candle, because we can find its standard luminosity by observing the flux when one explodes in a galaxy whose distance we already know.

There are two types of supernova used as standard candles. One involves the explosion of a particularly massive star (about 10 times more massive than the Sun, or more). The only significant X-factor in this case is the assumption that these supernovae are all alike, because so far, only theoretical computer models of the stars and their explosions can give us that information, and there may be unknown components that need to be included in the models. Still, this method has produced results that closely agree with Cepheid distance measurements, so modeling such explosions is at least on the right track.

The other type of supernova involves less massive stars; it's a more promising standard candle, called a "type Ia" supernova. (The massive-star version is a "type II." The unimaginative naming scheme refers to the observed spectrum of the supernova: Type II supernovae contain hydrogen lines; type I do not, and type Ia contain silicon, but not hydrogen.) We have strong evidence that type Ia supernovae are standard candles—evidence based not on computer models, but rather on observations of the light output of many such sources.

The case for type Ia supernova standard candles stems from their light curves. The supernova glow spikes up in the first week or two and then fades away over the next two months. What's so compelling here is that the light curves for many different type Ia supernovae look very much alike. Their shapes do not change from one supernova to the next. This implies that type Ia supernova explosions are, in some ways, all alike. Studies reveal that the rate at which the light fades away after it peaks is closely correlated with its luminosity at the peak. So the method works like this: You measure the light curve, and from that, you deduce the luminosity.

The evidence for this relationship between the light curve and the luminosity is displayed in Figure 3.14, in which all four frames are light curves. Frame (a) is the hypothetical light curve of the yellow light on a traffic signal, to help demonstrate the light-curve concept with something more familiar than an exploding star. Frame (b) shows the light curve for a Cepheid variable star. Frame (c) is a type II supernova. Frame (d) has several type Ia supernovae light curves. Notice the key to the whole type Ia supernova technique in frame (d): All the curves follow a distinctively shaped pattern.

The real merit of supernova standard candles is their tremendous brightness. A typical supernova shines so brightly that one star, all by itself, emits enough light to approach the light output *of the entire galaxy* it inhabits—a galaxy containing hundreds of billions of stars (several times 10^{11})! (Curiously, this statement is true for either type of supernova, even though the two explosions have completely different causes.) You can see this brightness in Figure 3.15. An exceptionally bright standard candle can be used to measure exceptionally great distances. Type Ia supernovae are now being used to measure distances of several billion (10^9) light-years. Remember that we see objects at great distances not as they *are* today, but as they *were* long ago, when the light that is arriving here now was originally emitted. (Recall Section 2.5.) Because of their great brightness, then, supernova standard candles give us knowledge of the ancient universe, back to a time when it was only about half of its current age.

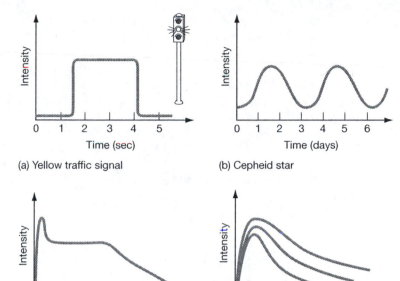

(a) Yellow traffic signal

(b) Cepheid star

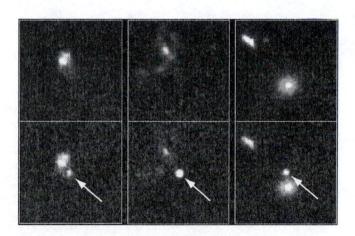

(c) Type II supernova

(d) Type Ia supernova

FIGURE 3.14 Supernova Light Curves
Light curves (brightness vs. time) are shown for different sources of light. Each frame is discussed in the text. Of particular importance here are frame (c), type II supernova, and frame (d), type Ia supernova. In frame (d), you can see how the shape of the light curve follows a pattern from one supernova to the next, letting us use these supernovae as standard candles.

FIGURE 3.15 Distant Supernovae Images of three very distant galaxies before (top) and after (bottom) a type Ia supernova explosion (denoted with arrows) within the galaxy. The supernova is nearly as bright as its entire host galaxy! (◆ Also in Color Gallery 2, page 27.)

The standard-candle technique is extremely important to our understanding of distances in the universe, near and far, and is therefore extremely important to our understanding of the universe itself. In the next section, we will organize that distance information and use it to build an appreciation for the *cosmic* sizes and scales that will occupy our attention for the rest of this book.

3.3 The Scale of the Universe

Much of cosmology relies on distance measurements, so we'd like to construct our distance scale piece by piece, adding each piece with a connection to what we already know, and with an understanding of how we extended our knowledge to the next step. (This technique is often called "bootstrapping," since we gain knowledge by "pulling ourselves up by our bootstraps." You can feel free to use this literary metaphor; it's officially sanctioned by professional astronomers.)

Your understanding of the history and structure of the universe is greater when you understand how that information was obtained. At least in principle, you could repeat the measurements yourself. In actual astrophysical research, the techniques involve sophisticated instrumentation that can seem mysterious and detached from your experience, but all are based on principles of nature that operate all around us and that we use in our daily lives without even thinking about them.

The Astronomical Distance Ladder

Most methods for determining distance need to be calibrated on nearby objects. When we calibrate a measurement system, we adjust its readings by comparing them with readings from a known standard. (Imagine calibrating a thermometer by making sure that it reads 100 °C in boiling water.) The distance to a supernova, for example, depends on first learning how to interpret its light curve. To do that, we start with light curves from nearby supernovae, for which we have already figured out the distance by some other technique, such as using Cepheids as standard candles. But what if we've made a mistake in our application of Cepheids as standard candles? Then the supernova measurements would be wrong, too.

We've already seen several methods for measuring distance; they're summarized next. But this time, as you read, notice the calibration (bootstrapping) needed for each, highlighted with italics.

- *Radar* (Section 2.5) uses the known speed of light to convert radio signal travel times to distances. This "brute-force" approach is useful and reliable, but only within our solar system.

- *Parallax* (Section 2.5) uses pure geometry to look for shifts in the sky position of relatively nearby stars in our Galaxy with respect to more distant ones. The method relies on a known baseline distance—the diameter of Earth's orbit around the Sun—*obtained by radar measurements.*

- *Main-sequence fitting* (Section 3.2), useful within our Galaxy, is a standard-candle technique and therefore exploits the inverse-square-law behavior of light. The method employs a catalog of spectroscopically classified stars of known luminosity, obtained by applying the inverse-square law to nearby stars *whose distances were measured by parallax.*

- *Cepheid variable stars* (Section 3.2), useful for measuring distances to relatively local galaxies, are also standard candles. Therefore, they must be calibrated as well, through studies of Cepheid stars in the Milky Way *whose distances are known by main-sequence fitting.*

- *The Tully–Fisher and Faber–Jackson relations* (Section 3.2) are also standard candles, used for middle-range galaxies and *calibrated with galaxies whose distances are known by Cepheid measurements.*

- *Supernovae* (Section 3.2), especially type Ia for measuring very distant galaxies, are standard candles, which again are *calibrated by Cepheids.*

Another important technique that we'll be adding later in the book, when you have adequate background,

- *The Hubble law* (to reappear in Section 5.1) is an *empirical* technique, which means that (as with other laws) we know that it works because our measurements consistently verify it. This method is capable of measurements to great distances across most of the observable universe. However, it is *calibrated by Tully–Fisher and Cepheid measurements.*

Taken together, these techniques, along with a handful of others, make up a system of measurement known as the **astronomical distance ladder.** Figure 3.16 depicts this ladder—another comprehensive astronomy metaphor—upon which you can climb to ever-greater heights (i.e., measure ever-greater distances). To get to the upper rungs, you need all the rungs below you, since each technique is calibrated by others below. More calibrations means more places for errors to creep in, so it would be nice to confirm our distance ladder measurements by some *other* type of measurement. We'll introduce one of these consistency-check methods now, and two others later in the book, when you know the physics you need.

Standard rulers   This technique is very much like the standard candle: Something is known, something is measured, and distance is calculated from the two with a mathematical formula. With a standard candle, you know the luminosity, measure the (apparent) flux on our sky, and compute the distance with the inverse-square law. With a standard ruler, you take something whose *physical size* you somehow know, you measure the *apparent* size of that object on the sky, and you compute its distance away with trigonometry.

Here's a simple illustration for a spiral galaxy. Suppose that we use some method to measure the diameter of the spiral disk of our Galaxy and a few nearby galaxies, and they all come out around the same number. Then we measure the *apparent* size of a more distant spiral galaxy (an *angle*), and we *assume* that its actual physical diameter is the same as the others we've measured. We can then calculate its distance with trignometry. A more distant galaxy would have to be larger to occupy the same angle.

The same sort of overall relationship works for rulers as for candles: There are three relevant properties—physical size, apparent size, and distance—and if you know any two, you can always calculate the third. But unlike apparent brightness used for candles, you can't always measure the apparent size (angle), because if an object is too distant, it will appear as just a dot.

In practice, standard rulers aren't very commonly used, because standard candles usually accomplish the same thing with greater accuracy. Standard rulers do, however, provide an important consistency check on our bootstrapped measurements. Since various ruler measurements agree with candle measurements, we know that the distance ladder is reliable.

Two other techniques, to be discussed in detail later on, are *gravitational lens time delays* (Chapter 6) and the *Sunyaev–Zeldovich effect* (Chapter 8). Both of these are indicators of *absolute* distance, which means that they do not need the lower rungs of the distance ladder for calibration. Instead, they yield a direct distance—with some experimental uncertainty, as always. These two methods provide an important verification of our *relative*-distance indicators (those which *do* require calibration) used in the distance ladder. Relative-distance indicators are generally easier to use, since absolute-distance indicators require that

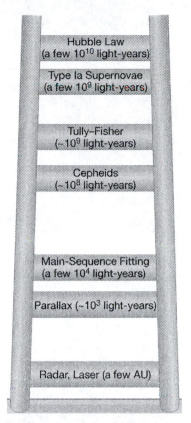

FIGURE 3.16 The Astronomical Distance Ladder   Each rung of the ladder is a technique for measuring distances that builds upon the previous steps.

you find a special kind of source. In the absence of such a rare source, the distance ladder is your only option.

At this point, we have a comprehensive cosmological distance measurement apparatus in place. Now let's use the distances we know to build up a picture of the universe.

A Universe of Galaxies

This is the home to about seven billion people:

FIGURE 3.17 Planet Earth   Our Earth is a measly 12,750 km in diameter.

Earth is much too small to measure in light-years. Its diameter is 12,750 km. That sounds big, but the odometer of your car probably had a higher reading than that after only one year of use.

The Moon's diameter is about four times smaller than the Earth's, and on the basis of high-precision laser measurements (Section 2.5), it lies 384,000 km away, on average. How do you appreciate these numbers? Simple: Divide them. The distance to the Moon divided by the diameter of the Earth is 384,000/12,750 = 30. This means that you could line up about 30 Earths, side by side, between here and the Moon:

Earth   Moon

FIGURE 3.18 Distance to the Moon   30 Earths and 1 one moon, to scale.

Using radar, we've measured the distance to the Sun. And by measuring its apparent (angular) size on our sky, as described earlier for standard rulers, we can compute its physical size. The Sun is 1,390,000 km in diameter and 150 million (150,000,000) km away. Dividing as before, you could fit 150,000,000/1,390,000 = 108 Suns between the Earth and the Sun:

Sun   Earth

FIGURE 3.19 Distance to the Sun   108 Suns and one Earth, to scale.

Or do this: 1,390,000/12,750 = 109 Earths that we could line up across the Sun:

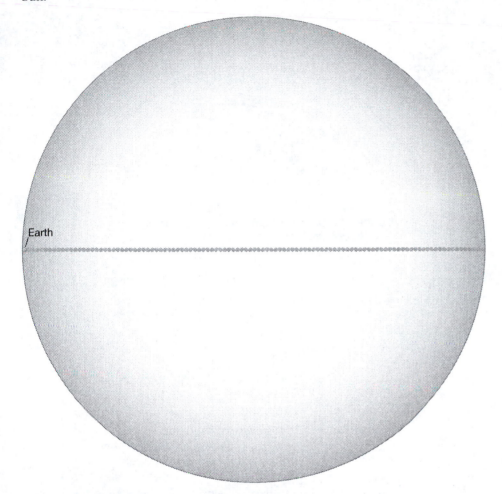

Earth

FIGURE 3.20 Size of the Sun   109 Earths and the Sun, to scale.

For that matter, 1,390,000/384,000 = 3.6. Which means the Sun is more than three times bigger than the Earth-Moon separation:

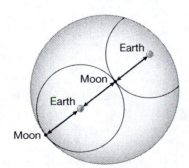

FIGURE 3.21 Size of the Sun, Again   The Earth–Moon system overlaid on the Sun, to scale.

The planet Neptune is the most distant planet in our solar system, not counting dwarf planets like Pluto. Its average distance from the Sun is 30 times that of the Earth:

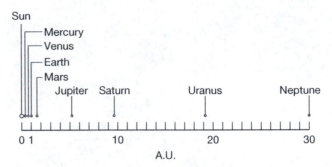

FIGURE 3.22 Distances between Planets   You could put 30 Sun–Earth systems between the Sun and Neptune.

The closest star to our Sun is called Proxima Centauri (meaning the closest star in the constellation Centaurus). By parallax, we know that it's 4.2 light-years away. To connect light-years to our previous numbers, recall that light travels one light-year in one year. The Earth–Sun distance is about 8.3 *light-minutes*, so sunlight takes about 8.3 minutes to get here once it leaves the Sun. The Earth–Moon distance is 1.3 *light-seconds*.

There are over 31.5 million seconds in a year (you should verify this!), so that's 102 million (31.5 million × 4.2/1.3) Earth–Moon distances to Proxima Centauri. A number as big as 102 million is hard to imagine, though, so let's instead compare it with the solar system distance out to Neptune. The result is that if you were to travel the distance to Neptune, but in the direction of Proxima Centauri, you'd have to repeat that amount of distance 8,800 times over again before you arrived!

There's no way around this: There's a leap in scale between solar system distances and *interstellar* distances (the distances between stars). If the distance between the two dots shown here represents the distance from our Sun to Proxima Centauri, then the sizes of the two dots are way too big for the scale model. That is, the left dot is too big to represent Proxima Centauri to scale, and the right dot is too big to represent *our entire solar system* to scale!

.   .

Proxima Centauri   Our solar system

Out in the disk of the Galaxy, where we live, the stars are typically separated by distances similar to the gap between us and Proxima Centauri. (Remember, all of this is well measured by the parallax technique.) We can imagine a local grid of stars, separated by a few light-years. But the arrangement isn't really a perfect grid, so, to be realistic, we should mess it up a little, as in Figure 3.23.

Consider next the thickness of our Galaxy. Parallax is not currently adequate for distances as large as the galactic disk, but main-sequence fitting is. So we are able to measure that thickness as roughly the distance to the most distant stars we

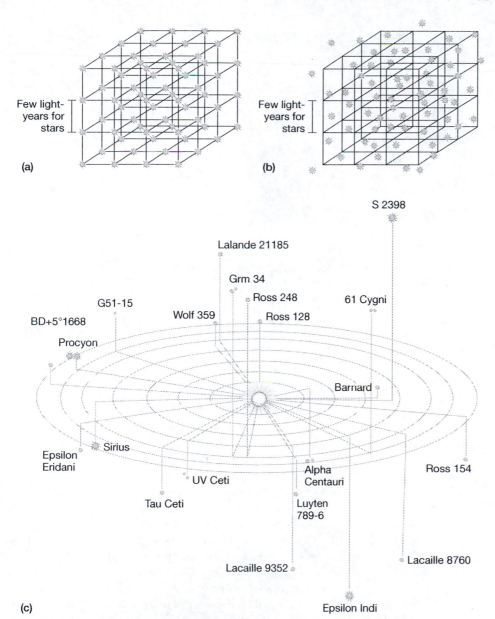

FIGURE 3.23 The Solar Neighborhood   Frame (a) is a local stellar grid. Frame (b) is the same, but with stars moved around a bit to be more realistic. Frame (c) is a representation of our solar neighborhood, showing the positions of neighboring stars relative to our Sun. (Star sizes not to scale.)

can find in the "galactic north" and "galactic south" directions; the thickness turns out to be about 2000 light-years.

Measuring the diameter of our Galaxy—or our location within it—is trickier. Most of what we know now is the result of observing in nonvisible frequencies, particularly radio and infrared. As for our location, in the 1920s an American astronomer named Harlow Shapley catalogued some important distances to help us with that; he measured the distances to globular clusters.

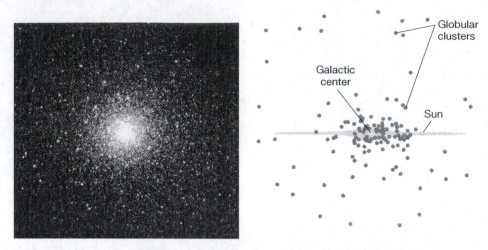

FIGURE 3.24 Globular Clusters   (a) The globular cluster M80 and (b) the distribution of globular clusters around our Galaxy. Each dot in (b) looks like the entire swarm of stars shown in (a).

Globular clusters are clusters of hundreds of thousands of stars (several times 10^5), all crowded into a region comparable to the typical spacing between individual stars in our part of the Galaxy. There are about 150 known globular clusters associated with the Milky Way, but they do not lie in the Galaxy's flat disk. When Shapley measured the distances to all these clusters, using variable stars as distance indicators, he discovered that they're clustered spherically around a central point, as in Figure 3.24.

As it turns out, that central point is located 28,000 light-years away from us. Apparently, we do not live near the center of our Galaxy. Not that it matters: The Galaxy itself is about 100,000 light-years in diameter, and wherever we find ourselves is bound to be incredibly far from everything else. (See Figure 3.25.)

By measuring distances to Cepheids in the neighboring Andromeda Galaxy, also in the 1920s, Edwin Hubble was able to measure the distance to that galaxy. To the great surprise of many, Andromeda, which looks much like a fuzzy nebula, turned out to be a whole other galaxy. Its Cepheid-based distance today is 2.5 million light-years. Dividing numbers as we did earlier, we see that 2,500,000/100,000 = 25 Milky Way diameters between our Galaxy and the next big one (Figure 3.26).

Astronomers have also discovered a number of other galaxies around us—45, and that number keeps growing as fainter and fainter ones continue to be discovered. In this **Local Group,** Andromeda and the Milky Way are the biggest and most massive. (We can feel superior to Andromeda if we presume, without evidence, that it lacks intelligent life. Of course, one could argue that the need to feel superior might undermine our own claim on intelligence.) Around us are many smaller galaxies, most of which are called **dwarf galaxies**. Two among them are visible to the naked eye in the Southern Hemisphere sky: the Large Magellanic Cloud (LMC) and the Small Magellanic Cloud (SMC), so named because their cloudlike appearance was noted by Portuguese explorer Ferdinand Magellan (1470–1521) during his career as a global circumnavigator. The Local Group, all together, occupies a region about 4 million light-years across.

With Cepheids and Tully–Fisher, astronomers have mapped out the locations of other, nondwarf galaxies around us. We find that the typical spacing between galaxies is a few million light-years. We can once again imagine the structure we

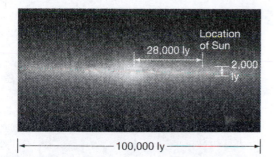

(a)

(b)

FIGURE 3.25 The Milky Way Galaxy   (a) Edge-on view of our Milky Way, seen in infrared light. (b) Popular depiction of a galaxy, with an arrow saying "You are here." No such picture of our Galaxy could ever have been taken by us from the outside like this. Even at the speed of light, it would take about a hundred thousand years to move into position to take the picture! And even if this were a picture of our Galaxy, we're not that far from the center!

FIGURE 3.26 Distance between Galaxies   Twenty-five Milky Ways and the Andromeda galaxy.

used for stars within our neighborhood, separated by a few light-years (a difference in scale of a million times between stars and galaxies), by inventing a grid and then messing it up a little, as before:

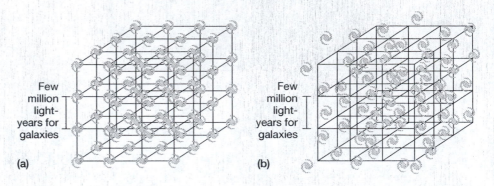

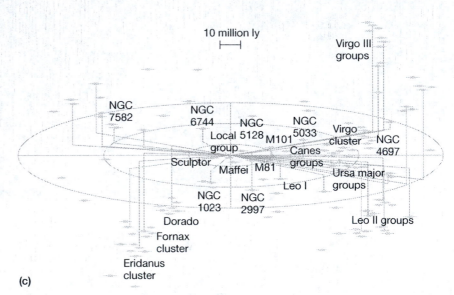

FIGURE 3.27 The Galactic Neighborhood.   Frame (a) shows a perfect grid of galaxies; frame (b) is the same, but with galaxies randomly scattered: frame (c) shows the actual local galaxy distribution. (Galaxy sizes not to scale.)

We also find that galaxies often collect into structures larger than our Local Group; these larger structures are called **galaxy clusters** (Figure 3.28). On top of that, galaxy clusters can cluster, just like stars and star clusters; a cluster of galaxy clusters is sometimes called a **supercluster**. Galaxy clusters can be 10 or 20 million light-years across, with hundreds or thousands of galaxies in them. Superclusters are more like 100 or 200 million light-years across. We live on the outskirts of the Virgo Supercluster, named for the constellation in the same direction on the sky as most of the supercluster's member clusters.

Galaxies and clusters, spaced and sized as described here, appear to dominate the universe in every direction, for as far as we can see. Galaxy clustering is an important part of cosmological research today; we'll consider the motion in clusters in the next

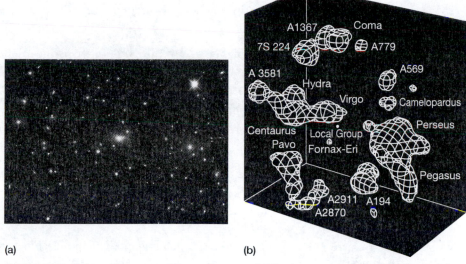

FIGURE 3.28 Galaxy Clusters and Superclusters   (a) The Coma galaxy cluster. (◆ Also in Color Gallery 1, page 14). (b) A three-dimensional map of the groups and clusters that make up the Virgo supercluster. The entire Coma cluster from frame (a) is just one of the smaller clusters in the supercluster (b).

chapter and the formation of clusters in Chapter 12. Large-scale structure, too, of which superclusters are an example, will show up again soon in Chapter 5. Now, however, is a good time to think about how impressive a picture of the universe we obtain from the distance ladder laid out earlier in this chapter—a ladder that was built up from near to far, with relatively ordinary techniques that you use every day without even thinking about them: depth perception with two eyes and standard-wattage light bulbs.

The Size of the Skies

Numbers are useful for comparing things. ("You only have 12 dollars in your pocket? Well, I have 40—a little over three times as much.") But how are you supposed to deal with huge, astronomical numbers?

There are two fairly direct answers.

FAIRLY DIRECT ANSWER #1: Choose to care about the *order of magnitude* (the exponent in scientific notation) first. For example, suppose you win the lottery, and receive a lump-sum check for 5.6×10^9. First look at the 10^9 part, and then look at the 5.6 part. The "9" tells you that you're a billionaire. The "5.6" just tells you the petty details: It doesn't distinguish between thousands, millions, and billions; it only says *how many* billions. Do you really care whether you're awarded 3.3 or 5.6 or 9.1 billion dollars, or would they all have essentially the same impact on your life? What really matters is the order of magnitude. Why? Because it immediately tells you things like "enough money to never work again," but "not enough to buy California and rename it after your dog." The only time you need to worry about the coefficient (the 5.6 part) is the day you file your tax return or when you meet some other billionaire with whom you feel a primitive, territorial need to compete over money.

FAIRLY DIRECT ANSWER #2: Memorize just a few typical numbers, so that you can use them as benchmarks whenever you encounter a new number. Some of the major ones are:

- solar system distances (between planets)—several AUs (an AU is the average distance between the Earth and the Sun), 10^5 times smaller than a light-year
- interstellar distances (between stars)—several light-years
- intergalactic distances (between galaxies)—several million (10^6) light-years
- cosmological distances (across observed universe)—several tens of billions (10^{10}) of light-years

- size of planets—thousands to tens of thousands (10^4) of kilometers (you could drive that far in less than a week)
- size of main-sequence stars, such as the Sun—tens, hundreds, or thousands of Earths
- size of star clusters—tens of light-years, much bigger than an individual star (by ~10^7 times!)
- size of spiral galaxies—10^5 light-years, made up of hundreds of billions (10^{11}) of stars
- size of galaxy clusters—tens of millions (10^7) of light-years, made up of thousands of galaxies

Suppose you read that some galaxy has been discovered at a distance of a billion light-years. You can immediately place this number by comparing it with the preceding ballpark reference numbers. Typical galaxy separations are millions of light-years, and because 10^3 (a thousand) $\times$ 10^6 (a million) = 10^9 (a billion), you know that this new galaxy is something like "the thousandth galaxy away in some direction." Is that very far? Yes (compared with the distances of many known planets, stars, galaxies, and clusters) and no (compared with the most distant objects ever observed).

So you see how memorizing a few numbers helps you place new information. Unfortunately, to *visualize* these numbers, you'll have to refer back to the figures in this chapter, with Earths lined up from here to the Moon, etc. But the most important point for your study of cosmology is this: You don't have to be intimidated by astronomical numbers. Believe us, they're not intimidated by you. You have to look at them a little differently than you do the numbers of daily life (e.g., look at the exponent first), and that takes some practice, but once you know the trick, it becomes quite natural. In no time, you'll find yourself saying things like "A *million* light-years? That's *way* too small!"

Quick Review

In this chapter, you've learned more about the properties of light, as well as the matter that light interacts with. You've seen how astronomers have identified and catalogued certain types of light sources in order to use them as standard candles. With this understanding of light and its sources, astronomers have measured the distances to many of the objects in the sky, revealing the remarkable vastness of the universe.

Try the following crossword puzzle to test your knowledge of key terms and concepts from Chapter 3:

ACROSS

2 Type of energy associated with random motion
6 Mathematical way to use spiral galaxies as standard candles
8 Closest supercluster of galaxies
9 Tiny, positively charged object at the center of an atom
11 Type of energy associated with ordered motion
13 Astronomical distance _____
14 Used to measure very local distances
15 Used to characterize the average speeds of a collection of particles
16 The glow of a hot object
17 Gas of ions and electrons
18 Negatively charged particle found in atoms
20 Construct made from atoms chemically bonded together

DOWN

1 Uncharged particle inside atomic nuclei
3 Type Ia supernovae produce these very predictably
4 Object of known luminosity
5 Type of star that's useful for measuring extragalactic distances
7 Neighboring galaxies

10 Positively charged particle found inside atomic nuclei
12 Atom that's missing one or more of its electrons
17 Particle of light
19 Amount of light crossing a particular area in a particular time

Further Exploration

(M = mathematical content; I = integrative—builds on information from more than one chapter; P = project idea; R = reflection; D = suitable for class discussion)

1. (M) On *Star Trek*, the starship *Enterprise* travels at "warp speed," a futuristic faster-than-light technology. Warp speed scales with powers of two; for example, "warp 3" means "$2^3 = (2)(2)(2) = 8$ times the speed of light." The speed of light is 300,000 km/sec, or 1 light-year/year.

   On the show, the *Enterprise* appears to be capable of warp 9-point-something, but not warp 10. So, how long would it take to reach each of the following locations, starting from Earth, at warp 9?
   (a) The Sun
   (b) Proxima Centauri
   (c) The center of the Milky Way
   (d) The Andromeda Galaxy

   Finally, comment on a realistic goal for the exploration of our own galaxy in the next few hundred years (between now and the various *Star Trek* program eras). Assume that we can't exceed the speed of light, as our current understanding of physics informs us. What percentage of our Galaxy might we realistically hope to explore in that time?

2. (R) Explain sequentially how to measure the distance to a spiral galaxy with the Tully–Fisher relation. Omit nothing! Exactly what do you measure, and how do you measure it? What don't you measure, and how do you know it without measuring? How do you get the distance? Be specific. Finally, how did Tully and Fisher formulate their relation to begin with? That is, how did they know the intrinsic luminosity of any galaxies, in order to correlate that luminosity with rotation speed?

3. (RI) Briefly describe the emission and absorption of light, from Chapter 2, in terms of photons.

4. (R) The Sun's central temperature is 15 million kelvins, and its surface is 5800 kelvins; the temperature drops smoothly from 15 million to 5800 in between the center and the surface. In light of this temperature difference, and in light of Figure 3.4, try to reconcile the following two correct, but seemingly disparate, statements:

    • Before emerging from the Sun's interior, a photon interacts with matter particles all over the interior, thus distributing energy throughout the Sun.
    • Different parts of the Sun have different, but constant, temperatures.

5. (M) (a) Estimate the fraction of your body's volume occupied by electron clouds and the fraction occupied by nuclei. (b) Do the same for mass. (*Hint*: Most of your body's mass is from atoms with equal numbers of protons and neutrons; for example, oxygen, the heavy part of H_2O, has eight protons, eight neutrons, and eight electrons.

6. (R) Argue that the hydrogen molecule, H_2, has less energy than would its two H atoms if they were separated from one another. To do this, think about how energy would be involved in splitting up the atoms.

7. (R) (a) Describe a beam of white light as completely as you can. (b) Suppose the light is created by a blackbody emitter, such as an incandescent light bulb. Now what can you add to your description from part (a)?

8. (R) A star gets dimmer as seen on the sky. Which of the following, if true, could explain why?

    • Its intrinsic brightness decreased.
    • Its apparent brightness decreased.
    • Its distance from Earth decreased.

9. (R) Which has more luminosity and which has more flux?
    (a) A laser pointer or 100 laser pointers held together and shining parallel beams?
    (b) A burning match or 100 burning matches grouped together?

10. (R) It has been proposed that an advanced alien civilization might construct a "Dyson sphere"—a huge spherical shell surrounding its parent star. Explain, in terms of energy, what building a Dyson sphere might accomplish for the alien civilization.

11. (R) What is a light curve, and why is it important to the type Ia supernova distance indicator?

12. (RD) Give some examples of how you calibrate the standard-candle and standard-ruler distance measurements you subconsciously make every day.

13. (MR) Describe to an English-speaking alien, on a planet with no mountains, how big a mountain is.

14. (MR) Suppose we randomly identify a cube-shaped region a few million light-years on a side in the universe. How many galaxies are likely to be in this region? How many galaxies could possibly be in it?

15. (RM) About how many star systems could have inhabitants who could possibly know that you have been born? (*Hint*: Imagine a sphere centered on Earth and whose radius is as many light-years as you are years old. Refer to Figure 3.24 to estimate how many stars fall within this sphere.)

16. (R) Explain what heat flow is in terms of the particle description of matter.

17. (RID) How does the particle model of light and matter help explain the blackbody spectrum?

18. (R) Describe, in your own words, the inverse-square law for light.

19. (RIP) Make your own imaginary scale model to help you understand distances in the universe: Pick an object to represent the size of our solar system (out to Neptune). If all distances in the universe were scaled down in the same way as the solar system, what would be the distance to nearby stars, nearby galaxies, and, finally, the most distant galaxies in the observed universe?

20. (RI) On the basis of the material covered so far in this book, and in the spirit of the quote that follows, list 5 to 10 examples in which "the beauty of nature appears in its logic." below. For each example, write a sentence or two explaining why you chose it.

> "*A painter paints the sunset, and a scientist measures the scattering of light. The beauty of nature appears in its logic as well as appearance. And we delight in that logic: The square of the orbital period of each planet equals the cube of its distance from the Sun; the shape of a raindrop is spherical, to minimize the area of its surface. Why it is that nature should be logical is the greatest mystery of science. But it is a wonderful mystery.*"
>
> —Alan Lightman, *Great Ideas in Physics*

The Universe We Discover *through* Motion and Gravity

"...when you have eliminated the impossible, whatever remains, however improbable, must be the truth."

—*Sherlock Holmes*

In the previous chapter, we saw how learning about light enables us to construct an understanding of what's out there. What people have learned by looking at distances alone, using the inverse-square law for light, can be regarded as a revolution in human thought. Just getting a sense of the scale of the universe, compared to the scale of the human world, demands a considerable shift in perspective.

Even as we have worked to learn about the sizes and arrangement of stars, galaxies, and clusters, we have so far ignored the motions of these objects. What we learn from these motions will revolutionize our thinking yet again. Another inverse-square law comes to our aid in this chapter, with another series of implications for the universe. How do galaxies hold together, and why do some of them develop a spiral shape? What's really behind their organization and behavior? We'll see that a little study of gravity reveals a lot about galaxies, including a profound and unexpected discovery.

4.1 Enter Gravity

Now that we're armed with a sense of the scale and arrangement of the objects in space, it's time to add another element to the picture: how those objects interact gravitationally. We'll begin with our solar system and then expand far beyond it.

Kepler and Newton

Planets move across the sky over time because they are traveling through space. The path a planet follows when moving around the Sun is called its **orbit.** The term "orbit" applies to other objects as well, such as the Moon orbiting Earth. How do orbits work?

By 1619, German astronomer and mathematician Johannes Kepler had published his now-famous three laws of planetary motion. His laws were based on a careful analysis of data obtained chiefly from Danish astronomer Tycho Brahe's observations of the planets. "Not all the planets are borne with the same speed, as Aristotle wished," wrote Kepler. Instead, he discovered that planetary motion is more complex:

> "... it is of irregular speed in its parts; and it makes the planet in one fixed part of its circuit digress rather far from the sun, and in the opposite part come very near to the sun: and so the farther it digresses, the slower it is; and the nearer it approaches, the faster it is ..."

> —Johannes Kepler, *Epitome of Copernican Astronomy*

Kepler's observation-based laws paved the way for others to gain a deeper understanding of *why* planets move as they do. But first he had to describe *what* they do. He correctly claimed that:

1. Planetary orbits are *ellipses*, with the Sun located at one *focus*.

Prior to Kepler, heavenly bodies were assumed to orbit in perfect circles. That was Aristotle's idea, and its philosophically pleasing description dominated Western thinking for nearly 2000 years before Kepler showed it to be wrong. An ellipse is a mathematically described shape that looks like a stretched-out circle and is commonly referred to as an oval. (See Figure 4.1.) The figure also shows what the word "focus" means in this context: An ellipse is like a circle, except that it has two "centers," each called a focus.

The long axis of an ellipse—the longest straight line you can draw through the ellipse, like the diameter of a circle—is called the *major axis*. Some elliptical orbits with differing major axis lengths are shown in Figure 4.2. We can say that the ellipses shown have varying *eccentricities*. The most eccentric ellipses are the most stretched out, with an eccentricity of nearly 1. Under Kepler's first law, a perfectly circular orbit is possible, too, with an eccentricity of precisely zero.

Kepler determined that orbits are ellipses, although he didn't know why they were. Elliptical orbits are simply what the data revealed to him. (He knew the precision of Tycho Brahe's data and realized that trying to force a perfect circle to fit those data, particularly with regard to the motion of Mars, would be "cheating.")

Kepler's second law is

2. Planetary orbits sweep out equal areas in equal times.

This "sweeping out areas" business is shown in Figure 4.3. The first law tells us what the path of an orbiting planet is; this second law tells us about the speed of the planet in its orbit. If the planet is far from the Sun, then the planet moves slowly, because little motions sweep out big areas (i.e., it will take the same amount of time to cross the distance covered by the three arrows in Figure 4.3). In order to take the same amount of time to sweep out a certain area when a planet is far from the Sun as when it is near, the planet has to slow down when it's far away and speed up when it's close.

We might suspect that a description of gravity would help here: We'd like to know *why* planets orbit faster when they're close to the Sun. But first, let's complete the list of Kepler's laws:

3. The size (*a*) and period (*P*) of a planetary orbit around the Sun are mathematically related by

$$P^2 = a^3 \,,$$

(eq. 4.1)

*Tycho Brahe (1546–1601), from Denmark, and **Johannes Kepler** (1571–1630), from Germany, revolutionized our understanding of the motions of the planets in our solar system. Tycho made comprehensive, detailed observations of the positions of planets from night to night, at a remarkable precision (at that time) of nearly 1% of 1 degree on the sky. His results led Kepler to construct an accurate scientific description of how the planets move, now collectively known as "Kepler's laws" of planetary motion. The work of both of these astronomers showed convincingly, for the first time in history, that the Sun, and not the Earth, was the center of the solar system.*

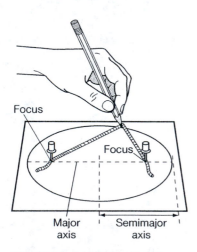

FIGURE 4.1 An Ellipse An ellipse is the shape you get when you draw a figure with fixed total distance from two foci (plural of "focus"). Here, this fixed total distance is enforced by a length of string held down by two tacks, one at each focus point.

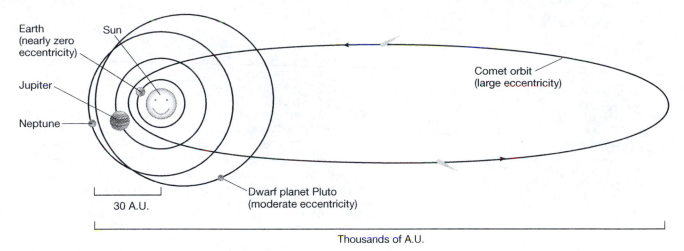

FIGURE 4.2 Some Elliptical Orbits in the Solar System   Orbits can range from nearly perfect circles to highly elongated ellipses with large eccentricities.

or, essentially,

$$(\text{how long the orbit takes})^2 = (\text{how big the orbit is})^3 .$$

Don't let the squaring and cubing bother you. The point is much simpler than that: Orbits are all alike, in that there's a mathematical recipe connecting their size scale to their time scale.

We use *a* to denote the length of the *semimajor axis*, which is just half the length of the major axis, measured in AU. *P* represents the time elapsed, in years, for a complete orbit. Watch how the law works:

For Earth, $P = 1$ year and $a = 1$ AU, so
$P^2 = (1)(1) = 1$, and $a^3 = (1)(1)(1) = 1$. Then for Earth, $1 = 1$.
(Woohoo! Kepler was right!)

For Neptune, $a = 30$ AU. By Kepler's third law,

$$P^2 = (30)(30)(30) = 27,000 .$$

Taking the square root of both sides gives $P = 164$ years, which is indeed Neptune's measured orbital period.

Now, Kepler's laws are observationally true, but they lack *motivation*. We can see that they are true, but it is human nature to want to know *why* they are true. A more satisfying explanation comes in the form of a theory of gravity. For this, we look to Sir Isaac Newton. It is difficult to overstate the importance of his contribution to the field of physics, although Queen Anne of England must have recognized it, since she knighted him in 1705, making him the first Briton to be knighted for scientific accomplishment. Newton's big achievement in the field of gravity (as contrasted with his other big achievements in the fields of calculus, kinetics, fluid mechanics, and optics!) was his recognition that gravity

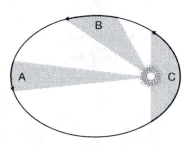

FIGURE 4.3 Kepler's Second Law   When a planet is far from the Sun (A), it travels relatively slowly, sweeping out a long, thin area. When it is near the Sun (C), it moves faster, sweeping out a short, fat area. B is an intermediate case. Kepler's second law states that planets slow down and speed up along their elliptical paths in just such a way as to ensure that all these (shaded) areas swept out in a fixed amount of time are equal.

Sir Isaac Newton (1643–1727) is often called "the father of modern science" because his discoveries powerfully reshaped our understanding of the universe. Among his many other scientific achievements, he constructed our modern understanding of motion and forces (including gravity) and he invented calculus. Because of the importance of Newton's work, one of his professors at the University of Cambridge actually resigned so that Newton could have his job. In 1705, Newton became the first scientist ever knighted.

could be invoked to explain *both* the everyday phenomenon of falling *and* the behavior of celestial bodies. In his landmark book, *Philosophiæ Naturalis Principia Mathematica* (*Mathematical Principles of Natural Philosophy*), or, simply, the *Principia*, Newton made the following argument that his gravity theory applies to *all* bodies:

> *"Finally, if it is universally established by experiments and astronomical observations that all bodies on or near the earth gravitate toward the earth, and do so in proportion to the quantity of matter in each body, and that the moon gravitates toward the earth in proportion to the quantity of its matter, and that our sea in turn gravitates toward the moon, and that all planets gravitate toward one another, and that there is a similar gravity of comets toward the sun, it will have to be concluded ... that all bodies gravitate toward one another."*

> —Isaac Newton, *Principia*

Newton published his theory of gravity, and other pioneering work, in 1687. More than 200 years later, Albert Einstein discovered that Newton's theory was incomplete, and in Chapters 6 and 7 we'll study Einstein's improvement upon it and how his correction to Newton's theory applies to cosmology. But Newtonian gravity is still highly useful. For one thing, it remains true for most astronomical systems; that is to say, Einstein's correction is often too small to worry about. Perhaps more importantly, Newton's description of gravity is so clear and direct, that the small investment you make by studying it will yield a large payoff in terms of understanding your universe.

Believe it or not, the easiest way to appreciate Newtonian gravity may be through an equation. **Newton's law of gravity** looks like this:

$$F = G\frac{mM}{r^2} \ ,$$

(eq. 4.2)

where

- *F* is the gravitational *force*, measured in metric units called newtons, abbreviated N as in "800 N."

- *G* is a number, called "Newton's constant" or "the gravitational constant of the universe," relating to the strength of gravity.

- *m* and *M* are the masses, in kilograms, of two objects that gravitate toward one another. Usually, the two objects have different masses (Earth and Sun, for example), in which case the smaller one is *m* and the larger one is *M*. (We'll see in Chapter 9 that mass is actually a complicated concept. However, for now, its more familiar meaning will suffice: Mass is a measure of the "amount of matter" something is made of. Here on Earth, we can find something's mass from its weight: Heavier things are made of more matter. In outer space, a freely floating object is weightless, but still has mass.)

- *r* is the distance, in meters, between the two gravitating objects. Usually, the separation between two objects is the radius of a circle; for example, the radius of the Earth's orbit is also the separation between the Earth and the Sun. That's why the letter *r* is used. But be aware that this is somewhat misleading notation, and there need not be any circles involved for Newton's gravity law to apply. (We could say *d* for "distance" instead, but the *d* might remind you of "diameter," so let's stick with *r*.)

You may have recognized that this is another inverse-square law, with the distance squared in the denominator, just like the inverse-square law for light. The inverse nature has the same effect in both cases: Gravity and brightness each get *weaker* at larger distances. And because *r* is squared in both cases, a *small* increase in distance generates a *large* decrease in gravity (or brightness).

On the other hand, the masses are in the numerator. If one or both masses increase, then the whole thing increases, so the gravitational force gets stronger. In words, Newton's law of gravity goes like this:

The strength of the natural gravity force =

         a number that has been measured and is always the same ×
         the mass of one object × the mass of another object/
         the distance between the objects, squared .

To put all of this together, you have to view gravity as Newton did, and *not* how you normally do in your daily life. Normally, you might think of gravity as a force holding all of us to the Earth. That's true, but it's not the complete picture. Instead, think of it as a force between *any* two objects anywhere, such as the Sun and a planet, isolated in space. These two such objects are acted upon by a gravity force, which is always *attractive*. Newton's equation says that the force will increase if either mass increases. But Newton's equation also says that the force will decrease if the distance grows. These properties of the gravity force are illustrated in Figure 4.4.

And how do we handle more than two objects? Simple: Just use Newton's law for each pair. Then each object gets a force arrow from every other object present, one by one, and is pulled simultaneously in the direction of all the arrows, as in Figure 4.5.

In frame (b) of Figure 4.5, a spaceship is acted upon by all the little gravitational forces from each star in a distant cluster of stars. In frame (c), we see that we can represent this effect with one big force coming from all the stars in the cluster combined. A similar effect is operating on you right now. The reason that gravity holds you to the Earth is that your body (m) and the Earth (M) are experiencing an attractive force at a distance (r) that runs from the middle of the Earth to the middle of you. You are pulled toward each bit of the Earth with a force that varies with the inverse square of how far away that bit is from you. The net effect of all those little forces, as shown in Figure 4.6, is to pull you with a fairly strong force directed toward the center of that collection of bits (i.e., toward the center of Earth).

At this point, you might be asking yourself why gravity pulls you toward the Earth, but not toward every other object you encounter. "Self," you might be saying, "if Newton was so smart, then how come books don't fly off their shelves and hit me when I walk by? Why don't I gravitate into large buildings when I'm strolling along the sidewalk?"

The answer lies in the fact that gravity is by far the weakest natural force that scientists have discovered in the universe. It may not always seem that way, especially when the elevator is out of order and you have to hike up to the 11th floor in July with 94% humidity, but the very fact that you *can* climb stairs tells you how weak gravity must be. Think of it: All the mass of the entire planet is trying to hold you down, yet you come out on top.

Newton built the weakness of gravity into his equation through the gravitational constant. It is a very small, well-measured number: $G = 6.67 \times 10^{-11} \, \text{m}^3/\text{kg s}^2$

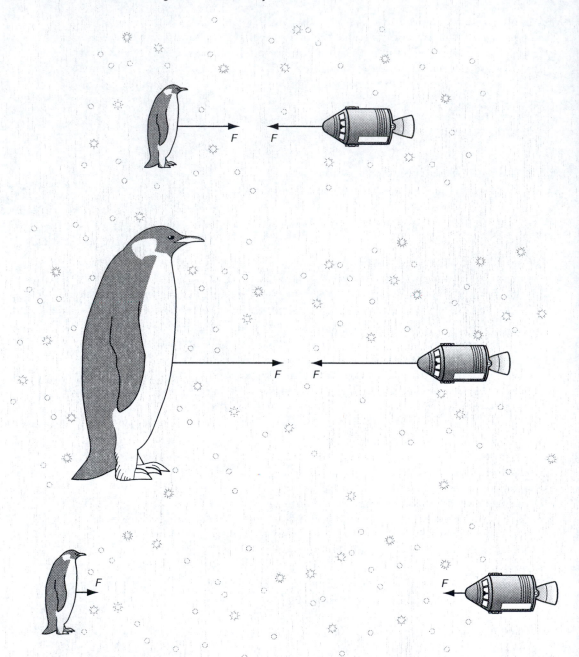

FIGURE 4.4 Gravitational Attraction Both the space penguin and the spaceship have mass, so they are pulled toward each other by their mutual gravitational force (indicated by arrows). There is only one equation for this attraction, and not a different equation for each object, so the force that is calculated is the same for both; thus, both force arrows are drawn the same length. The force of gravity increases if *M* or *m* increases and also increases if the masses get closer together, but decreases when the masses move farther apart.

(SI units). With *G* that small, we need *M* to be huge, or else *F* is too small to notice. Books and buildings and other people don't have enough mass to pull you —or rather, they *do* pull you, but not hard enough for you to notice. You only feel Earth's gravity because Earth's large mass makes up for *G*'s smallness.

In any case, gravity *is* strong enough to affect massive celestial bodies. If this were the whole story, however, then the space aliens in movies wouldn't *want* to colonize our planet, because the attractive force, left to its own devices, would pull our planet into the Sun, where it would incinerate. But if our planet were set in motion first, then there is a special path that it could take—a path that is subject to the continuous inward pull of gravity, yet doesn't head into the Sun. That path is an orbit. (See Figure 4.7.)

Now, there are good reasons to expect that the Earth and other planets all started out in a state of motion, as though heading alongside the Sun. We won't go into those reasons right now, because the formation of the solar system is not our present concern. Rather, we wish to demonstrate how it is possible to have stable orbits, instead of gravity sucking all the planets into the Sun. (But if it bothers you to imagine a planet "starting out" moving at some speed, then ask yourself this: In outer space, where there is no air resistance and no friction, what would cause something to sit still?)

The way to interpret Figure 4.7 is this: Because gravity's force is always pulling the planet inward, it never gets to just fly away—which is what it would do if there were no gravity. Instead, it's constantly *arcing toward the Sun*. Yet if the planet's moving with enough speed, then it's always in the process of *moving past the Sun*, too. The two effects *combined* create a stable orbit. You can think of it as a win–win compromise (there are no thought police to stop you): Gravity wins by making the path curve around the Sun, and the planet's motion wins because it never stops moving or falls into the Sun.

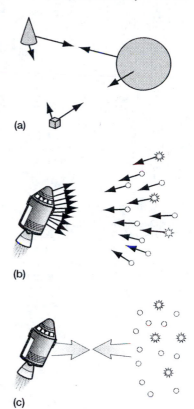

(a)

(b)

(c)

FIGURE 4.5 How Newtonian Gravity Works for Multiple Objects
(a) Three objects gravitate toward one another pairwise. That is, Newton's law of gravity can be used to compute the gravity force between each pair of objects. Notice how the pairs of arrows work, with two objects exerting the same force on each other, but each object being acted upon by different forces from the other two objects. The small, bottom object, for instance, is being pulled gently up and left, and strongly up and right. When all the forces are taken together, the bottom object is being pulled mostly upward. (b) A spacecraft near a cluster of stars is acted upon by a collection of forces caused by (and directed toward) each star. (c) The combined effect of all the forces can be understood by viewing the entire star cluster as a single massive object.

EXERCISE 4.1 Earthbound

If you had all-encompassing power over the world, and you wanted to quadruple the gravity force felt by the rest of us lowly earthlings, just to amuse yourself, how would you do it? Specifically,

(a) Could you do it by changing the mass of the Earth only, but keeping Earth the same size? If so, how much would you change the mass? Would you halve it? double it? quadruple it?

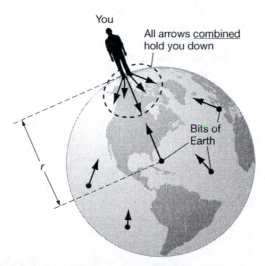

FIGURE 4.6 Gravity at the Earth's Surface   You are pulled toward the center of the Earth with a force that is due to the combined forces of all the bits of mass that make up the Earth.

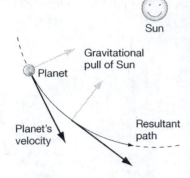

FIGURE 4.7 The Cause of Orbits
A planet such as the Earth would "try" to continue on past the Sun in a straight line if it were moving in that direction to begin with. But the gravitational force "tries" to pull the planet toward the Sun. The resulting compromise is an elliptical orbit around the Sun.

(b) Could you do it by changing the size (radius) of the Earth, but leaving the mass alone? If so, how much would you change the radius? By a quarter, half, double, etc.?

(c) If some *other* all-powerful individual—your sister perhaps—suddenly makes the Earth nine times less massive (but leaves the size alone), and you still want to quadruple the gravity force without undoing her action, then how much would you have to change the radius?

Gravitational Binding Energy

It's nice to describe gravity in terms of its *force*, since that's what you can actually feel. But it's often useful to characterize it as an *energy* instead. That way, we can directly compare gravitational energy with other forms of energy. The SI units for energy are "joules," abbreviated J as in "3000 J."

The energy associated with gravity is called a **binding energy**, because gravity binds us to the Earth, and binds Earth to the Sun, and so on. The idea of being *bound* is simple: It means that, without acquiring some additional energy, you can't escape. Without additional energy provided by a large rocket engine, your body is bound to the Earth. (Of course, your *mind* can go higher by studying cosmology.) We can quantify *how bound* you are with a single number, called the gravitational binding energy.

There are other binding energies, too. *Electrostatic* binding energy quantifies the binding between electrons and nuclei in atoms. *Nuclear* binding energy quantifies the binding between protons and neutrons inside atomic nuclei. One interesting thing about these binding energies is this: We assign them a *negative* number value. For gravity, the equation for the binding energy looks similar to the equation for its force:

$$E_g = -G\frac{mM}{r} \quad .$$

(eq. 4.3)

Why the minus sign? If we start with two objects infinitely far apart, then they share no gravitational binding energy, because they're too far apart to feel each other's gravity. You can see this in Equation 4.3 (and in Equation 4.2), because dividing by an infinite *r* gives zero energy (and zero force). But if two objects start out close to one another, then we would have to *add energy* in order to separate them infinitely—with rockets again, perhaps. Mathematically, it follows that the binding energy must be negative, because adding a positive (rocket) energy can bring it up to zero.

All the letters on the right side of the equation are the same as with the gravitational force, but their meanings here reflect two differences. The first difference is the minus sign, to indicate that we have a binding energy, which is always a negative value. The second difference is that *r* is no longer squared in the denominator. But since the same quantities are on top and bottom in this equation as in the force equation, they do at least act in the same direction. More mass means more force *and* more binding. More separation distance means less force *and* less binding—except that when we say "less binding," we mean "a less negative binding energy."

As promised, the benefit of having learned all this is at hand, because now we can compare *gravitational energy* with *kinetic energy* (recall Section 3.1), and from that comparison, we gain a strong understanding of motion in space. The formula for kinetic energy—the energy of motion—is

$$E_k = \frac{1}{2}mv^2 \, , \qquad\qquad \text{(eq. 4.4)}$$

Where, as before, m is mass, and now v is velocity or speed. First, just read the equation, and notice how the squared velocity makes it count twice as much:

kinetic (or motion) energy = one-half × mass of moving object
× speed of moving object, squared

Notice that if m or v increases, or if both increase, then E_k increases, too. And because of the square, changes in v have a greater effect than changes in m. Doubling m doubles E_k, but doubling v *quadruples* E_k, because 2^2 is 4. Still, even without this level of detail, we can very quickly see that kinetic energy depends on what we would expect it to depend on: faster moving and/or more massive objects have more energy when they move than do slower and/or less massive objects.

One more thing to notice about kinetic energy is that it has no minus sign. It's *not* a binding energy. In fact, if you think about it, it's more of an *un*-binding energy: More mass and more speed make for more ability to break free from some sort of binding. And this is exactly what we need to consider.

In a typical orbit, the total energy is the kinetic energy of the orbiter plus the gravitational binding energy between the orbiter and the orbitee (yes, we just made that word up). We can often consider one object to be stationary while the other moves (like the Earth around the Sun); this is never literally true (the Sun moves a little, too), but it's usually accurate enough if one object has significantly more mass than the other. So we have

$$E_{\text{total}} = E_g + E_k = -G\frac{mM}{r} + \frac{1}{2}mv^2 \, . \qquad\qquad \text{(eq. 4.5)}$$

In words,

Total energy = gravitational energy, based on mass and distance (negative because it binds things) + kinetic energy, based on mass and velocity (positive because it sets things free) .

Now, since one term in this equation is negative and the other is positive, the total energy could go either way. In other words, there is a battle between the gravitational and kinetic energies, and whichever there is more of wins. We already know what it means to "win": More gravitational energy means that the total energy of the system is negative, so the system is bound; more kinetic energy means that the system is not bound and the objects will escape from one another. (See Figure 4.8.)

We can calculate the outcome of any encounter in space—say, between a planet and an asteroid—with this equation. If the total energy comes out positive, then the asteroid is not bound, and it flies away. If the total energy comes out mildly negative, then the asteroid ends up bound in a new orbit around the planet.

FIGURE 4.8 Bound and Unbound. This is a simpler version of the binding and unbinding discussed in the text. (a) A ball rolling fast enough to have more positive kinetic energy than negative gravitational energy is not bound and will escape the dip. (b) However, if there is more negative gravitational energy than positive kinetic energy, the ball is gravitationally bound and will not escape.

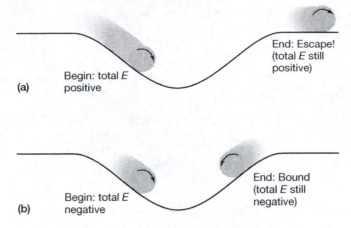

(a) Begin: total *E* positive

End: Escape! (total *E* still positive)

(b) Begin: total *E* negative

End: Bound (total *E* still negative)

Several moons in our solar system are asteroids captured in this way. If the total energy is *very* negative, then the asteroid crashes into the planet. Notice that the asteroid can collide or be captured into an orbit; in both cases it's bound, because it'll never escape. It's just a question of in which case it is *more* bound. Various bound and unbound trajectories (orbits) are shown in Figure 4.9.

Exercise 4.2 is intended to help you work on your intuition for motions and encounters in space. You only need E_g and E_k; together, they can answer everything for you.

FIGURE 4.9 Bound and Unbound Trajectories Circular and elliptical orbits are bound. *Hyperbolic* paths are not bound. *Parabolic* paths are not bound, in that something following a parabolic path away from the Sun never returns. However, a trajectory is parabolic when there's zero total energy—right on the boundary between bound and unbound.

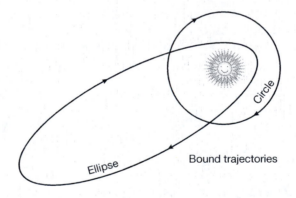

Circle

Ellipse

Bound trajectories

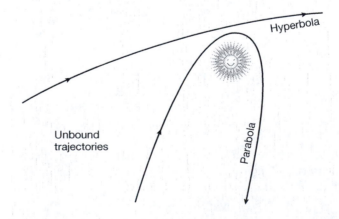

Hyperbola

Unbound trajectories

Parabola

EXERCISE 4.2   To Bind or Not to Bind?

So you think you understand all this energy stuff? Prove it! Try to hold your ground against the following counterintuitive examples.

(a) If you're piloting the space shuttle in orbit, and you want to return to Earth, you might try thrusting the shuttle in the direction of the Earth. This will not work! Chances are, you'll miss the planet altogether. (Or if you hit it by these means, that's not good either!) Suggest what will actually happen, and then use the total-energy formula to demonstrate that this approach will tend to unbind you from the Earth—just the opposite of what you want. Finally, figure out what you should do instead in order to get home.

(b) Suppose some astronomers announce that they just discovered a deadly asteroid that's expected to come dangerously close to the Earth. Demonstrate that the faster the asteroid is moving, the *better* for all of us.

(c) Suppose that the asteroid from part (b) turns out to be less massive than the astronomers initially realized. But don't relax yet! Sure, if it hits the Earth, it'll do less damage. (Explain why.) But does less mass mean that it's less likely to hit us? After all, less mass means less gravitational force and less negative gravitational energy. (*Hint:* Decide carefully which mass, m or M, is in question here, and look over the whole equation for E_{total} before answering.)

(d) Air molecules on Earth have a variety of speeds. For some, their kinetic-plus-gravitational energy comes out positive. Does that mean that our atmosphere is slowly bleeding into space? Is it possible that the Moon, which is about 100 times less massive and 4 times smaller than the Earth, had an atmosphere in the past, but then lost it in this manner?

(e) Flip back to refresh your memory of *thermal energy* from Section 3.1. Which is there more of on the surface of the Sun, thermal energy or gravitational energy? In other words, would $E_g + E_{th}$ be positive or negative?

As you know from Chapter 2, we can use the Doppler effect to measure the speeds at which objects in the universe move toward us or away from us. Now, with gravity in the mix, we can learn not only how fast an object is moving, but also *why* it's moving. Is it in orbit? If so, can we find out how massive an object it orbits around? It turns out that, yes, we can.

We can also learn to recognize bound and unbound systems. For example, in Figure 4.10, is the knot of stars at the left a star cluster? That is, is it gravitationally bound, so that it will stay together, or is it just a clumpy arrangement of unrelated stars on the sky, which will disperse over time? In the case shown in the figure, the stars *look* bound, in that they look similar to other bound star clusters, but actually they are not. On the basis of Doppler measurements of their speeds, they have too much kinetic energy to be bound.

In order to recognize gravitational motion in our astronomical and cosmological observations, we must now define two particularly relevant speeds, which we can then compare against Doppler measurements.

PARTICULARLY RELEVANT SPEED # 1: If E_g exactly equals E_k, then the system is on the borderline between bound and unbound. The speed at which this happens, then, can be thought of as the bare minimum speed needed for an object to *escape* from being bound. For example, between you and the Earth, it's the speed at which you'd need to jump so that you'll never come back down.

FIGURE 4.10 Combining Doppler Measurements with Knowledge of Gravity   In this region called NGC 2467, the grouping of stars at the left appears to be a star cluster—a gravitationally bound system—because its appearance is similar to that of many other bound star clusters. In this particular case, however, the appearance is deceiving; their Doppler shifts show that these stars are moving too fast to be bound as a group. (◆ Also in Color Gallery 1, page 9.)

We define this speed as the **escape velocity** v_{esc}. If we set the total energy to zero in Equation 4.5, then some quick algebra reveals that

$$v_{esc} = \sqrt{\frac{2GM}{r}} \; .$$ (eq. 4.6)

Remember: M is the mass of the object being orbited, not the mass of the orbiting object itself.

PARTICULARLY RELEVANT SPEED # 2: For a bound, circular orbit, we find that the magnitude of E_g is exactly double that of E_k. It's useful to denote a **circular orbit velocity** v_{circ} whose value is

$$v_{circ} = \sqrt{\frac{GM}{r}} \; ,$$ (eq. 4.7)

which is about 1.4 times smaller than v_{esc}. The importance of this speed may not be immediately obvious, since you have learned that orbits can be elliptical instead of circular, in which case the orbital speed is not v_{circ}. But many orbiting systems in the universe, like the majority in our solar system, *do* have nearly circular orbits, so their speeds will be at least close to v_{circ}.

Astronomers can use both the escape and circular orbital speeds to understand the nature of many objects in the universe. Imagine that we observe a galaxy cluster. For each galaxy, one by one, we can compare these two calculated speeds with its Doppler-measured speed to learn which galaxies are bound in orbit and which are just passing through before they escape. Or suppose we observe an object moving near another massive object, after an explosion, perhaps. We can now determine whether one object is orbiting the other, or whether it was set in motion by the explosion instead. In other words, understanding gravity allows us to understand what caused the motions we see.

Next, in Section 4.2, we'll do just that: use our knowledge of gravity and distance to interpret the motions of objects we see in the universe. But we can do even better than that, because in Section 4.3, we'll apply our knowledge to material in the universe that we *can't* see.

4.2 Stars and Galaxies in Motion

In our solar system, planets orbit the Sun in nearly circular paths. Each planet travels at a speed close to v_{circ}, which is a different speed for each planet. Why? Because at different distances from the Sun, the gravitational pull on each planet is different—or mathematically, because r appears in the equation for v_{circ}. The resulting solar system speeds are in the vicinity of a few tens of kilometers per second; Earth orbits at 29 km/s.

We can also zoom out and consider motions in the "solar neighborhood": the motions of the Sun and other nearby stars. In Figure 2.31, we saw random motions of nearby stars. But with our knowledge of gravity, we can do better now. Our Galaxy is a massive structure, and we can view the Sun and its stellar neighbors as orbiters, circling the center of the Milky Way.

Astronomers measure orbital speeds within our Galaxy by observing Doppler-shifted lines in radio emission by galactic hydrogen gas, and by making Doppler measurements of many globular clusters. Globular clusters are useful to look at, too, because they do not orbit within the disk of the galaxy as we do; therefore, we can determine how much we appear to be moving relative to the center of their overall population. These techniques tell us that stars at the same distance from the galactic center as our Sun travel at v_{Sun} = 220 km/s. That's about 10 times faster than the typical orbital speeds in the solar system. Figure 4.11 shows our planet moving around the Sun, as the Sun and planets move around the Galaxy.

From our perspective, the Sun's neighbor stars appear to move at about 10 km/s, every which way. Now let's reevaluate that appearance from a more Galaxy-centered point of view. If local stars all move at *nearly* 220 km/s, in *nearly* the same direction, then, from our moving viewpoint, it will just look like random scatter around us. Faster stars will pull away from us. But slower stars will also move away from us, because we're pulling away from them! (See Figure 4.11.)

We should also apply Newtonian gravity to other systems in space. First, consider the stars within a globular cluster. They orbit in all possible directions, like a swarm of bees (or perhaps a swarm of less scary gnats). How should we interpret

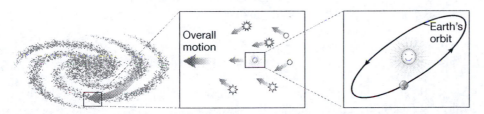

FIGURE 4.11 The Solar System within the Galaxy   The main figure shows our location within the Milky Way. The first inset shows that our solar system and neighboring stars have a little scatter in their motions, but are all moving in the same general path around the center of the Galaxy. The second inset zooms in further to reveal that our solar system is inclined relative to the Galaxy.

that motion? To answer this question, just look at any *one* of the 10^5 or 10^6 stars in a typical globular cluster. That one star is acted upon by the gravity forces *from all the other stars*. So one star orbits the center of the cluster as a result of being pulled by every other star, most of which are concentrated near the center of the cluster. Whatever direction the star happens to be moving initially, coupled with the gravity of all the other cluster stars, will define the star's orbit.

Many other systems behave this way, too. Like a globular cluster, elliptical galaxies have "swarming" stellar orbits, although ellipticals are about 10 million times more massive than globulars. The central "bulges" of spiral galaxies behave this way as well—like miniellipticals bulging out from the center of the spiral disk. Even galaxies swarm around each other in a similar way, responding to the mass of other galaxies in small collections, such as our local group, and in massive galaxy clusters, where individual galaxies travel at about 1000 km/s. Larger scales appear to imply larger speeds, from planets up to galaxy clusters.

Globular clusters swarm around our Galaxy, too, although they respond to the large gravitational pull of the Galaxy much more than they respond to one another. And in the same spherical region surrounding our galactic disk, there also exists a small population of individual stars called "halo stars," orbiting the Milky Way in all directions. In the next subsection, we'll take a closer look at the systems that make up a spiral galaxy like ours.

EXERCISE 4.3 Virgocentric Infall

At the end of Section 3.3, we encountered the cosmically nearby Virgo Supercluster, composed of many galaxy clusters. Our local group, including the Milky Way, and many other galaxies in our larger neighborhood are moving toward the Virgo supercluster. Explain why. Then speculate on the eventual outcome of this process, billions of years from now.

Before you answer, consider *Hint # 1:* The more matter that accumulates in the supercluster, the stronger its gravity will be.

But before you assign too much weight to *Hint # 1*, consider also *Hint # 2:* Clustered objects appear to be relatively stable; galaxies, star clusters, galaxy clusters, and superclusters can keep on swarming without any great tendency to crush down into one superobject.

Why Disks?

Figure 4.12 depicts the Milky Way and highlights the motions of various parts of it; refer to it as you learn the official names for the parts of the Galaxy. The elliptical or spherical components include the **bulge** and the much larger **halo,** both of which feature the swarming motion we've been analyzing. The **disk** component includes our Sun, other stars, dust, and a large amount of gas. The gas is mostly hydrogen, and its contribution to the mass of the disk is comparable to that of all the stars in the disk. The material in the disk orbits the galactic center in a plane, in a more organized way than the "swarming components" (the bulge and the halo), and is generally arranged into a pattern of **spiral arms.** Figure 4.13 depicts the overall stellar motions in each of these locations.

The elliptical, swarming components of various objects—elliptical galaxies, globular star clusters, spiral galaxy bulges, and galaxy clusters—tend to have very little gas in them. (We learn this by looking for a particular type of radio emission from hydrogen.) The stars in these regions are generally yellow-orange, a color associated with relatively old stars. Younger populations of stars usually contain

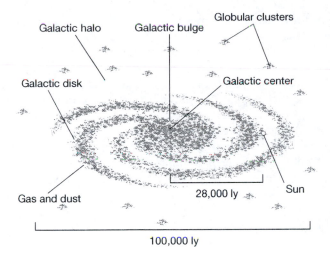

FIGURE 4.12 Galactic Anatomy The structure and dimensions of our Galaxy.

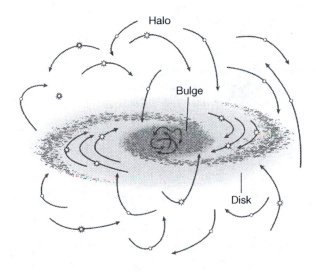

FIGURE 4.13 Stellar Motions in a Disk Galaxy The motions of bulge, disk, and halo stars are shown here. Disk stars orbit within the plane of the disk (and even if they wander slightly above or below the disk, the gravity of the disk always pulls them right back). Halo and bulge stars swarm around in all three dimensions and can repeatedly pass through the disk (which is mostly empty space) without incident.

several bright blue stars, making the population of stars look blue overall. (We've already seen this aspect of stars: Remember the blackbody concept from Section 3.1, which tells us that the hottest stars will be both the bluest and the brightest.) You can see these galactic colors yourself in any of the various color images of galaxies and galaxy clusters in the color galleries.

Putting all this together, we observe that

- The disks of spiral galaxies like ours contain a lot of gas, which can form new stars; this makes sense because the blue color indicates the presence of young stars that must have formed recently from gas.

- The elliptical, swarming systems, such as globular clusters or elliptical galaxies, contain little gas and therefore can't produce very many new stars; this makes sense because the yellow-orange color indicates the absence of young stars.

Here again, as with the rest of this chapter, we can combine our knowledge of physics (blackbody colors, the Doppler effect, hydrogen lines, and Newtonian gravity) with our observations in astronomy (colors and shapes of stellar populations and the relative abundances of gas and stars), to comment on how galaxies and clusters form. In particular, which came first? Did a large quantity of gas assemble and then fragment into separate objects (like stars), or did the gas form stars first, which *then* assembled into a system (like a galaxy)?

It's the gas that first forms the disk of a spiral galaxy, and that gas forms into stars later. Unlike stars, clouds of gas interact strongly with each other. One cloud can't easily pass through another cloud; they tend to mix and merge. A spinning system of gas, then, is more "connected" than a swarm of individual, pointlike stars. The gas system can become flattened into a disk, like tossed pizza dough. (An analogy for stars might be trying to toss and spin a collection of distinct, hardened blobs of pizza dough and having them just scatter all over the kitchen.)

Imagine following a large gas cloud trying to "wobble" its way through our Galaxy, as in Figure 4.14. With each attempt to pass through the disk, the cloud is obstructed. The only direction in which it encounters no such obstruction is the direction in which the disk material is already moving. Now, you may think this explanation sounds like we're cheating, since it started with a disk already in existence. But any self-gravitating system, like a galaxy that is still forming, will tend to have some net spin direction, compared with the larger surrounding medium. Everything moves and spins in space, where there's no friction to make things sit still. In fact, when a galaxy begins to form, its gas pulls itself inward by gravity and therefore spins *faster*, like a figure skater pulling her arms in. It follows that there's always a preferred orbital direction and that all other directions will be more obstructed. Think of a swirling gas system like a tornado: There's definitely a preferred direction of motion, and it's harder to move straight up or down, or against the circulation. A disk therefore *forms automatically*. The point is, if the gas is already orbiting in a disk by the time the individual stars form, then those stars will stay in the disk: There is nothing to cause them to turn away.

A spiral galaxy, evidently, has a complicated formation history: It contains two separate elliptical components—a large halo and a small bulge—which formed differently than its disk, the part that's active with new star formation. Stars in the elliptical components evidently formed earlier, or else the gas they formed from would have merged into the disk. This situation is shown in Figure 4.15. You might be tempted to say that such an intricate formation process should occur less frequently than the seemingly simpler assemblage of stars in elliptical galaxies. But actually, it seems to be the other way around: Disks seem to emerge quite naturally, as a result of gas existing in a galaxy or even new gas entering a galaxy.

Elliptical galaxies, however, do not appear to have much gas, and they have a rounded shape rather than a flattened disk. Perhaps they form from gas clouds with

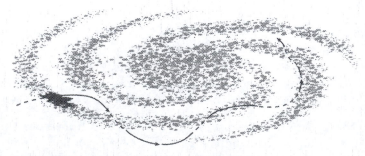

FIGURE 4.14 How Gas Clouds Conform to the Motion of a Disk A blob of material moving through the galaxy will fall into line with the other material that's there, because that's the path of least resistance.

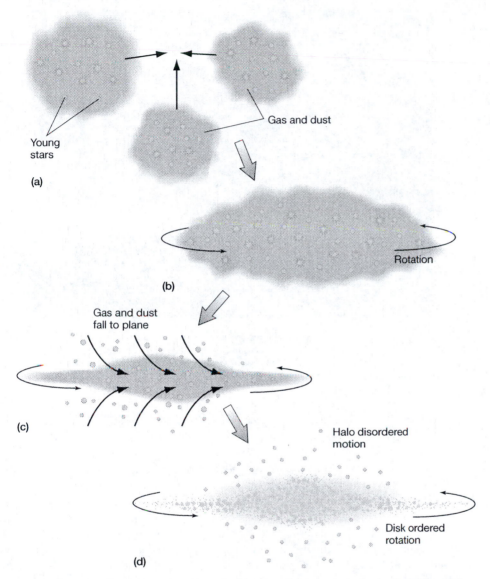

Young stars

Gas and dust

(a)

Rotation

(b)

Gas and dust fall to plane

(c)

Halo disordered motion

Disk ordered rotation

(d)

FIGURE 4.15 Spiral Galaxy Formation   The formation of spiral galaxies under the influence of gravity and rotation generates separate components. Shown here, the disk forms from a rotating gas cloud, which later fragments into stars. The halo appears to be composed of stars that formed prior to the disk; now the swirling gas in the disk is not able to "push" halo stars into flat orbits.

very little spinning motion, or maybe their stars just formed so quickly that the gas they formed from didn't have enough time to reorganize into a disk. Or maybe they form from preexisting collections of stars. In galaxy clusters, there are many more ellipticals than spirals. But in *the field*—that is, anywhere other than galaxy clusters—there are many more spirals than ellipticals. This difference has led many astronomers and cosmologists to speculate that a galaxy forming on its own is perhaps most likely to form as a spiral, but that the frequent interaction between galaxies in clusters tends to convert spirals into ellipticals. You can see for yourself, in Color Figures 27 and 30 on pages 13 and 14 of Color Gallery 1, that collisions between galaxies do occur, and complex arrangements can result.

FIGURE 4.16 The Abell 1185 Galaxy Cluster   Here we see a galaxy cluster, rich with a variety of member galaxies: spirals, ellipticals, and even a guitar-shaped collision event in progress (on the left). The collision is visibly changing the shape of the colliding galaxies. Perhaps when it's done, the inter-action will have created one or more new elliptical galaxies. (◆ Also in Color Gallery 1, page 13.)

In Figure 4.16, you can see a cluster of galaxies that has it all: spirals, ellipti-cals, and even a guitar-shaped region where at least two galaxies are interacting with one another. Gravitationally, the "guitar galaxies" are pulling each other apart, distorting whatever shape they had before. We might envision two spiral galaxies colliding and their disks being ruined by the collision. The stars are gravi-tationally pulled into a swarm of nondisk orbits. The result is an elliptical galaxy. In the merger of the "Antennae Galaxies," (See Color Figure 27(b) in Color Gallery 1, page 13.) you can see that a merger tends to separate out the stellar and gaseous components of the interacting galaxies. Although stars are too widely sep-arated to collide with each other, massive clouds of gas and dust do collide with each other. A burst of new star formation caused by the collision can lead to super-nova explosions that eject gas from the colliding galaxies. Thus, mergers can sepa-rate the stars from the gas and reshape spirals into ellipticals.

Spiral Structure

So why do disk galaxies develop spiral arms? This, too, is answerable in terms of gravity and measurable in terms of the Doppler effect, although it is complicated.

When astronomers survey the motions of stars near us in the disk, they find a pattern to the Doppler shifts. Figure 4.17(a) shows this pattern, in which stars closer to the galactic center outpace us and therefore become blueshifted as they approach and redshifted as they move away. The reverse is true of stars outside our galactic radius: we outpace them, so they appear blueshifted as we approach and redshifted as we pull away. The pattern is purely geometric. If you've ever run (or seen) a track race, then you know the advantage in taking the inside track around a curve. The inside track is shorter than the others. (See Figure 4.17(b).)

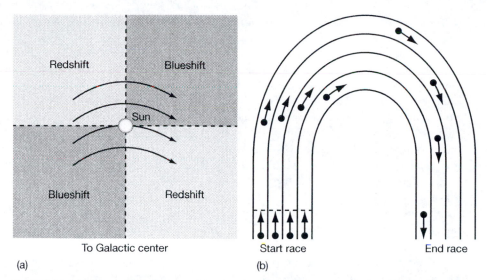

To Galactic center   Start race   End race

(a)   (b)

FIGURE 4.17 Differential Rotation   (a) Using the Doppler effect, we can identify distinct regions where other stars in our Galaxy appear to be coming toward us (and thus are blueshifted) or pulling away from us (and thus are redshifted). Inside our orbit around the galactic center, other stars are passing us; outside our orbit, other stars are being passed by us. (b) A race with equally fast runners tied as they approach a turn, but no longer tied after the turn, because the inside track is shorter.

On a galactic scale, this phenomenon is called *differential rotation*, and a spiral galaxy is said to be rotating *differentially*. Differential rotation is merely the result of stars traveling around smaller circles (inner orbits) in less time than those traveling around larger circles (outer orbits). Notice that we said "in less time" and not "at a faster speed." This is an important distinction to which we will return with great emphasis in the next section.

Let's get an idea of the time scales involved. The Sun's trip around the Galaxy, and therefore your trip as well, takes about 225 million years. Astronomers calculate that number from the size of our orbit, which is based on the distance ladder, and the speed of our orbit, which is based on Doppler measurements.

To get started, let's propose that differential rotation causes the spiral arms to form and sweep back. By the scientific method, the next step is to ask what predictions this model makes, so that we can see if they're supported by observations of galaxies. The primary prediction in this case is how the spiral arms would change over time. Because if the arms move together with the stars and gas in the disk, then differential rotation would spin them faster as you approach the center. Imagine

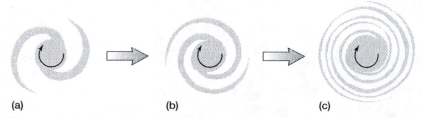

(a)   (b)   (c)

FIGURE 4.18 Galactic Winding   If spiral arms were fixed to the stars in the galaxy, then differential rotation would tend to wind them into an ever-tighter spiral (shown in frame (c)), which we do not observe.

this, first in your mind's eye, and then in Figure 4.18. The result is inescapable: A galaxy would wind up into an ever-tightening spiral.

Does this really happen? Proceeding with the scientific method, we need to test the prediction with evidence from space. This would seem hard to do, since galaxies take so long to rotate, and few galactic astronomers are willing to wait hundreds of millions of years for an answer. Fortunately, there's another way. If we survey *many* galaxies, of many different ages, we should find a wide variety of "winding-tightnesses." Furthermore, galaxies with arms, say, half a billion years older than ours should have no spiral arms left at all, since that's the duration of a few galactic orbits. In other words, spiral galaxies should be rare, since the spiral pattern lasts only for a half billion years or so, and that's not very long: Our own Sun is about nine times older than that, and the universe is about 27 times older! As it turns out, astronomers do observe *some* variety in spiral windings, but nothing like Figure 4.18(c). More to the point, they see *far* too many spiral galaxies to be consistent with a model that assigns them such a short lifetime.

We conclude that the spiral pattern *must not be rigidly connected* to the stars and gas in the disk. And if the spiral pattern moves at all, it must move much more slowly than the stars. In other words, orbiting stars pass *through* the spiral arms and reemerge on the other side. Once we accept that idea, another explanation for the spiral pattern becomes credible: In order for the spiral pattern to persist, stars need not stay permanently in the arms; they need only *linger* there a little longer than they do elsewhere in their orbits.

Astronomers often explain spiral structure by analogy to a traffic jam: You drive through it and eventually emerge in front of it, but you spend a long time stuck in it. Now, imagine that some evil space aliens who have never seen a traffic jam take a satellite photograph of one of ours with their undetectable spy satellite. They'll see a clump of cars, but they won't realize that the clump itself is *not* moving with the cars; in fact, the clump pattern could be moving forward or backward. Usually, after an accident has been cleared off the road, the front of the traffic jam dissolves while the back continues to pile up; that is, the clump itself moves backward even though the cars in it are all moving forward. But even if the clump moves forward, the cars in it must move faster, since they eventually escape from it.

Why is there a galactic traffic issue? Imagine that a star's galactic orbit is circular, but that the existing spiral arms tug on the star every time it passes by, because the arms have extra mass and therefore extra gravity. This tug might stretch the star's orbit a bit, pulling it outward into the spiral arms. Because the star now spends more time along the spiral arm—being delayed there by the arm's gravity—the arm will be one star brighter. Furthermore, the arm will now be one star more massive, enhancing its gravity for the next star that comes along. If many stars behave this way, then the arm will always be brighter than the rest of the disk.

This explanation comes from *density wave theory*. The theory itself is beyond the level of our discussion here, but it is based on modeling the gravitational intricacies involved in a center-weighted disk of particles (stars) rotating at different speeds. This model is shown in Figure 4.19, in which, if you look closely, you'll see a pileup of individual orbital paths forming the shape of a spiral! Therefore, there's always an excess of starlight coming from along the spiral arms. Individual stars move through the arms, but at any moment, there's always a concentration of stars in them (similar to cars on a congested highway).

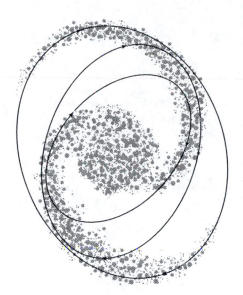

FIGURE 4.19 Why Galaxies Appear to Have Spiral Arms Density wave theory explains the existence of spiral arms with stellar orbits that twist more and more as you look farther and farther from the center. This increasing twisting creates an overlap of orbits along the familiar arc of a spiral arm, so that's where a majority of the starlight comes from. Notice, however, that stellar orbits don't follow the arms forever; individual stars move *through* them.

To enhance the spiral pattern, consider another aspect of the theory: When orbiting gas enters the arm (gas orbits, too) it is compressed. Compressed gas often collapses by its own gravity and thereby forms new stars. We saw recently that new star populations tend to appear bright and blue, just like spiral arms, because the brightest stars emit the bluest blackbody light and the overall glow is dominated by those brightest stars. But the brightest stars are also the hottest blackbodies, so they burn out the fastest, and by the time they cross the spiral arm, they've gone dark. So that's another reason for the prominence of spiral patterns in observed galaxies.

Interestingly, density wave theory says that even when there isn't an arm there to begin with to distort stellar orbits, one will form naturally. So if you start with a disk, a spiral pattern will automatically emerge, just from gravity. Quite clearly, nature "knows it's supposed to" form disks and then "knows it's supposed to" form spiral patterns in those disks, all over the universe. It's really quite remarkable how much complexity and beauty stems from Newton's seemingly simple recipe for gravity.

As this section draws to an end, it's valuable to think about the following pervasive message laced throughout the chapter, which is: People are good at this stuff! The scientists who endeavor to explain our observations of stars and galaxies, using known concepts such as the inverse-square laws for light flux and gravity, have met with astonishing success. We hope you will agree that you now have a good sense of what's out there, how large and distant it is, and how it's arranged in clusters upon clusters, etc. You even know the principles behind a galaxy's ability to generate its own complex and beautiful spiral shape.

We make a big deal about all this so that you will actively decide how confident you are in the scientific methods used to get us here, because the next section may test your confidence. What happens when your highly successful, tried-and-true methods reveal something totally unexpected, something that exerts great influence over the entire universe, but proves to be completely invisible to direct observation by every astronomical instrument in the world? What happens when every test confirms exactly what you didn't expect? At what point do you start to believe the data? (Remember Kepler and the elliptical orbits of planets: Knowing when to believe the data can lead to great advances.)

EXERCISE 4.4   Last Chance for Doubt!

Write a few hundred words about the techniques applied in this chapter to interpret our observations of the universe. Think about blackbodies, photons, starlight, and temperatures; think about standard candles, the distance ladder, and the measured large-scale arrangement of galaxies in space; think about Newtonian gravity, Doppler speeds, and the motions of celestial objects. Finally, try to imagine how we might be wrong about any of this; is there anything we're assuming without adequate justification? Could we delete any one principle without ruining the rest of the picture?

4.3 Dark Matter

The velocity of an object in orbit is

$$v_{\text{circ}} = \sqrt{\frac{GM}{r}} \, , \tag{eq. 4.8}$$

for an orbit at a radius r from the large central mass M. This equation is exactly true for a circular orbit and approximately true for a nearly circular orbit.

In practice, the easiest quantity in this equation to obtain from measurements is the velocity, since we can usually get that from the Doppler effect. We can also measure sizes and distances like r, from measurements based on the distance ladder. The mass of an astrophysical system—which isn't as directly measurable—can be calculated from the speed and distance of the objects orbiting it. The mass of a galaxy, for example, can be calculated from the speed and distance of the stars orbiting it. So let's flip the equation around algebraically and rewrite it in terms of the mass, to see how cosmologists determine the amount of mass out there:

$$M = \frac{rv^2}{G} \, . \tag{eq. 4.9}$$

In words, we're saying that

The mass being orbited by something = the radius of that orbit × its speed, squared / Newton's gravity constant (just a number)

This is the mass enclosed by the orbit—the mass located inside the orbit's circle.

A Glimpse of the Problem

At this point, we're ready to start "weighing" galaxies and other systems (i.e., calculating their masses), because we have all the measurements and we have the equation we need to convert them into mass. Before we do that, however, it's worth pointing out that we could simply add up the mass of a galaxy instead, by counting the quantities of gas and stars within it, both of which are measurable as well. The gas is detectable in spectral lines, and the stars obey well-known models, allowing us to connect the color and quantity of the starlight to the amount of stellar mass that generates it. That is to say, we have *two* ways to weigh a galaxy—which would be great if only they agreed with each other!

Using the Sun's 220 km/s orbital speed in Equation 4.9, we measure 95 billion solar masses ($\sim 10^{11}$) inside the Sun's orbit. By measuring light, scientists count about 15 billion solar luminosities inside that same region.[1] These two measures are consistent only if the matter inside the Sun's orbit is typically less "productive" than the Sun; that is, per solar mass, the matter inside the Sun's orbit produces about six times less light than the Sun (because 95/15 = 6.3)

A cosmologist would describe all this by saying that the **mass-to-light ratio** in the inner Galaxy is 6.3 solar masses per solar luminosity. (A solar mass is the mass of the Sun, and a solar luminosity is the luminosity of the Sun, so the mass-to-light ratio of the Sun itself is 1 solar mass per solar luminosity.) A mass-to-light ratio of 6.3 is a few times higher than astronomers expect for stellar populations in both the disk and the bulge, based on their understanding of stars. A high ratio like that might result from a large number of small, dim stars, which add mass without adding much light. A globular cluster, for example, is an assemblage of such stars, but even so, its mass-to-light ratio is only about 2 solar masses per solar luminosity.

Outside the Sun's orbit, things get even more dicey. If we use Equation 4.9 for orbits at the edge of the observable galactic disk, we obtain a mass of 200 billion solar masses (a little more than double what's inside the Sun's radius). Figure 4.20 shows a spiral galaxy like the Milky Way, with dotted circles at the Sun's orbit and at double that radius, encompassing all of the visible Galaxy. You can see clearly in the figure how strange it is that doubling the radius doubles the mass inside, because the vast majority of the visible material in the Galaxy is well inside the *inner* circle. So making the circle bigger shouldn't change the total mass figure so much. But the real mystery arises when cosmologists look at orbiting material farther out, beyond the edge of the galactic disk of stars. Even without any stars there to observe, it is possible to get the Doppler speed of hydrogen gas orbiting at much larger galactic radii. This material extends out beyond the stars, and spectroscopic instruments can often measure its velocity at distances of three to five times the size

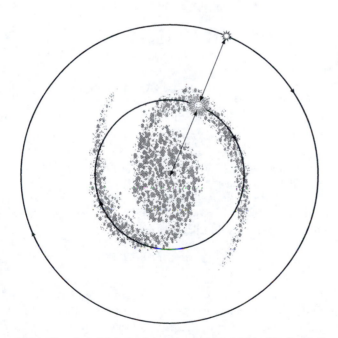

FIGURE 4.20 The Mass Enclosed by Two Orbits The amount of mass inside an imaginary sphere keeps increasing as the sphere gets bigger, even when we can't see much more mass to include. Pictured here, the Sun's orbit appears to enclose the majority of the mass enclosed by another star's orbit, twice as far from the center of the Galaxy. It looks like mostly empty space between the two orbits. Yet that star's galactic orbit is just as fast as the Sun's, which means (mathematically) that it must enclose *twice* as much mass as the Sun's orbit does.

[1]Jeffrey Bennett, Megan Donahue, Nicholas Schneider, and Mark Voit. *The Cosmic Perspective*, 4th ed. San Francisco: Pearson Addison Wesley, 2007.

of the visible part of a galaxy. Oddly, measurements indicate that there's three to five times more mass there than what's located within the radius of the visible galaxy. To be clear, *the farther away from a galaxy's center you go, the less stuff you see, but the more mass you detect*, on the basis of its gravitational influence.

Cosmologists can verify the orbital speeds of globular clusters and dwarf galaxies orbiting larger galaxies, including our own. Measurements in the Milky Way are a bit more difficult than in external galaxies, just because we're observing from within, but the results are the same here as in all the other galaxies surveyed: The mass-to-light ratio for a typical galaxy is 10 to 20 (usually closer to 20 and sometimes more than 50!). For some local dwarf galaxies, the ratio is several *hundred*. But either way—tens or hundreds—these numbers are far outside the range of any reasonable distribution of stars and gas.

THE PROBLEM WITH GALAXIES: Galaxies typically have >10 times more matter than we can account for in any simple way.

That's just galaxies. Healthy skepticism demands that we look at other large-scale gravitationally bound systems, too. After all, we don't need any extra mass to explain the orbits of the planets, so we want to know if it's just galaxies.

Accordingly, we turn our attention to larger scale bound systems: galaxy clusters. We can make the same measurements on clusters as on individual galaxies, but we measure the velocities of individual galaxies in the cluster, instead of individual stars in a galaxy. We again measure these velocities by Doppler methods, and we again measure the radii of the orbits with distance ladder techniques. Everything follows the same pattern we saw with galaxies, except that all the numbers are bigger, including, it turns out, the mass-to-light ratio: several *hundred* solar masses per solar luminosity. That is, we find exactly the same story in clusters as in individual galaxies: There's much more matter than we can see. And it's much more than we could account for even if we use all the mass in all of the member galaxies combined.

THE PROBLEM WITH GALAXY CLUSTERS: Galaxy clusters typically have >100 times more matter than we can account for in any simple way. (But it is possible to account for *a little* of it in a *semi-simple* way; keep reading this section.)

The numbers just cited—more than 10 times for galaxies and more than 100 times for clusters—are based on mass-to-light ratios and therefore include the mass of some ordinary kinds of matter, such as gas and dust, that don't shine like stars. But gas and dust account for only a small fraction of the excess mass; the vast majority of it is something else.

As a student of cosmology, confronted with these unexpected measurements, you probably want a little more detail. First, we'll map out where exactly this excess matter lives, and then we'll ask what it might be.

The Dark Halo

The most important tool for studying the amount of matter and where it lies within a bound system is called a **rotation curve**—a graph that displays the orbital speed (v) on the vertical axis and the distance (r) from the center of the system on the horizontal axis. To get a sense of what a rotation curve looks like, we can plot

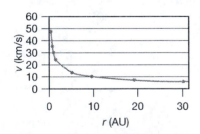

FIGURE 4.21 Solar System Rotation Curve   The velocity of an object in the solar system decreases with distance away from the Sun, because most of the mass (responsible for the gravitational force) is located right at the center, in the Sun. This is the behavior predicted by Kepler's and Newton's laws.

one for the planets in the solar system. (See Figure 4.21.) The dots on this graph display measured orbital data for the planets; the line displays the values predicted by Newton with Equation 4.8. You can see how well that equation accounts for our solar system's motions when we use the mass of the Sun for M. The dots all sit on the Newtonian line.

Get a good look at the solar system rotation curve and the way it arcs downward. When you look at a galaxy, the structure seems similar, so you might expect a similar rotation curve. And though you may be wise to expect a similar curve— after all, galaxies appear to have their mass lumped primarily in the middle, as the solar system does—you would be wrong.

Largely during the 1970s, astrophysicist Vera Rubin of the Carnegie Institution of Washington, DC, and her colleagues observed a large variety of galaxies and constructed their rotation curves. A small sample of these curves, including one for the Milky Way, is plotted in Figure 4.22. As you can see, the shape of galactic rotation curves is nothing like that of the solar system rotation curve we just saw in Figure 4.21. (Go ahead and check!). Rubin published the following account of her discovery:

> *"Available optical instrumentation now permits the detection of emission across a very large portion of the disks of spirals, well beyond the 'turnover point' in the rotation curves. We have initiated a program to obtain spectra of spiral galaxies at high velocity resolution and at a large spatial scale, in order to study their properties as a function of HT [Hubble Type]. ... The major result of this work is the observation that rotation curves of high-luminosity spiral galaxies are flat, at nuclear distances as great as $r = 50$ kpc [160,000 light-years]. These results take on added importance in conjunction with the suggestion...that galaxies contain massive halos extending to large r."*
>
> —Vera C. Rubin, W. Kent Ford, and Norbert Thonnard, *Extended Rotation Curves of High-Luminosity Spiral Galaxies, IV. Systematic Dynamical Properties, Sa. ØSC*

Galactic rotation curves are essentially *flat*. Look at the axes to understand what that means: As you go farther from the center, where we would naively expect gravity to fall away, the stars always orbit at about the same speed. For galaxy after galaxy studied—spiral, elliptical, whatever—velocities are flat. This is true even out to distances several times beyond the radius of the visible galaxy. Rubin quoted results out to 160,000 light-years, and since then, rotation curves have been shown to remain flat several times farther out than that!

Let's pause to evaluate what all this means. The visible part of a galaxy is typically about 100,000 light-years in diameter, or about 50,000 light-years in radius. The flat part of the rotation curve starts at a radius of about 10,000 light-years and often extends outward to about 200,000 light-years (where it is measured in gas motions, not stellar motions). The interesting conclusions apply everywhere outside the inner 10,000 light-years or so, which is a pretty small region by galactic standards: Stars and gas travel *too fast to be gravitationally bound* by the visible amount of mass. Unless there's at least 10 times more hidden matter than what you can see, the galaxy wouldn't exist. It would spin itself apart, flinging stars off into intergalactic space. The fact that we find galaxies held together tells us that the mass must be there somewhere.

So where is the mass? The rotation curve answers that as well. Because the curve doesn't descend like the solar system curve for a central point mass (the Sun), we can take what the curve *does* do, which is remain flat, and infer the underlying

Vera Rubin *(1928–present) has made many important contributions to astrophysics and cosmology, the most revolutionary of which was the discovery that galaxies are flooded with far more matter than can be directly seen. The excess "dark matter" turns out to be much more abundant in the universe than normal matter. Her discovery completely changed our view of the universe.*

(a)

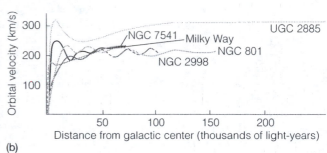

(b)

FIGURE 4.22 Galactic Rotation Curves Unlike the velocities of objects in the solar system, the velocities of stars do not decrease as one observes stars that are farther away from the center of their galaxy. This implies that the mass of the galaxy extends well beyond the stars whose motions we are observing. (◆ Also in Color Gallery 2, page 18.)

mass distribution. There are two important aspects to the distribution of unseen mass:

IMPORTANT ASPECT # 1: The matter distribution is densest at the center and spreads out more and more as you go away from the center. But it does so relatively slowly, such that with each step away from the galactic center, the amount of mass you "step over" is just enough extra to counter your extra distance away, leaving your orbital speed unchanged.

The mass is distributed over a wide distance from the galaxy's center. It's not just a blob in the middle; it's a smear that continues to affect objects well beyond the visible galaxy.

IMPORTANT ASPECT # 2: The matter distribution is spherical, not flat. This is true of both elliptical and disk galaxies. Globular clusters, satellite dwarf galaxies, halo stars—these things all orbit around galaxies at the same rotation speed (around 220 km/s for the Milky Way), but *outside* the plane of the disk. The constant rotation curve holds true even if you turn the galaxy sideways.

So the primary mass constituent of a galaxy is more massive and more spherically spread out than the visible galaxy appears. It is also something *dark*. Most of a galaxy's mass is dark, in the sense that it diminishes the mass-to-light ratio for the galaxy. It has mass, but doesn't shine like stars, and it doesn't absorb spectral lines like gas. We can't *see* it. To be precise, it doesn't interact with light—either producing it or blocking it—in any direct or easily measurable way. This is what is meant by **dark matter**.

All of this mass data *compels* us to adopt a radically new view of galaxies (and clusters). We thought galaxies had a certain mass, but it turns out that they have 10 times more. We thought the mass was localized to the visible radius, but the mass-containing radius is actually much larger. We thought spiral galaxies were basically two-dimensional disks, but most of their material occupies a three-dimensional sphere. And we thought that galaxies were made of stars and gas, but the vast majority of the mass is made of something else.

Say hello to the revised, modern galaxy shown in Figure 4.23.

The large spherical component is called the *halo*. We saw that term earlier in the chapter, but now we recognize that (a) it's much bigger than the visible galaxy, and (b) it's almost entirely made of dark matter. Therefore, we usually refer to the **dark matter halo.** As measured by size and mass, the dark matter halo is much greater than the luminous part of the galaxy. For example, a "spiral galaxy" is really a big dark halo, with a little, luminous spiral structure in the middle. (People sometimes refer to *luminous matter*, which literally means "matter that emits light." In practice, it refers to everything other than the dark matter.) Figure 4.24 shows an example of the gravitational influence of an unseen mass of dark matter.

Cosmologists have no real way to detect the edge of our Galaxy's halo, nor is there any reason to expect a sharp boundary; the dark matter distribution probably just fizzles away, blending with the ambient dark matter density beyond our Galaxy. Still, current estimates suggest that the dark matter halo is typically about 10 times the size (radius) of the visible galaxy!

In order to think more cosmologically, you might try imagining the universe as a network of galaxies—as in Section 3.3 and Figure 3.28—but without the galaxies!

FIGURE 4.23 A Complete Galaxy
The visible part of any galaxy (shown as a spiral galaxy here) is embedded in a much larger dark matter "halo." This halo's mass of dark matter is concentrated toward the center and gradually diminishes outward.

FIGURE 4.24 The Influence of Dark Matter The long strip of material streaming out to the right from visible galaxy UGC 10214 has been torn loose by something unseen, probably a large mass of dark matter nearby. (◆ Also in Color Gallery 2, page 19.)

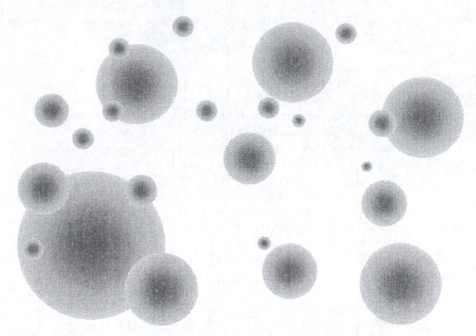

FIGURE 4.25 A Universe of Halos Halos of dark matter are the primary gravitating structures in the universe, because they're much more massive than the galaxies that live in them. Therefore, as a decent approximation to the real universe, you might just imagine a fairly regular arrangement of dark matter halos and ignore everything else.

In other words, map out the universe in your mind as vast arrangement of dark halos. Just ignore the luminous matter for now; after all, relatively speaking, there's not much of it out there anyway. Then the universe looks like Figure 4.25.

All the previous rules apply. Galaxy halos are spaced by a few million light-years, on average, forming something like a grid, except in more closely packed clusters and groups. Dark matter halos aren't solid (and neither are the luminous parts of galaxies), so they can merge, or pass through one another (as can the stars).

Now that you have the "halo universe" properly set up in your mind, you can go ahead and put the luminous matter in there too, as in Figure 4.26. In fact, a relatively accurate way to view the universe might be as follows: The universe is a network of dark matter halos sprinkled with luminous matter as *tracers*, to show us where the dark matter is. Looking at bread crumbs tossed onto a river in order to observe the current is like looking at luminous galaxies within dark halos in order to observe the mass distribution. Whenever you see a galaxy, through a telescope or in photograph, try to imagine the larger halo around it. After all, the halo is the bulk of the mass, and therefore of the gravity; it's the part that mainly influences the motions of other objects in space.

What Could It Be?

Just what is dark matter?

Throughout large galaxy clusters, we are able to observe x-ray emissions. These x-rays come from the very hot *intracluster medium* (ICM), a gas that pervades the cluster, as seen in false color in Figure 4.27. The spectrum of the emission indicates that the gas has a temperature of around 10 *million* kelvins and is usually found in quantities a few times greater than the luminous mass of the cluster. But remember that the mass discrepancy in galaxy clusters is typically a factor of a hundred or

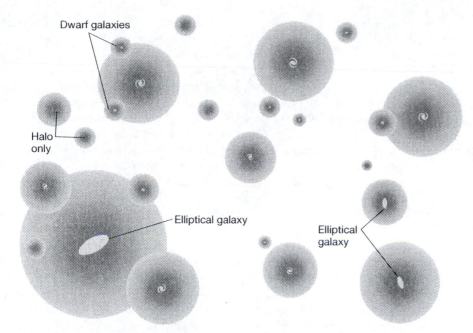

Dwarf galaxies

Halo only

Elliptical galaxy

Elliptical galaxy

FIGURE 4.26 Matter in the Universe   The sketch shows the arrangement of matter in the universe: halos of dark matter "decorated" with luminous normal-matter galaxies at their centers.

FIGURE 4.27 The Intracluster Medium (ICM)   X-ray emissions from the intracluster medium (which looks like a fuzzy cloud) of a rich galaxy cluster, overlaid upon an infrared image of the cluster. The mass of x-ray-emitting gas is generally more than the mass found in all the visible galaxies combined, but there is still much more (dark) matter unaccounted for.

more. So the ICM is a nice start, but there's still lots more dark matter in clusters yet to be found.

In February 2005, researchers using NASA's Chandra X-ray Observatory reported evidence of two giant clouds of diffuse gas. These clouds lie in intergalactic space (the space between galaxies), and if there are many more like them, then they could reasonably account for up to 10% of the dark matter we measure. This gas has been dubbed the "warm–hot intergalactic medium" and has even earned its own acronym, WHIM (which, believe it or not, is *not* the most ridiculous acronym in this section—keep reading).

Some of the dark matter seems to take the form of failed stars called *brown dwarfs*. Brown dwarfs are like big Jupiters: giant gaseous bodies that are a little too small to become stars. Another contribution to the dark matter is probably old, dead stars. These stars belong to a class of *compact objects*, of which white dwarfs, neutron stars, and black holes are members. Compact objects are remarkable entities in their own right, apart from their potential role as dark matter. Among their special properties are their extreme surface gravity (the *weakest* is a million times stronger than Earth's!) and their compact size (the *largest* is about the size of the Earth—about 1% the radius of the Sun and a millionth of its volume!).

Astrophysicists call all of these mass-hiding, halo-dwelling, small stellar drifters *massive astrophysical compact halo objects*—that is, MACHOs. It is a great observational challenge to detect them, because they're both dark and small! But we'll see in Chapter 6 that Einstein gave us a very clever way to search for them, and that way has so far revealed that roughly 20% of the dark matter in the Milky Way halo comes from MACHOs.

So the ICM and the WHIM have actually been found, although one could argue that neither gas really qualifies as *dark* matter, since each is seen in x-rays. MACHOs have been found too, but they can be detected only around the Milky Way so far, and their measured abundance is highly approximate. All of these combined—the ICM, the WHIM, and MACHOs—are still much less than half of the total amount of dark matter. From here on, we have to *speculate* on what the remaining dark matter is. But first let's talk about what it's *not*.

WHAT DARK MATTER IS NOT: Gas at other temperatures (temperatures other than what the ICM and the WHIM have been measured at) is usually dismissed as a candidate for dark matter. The traditional reasoning would be as follows: Either the gas is hot, so that it emits line radiation, or it's cold, so that it absorbs line radiation from the starlight coming from behind it. Either way, scientists would notice its presence. Since they have not, it must not be there. So we've got "very hot" intracluster gas and "warm–hot" intergalactic gas, and we've argued against "hot" and "cold." It seems that we're running out of gas temperatures at which to hold out hope for a new dark matter discovery! Let's consider some alternatives.

Might dark matter take the form of objects like asteroids, or even pebbles, or dust grains, drifting through galaxies? The answer to the possibility of such solid objects making up the dark matter is a solid "no." Astronomers have a highly reliable explanation for how various chemical elements are produced astrophysically, which we'll talk about in Chapter 9. The relevant result for us here is that the heavy metals constituting those objects are far less abundant than hydrogen and helium, and stars are generally around 98% hydrogen and helium. (In astronomy, every element above helium, element number two in the periodic table of elements, is called a "metal"—it's just a quirk of the field.) So, for every bit of iron or carbon used to make asteroids, there is a corresponding and much more massive "bit" of hydrogen and helium used to make additional stars, which wouldn't be dark. Since the bulk of the elements in the universe are not metals, there simply isn't enough metal in the universe to contribute to the dark matter mystery.

One proposal for some of the dark matter is simply that it takes the form of dim stars. Low-mass stars—say, 10% of the Sun's mass—produce only about 0.1% of the Sun's light. Thus, these stars could add mass, but not much light, to a galaxy, thereby boosting the mass-to-light ratio. Because such stars are cooler than the Sun, their blackbody nature gives them a dim red glow, and they are called *red dwarfs*.

However, observations of globular clusters by the Hubble Space Telescope show a distinct absence of these red dwarfs. The data suggest a cutoff of approximately 20% of the Sun's mass, below which few stars seem to exist. At present, these data are probably insufficient to *rule out* red dwarfs as a significant contribution to the dark matter, although that data may be coming soon. For now, scientists certainly have adequate cause to *doubt* that they exist in sufficient quantity to account for dark matter.

The dark matter story continues on and on, just like this. Could it be X? No. Could it be Y? No. And needless to say, Z isn't looking good either. This story of the failure to hammer down exactly what this dark matter consists of is particularly interesting in light of how old the problem is. The first major evidence for dark matter in galaxy clusters was presented in 1933. By 1980, the case for dark matter in galaxies was overwhelming, too. The dark matter mystery has been open for over three-quarters of a century, but most of the mystery remains. Researchers just chip away at it, learning more and more about what dark matter is *like* and contemplating what it might be.

Fundamentally, the undiscovered dark matter is something that doesn't interact significantly with light. It doesn't emit light, or we'd see it. And it doesn't absorb light or obscure it, because we'd see that, too. It lives in galaxies, but it also lives between galaxies, in clusters. And on smaller scales, like the solar system, it has virtually no effect. Strangely, as cosmologists learn more about dark matter, they fail to demystify it. Instead, they learn how special it must be and thereby make it more exotic!

So what might the dark matter be? Although no one knows what it *is* yet, cosmologists are actually quite good at answering what it *might* be. There is now, and has long been, a major contender:

WHAT DARK MATTER MIGHT BE: Most of it probably isn't made of "normal" matter at all, but rather some form of *exotic* matter. Think about this: We are looking for something that pervades the universe on large scales, but has no effect on small scales. We're looking for something very massive that doesn't emit light or interfere with light. So why not just invent something that actually fits that description, rather than try to figure out ways to cram normal matter into that mold? Instead of large objects, we could imagine dark matter as a vast agglomeration of individual *particles*. And it's not the usual particles—protons, neutrons, or electrons—because, to fit the bill, we need something much *more* massive and much *less* prone to interactions that we could observe. In other words, what's needed is a *weakly interacting massive particle*—a WIMP (yes, really).

WIMPs could serve as spherical galactic halos by swarming gravitationally, just as halo stars or globular clusters do. They could also serve as "galaxy cluster halos" in the same way. A universe of gravitational structures could be built from a network of WIMP particles. Look at Color Figure 41 in Color Gallery 2, page 19 to see visual evidence of weakly interacting dark matter. This image shows two colliding galaxy *clusters*, in which the ICM of one cluster (the x-ray-emitting intracluster medium, making up most of a cluster's "normal" matter) has "bumped into" the ICM of the other cluster, causing it to remain behind as the two clusters pass through each other. If the majority of the dark matter is weakly interacting, then one cluster's swarm of WIMPs would *not* "bump into" the other, but would just pass through with the galaxies instead. The figure shows exactly that: The primary (exotic) gravity sources have not remained behind the way the normal-matter ICM has.

So WIMPs fit the bill perfectly, and it turns out that physicists do know of one well-measured weakly interacting particle called a neutrino. Although neutrinos have too little mass to account for the amount of dark matter we seek, their existence supports the idea that it may be reasonable to invent another weakly interacting particle. The challenge here, again, is: How do you measure such a thing—a particle that, by definition, interacts too weakly to measure (other than by its gravitational influence)? But, as was the case with MACHOs, we need not give up on measuring WIMPs; we just need further study—Chapter 9 in this case.

So after many decades, during which some researchers have been searching for dark matter and others have been trying to disprove its existence (see the box "Gravity on Large Scales"), it's evidently here to stay. Although the saga of dark matter may seem like a quirky side story, it has profound importance to cosmology. As you continue your study of the universe throughout this book, dark matter will repeatedly reemerge. (There's probably some dark matter *in this book right now*, making it just a tiny bit more massive.) You'll read more about MACHOs and WIMPs, for example, when you gain the relevant background science for each. But you'll also see how dark matter enters other cosmological calculations and measurements, and how it plays into the overall evolution of the universe.

Interestingly, we don't have to know *what* dark matter is to know *that* dark matter is. In this sense, the "dark matter problem" is not a problem. There is no rule that forbids the universe from challenging our expectations.

Gravity on Large Scales

A small number of researchers have suggested that there is *no* excess of mass. They correctly note that there is another possibility: Scientists might simply be mistaken about how gravity makes things move on large scales. Newton's equations for motion under the force of gravity are a great success in the solar system, but does that mean that they work for entire galaxies? An alternative, called MOND (MOdified Newtonian Dynamics), proposes a change to the orbits of Newtonian gravity—a change that is expressly designed to explain the (presumably deceptive) appearance of larger and larger mass-to-light ratios corresponding to larger and larger size scales. MOND's ability to accomplish this makes it tempting to consider simply changing the laws of physics affecting orbits, rather than adding dark matter to our inventory of the universe. And it's always healthy to question one's assumptions.

There is no way to be absolutely certain—as is always the case in science—but there are two fairly strong arguments against MOND (or any theory like it) and in favor of dark matter. One is based on data: Some systems are wildly outside of the predictions of MOND, but they are still consistent with the presence of dark matter. In particular, many dwarf galaxies with small size scales have huge mass-to-light ratios, approaching 500 solar masses per solar luminosity! Alternatives to gravity or motion laws aren't equipped to handle this sort of variation, since one law would have to work for many different systems. Dark matter, by contrast, accounts for such variations quite naturally: You just vary the amount of dark matter from one system to the next. There are many prosaic ways to imagine such variations arising. Just as the visible parts of galaxies come in a wide variety of masses, it seems reasonable to expect that the quantity of dark matter would vary as well.

Color Figures 40 and 41 on page 19 in Color Gallery 2 are also examples of systems that don't fit the MOND paradigm. In these figures, not only is the gravity greater than it should be for normal matter, but it emanates from regions that do not appear to contain the normal matter. So even if one changed the laws of physics to enhance the effect of gravity, it wouldn't work because then the enhanced gravity would occur where the normal matter is, rather than where the gravity's effects are actually observed.

The other argument favoring dark matter has to do with calculations. The main benefit of MOND is its ability to explain, apart from some dwarf galaxies, the behavior of many different systems at different scales with one compact mathematical expression. This is interesting, but not compelling, because it has since been calculated that the effect of dark matter fits that formula, too, varying with different size scales *in the same way* as MOND does. The calculation for dark matter isn't as simple as that for MOND, but it is a natural consequence of multiple standard elements of modern cosmology taken together. The calculation didn't have to work out, so the fact that it did provides strong support for dark matter cosmology. By contrast, the MOND calculation *did* have to work out, because it was constructed for just that purpose.

The State-of-the-Universe Address

To our fellow wonderers, students of cosmology, and citizens of the universe:

We are entering a new chapter in our history. In this new chapter (5), we will expand our observations beyond individual systems—stars, galaxies, and clusters—and delve into the truly cosmic. We will observe large-scale phenomena that appear in every place and in every direction. So far, we have constructed a powerful basis for understanding the world of deep space. We know how stars shine as blackbodies and how swirling galactic disks emerge. We know how light and gravity behave. We know how to measure great distances and how to interpret what we see there. But before we move on, let's contemplate the significance of what we've already accomplished.

Long ago, people believed that the Earth sits still at the center of the universe. But the observations of Galileo and Brahe, the laws of Kepler and Newton, and the entire distance ladder firmly place our world within a vast arrangement of orbits upon orbits. It may seem like we're closing in on the truth, but along with such truth comes a price in perspective. The more we learn about the universe, the more foreign are the surroundings that we find ourselves in. The Earth is not the center; the Sun is. The Sun is not the center; the bulge of the Milky Way is. The Milky Way is not the center … does it seem likely that the Virgo Supercluster is?

Observations of the 20th century place us in a galaxy far bigger than anything our species can reasonably hope to explore by any foreseeable transportation technology. Galaxies drift through wide-open space, separated by millions of light-years, which we've observed at tens of thousands of such separations away from us. The universe is vastly larger than any known human purpose could justify.

And then there's the content and the complexity. The universe is a shell of stars around us! Then, no: It's a disk of stars *including* us, and a fleet of galaxies beyond. The galaxies are the same as nebulae! Then, no: Galaxies are huge, isolated stellar communities. They come in a narrow variety of shapes, sizes, and colors. Disk galaxies arrange themselves into beautiful spiral patterns. Sometimes we find thousands of galaxies huddling together, swarming in clusters that aggregate as well.

Your universe would be a network of galaxies and clusters of galaxies, except that each galaxy turns out to be a relatively tiny visible speck, centered in a far grander halo. It is halos, then, of various sizes that pervade the universe, *decorated* by pretty galaxies. And what are halos made from? Probably a new form of matter, which we already know something about, yet we still struggle to discover its identity. After decades of alternatives and intellectual challenges to dark matter, the challengers have fallen and the dark matter remains. In other words, today we know very well that the universe is predominantly made of something that we don't know very well.

Quick Review

In this chapter, you've learned how astronomers can "observe" the universe by means other than looking at light. You applied this knowledge to understand orbits and the structure of bound systems, such as galaxies. The force of gravity affects the motion of all matter and reveals the presence of material that is invisible to our eyes and telescopes. Through an understanding of gravity, you've learned that the material we can see makes up only a small fraction of all the mass in the universe.

Try the following crossword puzzle to test your knowledge of key terms and concepts from Chapter 4.

14 Shape of orbits
16 Small, dim star
18 Graphical tool for studying dark matter
19 Trajectory for a gravitationally bound object moving in space

DOWN

1 Massive spherical region that a galaxy lives in
3 Useful ratio for studying dark matter
4 Speed needed to leave and never return
5 Emitted by the intracluster medium
8 Large central feature in a spiral galaxy
9 Stars display _____ rotation about the center of a spiral galaxy.
10 Theoretical alternative to dark matter
12 Structure formed by density waves
13 When two galaxies meld together
15 Natural force that depends on mass and distance
17 A particle of dark matter

ACROSS

2 Doesn't significantly produce or block light
6 One of two special points in an ellipse
7 Massive astrophysical compact halo object
9 Flat component of a spiral galaxy
11 Type of energy associated with particles that are not free to leave

Further Exploration

(M = mathematical content; I = integrative—builds on information from more than one chapter; P = project idea; R = reflection; D = suitable for class discussion)

1. (R) Newton's law of gravity can be used to predict Kepler's laws. In this exercise, just worry about Kepler's first two laws. Suppose that a huge asteroid collides with the Earth at an angle that causes the Earth to slow down in its orbit without changing directions. Assume that before the collision Earth's orbit is perfectly circular.
   (a) Consider the immediate effect, in terms of the balance between gravity pulling the Earth inward and the Earth's motion "past" the Sun. Which effect(s) is (are) changed by the collision? Argue that the overall result should turn the Earth inward more sharply than before.
   (b) Show that the gravity force increases as the Earth is pulled inward.
   (c) Because of this increased pull, Earth will speed up. Demonstrate with words and pictures that this speedup should prevent the Earth from continuing inward toward the Sun and that a new, elliptical orbit might result, with the Sun at one focus—or at least, with the Sun no longer as close to being centered inside Earth's orbit (as in Kepler's first law). (*Note:* You are not *proving* that orbits are elliptical, as Kepler did; you are simply showing that an ellipse makes sense in this context.)
   (d) Show further that, in this new elliptical orbit, the speed is fastest when Earth is near the Sun and slowest when it is far away (in accordance with Kepler's second law). To do this, think about how Newton's gravity would predict some situations that lead to a speedup, and others that lead to a slowdown, in Earth's orbital velocity.

2. (RD) In a few sentences each, summarize (a) the evidence for dark matter, (b) the location of dark matter, (c) what dark matter can't be, and (d) what dark matter might be. Then, (e) write a paragraph on how your answers to (a) through (d) affect your universe as you understand it now, as contrasted with what you believed back in Chapter 1.

3. (M) (a) An alien ship is in a circular orbit of a planet at 100 joggles per blag. (Back on the alien home world, a joggle is a unit of distance and a blag is a unit of time.) If the aliens on the ship wish to break orbit and fly away from the planet, how fast will they need to be going (at a minimum)? (b) If another ship, with three times as many sizbars of mass as the first ship, does the same—break orbit by accelerating from 100 joggles per blag—what must its speed become?

4. (M) Calculate E_g and E_k for the Earth's orbit around the Sun, and use these quantities to verify that we're really bound to the Sun. (You'll sleep better afterwards.) The Sun's mass is 2.0×10^{30} kilograms, the Earth's mass is 5.98×10^{24} kilograms, the Earth's radius is 6.37×10^6 meters, the radius of its orbit is 1 AU = 1.5×10^{11} meters, and you should verify that its speed is 29,000 meters/second.

5. (R) Regarding disks,
   (a) Explain why disks form in some galaxies.
   (b) Speculate on why the solar system is disk shaped.
   (c) Why would dark matter be arranged in a spherical halo rather than in a disk shape? What does this tell you about the nature of dark matter?

6. (RI) (a) How would you expect the spectrum of a disk galaxy to differ from that of an elliptical galaxy? Answer in terms of blackbody properties *and* spectral lines.
   (b) From the spectral information from these two types of galaxies, could we use ellipticals or spirals as standard candles? Why or why not?

7. (R) How do we know that stars move *through* spiral arms, rather than with them?

8. (MR) In terms of v_{circ} and v_{esc}, why do flat rotation curves imply the existence of dark matter? What would the rotation curves look like without it?

9. (RD) For each of the following candidates for dark matter, explain what makes it "dark":
   - MACHOs (e.g., "Jupiters")
   - WIMPs (i.e., "invisible particles")
   - red dwarf stars
   - asteroids

10. (MRI) Use what you learned in Section 3.3 to help you decide whether or not galaxy halos typically overlap, for galaxies *in the field* (i.e., galaxies not in clusters).

11. (R) Suppose that dark matter is made primarily of WIMPs. How could a bunch of tiny particles generate any appreciable gravity force on larger objects such as stars? Answer in terms of Newton's gravity force.

12. (RI) Compare gravitational binding energy with *electrostatic binding energy*, which keeps electrons bound to atoms and holds atoms together in molecules. Would electrostatic energy be a negative quantity, like gravitational energy? Imagine the solar system as if it were an atom, and describe the energy involved in "ionizing" it. (This is a warm-up for Chapter 8.)

13. (RI) Suppose you took Doppler shift measurements of 45 local group galaxies. What results would you expect to obtain? What range of speeds would you expect? (Give numbers.) Would the shifts be red, blue, or both? Finally,

given the speeds you estimate, why don't we observe proper motions—such as the Large Magellanic Cloud galaxy zooming across the sky at night—with our naked eyes?

14. (RD) Come up with five scientific discoveries, throughout history and from multiple different scientific disciplines, that are similar to the discovery of dark matter in the sense that the scientists were led to believe in the existence of something they could not directly see at the time. For each scientific discovery you cite, comment in a few sentences about *how similar* that discovery is to the discovery of dark matter. For example, did that discovery depend on trusting some equations? Did it overturn the conventional thinking on its particular topic? Is it still at least somewhat unsolved?

Clues about the Cosmos

> "But nature gives most of her evidence in answer to the questions we ask her. Here, as in the courts, the character of the evidence depends on the shape of the examination, and a good cross-examiner can do wonders."
>
> —C. S. Lewis

In Chapters 2 through 4, we observed celestial objects and learned what we can conclude about our cosmic context—our place in the universe—from those observations. Thus, the stage is set: No longer will we be content simply to study the objects that the universe *contains*; now we will focus on those observations which hint at the nature of the universe itself.

The human quest to understand the universe unfolds like a detective novel. As with any good mystery, we begin by gathering clues and following leads. The leads are strange and may seem disconnected from each other at first. But deeper investigation will reveal what each clue means, and eventually, the full plot will be exposed.

The clues are in this chapter. The deeper investigation follows in the next four. Chapter 10 reveals the full plot. For now, just enjoy the beginning of the story. Get a feel for the type of leads we're working with. You can even start guessing the ending as you read. Twentieth-century science gave us this collection of cosmic clues, and professional astrophysicists continue to kneel over these footprints with their magnifying glasses today. However, the story is not only for them, and if you trace the evidence far enough back, some of the footprints are yours.

5.1 Redshifts of Galaxies

The first major cosmic clue comes from spectroscopic observations of light from many galaxies. To understand this evidence, begin by recalling the following information from Chapter 2:

- Light from an object can be represented with a *spectrum*—a graph that displays the intensity of light shining at each wavelength (or, loosely speaking, at each color).

- The spectra of light from stars and other objects include gaps at specific wavelengths. For example, sunlight, when dispersed with a prism, looks like a rainbow with several thin dark bands—dark gaps in the rainbow—where certain shades of certain colors are missing. These gaps are called absorption lines.

- Light from an object that is moving either toward us or away from us exhibits a *Doppler shift*, which tells us how fast the object is moving. The object's light arrives here with its color shifted to shorter wavelengths (if it's coming toward us) or longer ones (if it's moving away). The more the color is shifted, the faster the object is approaching or receding.

A galaxy's spectrum comes from the combined light of its many stars. It contains absorption lines that scientists recognize for two reasons. First, they can re-create these lines in a laboratory and thus determine which elements produce them. In this way, they can figure out that a galaxy's spectrum contains, at specific wavelengths, features that are caused by specific elements in the atmospheres of its stars. Second, the lines are familiar to astronomers because they appear in the spectra of nearby stars, including the Sun.

For example, the element hydrogen has several well-identified lines, as we discussed in Chapter 2. A hydrogen atom produces lines at specific wavelengths. (Recall Figure 2.32.) If we observe the spectrum of a galaxy and see that it includes these lines, then we can conclude that that galaxy contains hydrogen.

Now, suppose the galaxy is moving *away* from us. Then, because of the Doppler shift, every ray of light will undergo a *redshift*. This means that every wavelength becomes longer by the same percentage. Suppose we measure hydrogen lines on the galaxy's spectrum and find that they are redshifted by 0.01% (i.e., they are longer than the original wavelength by 0.01%). Then the colors we expect to see will be shifted to longer wavelengths, as shown in the top frame of Figure 2.32. Similarly, the spectrum of a galaxy headed toward ours will be *blueshifted*.

The observation of many lines shifted together in a galaxy's spectrum is quite compelling. For example, if all the hydrogen lines appear in a spectrum, all redshifted by the same percentage, we can be pretty confident that the redshifts are caused by the whole galaxy moving away from us, at 0.01% of the speed of light (or 300 km/s) in this case. Note that if a line is already in the red region of the spectrum and its wavelength is increased, we still call it "redshifted" even though it actually moves *away* from red. Astronomers abuse the term "redshift" to indicate any shift toward a longer wavelength.

Figure 5.1 shows an example of a redshifted galaxy spectrum. In this case, we have many lines, from many different elements, to work with. If all of the lines are redshifted by the same amount, this gives us great confidence in our measure of the redshift of the galaxy emitting the light.

Under the assumption that the redshift is a Doppler shift, the spectrum tells us (1) that the galaxy is moving away from us and (2) how fast it is moving away. In order to assign an actual numerical value to the galaxy's speed, we need the formula for redshift (usually represented by z):

$$z = \frac{\Delta\lambda}{\lambda} = \frac{\lambda_{\text{observed}} - \lambda_{\text{emitted}}}{\lambda_{\text{emitted}}} . \qquad \text{(eq. 5.1)}$$

The value of $\lambda_{\text{observed}}$ is the wavelength we actually observe for a particular feature in the spectrum of a celestial object, and λ_{emitted} is the nonredshifted wavelength

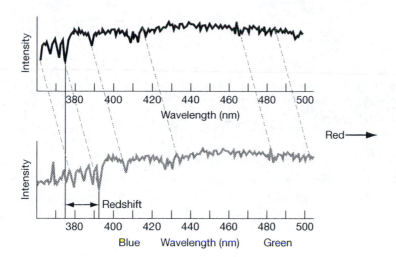

FIGURE 5.1　A Redshifted Galaxy Spectrum
The top frame shows a galaxy spectrum that is not redshifted (not moving away from us), and the bottom frame shows a spectrum that is redshifted. Notice that there are several dips, representing absorption lines caused by particular elements. The redshift of these dips indicates that the galaxy is moving away from us.

of the line, as we would measure it in the lab. The symbol $\Delta\lambda$ gives the nuer of nanometers that the wavelength has shifted. (Recall that a nanometer is 10^{-9} meter.) Therefore, a large $\Delta\lambda/\lambda$ means a large redshift z, which indicates a large speed for a galaxy traveling away from us. For speeds that are small compared with the speed of light, the speed of the galaxy is just the redshift times the speed of light, or

$$v = cz \quad \text{for } v \ll c .$$ (eq. 5.2)

That is, when due to the Doppler effect, redshift measures the speed at which a galaxy is moving away from us, expressed as a fraction of the speed of light. In our previous example of a galaxy moving away from us at 0.01% of the speed of light, we'd say that $z = \Delta\lambda/\lambda = v/c = 0.0001$. (The two extra zeros just convert the percent into a fraction).

By itself, and for a single galaxy, this information is not terribly *cosmic*, since it tells us nothing about the nature of the entire universe, nor is it terribly profound. Redshifts become cosmologically interesting only when we sample *many* galaxies. If galaxies just move around at random, then we should expect to find about half of them moving away from us (redshifted) and the other half moving toward us (blueshifted) at any given time. That's what everyone expected, which is why the actual spectra came as such a surprise.

Beginning in 1912, Vesto Slipher, at Lowell Observatory in Arizona, began measuring the spectra of galaxies. Remarkably, out of 25 galaxies whose spectra he measured and reported on in 1917, only 4 were blueshifted, while the remaining 21 showed a redshift. Some of the nearby ones were coming toward us, like the Milky Way's next-door neighbor, the Andromeda galaxy. But modern observations confirm that *all* galaxies outside of our local neighborhood are redshifted. The universe, in other words, is full of galaxies moving *away from us*. What's so special about *us*? Why are *we* so unpopular?

There's more. In the late 1920s, American astronomer Edwin Hubble, working at the Mt. Wilson observatory in California, combined Slipher's redshift data with his own measurements of distance to the galaxies (based on

*Vesto Slipher of Lowell Observatory in Arizona (1875–1969) and **Edwin Hubble** of Mt. Wilson Observatory in California (1889–1953) are together responsible for discovering that other galaxies are receding from ours. For a large part of the 20th century, this observation alone enabled astronomers to piece together much of our present understanding of the universe.*

Cepheid standard candles discovered by Leavitt, as discussed in Chapter 3). Hubble wrote:

> "The results establish a roughly linear relation between velocities and distances among nebulae [galaxies] for which velocities have been previously published [by Slipher], and the relation appears to dominate the distribution of velocities."
>
> —Edwin Hubble, "A Relation Between Distance and Radial Velocity Among Extra-Galactic Nebulae"

In other words, the data revealed a relationship between the distance to the object and the shift in its spectrum: The amount of *redshift increases linearly with increasing distance*. So if we compare two galaxies, one twice as distant as the other, the more distant one appears to be moving away from us twice as fast.

In 1929, Hubble put together the first famous plot of this relationship, which is referred to as the *Hubble diagram*. The relationship the diagram depicts can be expressed in words by saying that the amount of redshift is directly proportional to the distance to the galaxy. This relationship is graphed in Figure 5.2 (the Hubble diagram) and can be summarized with the following formula, called the **Hubble law**:

$$zc = H_0 d \ . \tag{eq. 5.3}$$

In this equation, z is the redshift of a galaxy (as defined in Equation 5.1), d is the distance to that galaxy, c is the speed of light as usual, and H_0 (the slope of the line in Figure 5.2) is referred to as the **Hubble constant**. You can see from the equation that as d goes up, so does z. How much z goes up when d goes up depends on the value of H_0. This relationship has been confirmed for a vast sample of galaxies and therefore must be a key feature of any successful model of the universe. In Chapters 7 and 10, we will explore *why* galaxies all over the sky are coordinated in such a way that their redshifts are proportional to their distance from us.

If the redshift is due to the Doppler effect (which is how it was originally interpreted), z is a measure of the velocity at which the galaxy is moving away from us, called a *recession* velocity, as described by Equation 5.2. In this case, we can rewrite Equation 5.3 in simpler form to show that velocity is directly proportional to distance away from us:

$$v = H_0 d \ . \tag{eq. 5.4}$$

This is often the way you will see the Hubble law written, graphed, and discussed: The farther away a galaxy is, the greater is its recession velocity. But it's worth remembering that the velocity is an *interpretation*. The direct observational statement is the relationship between *redshift* and distance from us, not between velocity and distance. That is, if the redshift were caused by some other agent, rather than the Doppler effect, then it might be inaccurate to claim that "redshift" equals "moving away from us." For the time being, we're going to ignore this issue and conclude that galaxies really are getting farther away; we'll need to double-check this point later, in Chapter 7.

The current best measurement of the Hubble constant is $H_0 = 71 \pm 4$ (km/s)/Mpc. The units of the Hubble constant indicate its meaning, linking a galaxy's recession speed to its distance, as in "Galaxies recede at a speed of 71 km/s

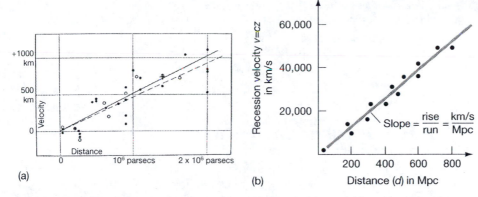

FIGURE 5.2 Hubble Diagram   (a) Hubble's original Hubble diagram and (b) a more modern one. A Hubble diagram is a graphical illustration of the Hubble law: The more distant the galaxy, the larger is its redshift. Redshift times speed of light gives units of velocity, as indicated in the figure. If interpreted as a Doppler shift, the redshift is a measure of the galaxy's velocity as it moves away from us. Notice that this trend emerges only for galaxies beyond about a million parsecs (a few million light-years) from us. Hubble's original plot, (a), would occupy only a tiny rectangle in the bottom left corner of the modern plot. Some of these nearby galaxies are actually moving toward us; can you tell which ones?

for every Mpc (megaparsec) they lie away from us." (Recall from Section 2.5 that a parsec is 3.26 light-years, so 1 Mpc = a million parsecs = 3.26 million light-years; it just happens that the units cosmologists use for the Hubble constant involve megaparsecs rather than light-years.) The units of the Hubble constant nicely imply what Equation 5.4 says: Galaxies 2 Mpc distant should recede at $71 \times 2 = 142$ km/s, galaxies 3 Mpc distant should recede at $71 \times 3 = 213$ km/s, etc., although you really have to look farther away than a few Mpc before the Hubble law behavior emerges.

Galaxy redshifts may seem mysterious. It's hard to see how such widely separated galaxies could all conspire to move in such an organized way. However, don't let that puzzle diminish the significance of galaxy redshift measurements in your mind. These discoveries—that nearly all line shifts are red and not blue, and that more distant galaxies have greater shifts than nearby ones (the Hubble law)—are deeply important clues to the nature of space and time in our universe. We'll need Einstein's theory of general relativity to address them adequately (Chapters 6 and 7). For now, just begin to appreciate how pervasive the galaxy redshift mystery is, starting with Exercise 5.1. Then continue through this chapter to assimilate the other major clues to the nature of the universe.

EXERCISE 5.1 The Relationship between Redshift and Distance

Modern versions of the Hubble law are based on a vast sample of galaxies, far more than Hubble originally had to work with. The Sloan Digital Sky Survey (SDSS, a collaborative effort aimed at collecting and studying an enormous sample of galaxies around us) contains images and spectra of hundreds of thousands of galaxies, available in a database you can access on the Web at http://cas.sdss.org/dr3/en/proj/advanced/hubble/. This site takes you to a part of the database with a tutorial for making your own Hubble law plot. You can also venture off into the database on your own to look for galaxies in particular regions of the sky. If you pick out similar-looking galaxies, their distances can be estimated from their apparent brightnesses by a rough version of the standard-candle technique which assumes that all the galaxies you choose have the same luminosity. The redshift can be obtained as we described earlier (see Figure 5.1), by looking at features in the spectrum of each galaxy and measuring how far these features are displaced from where they would be if the light source were at rest relative to us. (This has been done by others using the spectra you see, so you can just read off the value for *z* listed on the plots, or you can check it for yourself.)

Follow the procedure outlined on the SDSS website to generate your own plot of the relationship between redshift and distance for galaxies in your sample. What value do you obtain for the Hubble constant?

5.2 The Distribution of Galaxies

The Hubble law today is based on observations of far more galaxies than are shown in Figure 5.2. Scientists have now measured spectra for nearly a million galaxies. So when we learn that nearly all galaxies are redshifted and that the redshift is greater the farther away the galaxy is from us, we're talking about extremely well established knowledge. And if the redshifts represent velocities in the familiar sense, then galaxies all over the universe are speeding away from our Galaxy. This behavior is not something that can be easily explained away. It seems to be going on everywhere in the universe, in every direction we look.

The best measurement technique for acquiring this type of data is a "galaxy survey," also called a "redshift survey," because it samples galaxies over a range of redshifts. (By the Hubble law, that means a range of distances, near and far.) For various directions in space, these surveys create a catalog of all the galaxies they find, including spectra and images that show the brightness of the galaxies in various colors. They thus provide a detailed description of our surroundings. The Sloan Digital Sky Survey (SDSS) is the largest redshift survey carried out to date.

Since the Hubble law—the equation connecting a galaxy's redshift to its distance—is well demonstrated by a large sample of galaxies, why not turn it around and use it? We can measure a galaxy's redshift directly by taking a spectrum (as in Figure 5.1) and then apply the Hubble law to discover how far from Earth the galaxy is located. In other words, the Hubble law can help us to measure distances.

A redshift survey therefore provides more than just a catalog of celestial objects: It gives us a three-dimensional map of the universe around us. Such a map is depicted in Figure 5.3. This map, from the SDSS, shows nearly 67,000 galaxies, each represented with a dot. We are located at the point where the two lines meet, so the most distant galaxies in the figure are at the outskirts. The segment shows a wedge of the night sky extending away from the Milky Way. It's shaped like a slice of pizza; we inhabit the tip of the first bite.

The most distant galaxies in the figure are a little over 3 billion light-years away. For comparison, the closest galaxy to us—or rather, the closest one comparable in size to the Milky Way—is the Andromeda galaxy, about 2.5 million light-years away. So this survey covers over 1000 times the distance from us to Andromeda.

At first glance, the large-scale distribution of galaxies may appear random. But these data offer more information than that when we know how to appreciate what they're telling us. First, we need to understand what they're *not* telling us: Nothing in this map suggests that we are at the center of the universe. It can appear that way, partly because we're at the center of the survey itself and partly because the galaxies appear to dwindle away as you travel farther from us. Neither effect is real, though. We're at the center of the picture because we have no alternative reference point apart from our own galaxy (which is our reference point only because we happen to live here). And there is a very natural reason for the dwindling number of galaxies

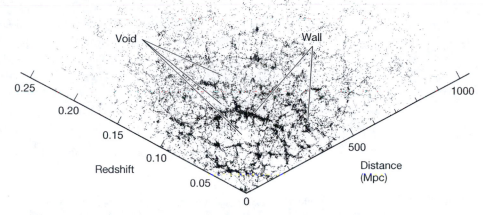

FIGURE 5.3 Distribution of Galaxies These data from the Sloan Digital Sky Survey show nearly 67,000 galaxies, each represented with a dot. We are located at the vertex at the bottom. The scale outward is listed both in redshift and in Mpc. The redshift is the direct measurement and the Mpc scale is obtained from the redshift via the Hubble law equation. Several walls (rows of galaxies) and voids (regions without many galaxies) can be seen in the data. (◆ Also in Color Gallery 2, page 20.)

far from us: The farther away a light source lies, the dimmer it appears, which makes it harder to see. When the dimming is taken into account, there appear to be just as many galaxies 2.5 billion light-years away as there are nearby; we just can't see them as well.

So what *does* the survey tell us? When researchers Valérie de Lapparent, Margaret Geller, and John Huchra conducted the first such redshift survey in the 1980s, they reported the following:

> "Several features of the results are striking. The distribution of galaxies in the redshift survey slice looks like a slice through the suds in the kitchen sink; it appears that the galaxies are on the surfaces of bubble-like structures with diameter 25–50 h^{-1} Mpc [110–230 million light-years].... The galaxies appear to be distributed in elongated structures which surround the empty regions. Most of the galaxies belong to one of these large structures. The voids have a typical diameter of ~25 h^{-1} Mpc."
>
> —Valérie de Lapparent, Margaret J. Geller, and John P. Huchra, "A Slice of the Universe"

Since that first survey, more detailed and far-reaching surveys (like SDSS) have solidified these conclusions. This line of research has revealed much about the large-scale structure of the universe:

- On large scales, galaxies are arranged in dense clusters (dark regions on the map). Clusters are arranged into vast *walls*—sheets or bubbles that interlace with one another to form a spongelike shape.

- In between the walls are *voids*—holes in the sponge that galaxies do not seem to inhabit. These voids are about a hundred million light-years in size.

Margaret Geller (born 1947) and John Huchra (born 1948), from the Harvard–Smithsonian Center for Astrophysics, discovered that galaxies are arranged throughout the universe in a pattern that resembles, as they put it, "suds in the kitchen sink." Together with their graduate student, Valérie de Lapparent, they used telescopes on Mount Hopkins in Arizona to discover the "bubbly" nature of our universe. Where this structure comes from is one of the great stories of modern cosmology.

- The pattern is not random: Clusters, walls, and voids are generally the same size everywhere. These features appear to have a maximum size.[1]
- There is no evidence from optical galaxy surveys of any structure larger than these walls and voids.[2] In other words, it appears that if we could somehow zoom out and see a wider field of view, we would just see more of the same, not forming any special shape beyond what we can already see in this map.
- There is no evidence of a center and no evidence of an edge. The universe appears the same in every location (it is **homogeneous**) and in every direction (it is **isotropic**). There is no particularly interesting pattern to the clusters, walls, and voids located on the right-hand side of the slice that you can't also see on the left.

We noted in the previous section that any successful theory of the evolution of the universe will have to explain Hubble's law and the observation that nearly all galaxies have redshifts and not blueshifts. Now we can add that the successful theory will also have to consistently explain all the observations we just listed from redshift survey data. It will have to explain the origin of walls and voids, and why the voids have a characteristic size of about 100 million light-years. And it will have to explain the universe's apparent lack of a center or an edge. (Or, at the very least, it will have to explain how these things manage to elude human detection if they do exist).

EXERCISE 5.2 Check the Status of Your Universe

Compare and contrast each item in the preceding list with your responses in Exercise 1.7. To what extent do these survey data answer your questions from Exercise 1.1?

5.3 Microwaves from Every Direction

Robert Wilson (born 1936) and Arno Penzias (born 1933) discovered a microwave radio source that appears remarkably identical in every direction in the sky. It evidently floods the entire universe. This cosmic microwave background (CMB) has provided some of the most important data in the history of cosmology and, perhaps, in the history of all science.

One of the truly great all-time scientific discoveries—possibly the single biggest clue about our universe ever found—was a complete surprise to its discoverers. During the early 1960s, Arno Penzias and Robert Wilson at Bell Labs in Holmdel, New Jersey, were testing a radio antenna for an early satellite communication system. They were struggling to eliminate a soft radio hiss from the antenna. The hiss was like what you hear when you tune an ordinary radio to a frequency where there is no radio station. Nothing worked. The noise remained, no matter what part of the sky they pointed the antenna toward. That meant one of two things: (a) The noise wasn't real, and the antenna was just broken, or (b) the noise was real, but its source was somehow everywhere at once. Penzias and Wilson pursued option (a) for a while. They even tried scraping off "a white dielectric material" deposited on the receiver by a family of pigeons that had made their home there. It didn't help, though, because there was nothing

---

[1]Just as this book went to press, new results from radio astronomy measurements were announced that *imply* the existence of a single, gigantic void, about ten times larger than all the other voids. Additional research is needed before these observations can be genuinely understood. For the time being, it is probably reasonable to treat this one giant void as a completely different kind of entity, while we continue to learn about all the other "normal" voids.

[2]See footnote 1.

wrong with the antenna. It turned out that they couldn't get rid of the radio noise, and they couldn't find a unique source for it, because it floods the entire universe. In this sense, they had discovered a truly *cosmic* signal, without even knowing what they were looking for. They published the following account of it:

> "Measurements of the effective zenith noise temperature of the 20-foot horn-reflector antenna at the Crawford Hill Laboratory, Holmdel, New Jersey, at 4080 Mc/s [megahertz] have yielded a value about 3.5° K higher than expected. This excess temperature is, within the limits of our observations, isotropic, unpolarized, and free from seasonal variations (July, 1964–April, 1965)."
>
> —Arno A. Penzias and Robert W. Wilson, "A Measurement of Excess Antenna Temperature at 4080 Mc/s"

As it turned out, only 30 miles away, in Princeton, New Jersey, a group led by Robert Dicke was *building* a radio antenna to look for just such a cosmic signal. This signal from every direction in the sky had been predicted by Dicke's group (as well as by astrophysicist Yakov Zel'dovich, in Moscow, and others) as part of a theory to explain the galactic redshifts we discussed in Section 5.1. The theory predicted that the signal should be in the form of microwaves, which are waves from a region of the electromagnetic spectrum at the short-wavelength end of radio and longer than infrared. (See Color Figure 13 in Color Gallery 1, page 7.) In 1965, after Penzias and Wilson met with the Princeton group, the pair reported their observation of microwaves from the sky.

So the basic observation is this: Microwaves are picked up by a receiver pointed at the sky in any direction, even when the receiver is just pointing toward empty space. The radiation is remarkably *isotropic*—the same in all directions. It's as if the radiation is all coming from the same source, except that the source is in every direction. This is a puzzling and important clue to the nature of the source. Because of its universal nature, this glow from everywhere is called the **cosmic microwave background (CMB)**.

Microwaves are the same kind of radiation used for all sorts of modern technology, such as radios, cordless telephones, cell phones, wireless computer networking, satellite communications, and even television. If your TV has an antenna and receives stations through the air, you can even use it to detect cosmic microwaves. Just adjust your remote or turn the knob (TVs used to have knobs before they had remote controls—it was a very dark time in human history) to a channel which comes in only as fuzzy "snow." Some of the snow on the screen is the CMB—the radio noise that Penzias and Wilson couldn't eliminate. (If you have cable or satellite TV, unplug your television from the converter box, and then turn to a fuzzy channel to watch the cosmic microwaves.) We didn't say the CMB would be exciting to watch, just that you could see it.

EXERCISE 5.3   Cell Phone Microwaves and Cosmic Microwaves (Infinite Anytime Minutes for Free!)

The cosmic microwaves cover a range of wavelengths in the electromagnetic spectrum, but their peak intensity is at about 150 gigahertz. (See Figure 5.4.)

(a) How much faster are these microwaves oscillating, compared with the transmission frequency of your cell phone?

(b) How does the CMB peak frequency compare to the frequency of your favorite radio station? (See Color Figure 13 in Color Gallery 1, page 7.)

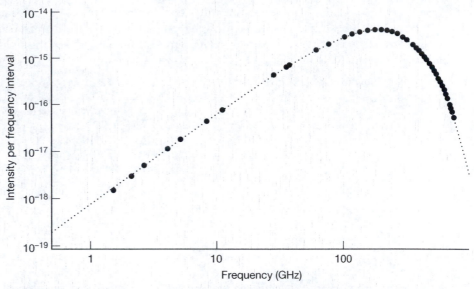

FIGURE 5.4 CMB Spectrum This is the CMB spectrum, in which most of the dots represent measurements taken by the COBE satellite. The vertical axis gives the energy flux per unit frequency interval. The horizontal axis is frequency in gigahertz. Both axes use a logarithmic scale, which compresses the data in order to fit more on the plot. The dotted line is a theoretical spectrum for a blackbody with a temperature of 2.725 K. This theoretical (dotted) line fits the data remarkably well: It is not just a connect-the-dots; rather, the measured points happen to line up almost perfectly with the spectrum for a perfect blackbody.

More detailed observations of the CMB reveal further clues. The radiation covers a range of frequencies with different intensities. By measuring the intensity of the radiation at various frequencies, we can plot a spectrum of the CMB, just as we did for other colors of light in Chapters 2 and 3. Figure 5.4 shows a spectrum of the CMB obtained by the COsmic Background Explorer (COBE) satellite in 1989.

The plot in Figure 5.4, combined with our study of temperature and blackbody spectra in Section 3.1, gives us a new way to interpret the CMB. Remember, a blackbody glows with a precise spectrum that is mathematically determined by its temperature, like a red-hot coal. This cosmic radiation is a nearly *perfect* blackbody. It seems to indicate that the temperature of the universe is only a little bit above absolute zero: 2.725 K. It has a *far* smoother blackbody spectrum than that of *any* other known source; that makes the CMB special in some way. Eventually we'll demand to know in what way, but for now we're just identifying the cosmic clues.

We can express the isotropy of the CMB by saying that the temperature is the same in every direction, to very high precision. We can display isotropy graphically with an all-sky plot of the CMB temperature. In order to fit the whole sky onto a flat page, we'll map it onto an oval (just as in Figure 5.5). As seen by a detector that's incapable of sensing variations finer than 0.01 K, the temperature is exactly the same in all directions, as shown in Figure 5.6.

So this is our updated understanding of the microwave clue to the cosmos: The universe glows in a way that indicates a temperature, but it's not the temperature of the stars or other objects *in* the universe. (Those objects are much hotter

FIGURE 5.5 Temperatures on Earth   For comparison, here is a temperature map of the Earth's ocean surface projected onto an oval. Temperature variations from 0 °C to over 30 °C are represented as different shades. In both this map and the following CMB map (Figure 5.6), a full sphere of temperature information is shown on a flat oval. (◆ Also in Color Gallery 2, page 21.)

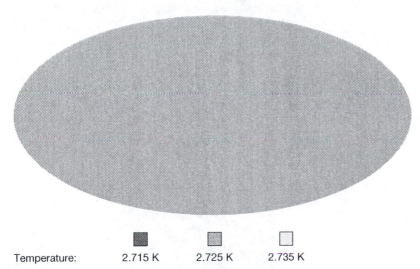

Temperature: 2.715 K   2.725 K   2.735 K

FIGURE 5.6 Isotropic Temperature of the CMB   This is what the CMB all over the sky would look like to a detector sensitive to 0.01 K temperature differences. Any temperatures less than 2.72 K or greater than 2.73 K would have shown up as a different shade of gray (according to the temperature scale on the figure), but the sky is uniform at this sensitivity level. Compare this figure with the temperature map of Earth in Figure 5.5, which shows much wider variations in temperature at different locations. (◆ Also in Color Gallery 2, page 21.)

than 2.725 K, and their spectra are much less smooth.) It's a temperature somehow associated with the universe itself.

In order to unravel the mystery of the CMB clue, we need to investigate the microwave radiation more carefully. When the temperature of the CMB in different directions is measured with greater sensitivity, we begin to notice slight variations from place to place on the sky. These variations reveal an intriguing pattern, displayed in Figure 5.7.

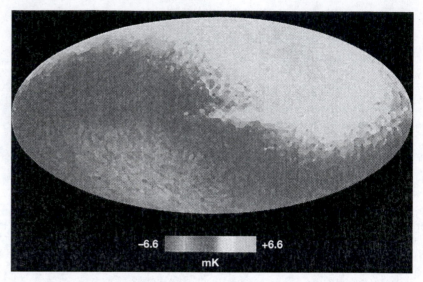

FIGURE 5.7 Temperature variations in the CMB The shading in the plot shows small temperature variations of a few thousandths of a Kelvin (a few mK, or millikelvins) away from the average of 2.725. (◆ Also in Color Gallery 2, page 21.)

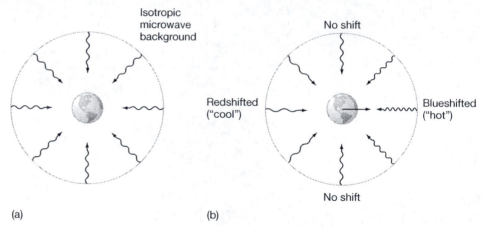

FIGURE 5.8 Moving through the CMB Our motion causes an otherwise isotropic distribution of background radiation to appear hotter in the direction we are moving toward and cooler in the direction we are moving from. (◆ Also in Color Gallery 2, page 21.)

The microwave radiation is very slightly *hotter* in one direction of the sky and slightly *cooler* in the opposite direction (180° away on the sky). Remember that hotter corresponds to a larger peak frequency in the blackbody spectrum, while cooler means a smaller peak frequency. What could cause such a symmetrical shift in frequency, toward higher frequency in one direction and toward lower frequency in the exact opposite direction? This effect brings to mind our discussion of Doppler shift in Sections 2.4 and 2.5, observed whenever a source and observer are physically moving toward or away from each other. So, is the CMB source moving in this specific way? No: It's far more likely that the Earth is the one moving. (See Figure 5.8.) Not only can we detect the temperature of the universe, but we can actually measure our motion relative to this background glow.

When the effect of Earth's motion is subtracted from the CMB temperature measurements, the uniformity of the temperature in all directions becomes even more remarkable. Once we take into account the Earth's motion, we can now increase the sensitivity to notice variations smaller than 0.0001 K, and still Figure 5.6 would be a featureless gray with no variations in shading. But be forewarned that hidden within the CMB, at even greater sensitivity, a wealth of cosmologically important information awaits. We'll return to the CMB in Chapter 8; for now, you only need to recognize that a source this special probably has an equally special origin.

5.4 Common Ingredients in the Universe

This chapter deals with *cosmic* clues: astronomical information that pertains to the entire universe, rather than just to one category of object (like "planets"). One such clue comes from identifying the ingredients common to *all* the objects in the universe.

For example, *you* are an object in the universe, and we could list *your* ingredients—brain, skin, bones, etc.—and then break them down further; for example, what is skin made of? The answer is skin cells, which contain various structures you can see with a microscope. Most of the material in these cells is water, or H_2O. That structure, a molecule, is made from two hydrogen atoms and one oxygen atom, which are in turn made of electrons and nuclei. Nuclei are made of protons and neutrons. (Review Section 3.1.) All things considered, most water molecules contain 28 subatomic particles (10 protons, 10 electrons, and 8 neutrons, it turns out).

We could go through a similar zooming-in procedure to list the ingredients of *any* familiar objects: rocks, cell phones, water bottles, penguins, planets, etc. When we do this, we find two very interesting observations:

1. They *all* can be broken down into the *same* three basic building blocks: protons, neutrons, and electrons.

2. The protons, neutrons, and electrons cluster together to form atoms, and there are only about a hundred different types of atoms, called *elements*. Each atomic element is defined by its *atomic number:* the number of protons in its nucleus. Each element (corresponding to a particular atomic number) also has a name, ranging from the familiar oxygen and hydrogen to less familiar ones such as francium. A listing of all the elements is usually displayed as a periodic table like the one in Figure 5.9.

At the atomic level, we find that everything can be reduced to a limited set of elements as listed in Figure 5.9. People, trees, rocks, and planets are different combinations of these elements, just as, in baking, we can make bread, cookies, or pasta out of different combinations of flour, salt, sugar, etc. The top 10 ingredients that make up *you* are listed in Table 5.1.

As previously mentioned, the atomic elements are made up of more basic particles: protons, neutrons, and electrons. Although we share these same three basic ingredients in common with everything else in our surroundings, this does not make us the same as everything else, because how the particles are arranged is all important. Every part of your body is made of protons, neutrons, and electrons, but it's the *sequences* in which these particles are organized that results in the variety of protein and fat, tooth and pancreas.

Legend:
2
He
4.003
Helium
— Atomic number
— Symbol of element
— Atomic weight
— Name of element

1 **H** 1.0080 Hydrogen																	2 **He** 4.003 Helium
3 **Li** 6.939 Lithium	4 **Be** 9.012 Beryllium											5 **B** 10.81 Boron	6 **C** 12.011 Carbon	7 **N** 14.007 Nitrogen	8 **O** 15.9994 Oxygen	9 **F** 18.998 Fluorine	10 **Ne** 20.183 Neon
11 **Na** 22.990 Sodium	12 **Mg** 24.31 Magnesium											13 **Al** 26.98 Aluminum	14 **Si** 28.09 Silicon	15 **P** 30.974 Phosphorus	16 **S** 32.064 Sulfur	17 **Cl** 35.453 Chlorine	18 **Ar** 39.948 Argon
19 **K** 39.102 Potassium	20 **Ca** 40.08 Calcium	21 **Sc** 44.96 Scandium	22 **Ti** 47.90 Titanium	23 **V** 50.94 Vanadium	24 **Cr** 52.00 Chromium	25 **Mn** 53.94 Manganese	26 **Fe** 55.85 Iron	27 **Co** 58.93 Cobalt	28 **Ni** 58.71 Nickel	29 **Cu** 63.54 Copper	30 **Zn** 65.37 Zinc	31 **Ga** 69.72 Gallium	32 **Ge** 72.59 Germanium	33 **As** 74.92 Arsenic	34 **Se** 78.96 Selenium	35 **Br** 79.909 Bromine	36 **Kr** 83.80 Krypton
37 **Rb** 85.47 Rubidium	38 **Sr** 87.62 Strontium	39 **Y** 88.91 Yttrium	40 **Zr** 91.22 Zirconium	41 **Nb** 92.91 Niobium	42 **Mo** 95.94 Molybdenum	43 **Tc** (99) Technetium	44 **Ru** 101.1 Ruthenium	45 **Rh** 102.90 Rhodium	46 **Pd** 106.4 Palladium	47 **Ag** 107.87 Silver	48 **Cd** 112.40 Cadmium	49 **In** 114.82 Indium	50 **Sn** 118.69 Tin	51 **Sb** 121.75 Antimony	52 **Te** 127.60 Tellurium	53 **I** 126.9 Iodine	54 **Xe** 131.30 Xenon
55 **Cs** 132.91 Cesium	56 **Ba** 137.34 Barium	71 **Lu** 174.97 Lutetium	72 **Hf** 178.49 Hafnium	73 **Ta** 180.95 Tantalum	74 **W** 183.85 Tungsten	75 **Re** 186.2 Rhenium	76 **Os** 190.2 Osmium	77 **Ir** 192.2 Iridium	78 **Pt** 195.09 Platinum	79 **Au** 197.0 Gold	80 **Hg** 200.59 Mercury	81 **Tl** 204.37 Thallium	82 **Pb** 207.19 Lead	83 **Bi** 208.98 Bismuth	84 **Po** (210) Polonium	85 **At** (210) Astantine	86 **Rn** (222) Radon
87 **Fr** (223) Francium	88 **Ra** 226.05 Radium	103 **Lw** (257) Lawrencium	104 **Rf** (261) Rutherfordium	105 **Db** (262) Dubnium	106 **Sg** (263) Seaborgium	107 **Bh** (262) Bohrium	108 **Hs** (265) Hassium	109 **Mt** (266) Meitnerium	110 **Uun** (269) Unununium	111 **Uuu** (272) Unununium	112 **Uub** (277) Unununium	113 **Uut** (undiscovered) Ununtrium	114 **Uuq** (285) Unununquadium	115 **Uup** (288) Unununpentium	116 **Uuh** (292) Unununhexium		118 **Uuo** (294) Unununoctium

57 **La** 138.91 Lanthanum	58 **Ce** 140.12 Cerium	59 **Pr** 140.91 Praseodymium	60 **Nd** 144.24 Neodymium	61 **Pm** (147) Promethium	62 **Sm** 150.35 Samarium	63 **Eu** 151.96 Europium	64 **Gd** 157.25 Gadolinium	65 **Tb** 158.92 Terbium	66 **Dy** 162.50 Dysprosium	67 **Ho** 164.93 Holmium	68 **Er** 167.26 Erbium	69 **Tm** 168.93 Thullium	70 **Yb** 173.04 Ytterbium
89 **Ac** (227) Actinium	90 **Th** 232.04 Thorium	91 **Pa** (231) Protactinium	92 **U** 238.03 Uranium	93 **Np** (237) Neptunium	94 **Pu** (242) Plutonium	95 **Am** (243) Americium	96 **Cm** (247) Curium	97 **Bk** (249) Berkelium	98 **Cf** (251) Californium	99 **Es** (254) Einsteinium	100 **Fm** (253) Fermium	101 **Md** (256) Mendelevium	102 **No** (254) Nobelium

FIGURE 5.9 Periodic table of the Elements All of the material we're familiar with on Earth is built out of these elements.

Table 5.1 Top 10 Elements in Your Body

The Ingredients of People. You are made up mostly of hydrogen, oxygen, carbon, and nitrogen, along with tiny amounts of other elements.[3]

Element	Percent by number of atoms
Hydrogen (H)	61.56%
Oxygen (O)	26.33%
Carbon (C)	9.99%
Nitrogen (N)	1.48%
Calcium (Ca)	0.24%
Phosphorous (P)	0.20%
Sulfur (S)	0.06%
Sodium (Na)	0.06%
Chlorine (Cl)	0.04%
Magnesium (Mg)	0.03%

Since the properties of objects depend on how their subatomic particles (particles within atoms) are arranged, the interesting changes in the world—changes that occur when you breathe, eat, run, turn on a light, or drive a car, for example—involve reorganizing these particles. There are two distinct kinds of reorganization that can occur: **nuclear** and **chemical**. The difference between them has to do

[3]This table is based on an activity from NASA's "Imagine the Universe" project: http://imagine.gsfc.nasa.gov/docs/teachers/elements/elements.html

with the structure of an atom. Recall from Section 3.1 that the nucleus is very small—about 100,000 times smaller than the atom itself—and tightly bound. (The exact size depends on the type of atom.) The electrons are much more loosely bound, the electron cloud defining the outer reaches of the atom. As a result, it's easy to make changes that shuffle around the distribution of electrons, *but leave the nucleus unchanged*. This is chemistry. You're doing it right now when you inhale oxygen (O_2) and exhale carbon dioxide (CO_2). Due to a rearrangement of their electrons, the oxygen atoms you breathe out are bonded to carbon atoms, rather than to each other.

By contrast, the tightly bound nuclei are much more difficult to rearrange. They remain intact when you breathe, for example. Before and after you exhale, the carbon is still carbon (defined by its atomic number, the six protons in its nucleus) and the oxygen is still oxygen (eight protons). Despite the hopes of ancient alchemists who dreamed of becoming wealthy by converting lead into gold, most of the transformations that occur in our everyday world—including even fire—are *chemical* and do not change the identity of the elements involved. That would require rearranging the *insides* of nuclei—the protons and neutrons—and would be called a *nuclear* process.

The fact that most familiar transformations are chemical and not nuclear is the key to the cosmic clue provided by the amounts of different elements we see. In the absence of nuclear processes, the amounts of common elements *can never change*. This leads us to ask the question, *Why are different elements present in the amounts they are?* Why is there more oxygen than aluminum on Earth? Presumably, each nucleus (of oxygen in your skin, for example) must have been formed by a nuclear process sometime in the past, since *only* a nuclear process can create (or change, or destroy) a nucleus!

What can we learn about the history of such nuclear reactions by looking at the amounts of different elements in the universe? We can start with what's close by. The oxygen in your body accounts for a little over half of your body mass. Oxygen accounts for much of the Earth's mass, too. (For example, most of the mass of ocean water comes from the O in H_2O). But the Sun is mostly hydrogen and helium; only about 2% of the Sun is made of other material (including oxygen). In the Earth and Sun combined, then, most of the mass is still hydrogen and helium, because the Sun is so much more massive than the Earth. This remains true for the whole solar system because the Sun is much more massive than all the planets combined. In fact, for the entire Galaxy, and in the rest of the observable universe, hydrogen and helium constitute the vast majority of the matter we see.

This discovery appears to be true cosmos-wide: The major ingredients in space, from stars and galaxies to clouds of gas and dust, are the same everywhere. But why should everything be made of all the same stuff? That's part of the cosmic clue, because, however it happened, it evidently happened throughout the entire universe.

At this point, you may be justifiably skeptical of the sweeping conclusions we've just made. How can astronomers possibly claim with any confidence to know the composition of even our nearby Sun, much less the rest of the solar system, distant stars, and galaxies at the edge of the visible universe?

Scientists get a little help from meteorites that land on Earth, which enable them to directly measure the composition of a few objects beyond Earth. But, as with most things in astronomy, the greatest source of information is light. Recall from Chapter 2 that each element produces a characteristic pattern of emission or absorption lines in the spectrum of light we receive from an object containing that element. (Recall Figure 2.23, also in Color Gallery 1 on page 8.) So, even when we

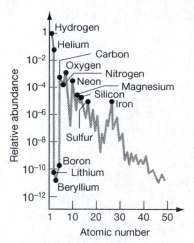

FIGURE 5.10 The Composition of the Universe This graph summarizes the results of the various abundance measurements discussed in the text. Hydrogen, the most common element in the universe, is assigned an abundance of 1, and the amounts of all other elements are expressed as a fraction of the amount of hydrogen. (◆ Also in Color Gallery 2, page 26.)

can't touch planets, stars, gas clouds, or galaxies, we can still learn their composition from the light we receive from them. These observations give astronomers an overall estimate of the abundance of elements in the universe. The results are shown in Figure 5.10.

Notice that the scale on the vertical axis is logarithmic; that is, it increases by factors of 10. So, although nitrogen (N) and oxygen (O) are plotted close to one another on the graph, there's about 10 times more O than N. Hydrogen is the clear winner in the universe, making up about 74% of the mass of the atoms overall. The runner-up on the graph is helium, which accounts for about 24%; all other elements combined amount to only 2%.

In order to understand the cosmic abundance clue, we need some additional information about nuclear structure. Most elements actually come in several different varieties, called **isotopes**. Two different isotopes of an element contain different numbers of neutrons, which makes the one with more neutrons heavier, but they're identical in most other ways. Different isotopes of copper, for example, would both have the same shiny metallic color and would both conduct electricity equally well. But one isotope would weigh slightly more, which leads to a subtle, but measurable, difference. The isotopes of greatest cosmological importance here are as follows:

Hydrogen (all isotopes have 1 proton and 1 electron)

- hydrogen-1, "ordinary" hydrogen: 0 neutrons
- hydrogen-2, called "deuterium" or "heavy hydrogen": 1 neutron
- hydrogen-3, called "tritium": 2 neutrons

Helium (all isotopes have 2 protons and 2 electrons)

- helium-3, which is relatively rare: 1 neutron
- helium-4, the most common variety: 2 neutrons

Lithium (all isotopes have 3 protons and 3 electrons)

- lithium-7, the most common variety: 4 neutrons

The numbering convention, called the **mass number** and often indicated by the number following the element name, is the total number of protons and neutrons *combined*. (Electrons are so lightweight by comparison, that they are ignored in the mass number.) Thus, helium-4 contains 2 protons and 2 neutrons, while helium-3 has 2 protons, but only 1 neutron. Since each element is defined by its number of protons, we only need to list the mass number next to its name or abbreviation in order to specify both the number of protons and neutrons. For example, when you see the term "carbon-14," you know it must have 6 protons in order to be called carbon (that's specified on the periodic table), and therefore it must have 8 neutrons to make its mass number add up: 6 + 8 = 14.

This is important in our current line of study, because nuclear reactions can change both the number of protons and the number of neutrons. In our accounting of abundances in the universe, we need to pay attention to which isotopes we find, because that will tell us something about which nuclear reactions formed them, which in turn will tell us something about what the universe must have been like at the time they formed. How much hydrogen-1 compard to hydrogen-2 we see in a gas cloud, for example, tells us something about how the universe formed that hydrogen.

Most elements are formed in the cores of stars, where it's hot enough for nuclear reactions. So, as time goes by and more stars live and die, more elements are formed. Therefore, when we look at older stars, we see smaller and smaller abundances of *most* elements—but not the six isotopes we've just listed! When astronomers measure abundances of these elements by looking at spectra of various objects, they always find that at least 23% of the mass is helium-4, whether they are studying stars, nebulae, or galaxies, and no matter how old these objects are! Although the abundances of other elements, such as oxygen and iron, keep decreasing as you look to older and older stars, helium levels off instead. Nearly all the remaining gas is hydrogen-1, but, as with helium-4, there is always a base amount of hydrogen-2 and helium-3 in every object we study, no matter how old. The base quantity of each of these two isotopes amounts to only a few grams for every 100,000 grams of normal hydrogen—a small, but very specific and measurable, abundance.

All of these quantities should be explainable in terms of the nuclear reactions that originally formed the isotopes. Nuclear reactions normally occur in stars, but even if we point our telescopes at regions of space with no stars and no galaxies, we still find these same abundances of hydrogen and helium isotopes, just drifting through intergalactic space. The pervasive, universe-wide aspect of such observations leads us to believe these light elements were formed somewhere other than in stars—in a nuclear event that is somehow common to the history of everything in the universe. The evidence for this conclusion becomes even more striking when we consider three particularly rare elements: lithium, beryllium, and boron.

Lithium, Beryllium, and Boron

Astronomers can also measure abundances of the light elements lithium, beryllium, and boron in the surface regions of stars. It's natural to look at these elements together, because all three are usually formed in the same way: by collisions that break up larger nuclei in interstellar space. By studying the spectra of stars, astronomers can relate the abundances to the age of the star. The abundances of all three elements diminish as we look at older stars. This finding is to be expected, since older stars formed long ago, before most of the interstellar collisions that make these elements had taken place. But while the amounts of beryllium and boron continue to decline as we search older and older stars, the abundance of lithium-7 levels off at about 3×10^{-10} compared with hydrogen (i.e., *three grams* of lithium-7 for every *ten billion grams* of ordinary hydrogen).

This is a very tiny amount, but it's measurable, and it's extremely significant. There exists no natural mechanism today for producing (or destroying) lithium any differently than the way beryllium and boron are produced (or destroyed). Yet, older and older stars contain less and less beryllium and boron, but in stars beyond a certain age, there's *always* the same abundance of lithium-7. To appreciate this finding, take a look at the graph in Figure 5.11, which compares the abundances of lithium, beryllium, and boron in stars of various ages.

Lithium-7, like deuterium and helium, evidently has some other cosmic formation process to explain why the same base amount exists everywhere, even in remote places where beryllium and boron never formed in comparable quantities. Somehow, long ago, lithium-7, along with hydrogen and helium formed *everywhere*. Astronomers know it was long ago because even the very oldest stars have the same base

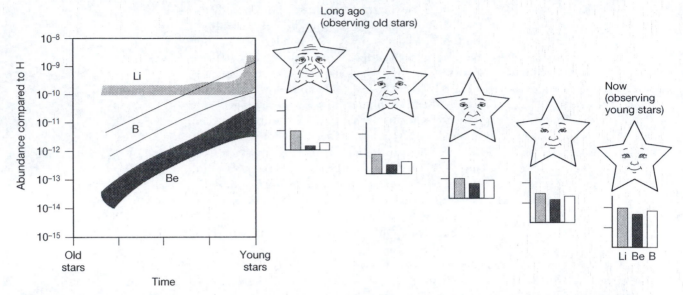

FIGURE 5.11 The Lithium Conspiracy   (a) Abundances of lithium (Li-7), beryllium (Be-9), and Boron (B-10 and B-11) plotted for stars of different ages. Shaded regions indicate the approximate range of abundance observations for each element. The abundance of light elements found in stars generally decreases as you look at older and older stars. But for lithium-7, the abundance levels off and never goes below a minimum of about 3×10^{-10} compared with hydrogen. (b) Cartoon depiction of the same: In the past, there's less and less Be and B, but Li had a steady base amount.

quantities of these isotopes. They know it was everywhere because the observation holds true in every direction.

Why is this "lithium conspiracy" so important? Because it's so specific! We now know that any scientific theory concerning the history of the universe must include an ancient nuclear process that manufactured lithium-7, in a precise amount of a few parts per 10 billion, but did *not* make any beryllium or boron. This is quite a challenge! Most nuclear environments tend to destroy lithium, beryllium, and boron, rather than assemble them. That's why their natural formation mechanism today requires splitting some larger nucleus in open space, where fragments like lithium nuclei will remain untouched once they have formed. But these collisions are not capable of producing lithium without also producing beryllium and boron. Evidently there must be a process that's more discerning: a nuclear reaction that makes lithium-7 specifically. This requires very special conditions, which currently do not exist anywhere in the known universe.

We will return to these observations after we've learned more in Chapter 9 about how to create an environment suitable for the nuclear reactions that produce elements. For now, the key point to remember is that the observations described in this section imply a single, brief era very long ago in which rapid nuclear processes generated all the base atomic material everywhere—material that would go on to form stars, planets, and people.

At this point, it's worth reminding ourselves what we're doing in this chapter. We're identifying a collection of cosmic clues—observations that appear to tell us something about the entire universe. But we're not trying to *explain* them yet, nor are we in any way trying to reveal how they relate to one another; we're just noting them. If you're struggling to make sense of all this stuff, don't panic; you *shouldn't* be able to make sense of it yet. So don't ask why the universe is like this or that, or what's the point of the abundances of the light elements. (Well, *ask* if you must, but we don't feel obliged to *answer* anything in this chapter.) The answers will come later, once we have a full appreciation of all the clues our answers must explain.

5.5 Comparing Astronomical Ages

One of the most important questions we can ask about the universe's history is how long it has existed at all: Has it always been here? As simplistic and inadequate as it might sound, a fairly useful starting point for this question is to ask the age of the oldest objects *in* the universe. This approach will prove useful only if we judiciously choose the right objects to study. For example, imagine that space aliens arrive here to study our planet. They might try to figure out how old human "civilization" is by somehow dating the oldest ruins of a fallen building on our planet. That might lead to a reasonable estimate of how long we humans have been civilized enough to dwell in buildings of our own construction—if the aliens would be satisfied with that. But they'd be pretty far off if they tried to figure out the age of human civilization by determining the age of the oldest person alive today.

Suppose that the aliens tried to measure how long human civilization has existed in North America by dating the oldest artifacts on that continent. They find jewelry, hand tools, pottery, dwellings, and human bones, all of the *same* approximate age, and correctly conclude that North American human civilization is at least as old as these artifacts, and perhaps not much older. We might hope for the same sort of agreement among the ages of stars, star clusters, and so on. There are a variety of independent methods for measuring ages of different objects in the universe. We describe some of the most useful ones here.

Radiometric dating of rocks and meteorites has been used to determine the age of the Earth and the solar system. Radiometric dating is a process in which scientists measure the abundance of some radioactive element inside an object and compare that present abundance with what they believe the abundance was when the object first formed. The reliability of the method therefore depends only on knowing (1) what the object's original composition was (and there are many circumstances in which that might be known) and (2) the radioactive decay rate (which is usually well measured). The decay rate is commonly referred to with the term "half-life," the amount of time it takes for the radioactive decay process to convert half of the original material into whatever it decays into.

Natural radiation from space (called "cosmic rays") alters our air by creating some carbon dioxide, CO_2, with a carbon-14 nucleus instead of the more common carbon-12. Plants "breathe" this gas, and animals eat these plants. As long as the animal is alive and eating, it will have a particular ratio of carbon-14 to carbon-12 that's common among all living animals. But when the animal dies and stops eating, no more carbon-14 is taken in, and the existing carbon-14 begins to decay. When scientists find the remains of the animal today, they need only compare the current ratio of carbon-14 to carbon-12 with the known original one to calculate how long the animal has been dead (or at least how long it's been fasting).

Astrophysical radiometric dating is similar to radiocarbon dating on Earth, except that it uses different isotopes of different elements, because the half-life of carbon-14 is only about 6000 years and we need something capable of dating objects billions of years old. Typical choices are uranium, potassium, rubidium, and lead. Measured in this way, our solar system (on the basis of samples from the Earth, the Moon, and meteorites) turns out to be 4.5 to 4.6 billion years old.

Radiometric dating can be employed to measure the ages of stars. From spectroscopic observations of long-lived isotopes in stellar atmospheres, astronomers get a radiometric age estimate for the star. The method begins by seeking very old stars, which, recall from the previous section, are distinctive because they are deficient in elements higher than helium in the periodic table. The spectra from these stars reveal relative abundances of isotopes. As we learned before, astronomers must choose to focus on isotopes which formed in a known way, so that they know their original abundance ratio.

The difficulty of performing this analysis solely from the information contained in stellar spectra, rather than with radioactive samples in the laboratory, leads to relatively large uncertainties. A 1999 study of the ages of the oldest stars found a best fit of 15.6 ± 4.6 billion years, using isotopes of uranium and thorium. A 2001 study found 12.5 ± 3.0 billion years with uranium. These numbers are extremely interesting, because astronomers know that some stars will live for *trillions* of years (Chapter 9), but the oldest stars out there are evidently only billions of years old. This is an important cosmological clue: Why are all the long-lived stars in the universe still in their infancy? Either the universe has been making stars only for the last 12–15 billion years, or maybe the universe has *existed* for only that long.

Astronomers can estimate the ages of star clusters because all stars in the cluster started forming at the same time from the same cloud of gas. Based on well-established models of stellar evolution, astronomers know that the least massive stars live the longest (more on this in Chapter 9). Further, the correlation between a star's mass and its lifetime has been firmly established. The star's luminosity, surface temperature, and spectroscopic features all change predictably when it nears the end of its life, and since all of these properties are easily observable, astronomers can isolate the age of presently dying stars in any particular star cluster. The result is approximately the age of the entire cluster. The oldest star clusters, measured in this way, are judged to be 12 ± 1 billion years old. Again, this number is much lower than it could be.

Astronomers can measure the ages of the oldest white dwarf stars on the basis of their cooling rates. White dwarfs are some of the oldest stars around; the Sun will become one when it's old enough. Astronomers can calculate how long they've been cooling from information they obtain on how hot they are now (measurable), when they formed (based on theoretical models), and how quickly they radiate energy away (measurable and well modeled). In 2002, the oldest white dwarfs were measured with good precision to be 12.7 ± 0.7 billion years old, with 95% confidence. A subsequent study of the same white dwarfs yielded a new measurement of 12.1 billion years, with a lower limit of 10.3 billion years, at the same confidence level.

The most significant aspect of all these astronomical observations is the remarkable concordance of ages measured by such different means. If they all came out different, we'd have to work much harder to figure out the real age of the universe. If some approaches generated numbers in the billions of years, but other results were in the millions, it would be difficult to print any of it in a cosmology textbook. Textbook writers would have to get day jobs on Wall Street instead. (On second thought, maybe not: If our stock predictions varied so widely, we'd soon be out on a different street!)

Fortunately, the real results turn out to be believable. Astronomers have many independent measurements that all generate ages in the vicinity of 12 to 13 billion years for the oldest astronomical objects. This sort of verification gives astronomers confidence in their methods for determining ages and gives us a hint about the age of the universe: If nothing in it is older than about 13 billion years, then it stands to reason that the universe itself might be almost that "young," too. This statement says a lot. If the universe has some particular age, then it might have had a *beginning*. Now, it's possible that the universe has been around forever but the objects in it appeared only 13 billion years ago; however, such a state of affairs seems a little contrived. Comparing astronomical ages, then, raises the possibility that the universe has been around only for a certain amount of time, much like those of us who live within it.

5.6 Olbers's Paradox

The last clue in this chapter is slightly different from the others. It is primarily a *thought experiment*. Unlike many of the clues we've discussed so far, this one does not require sophisticated equipment. The only observation involved is so obvious that it barely seems like an astronomical observation at all. In fact, the basic question in this section might sound too simple to be worth your time, but humor us anyway. The question is this: Why is the sky dark at night?

Start by considering an entirely different situation. You're looking at a distant mountain when it begins to snow. The snow comes down lightly at first, in flurries, but within a few minutes it's snowing hard. As the snowfall gets heavier, the mountain gets more difficult to see. It gets fainter and fainter, and whiter and whiter, until you can't see it at all; you see only white.

Before it started to snow, rays of light traveled from the mountain to your eyes. When the snow begins to fall, some of those light rays are intercepted by snowflakes. Light still leaves the mountain in all directions, including toward your eyes. Some rays reach your eyes, but others hit snowflakes instead, so they never get to you. As the snowfall thickens, fewer and fewer lines of sight go all the way from the mountain to your face without at least one snowflake getting in the way, somewhere in between. In heavy snow, whenever one snowflake moves out of your line of sight, another enters, so, effectively, there are no more light rays reaching your eyes from the mountain. Now you see only white, because you're actually looking at snowflakes.

This effect is illustrated in Figure 5.12. What you actually see is a collection of individual snowflakes. Because they're all at different (and constantly changing) distances from you, it's impossible to focus on all of them at once, and that causes the hazy appearance.

Now we're ready to apply this reasoning to the universe. It's littered with stars and galaxies—sources that shine light. Some of those light rays come toward the Earth, and if you stand outside and look up on a clear night, some of those rays will enter your eyeballs (i.e., you see stars, and even galaxies, if you know where to look).

Suppose that the universe is infinite and that there are an infinite number of galaxies in every direction (a possibility that is reasonably consistent with the galaxy survey data in Section 5.2). In that case, no matter where you look in the night sky, you should see a ray of light from at least one of these galaxies. (See Figure 5.12.)

Think about what that means. It implies that the entire sky should be glowing with light, even at night! It's similar to snowfall. If you can't see the mountain

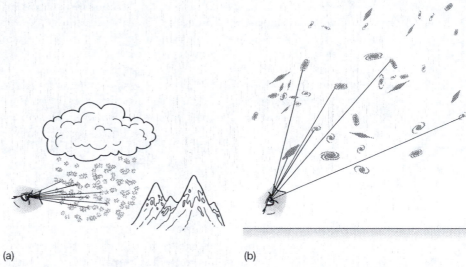

(a) (b)

FIGURE 5.12 Olbers's Paradox and Lines of Sight   Heavy snow puts a snowflake in the way of every attempted line of sight toward the mountain, and all you see is white. Similarly, in an infinite universe every line of sight eventually intersects a galaxy. It should be hard to sleep at night because the sky is so bright from the light of all those galaxies everywhere we look!

Heinrich Wilhelm Olbers (1758–1840) was an astronomer and physician in Germany. He popularized the "paradox" that the sky is dark at night (it was first noted by Kepler), and Olbers's paradox is still widely considered to be necessary background for studying the big bang theory.

because of intervening snowflakes along your line of sight, then you shouldn't see the dark sky because of intervening galaxies along your line of sight. Together with our contradictory observations of the dark night sky, this argument, which implies a bright nighttime sky, is called **Olbers's paradox**.

If you've been following the argument carefully, an objection may have occurred to you. "Wait a minute!" you might be saying. "The situation is more than *slightly* different from the snowflake example. In that case, all the light was reflected from the mountain a limited distance away, so each line of sight contained a light ray about as bright as every other line of sight. But for the universe, it's different. Some lines of sight reach galaxies so distant that they're too faint to see. I remember the inverse square law from Chapter 3, so I know that brightness falls off rapidly with distance. This explains why the sky is dark, and the paradox is resolved."

To see that there is still a paradox requires looking at the details. More distant galaxies are indeed fainter: We receive fewer photons from *each* galaxy at a greater distance. But there are also *more galaxies* at greater distances. Imagine dividing the universe into spherical shells centered on us. (See Figure 5.13.) If galaxies are uniformly scattered throughout all of space, then bigger shells will contain more galaxies. It turns out that each shell contributes the same number of photons to our night sky, regardless of its distance from us. (See Exercise 5.4 to investigate this phenomenon quantitatively.) Fewer photons from each galaxy reach our eyes, but the number of galaxies increases with distance in just the right way to keep the total number of photons hitting our eyes the same. Thus, Olbers's paradox remains.

Kepler was actually the first to put forth the paradox of the dark night sky. In 1826, Heinrich Wilhelm Olbers considered it again and brought it to the scientific forefront, leading to its renaming as "Olbers's paradox." (Others have reworked its math since, but the paradox remains.) Despite the fact that Olbers worked at a time when the existing technology didn't permit much observational cosmology, his paradox allows us to meaningfully question the overall nature of the universe.

The question becomes this: What does it mean that we *can* see dark patches of sky at night? Perhaps it means that the universe is not infinite, or that there are not infinitely many galaxies in it. It's not just that the light rays are being blocked, because if that were the case, then the light would heat up the blocking material, making it reradiate enough to replace all the light it blocks. Whatever the cause, the effect is happening on a very large scale. It is a clue to the nature of the universe, and in order for any cosmological theory to succeed, that theory will need to tell us why the sky is dark at night. It should explain the dark sky with clear, understandable logic, like that characterizing Olbers's paradox itself.

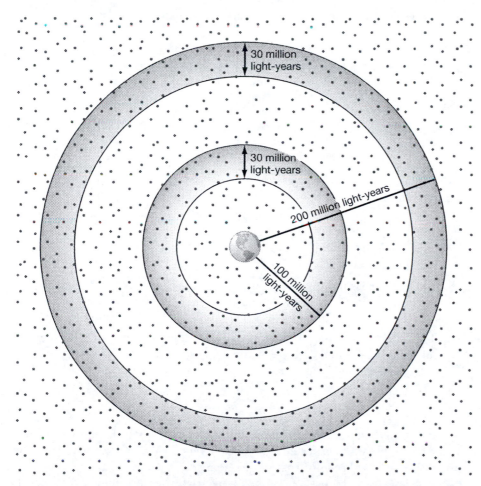

FIGURE 5.13 Geometry and Olbers's Paradox   The cross sections of two spherical shells, each 30 million light-years thick, are shown here. Both shells are centered on Earth, but one has a radius of 200 million light-years while the other is only 100 million light-years in radius. If the cosmos is uniformly filled with galaxies (dots in this figure), more of these galaxies will fall within the bigger shell, just because it has more volume. This increase in number of galaxies per shell as the distance to the shell increases exactly cancels the decrease in brightness per galaxy as we look at galaxies that are farther away. The result is that shells at 100 and 200 million light-years (or any other distance) contribute the same amount of light to our background sky. Thus, if the galaxies go on forever, our sky should be blindingly bright, even at night.

EXERCISE 5.4 Olbers's Paradox

(a) Using Olbers's assumptions (which ones?), we can see that the number of galaxies in any spherical shell around the Earth depends only on the volume of that shell; more volume equals more galaxies. Using the geometric formula for the surface area of a sphere, $4\pi r^2$, and assigning each thin shell the same thickness, which we'll call t, demonstrate that the total number of galaxies in a shell is $4\pi r^2 t \times$ {a constant number we'll call n}. What does the number n describe?

(b) Light from distant sources dims according to the inverse-square law from Chapter 3. Use this fact to show that each shell produces a measurable light flux on Earth equal to Lnt, where we'll choose L to represent the average luminosity of a single galaxy.

(c) Use your result from part (b) to demonstrate that *any* two shells will provide the same amount of light reaching Earth.

(d) Add up the light contribution for every shell in an infinite universe (one of Olbers's assumptions). How much light should we see in the night sky?

Just Stopping to Think

Observations of the universe from the last century reveal an unexpected strangeness:

1. All galaxies are moving away from us (apart from a small number of galaxies near enough to ours to be strongly affected by local gravity), and the farthest galaxies recede the fastest.

2. On large scales, galaxies are arranged in a spongelike network of walls and voids.

3. The universe is flooded with microwave radiation of astonishing uniformity. Its spectrum is that of an object glowing because of its own temperature, a few degrees above absolute zero.

4. The objects in the universe are made from a variety of chemical elements, but the abundances of the lightest isotopes appear to be fixed by some process that took place long ago and that spanned the entire universe.

5. The ages of the oldest objects in the universe are all in the vicinity of 12 to 13 billion years.

These five discoveries imply a universe that had a distinct beginning billions of years ago and that has experienced a few universe-wide leaps of evolution: Some process gave the whole universe its light elements, and another process created the CMB. These discoveries also point to a universe of galaxies, all of which somehow acquired a distinctive pattern of motion (the Hubble law) and a particular large-scale spatial arrangement. Such ancient history leaves relics that we can find and absorb into our knowledge base.

It's worth appreciating the fact that our society is capable of getting all this information about the universe, provided that it is willing to invest in the high-tech, expensive missions carried out by large-scale collaborations of hundreds of scientists, engineers, and technicians. But perhaps equally interesting is the result of a far simpler "mission":

6. The backyard observation that the sky is dark at night opposes the direct reasoning that an infinite universe full of galaxies should be bright in every direction.

Therefore, the universe we observe must be (a) not infinite, or (b) not full of galaxies, or (c) somehow capable of hiding light from galaxies. Before you ask yourself which option is the most plausible, take a moment to marvel at the options themselves. They are core, fundamental descriptions of the *entire* universe—all from a *thought* experiment! It took no extraordinary genius, no sophisticated technology, and no government spending to learn of these options. Olbers's paradox, and the meaning it delivers, is the result of nothing more than just stopping to think.

Which type of result is most believable to you, one that you can personally reason out by yourself in a quiet room with your eyes closed or one that's delivered to you in a technical magazine after having been sifted through expensive orbiting cameras, specialized software, and hundreds of Ph.D.s? (If you answered the latter, you might be surprised to know that some of the Ph.D.s themselves would say the former.) Although a genuine respect is appropriate for the big, professional scientific discoveries, it's equally noteworthy that science does leave room for meaningful interpretation by a thoughtful novice, with just a little independent study.

Quick Review

In this chapter, you've seen a collection of cosmic clues, selected for their importance in unraveling the mysteries of the universe. The Hubble law, the distribution of galaxies in space, the cosmic microwaves, the abundances of light elements, the ages of the oldest objects, and Olbers's paradox must all be explained by any good theory of the nature of the universe. The better you understand these six observations, the more you'll appreciate how much our theory of the universe accomplishes when you learn about it in the chapters to come.

Try the following crossword puzzle to test your knowledge of key terms and concepts from Chapter 5:

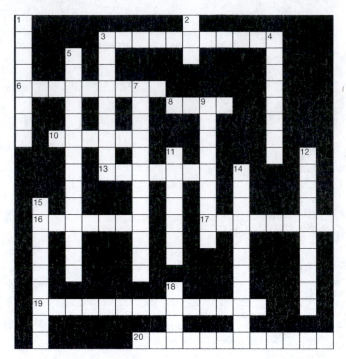

ACROSS

3 Techniques that use radioactivity to determine the ages of stars (and animal remains)

6 The more distant a galaxy is, the faster it appears to move away from us.

8 An enormous region in space that contains very few galaxies

10 24% of all normal matter in the universe

13 Reaction that changes nuclei from one type to another

16 Older stars have less beryllium and boron, but not less _____.

17 Same atomic number, but different mass numbers

19 Why is the sky dark at night?

20 Study to map out the large-scale arrangement of galaxies around us

DOWN

1 Observed in the spectrum of all distant galaxies

2 Faint, low-energy glow from every direction in space

3 Motion of distant galaxies

4 Reaction that leaves nuclei unaltered

5 Present-day expansion rate of the universe

7 Identifies a chemical element

9 Describes a universe that looks the same in every direction

11 The temperature of the CMB is 2.725 _____.

12 Type of star whose age is figured out from its cooling rate

14 Describes a universe that looks the same in every location

15 The spectrum of the CMB

18 Row of galaxies seen in redshift survey data

Further Exploration

(M = significant mathematical content; I = integrative—builds on information from more than one chapter; P = project idea; R = reflection; D = suitable for class discussion)

1. (MR) Get familiar with the redshift, z, by relating it to the recession speed it implies:
    - What's z for a typical local galaxy speed of $v = 1000$ km/sec? What does your answer tell you about the significance of the redshifts of distant galaxies?
    - At $v = 0$, what's z?
    - What's v for $z = 0.2$?
    - Calculate v for $z = 7$. Why does your answer indicate that Equation 5.2 (and therefore Equation 5.4 as well) is flawed for large redshifts like this one?

2. (RD) Sketch a hypothetical galaxy survey map of large-scale structure, (a) assuming that our universe has a center and an edge and (b) assuming neither center nor edge, but that the distribution of galaxies is totally random.

3. (RD) List some difficulties in trying to construct an artificial CMB-like radio source.

4. (R) Why should we expect the abundances of certain elemental isotopes in a star to correlate with the star's age?

5. (R) Summarize the Lithium-7 clue.

6. (RD) Suppose we have a large collection of computers, ranging from the current top of the line machines to the earliest household computer systems ever made, which

used televisions for their displays. For each computer in the sample, we measure

(a) the computation speed,
(b) the amount of memory, and
(c) the number of pixels it displays.

What does it tell us that all three measurements correlate with the computer's age, except that (c) remains constant for computers above a certain age while (a) and (b) keep getting smaller for older and older machines? In detail, relate this thought experiment to the measurement of the abundances of light elements in space.

7. (RD) How do we know that the source of the CMB isn't really. . .
   (a) in the Earth's atmosphere?
   (b) orbiting alongside our planet around the Sun, or alongside the Sun around the galactic center?
   (c) at the center of the Galaxy?
   (d) distributed throughout our Galaxy's halo?
   (e) created by stars—maybe even distant stars, all blended together?

   For each of the preceding, give as many reasons as you can.

8. (M) Which is increasing its distance from us faster, a galaxy 100 million light-years away or a galaxy 400 million light-years away? How much faster?

9. (RD) What is most noteworthy about the microwave radiation discovered by Penzias and Wilson?

10. (RD) List some typical ages of objects in the universe. Can you think of any objects that could be thousands of years old? Millions? Billions? Trillions?

11. (RD) Olbers's paradox asserts that it's puzzling that the is sky dark at night. Why *wouldn't* it be dark?

12. (M) Imagine that all the galaxies in our universe lie, for all practical purposes, in a two-dimensional plane, like most of the objects in our solar system. Modify the math of Olbers's paradox (as in Exercise 5.4) to fit this two-dimensional case, and figure out whether or not the paradox remains. Then do the same for a universe whose galaxies all lie on a one-dimensional line. (Hint: the sum of $(1/1) + (1/2) + (1/3) + (1/4) + \ldots$ is infinite, but the sum of $(1/1^2) + (1/2^2) + (1/3^2) + (1/4^2) + \ldots$ is finite.)

The Fabric of Spacetime

"Matter tells space how to curve. Space tells matter how to move."

—John A. Wheeler

To make sense of some of the observations presented in the last chapter, we are now led to investigate our understanding of space and time. We're so familiar with these concepts as the backdrop for our lives that we rarely have reason to question our preconceptions. We have an intuitive sense of space as something that gives us room to move around and keeps everything from happening all in the same place, but it's very difficult to pin down exactly what space *is* or what the world would be like without it. Furthermore, our knowledge of space is based on our experiences here on Earth. Are we justified in assuming that its properties are the same in deep space or near objects very different from our own planet?

Perhaps time poses an even greater challenge. Although the passage of time dominates our experience of life, we can't point to it or describe exactly how it "acts." Why does time seem to flow in one direction only? Why must we wait until a future event becomes the present in order to experience it? We can move either left or right in space, but we can only remember the past and not the future. Is this selective memory a property of time or a property of people? Our everyday notions of space and time are usually adequate here on Earth ("Meet me at the theater entrance at six-thirty"), but to characterize the universe properly, we need a framework that's valid everywhere. As you will learn in this chapter, it turns out to be more complicated to arrive at a black hole's entrance on schedule than at the theater entrance.

To get started, let's focus only on what we can *measure*. We can measure space by how many rulers it takes to cover the distance between two points, and time by how many "ticks" of a clock we must wait after one event and before the next. We know by intuition that space and time are related: We usually care about how far away something is because of how long it takes to get there. Motion, then, is related to the number of clock ticks that are counted as some number of ruler markings is traversed.

6.1 The Curved-Space Concept

A nice feature of the modern understanding of space and time—which is far from simple—is that it can be neatly described in terms of geometry. Usually, when people talk about geometry, they mean "Euclidean" geometry. This is the geometry we learned about in high school, involving triangles, parallel lines, and right angles. In Euclidean geometry, the following rules apply (see Figure 6.1):

- The shortest distance between two points is a straight line.
- Parallel lines never intersect.
- The circumference of a circle is 2π times its radius.
- The interior angles of any triangle add up to 180 degrees.
- The Pythagorean theorem relates the length of a right triangle's hypotenuse to the lengths of its other sides: $c^2 = a^2 + b^2$.

As we'll see momentarily, not all spaces need to obey these Euclidean rules, but those which do are termed **flat**. A word of caution here: The use of *flat* in this context is different from the everyday usage of the word. A flat space is not necessarily two-dimensional like a sheet of paper; to qualify as flat, it simply needs to obey the rules of Euclidean geometry.

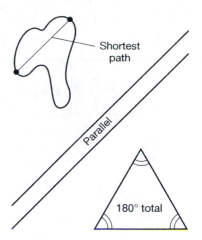

FIGURE 6.1 Some Properties of Geometry in a Flat Space (Euclidean geometry) A straight line is the shortest distance between two points, parallel lines stay parallel, and the interior angles of a triangle add up to 180°.

A common misconception surrounding the cosmological usage of the term "flat" arose at the beginning of this century, when a microwave experiment acquired evidence that our universe is (at least approximately) flat. Several news sources misinterpreted the terminology and incorrectly reported that we inhabit a "pancake-shaped" universe. (When the news media mess up reports of cosmological discoveries like this, they seem to prefer breakfast foods for their imagery. Almost 10 years earlier, the universe was incorrectly reported as being "egg-shaped", because the best CMB data set at that time was depicted in an elliptical map, similar to Figure 5.7 or Figure 5.5. Don't be surprised if the next major cosmological data set reveals that our universe is an Egg McMuffin®.) But "cosmological flatness" doesn't mean "paper thin" or "pancake-like"; it means that "Words like 'parallel' mean what you expect them to mean." Two perfect laser beams, fired parallel to each other in a flat universe, will always remain parallel. If the universe were not flat, the beams would eventually point in different directions.

Although most situations we're familiar with involve flat space, there are situations that don't. While a flat geometry works pretty well to describe spatial relationships over small distances (think of a standard city map), we need to use a curved geometry like that of a globe for larger scales. For example, the shortest airplane route from Boston to Moscow actually arcs substantially north of both cities.

Consider two people on the equator, one in South America and the other in Africa. Both face due north and begin walking. Convince yourself that at this moment their paths are perfectly parallel. But when they arrive at the North Pole, their paths cross. On a sphere, evidently, parallel lines do not remain parallel. Initially parallel lines converge in a (two-dimensional) space with

positive curvature. The surface of a sphere is one example of a positively curved space.

Now suppose that one of these adventurous hikers starts a new trek. (The other is eaten by polar bears.) She (the surviving hiker) walks due south without turning until she reaches the equator. There, she turns 90° to the right, now heading due west. By late afternoon, with the Sun in her eyes, she turns 90° to the right once again and proceeds straight back to the North Pole. In so doing, she traces out three sides of a closed figure, namely, a triangle. But two of the angles are 90°, so whatever angle her path makes at the Pole, the triangle's internal angles add up to more than 180° (possibly *much* more). This is another effect of positive curvature. (See Figure 6.2.)

We can also think about a space that's *hyperbolic*, rather than spherical. The example most commonly cited is the surface of a saddle. (See again Figure 6.2.) Parallel lines arc away from one another on a saddle, and a triangle's internal angles add up to *less* than 180°. This type of curvature is called **negative curvature**. (The Euclidean case, in which parallel lines stay parallel and so forth, is called zero curvature.)

Now, bear in mind that the examples we just mentioned—the sphere and the saddle—happen to be *two-dimensional* examples: Our hiker travels along the *surface* of the Earth and never tunnels underground. Spheres and saddles have two-dimensional surfaces that you can picture in your mind because you've seen them on three-dimensional objects within our three-dimensional world (e.g., a globe and a horse). But when astronomers talk about "the shape of space," they're considering the possibility of *three-dimensional curved space*, which is much harder to envision. In fact, the best way (and possibly the only way) to imagine a curved universe is by thinking about departures from Euclidean geometry. Think about what happens to parallel laser beams, or ask yourself where a spaceship will end up if its pilot makes a 90° left turn. In a curved space, parallel laser beams either converge or diverge, and a 90° left turn might eventually take you to the same place as a 90° turn to the right (think of a globe).

The shortest distance between two points in a curved space is *not* a straight line; consider trying to draw a straight line on a saddle. The shortest distance path is called a **geodesic**, which is straight only in a flat (Euclidean) space. In a curved space, the geodesic is like a flight path on Earth—the shortest route to travel in a curved world. It's worth looking at an example of this; spend a few moments examining Figure 6.3, which shows the flight path of a satellite.

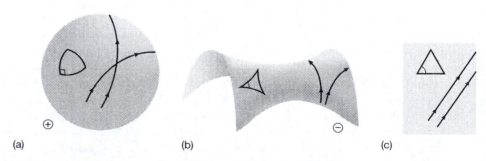

(a)   (b)   (c)

FIGURE 6.2 Different Spatial Geometries   Triangles and parallel lines behave differently, depending on the type of space they are embedded in. Shown here are (a) positive, (b) negative, and (c) zero curvature.

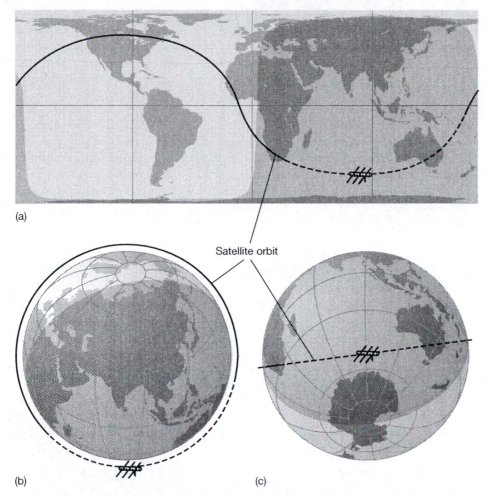

(a)

Satellite orbit

(b)   (c)

FIGURE 6.3 Statellite Geodesic   (a) Map of the world with a satellite orbit curve drawn on it. (b, c) The exact same orbit as in (a) (you should verify this!), no longer projected onto a flat map. The shading indicates which parts of the world are experiencing day and night at the particular instant shown.

6.2 Hints of a Deeper Gravity Theory

To see how geodesics and curved space offer insights into the nature of our universe, let's begin by considering three puzzling aspects of our physical world.

PUZZLE #1: *Objects with different masses fall at the same rate*, provided that air resistance is eliminated somehow. You can experiment with this phenomenon by dropping two objects of different mass at the same time—say, a baseball and a paper clip. (Neither is strongly affected by air resistance when dropped from a low height.) Heavy objects fall at the same rate as light ones.

But let's be more clear about what we mean by objects "falling at the same rate." More precisely, the Earth's gravity causes all objects to **accelerate** toward the ground at the same rate. That means many things. It means that both the baseball and the paper clip, when dropped from the same height, will take the same amount of time to reach the ground. It also means that the two objects will *always* be at the same height as they fall: By the time the baseball has fallen 3 inches, the paper clip has, too—exactly. Furthermore, at any moment, both objects will be falling at the same speed. They both start from zero speed when released, then gain speed in the same way, and finally hit the ground moving at the same maximum speed.

Physicists capture this concept compactly by saying that the two objects have the same acceleration, which can be expressed with a single number. Apart from very slight variations from place to place on our planet's surface, the acceleration due to Earth's gravity is 9.8 m/s^2, which can be thought of as "9.8 meters per second, per second." That means that after 1 second, the baseball (or the paper clip) is moving at 9.8 m/s. After 2 seconds, it has gained another 9.8 m/s, so now it's falling at a speed of $9.8 + 9.8 = 19.6$ m/s. Then, after 3 seconds, it's falling at 29.4 m/s, after 4 seconds, it's at 39.2 m/s, and so on.

Now contemplate for a moment just *how* the Earth moves different objects in the same way, because therein lies the curiosity. Could *you* move two very different objects—say, a flute and a piano—in precisely the same way? Would you be able to push each object across a frictionless floor with just the right amount of force (a gentle nudge for the flute, a heavy shove for the piano) to make them move together all the way? Gravity does this automatically, for every falling object. In a vertical race between any two objects (without air), gravity effortlessly produces a perfect tie every time. And it will be neck and neck the whole way.

PUZZLE #2: *Acceleration feels like gravity.* Whenever you step into an elevator, you get two chances to feel what it would be like on a planet with different gravity. Suppose you take the elevator to go up. For the first few seconds, as the elevator speeds up (accelerates), you'll feel slightly heavier. Then the elevator reaches its "cruise speed" and things feel normal. When the elevator slows down again (decelerates), you'll feel slightly lighter than usual, with a little more spring in your step. That is, upward acceleration and deceleration—changes in speed—feel the same as gravity does.

To Albert Einstein, this was more than just a curiosity. Figure 6.4 displays Einstein's elevator thought experiment. Imagine that you are riding in an elevator accelerating upward, not in a building, but through empty space, using a rocket engine attached to its bottom. If the rocket thrust were set to produce an upward acceleration of 9.8 m/s^2, then how would it feel inside the elevator? If you jump upward in the rocket–elevator, its floor rushes upward toward your feet at 9.8 m/s^2, because the rocket propels it that way. But if you jump upward in a different elevator that's just sitting still on Earth instead, your feet also reconnect with the ground at 9.8 m/s^2. Could you tell the difference?

Einstein realized that even if you had an entire scientific laboratory in the elevator with you, there exists *no experiment* you could perform to distinguish between a uniform upward acceleration and a uniform gravitational field. No arrangement of pulleys, chemicals, or lasers would do anything differently in the rocket–elevator than it would on the ground. Einstein wrote,

> "... all bodies are equally accelerated in the gravitational field. At our present state of experience we have thus no reason to assume that the systems... differ from each other in any respect, and in the discussion that follows, we shall therefore assume the complete physical equivalence of a gravitational field and a corresponding acceleration of the reference system."
> —Albert Einstein, "On the relativity principle and the conclusions drawn from it"

*No other physicist is as widely acclaimed for his scientific genius as **Albert Einstein** (1879–1955). In addition to making a host of major contributions to early quantum theory and to cosmology, Einstein constructed the theory of relativity, with which he radically changed the world's understanding of matter, energy, space, and time.*

This premise, that no experiment can distinguish between gravity and acceleration in a small, closed room like an elevator, is known as the **equivalence principle**. Gravity can be perfectly faked. A gravity force in one direction in a closed room can be faked by accelerating that whole room in the opposite direction. There's no other natural force like that.

PUZZLE #3: *Two different types of mass—gravitational mass and inertial mass—are always the same.* It takes more force to move a piano than to move a flute. But as we noted two puzzles ago, gravity moves both with the same acceleration. Therefore, gravity must somehow "know" exactly how much harder to pull on the piano. Expressing this thought like a pre-Einstein physicist, you could say that gravity's force on a musical instrument is proportional to the instrument's *gravitational mass*. Thus, gravity applies a large force to a piano and a small force to a flute, because the piano has more gravitational mass.

A different sort of mass also exists, called *inertial mass*. Objects have a tendency, called **inertia**, to maintain an initial state of motion and resist any attempt to change that state. For example, consider trying to stop a moving train or push one that's sitting still. The train's inertial mass quantifies its ability to resist changes to its state of motion. A large inertial mass requires a large force to affect its motion. But notice that inertia need not have anything to do with gravity, since we can use it to characterize trains, which move sideways. In fact, in the absence of gravity, inertia defines the *meaning* of an object's mass: It's how hard you have to push that object to move it around.

On the one hand, an object's gravitational mass determines the strength with which gravity pulls on it. On the other hand, the object's inertial mass determines the manner in which it responds to *any* force (not just gravity). But if the force in question happens to be gravity, then things get interesting. An object with a large gravitational mass is acted upon by a large downward force. (Think of anything heavy.) But that large gravitational mass is *precisely* canceled by the object's large inertial mass (which gives it a large resistance to forces), so the resulting falling motion is *always* the same: 9.8 m/s². This means that the two different types of mass—one connecting objects to gravity, the other completely independent of gravity—must always be the same number.

These three puzzles are really just three different ways to express one single puzzle: the special nature of gravity. It's difficult to explain gravity when we look at it as a conventional force, like the force from a spring or a magnet. You might take a minute to verify to yourself that any force you can think of, other than gravity, would behave very differently from gravity in the context of the three puzzles just presented. That's why Einstein's brainchild (well, one of his brainchildren, anyway) depicts gravity as a curving of space, and not a force. In the next section, we'll learn about this theory and how it handles the three puzzles.

EXERCISE 6.1   The Equivalence Principle

Using the equivalence principle, identify which way a helium balloon moves inside a car that's (a) accelerating from a red light and (b) making a right turn. To a physicist, both of these are changes from motion in a straight line at constant speed, and therefore both involve acceleration.

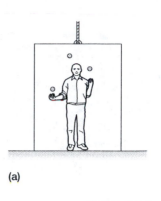

(a)

Rocket ship in outer space accelerates at 9.8 m/s² in this direction

(b)

FIGURE 6.4 Einstein's Famous Elevator Thought Experiment   In (a), the person standing in an elevator on Earth feels the pull of Earth's gravity, holding him to the floor. In (b), the elevator is out in space, far from any significant source of gravitational force. But the elevator is accelerating upward at 9.8 m/s². This produces the same effect as if a gravitational field of the same strength as the one at the Earth's surface were pulling him downward. The identical effect of acceleration and gravity is called the equivalence principle. Because of this equivalence, anything you can do in normal gravity, like juggling, would work exactly the same when you are accelerating upward at 9.8 m/s².

6.3 General Relativity

Einstein's theory of **general relativity** describes the nature of space and time. The theory characterizes the effect of gravity in terms of ordinary inertia, acting in a region of space that is *curved* due to the presence of matter and energy. That is, the effect of gravity is just that of ordinary motion through regions of curved space. Technically, general relativity involves curved *time* as well as space, but we'll focus on space for now to get a concrete grasp of the idea. Then we'll bring time into the picture.

An object's inertia reflects the fact that it has a natural state of motion through space, which it will follow automatically if left alone. In many familiar cases, this natural ("inertial") motion is motion in a straight line at unchanging speed, like a hockey puck sliding across the ice. But there's no reason this must always be the case. Consider what happens to a straight line drawn on the surface of a rubber balloon that is subsequently stretched into a random contorted shape: The straight line becomes curved. Einstein's insight was to explain the influence of gravity in a similar way: as a change in the geometry of space and time. He said that objects influenced by gravity are really just following their natural motion due to inertia, but they do so in a region of space that is no longer flat. Specifically, general relativity proposes the following two principles:

- *Empty* space is generally flat, in the geometric sense that we described in the previous section. But in the vicinity of matter or energy, such as a planet or a star, space is curved. *The presence of matter and energy causes the space nearby to bend.* When matter and energy are present, parallel lines curve in response to the bent space.

- The motion of matter and energy in a curved space follows the closest path to a straight line that the curvature allows. Just as a flight from New York to Paris follows the straightest possible path around a spherically curved world, natural motion in a space curved by matter or energy follows the straightest path it can. In other words, *objects travel through curved spaces by following geodesics.* Remember, a geodesic is the shortest distance between two points, but the shape of a geodesic depends on the shape of the space.

Stated compactly (as in the opening quote to this chapter), *matter and energy tell space how to bend, and space tells matter and energy how to move.* (Later in this chapter, we'll expand this concept to include both space and time.) To illustrate, consider the rubber-sheet example, depicted in Figure 6.5. Matter and energy tell space how to bend, as shown by Earth (the bowling ball) warping the otherwise flat, stretched rubber sheet (space). Both the Earth's matter and its energy (its warmth, for example) contribute to determining how deeply the rubber sheet bends.

The Moon (which warps its own local space as well, but not as much as the Earth does, since the Moon is about 100 times less massive than the Earth), "tries" to move through space in a straight line, but because the space around Earth is curved, the Moon is forced on a geodesic that traces out an orbit. According to general relativity, the gravity that causes the Moon's orbit is seen instead as the Earth bending the space around it and the Moon traveling through this warped space. Notice that in the simple case of the Moon's orbit, general relativity predicts essentially the same observable motion that Newton's theory of gravity (Chapter 4) does—and that makes sense, since any new theory of gravity must account for the observations we've already made.

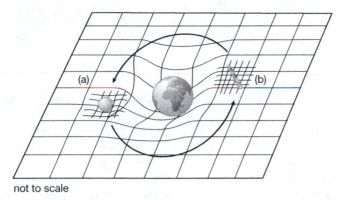

not to scale

FIGURE 6.5 The Rubber-Sheet Model   General relativity depicts gravity as a warping of space. (a) The Moon "tries" to travel in a straight line, but fails due to space being warped by Earth. (b) Any other object, even a flute, would "feel" the same curvature (since it's real!) and respond by moving in the same way.

If we were to replace the Moon with a much smaller object (like a flute or a piano), almost nothing would change! The flute still "tries" to take a straight path through the same Earth-curved space, so it follows the very same geodesic as the Moon. The only difference is that the flute warps space less than the Moon does, as shown in the figure. The Earth's "gravity" has the same effect on any object.

So: General relativity states that matter and energy cause space to curve and that the curvature influences the motions of other objects. But why does this matter? It matters because the science of the universe depends on the structure of space and time—the fabric of the universe. Over the next chapter and a half, we'll use general relativity to develop new techniques for studying the universe and to explain how the universe behaves in general. To do this, however, we must continue to gain mastery of the foundations.

Now let's look at another example. Imagine two asteroids drifting slowly through space. Suppose that they happen to be on parallel paths, headed for Earth, as shown in Figure 6.6(a). Normally, these asteroids would crash into the Earth's surface, but suppose that special tunnels have been drilled in the Earth to prevent this, as shown in the figure. Since Earth's gravity "pulls" the asteroids toward the center of the Earth, the initially parallel paths will actually converge there. This is a feature of geodesic paths in a positively curved space, as predicted by the theory of general relativity. [See Figure 6.6(b).] Remember that in a flat geometry, far from the Earth, geodesics are straight lines at constant speed, but this is no longer the case in the curved geometry near our planet.

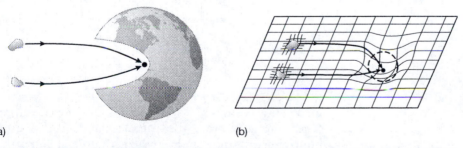

(a)   (b)

FIGURE 6.6 Asteroid Geodesics   (a) Two asteroids are slowly drifting toward our planet, and (b) if they could go through us, their initially parallel paths would converge at Earth's center because our space is curved in just that way. Note that this is true even if the two asteroids have different masses, like the Moon and the flute in Figure 6.5.

At this point, we have a confession to make: This rubber-sheet business isn't entirely *true*. For one thing, the rubber-sheet analogy includes only two of the familiar three dimensions of space, because objects move along the *surface* of the sheet only. The reality that general relativity aims to describe is actually *four* dimensional, including the three familiar dimensions in space *plus* time. Indeed, the passage of time is also "curved" in general relativity, and time intervals stretch or shrink in response to the presence of matter and energy. Note, however, that although we'll be adding the time dimension to our discussion soon, rubber-sheet diagrams can't really do it justice. Also, you have to imagine some sort of "extra" gravity pulling the sheets downward (i.e., toward a region below the sheet). A more accurate representation would somehow build that "extra" gravity into the appearance of space. Sadly, it was too expensive to print this book on four-dimensional paper, and the rubber sheets are the best we can do. Just be aware that they only offer *a way for you to visualize curvature* in general relativity; they don't capture the full curvature itself.

How does general relativity explain the puzzles we discussed in Section 6.2? Consider again the elevator thought experiment from Figure 6.4. When you're inside the rocket–elevator accelerating upward at 9.8 m/s² through empty space, you feel normal gravity as you would on Earth, because the floor is continuously being pushed up into your feet and toward anything you drop. It's the upward force from the floor that lets you feel your own weight.

But isn't the same true on Earth? Don't you feel the upward force of the floor pressing against your feet when you stand? Or the upward force from a chair, squishing your butt when you sit? Is this just an idle observation, or is there a real connection here?

Go back to the rocket–elevator in space, but this time turn off the rockets so that there's no acceleration. You're weightless inside, drifting between the elevator walls. Now try to come up with an analogous situation on Earth in a gravity field. Is there any way to truly simulate weightlessness on Earth? You may have heard that astronauts do some spacewalk training under water, but that doesn't really simulate weightlessness, because the astronaut feels pressed downward within her space suit, even though the space suit can be adjusted to hover at some particular depth in the water.

Now, what if the elevator were falling (Figure 6.7)? Imagine that your elevator, with you inside, is dropped from some great height. The elevator accelerates downward, and so do you, at exactly the same rate. So you don't move particularly toward the floor or the ceiling. You float inside, in a perfect simulation of weightlessness—but it's more than just a simulation. Weightlessness isn't just being in a region with no gravity sources; it's being *anywhere at all*, provided that there are no restraints preventing you from falling. According to general relativity, the simplest and most natural state is free fall. When objects are free to fall, they follow the geodesics that general relativity predicts.

We can now summarize how general relativity explains the three puzzles from Section 6.2.

NO-LONGER-PUZZLING #1: *Objects with different masses fall at the same rate.* Of course they do! Their acceleration is a product of two things, neither of which involves mass. One is the curvature of the space around them. The other is the state of motion of the observer, who stands on the hard ground and is therefore restrained from a natural state of free fall (the natural motion all objects follow unless something like the ground interferes). Since all observers on Earth

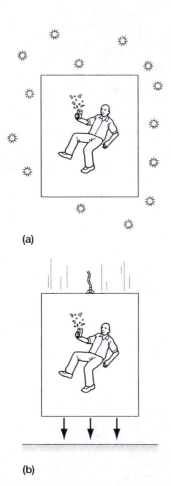

(a)

(b)

FIGURE 6.7 Free Fall (a) Floating in an elevator in space. (b) Floating in a free-falling elevator on Earth. Notice how the two situations appear identical from inside the elevator.

are restrained by the ground in the same way, we all agree on the acceleration of falling objects moving relative to us.

NO-LONGER-PUZZLING #2: *Acceleration feels like gravity*. Of course it does! The feeling of gravity is really caused when something like the ground keeps us out of free fall. That can be done equally well with a planet's hard surface or with rockets—in other words, by gravity or acceleration. In this way, the equivalence principle lives and breathes in every aspect of general relativity.

NO-LONGER-PUZZLING #3: *Two different types of mass—gravitational mass and inertial mass—are always the same*. Of course they . . . no, wait a minute: Gravitational mass doesn't even exist in the familiar sense in general relativity! Gravitational mass exists in Newton's theory of gravity in order to quantify the strength of the gravitational *force*, as in Equation 4.2. The bigger the gravitational mass, the stronger is the gravitational force that acts upon it. How much a falling object resists a "real" force (such as air resistance, wind, birds that get in the way, etc.) depends on its inertial mass. So, from the perspective of Newtonian gravity, the puzzle is why an object with more gravitational mass also gains inertial mass in the same proportion. But since gravity isn't a force in Einstein's theory, there's no need to talk about gravitational mass at all. From the perspective of general relativity, a falling object is just reacting to the shape of the space it moves through, so the puzzle goes away when we look at the situation from the perspective of general relativity.

Perhaps the most remarkable conceptual shift associated with general relativity is that space is no longer just the fixed, static backdrop for the real action of the universe. Space becomes tangible and dynamic. Whenever a galaxy moves or turns, its region of space changes shape, like a bowling ball rolling on a waterbed. Whatever an object does, it distorts space, and that distortion in turn changes how objects move. For the rest of this chapter, we'll use general relativity to study some important systems in the universe; in the next chapter, we'll use it to study the whole universe at once. Relativity theory is conceptually difficult, but your diligence will be rewarded with a link between the contents and the ultimate destiny of the universe.

EXERCISE 6.2   Broken Elevator

Let's take Einstein's famous thought experiment one step further. Suppose you get into an elevator at the top floor of a tall office building and press the button to descend to the bottom. The elevator doors close, and then. . . .

(a) First the elevator begins to accelerate downward at 2 m/s², and you accidentally drop your keys. Describe how they fall to the elevator floor, and how fast, compared with how things fall normally (not in a moving elevator). (*Hint:* Accelerating downward brings you a little closer to a state of free fall, so what would the keys be doing in that case?) Ignore air resistance throughout this exercise.

(b) Then the elevator reaches its "cruise speed" and stays at that speed; that is, the acceleration becomes zero. You accidentally drop your briefcase. (Face it, you're clumsy.) Describe how the briefcase falls.

(c) Then the elevator cable snaps, and your eyeglasses fall off your face. (You're nearsighted, too—bad genes.) Describe how the glasses fall. (*Hint:* Now you *are* in a state of free fall. You'd better hope that changes before you hit the cold, hard ground.)

(d) Miraculously, the bottom of the elevator shaft is filled with fluffy pillows, so when your elevator lands, it decelerates gently and comes to a stop. During this deceleration, your hearing aid falls out of your ear. (You're hard of hearing, too, from all the rock concerts you attended as a teenager.) Describe how the hearing aid falls while the elevator is decelerating through the pillows.

(e) For each of the four events above, (a) through (d), describe in two or three sentences what's going on in terms of general relativity and deviations from the natural state of free fall. How does the acceleration of the elevator enhance, diminish, or otherwise alter the effects of gravity?

6.4 Testing General Relativity with Photons and Gravitons

In the last section, we looked at some of the motivation for Einstein's theory of gravity. The theory explains several curiosities, expands our understanding of weight, and allows for curved space. But you'll recall from Chapter 1 that theory alone does not justify the acceptance of new science: Experiments must also show that the theory holds true in nature before we adopt it as our description of reality. Perhaps you think that the shape-of-space idea is interesting—but you might also wonder what's the point if it seems like just *another* way of thinking about gravity, rather than a *better* way.

When put to the test in the real world, it turns out to be a better way. In the next subsection, we'll look at direct evidence for the bending of space predicted by the theory of general relativity.

Gravitational Lensing

To *see* curved space in action, we begin by asking how it should affect what we actually see: light. Does gravity affect light?

According to Newtonian gravity, the answer is unclear. On the one hand, in Newton's theory gravity is a force, whose strength depends on the "gravitational mass" of light. That number is the photon mass, which is zero, so the force is zero. On the other hand, a force's effect on light depends on a photon's inertial mass (also zero). As a result, one doesn't *need* any force to move a massless photon. Then again, Newton's equations aren't intended for massless particles, so it's not obvious how seriously to take any of this reasoning.

In contrast, general relativity isn't so ambiguous. If space is actually curved, then *everything* passing by, massive or massless, will follow its (curved) geodesic. Light, then, is no exception. (There are no exceptions!) According to Einstein, a perfect laser, pointing perfectly horizontally in a lab, will deflect downward because of Earth's gravity. That is to say, *light falls*. (Clearly, if this is correct, then light must fall by only a tiny amount in Earth's curvature, or else we would see this falling in our daily experience. Flashlight beams would arc to the ground like water from a hose.) Try Exercise 6.3 to work out for yourself why general relativity predicts that light should be affected by gravity.

EXERCISE 6.3   Falling Light

In this section, we argue that the gravitational bending of light must occur, since the light is traveling through a curved space. Use the equivalence principle to construct another (and perhaps better) argument to demonstrate that the gravitational bending of light really happens. In other words, imagine an elevator accelerating upward through space as in Figure 6.4, far away from any gravity sources (like stars). Consider where a laser beam fired horizontally across the elevator will appear to go—where the laser dot will land on the opposite wall—when the elevator is accelerating (speeding up). Argue that the same must be true for lasers held still in the presence of a gravity source like the Earth. Answer in a few sentences or bullet points, accompanied by one or more sketches to illustrate.

Now, whether or not Newton's gravity bends light, his equations differ from Einstein's by a factor of two. That is, even if Newton's gravity force does bend light, Einstein's curved space bends it by twice as much. That means that if we could ever find a real-life example of light rays turning, we would simply have to compare how far they turn to decide which theory is correct. This is just what happened in 1919.

The world had its first opportunity to check Einstein's prediction that year, during a convenient arrangement of celestial objects. The Sun and Earth were aligned with a more distant star, such that from our point of view, the other star was behind the Sun. Figure 6.8 shows this geometry on a rubber sheet. From the figure, you can see that if Einstein's theory is correct and the Sun curves space, the starlight should bend around the Sun, which would make that star appear to move. Before and after the alignment, the star appears in its true position on the sky. But during the alignment, the star appears in a slightly different position, because its light is deflected.

In order to observe this effect, one needs to be able to see stars during the daytime, when the Sun is also visible. This is possible during a total solar eclipse, when the Moon is positioned just right to block out the light from the Sun. So the alignment was Earth–Moon–Sun–star, but because a solar eclipse is visible only from certain parts of the Earth, you have to go to wherever the eclipse is happening. The total solar eclipse of 1919 was visible over the Atlantic ocean mostly, reaching a little of Africa and a little of South America. The following is excerpted from the account of the scientists who conducted expeditions to Africa and South America

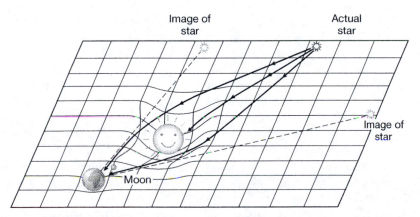

FIGURE 6.8 Starlight Bent by the Sun Light from a distant star follows a curved path due to the Sun's influence on the spacetime around it. The star then *appears* to be located in the direction that the light *appears* to be coming from, as indicated with the dashed lines. In this particular figure, we see that the same star can appear to be located in different places at the same time!

to observe this phenomenon:

> "The purpose of the expeditions was to determine what effect, if any, is produced by a gravitational field on the path of a ray of light traversing it. . . . The only opportunity of observing these possible deflections is afforded by a ray of light from a star passing near the sun. (The maximum deflection by Jupiter is only 0".017.) Evidently, the observation must be made during a total eclipse of the sun. . . . As totality approached, the proportion of cloud diminished, and a large clear space reached the sun about one minute before second contact. Warnings were given 58s., 22s. and 12s. before second contact by observing the length of the disappearing crescent on the ground glass. When the crescent disappeared the word 'go' was called and a metronome was started by Dr. Leocadio, who called out every tenth beat during totality, and the exposure times were recorded in terms of these beats...owing to a strike of the steamship company it was necessary to return by the first boat, if we were not to be marooned on the island for several months. By the intervention of the Administrator berths, commandeered by the Portuguese Government, were secured for us on the crowded steamer. . . . Thus the results of the expeditions to Sorbal [north Brazil] and Principe [an island in the Gulf of Guinea, Africa] can leave little doubt that a deflection of light takes place in the neighborhood of the sun and that it is of the amount demanded by Einstein's generalized theory of relativity, as attributable to the sun's gravitational field."
>
> —Frank W. Dyson, Arthur S. Eddington, and Charles Davidson, "A Determination of the Deflection of Light by the Sun's Gravitational Field, from Observations Made at the Total Eclipse of May 29, 1919"

FIGURE 6.9 A Common Example of Bending Light The pencil appears disjointed because the light from the underwater part of the pencil (bottom of picture) is bent onto a different path by the water. Gravitational lensing produces a similar effect, only in that case the bending of light is due to the curvature of space itself, rather than an intervening medium such as water.

The deflection of starlight was measured in 1919 and was widely considered a demonstration that space bends in the way general relativity predicts. However, the equipment of the time wasn't sensitive enough to rule out the possibility that Newton's theory of gravity was still correct and just needed to be extended to include a gravitational force on photons. Only later, with improved technology, was it confirmed that the amount of bending was numerically twice the amount calculated by Newtonian gravity and exactly what general relativity predicts.

The effect is similar to light passing through a glass of water, as shown in Figure 6.9. If you didn't realize there was water between you and the pencil, bending the path of the light, you would think the pencil had moved from its true location. Today we have some spectacular visual examples of the effect of the bending of light by curved space. (See Figure 6.10.) It's called **gravitational lensing,** because the light paths are being bent by gravity, as if through a lens. Gravitational lenses—large masses that perform gravitational lensing, like the Sun in Figure 6.8—magnify light, too, somewhat as optical lenses do.

Gravitational lenses also have other effects—distorting and multiplying images the way warped glass does—that can be both good and bad for astronomical study. It means that we have to be very careful how we interpret gravitationally lensed light (bad), but it also means that we get other information from the manner of warping that tells us about the lens itself, which is a massive object in space (good). Multiple imaging occurs when the light from a distant source bends around the lens along several different paths, somewhere between the source and Earth. This, too, can be good or bad: Multiple images tell us about how and by what the light is being bent (good), but they also force us to figure out spectroscopically whether two images really come from the same source in any given image (slightly bad).

One major benefit of gravitational lensing is that it aids in the study of dark matter, which was introduced in Section 4.3. Because dark matter, like any other matter or energy in the universe, causes space to curve, astronomers can use distant

FIGURE 6.10 Gravitational Lensing   Faint, blue spiral galaxies in the distant background are warped, magnified, imaged, and reimaged several times by the lens. (Color versions of this and other gravitational lenses are shown in Color Gallery 2.) The lens itself is the massive galaxy cluster Abell 2218 (seen here as the collection of elliptical galaxies). Multiple images of the spiral galaxy in the background surround the lens in a pattern of thin, concentric, circular arcs. (◆ Also in Color Gallery 2, page 17.)

sources (such as galaxies) that are warped and magnified by large dark matter lenses (like galaxy clusters) to study the amount and arrangement of dark matter in the lensing cluster, as was done in Color Figure 41 in Color Gallery 2, page 19. In that case, dark matter was "seen" by the gravitational lensing it caused, even thought it can't be seen in the usual ways.

Sometimes a lensing event is too brief or too faint to make out the detailed shape of the warping or to see multiple images. This is often the case when astronomers are watching a particular star and another object crosses their line of sight in front of the star. The result is a slight, but recognizable, brightening of the star, called *microlensing*. (The star's position should appear to shift as well, but with microlensing, the shift is generally too small to measure). Microlensing, too, can be very useful in the study of dark matter. Recall that one type of dark matter is the MACHO, an object like a dwarf star that's too dim to see directly. But with microlensing, astronomers see MACHOs indirectly when these objects pass in front of a background star and temporarily cause the background star to brighten. One of these detections is shown in Figure 6.11.

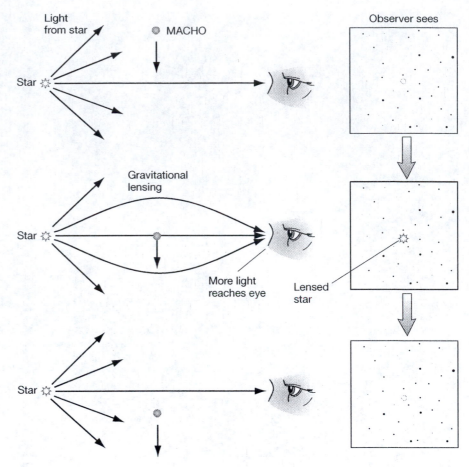

FIGURE 6.11 MACHO Experiment "Massive astrophysical compact halo objects" (MACHOs) are sought and found by gravitational microlensing. When a MACHO passes between us and a distant star, the star is briefly brightened by the gravitational bending of light, in a specific brightening and dimming pattern.

Today, large-scale MACHO searches use this very technique. Many such MACHO microlensing events have been seen to date, demonstrating that MACHOs are indeed a real form of dark matter. Continued monitoring of these microlensing events will tell us how many MACHOs are really out there—an important part of our attempt to understand dark matter.

Gravitational Waves

We just looked at how gravity affects light. As we saw in Chapters 2 and 3, light can be successfully modeled both as a collection of photons and as an electromagnetic wave. In the event of an electrical or a magnetic disturbance, like a spark or a flame, electromagnetic waves (photons) radiate away from that disturbance at the speed of light, carrying news about the disturbance with them.

General relativity says that gravity can also travel as a wave. You can picture this with the rubber-sheet analogy. (Recall Figure 6.5.) If the bowling ball moves around on the rubber sheet, the motion will temporarily change the distortions all over the sheet. But this change won't happen instantaneously; rather, it will travel outward from the bowling ball as ripples in the sheet. In the event of a gravitational disturbance, **gravitational waves** (a sort of "gravity transmission," which can also be described as particles called "gravitons"), like electromagnetic waves, radiate away from that disturbance at the speed of light, carrying news about the disturbance with them. But there are a few major differences:

- Gravitational waves are much weaker than electromagnetic waves; only major gravitational events send out waves energetic enough for our most sensitive instruments to notice.

- Electromagnetic waves superimpose electrical and magnetic fields on an existing space and time without significantly altering it. But since gravity is an expression of the curvature of space and time, gravitational waves carry a traveling "glitch" in spatial curvature, wounding and remending the shape of space as they pass through it.

- Electromagnetic waves interact only with electrical and magnetic objects, such as electrons. But since gravitational waves change the local shape of space itself, they interact with everything, even though the effect is usually too weak to notice.

During the next decade, two new gravitational-wave observatories will begin searching the universe for major sources of gravitational disturbances, like those produced when extremely dense objects called neutron stars and black holes merge. With two enormous detectors in the states of Washington and Louisiana, the advanced Laser Interferometer Gravitational-Wave Observatory (LIGO) should be able to detect the "curvature pulse" (gravitational-wave signal) of these mergers approximately once a year, from sources within the closest million or so galaxies, beginning in 2009. In addition, a joint mission between NASA and the European Space Agency, called the Laser Interferometric Space Antenna (LISA) and expected to launch in 2015 or later, will fly as a small constellation of separate spacecraft, orbiting the Sun. The antenna should be able to detect high-energy events from all over the observable universe. Both observatories will operate by comparing laser beams to see if a passing gravitational wave has compressed the distance that the beams travel. Over time, we will learn to interpret the details in these signals—their varying intensity, frequency, pulsing, etc.—and recognize exactly what type of event causes each detail. In addition, gravitational-wave astronomy holds the hope of testing a theory about the very early universe called "cosmic inflation" (Chapter 12); amazingly, that achievement will yield a fairly comprehensive understanding of our universe, all the way back to a time when it was less than a hundredth of a millionth of a trillionth of a trillionth of a second old (10^{-32} sec).

It has become clear that general relativity is a powerful tool for understanding space, time, and gravity. We turn next to two specific "objects" whose behavior is understood through the theory of general relativity.

6.5 The Black Hole and the Event Horizon

The theory of general relativity tells us that all matter and energy interact gravitationally, both by producing their own curvature and by moving on geodesics caused by existing curvature. As a result of this simple premise, you might wonder how any object can hold itself up against its own gravity. The Sun, for example, is a ball of hot gas and plasma, made of atoms and other particles. These particles follow geodesics all the time, so why don't all of them fall to the center of the Sun's curvature? (You can imagine this neatly with the rubber-sheet model that we've been using in this chapter; refer back to Figure 6.5.)

The answer is that the Sun *would* collapse to its own center if all of its particles were to slow down and if we could turn off their interactions with each other. That is, if we could "dial down" the temperature so that all the Sun's particles move slowly, then their geodesics would carry them to the center, the way the slow-moving asteroids we looked at in Figure 6.6 would meet at the center of the Earth. Even then, you couldn't cram all of the particles into the center of the Sun, because they generally repel one another (electrons, for example, electrically repel other electrons). But if you could also somehow turn off these interparticle forces or overpower them, then the Sun would collapse.

Like the Sun, most stars are supported against the force of their own weight pulling inward by hot, fast-moving particles that provide an outward push. And other stars are supported by interparticle forces even when those particles slow down. White dwarf stars, for example, are supported by forces between electrons. If those interelectron forces were overpowered by strong incurving gravity, then white dwarf stars couldn't exist, but neutron stars could, because they are smaller stars that are supported by forces between neutrons instead. And in 2002, astronomers published some interesting evidence for the existence of *strange stars*, which are even smaller than neutron stars and are supported by forces between quarks—the particles that hide inside protons and neutrons. Maybe someday high-energy physicists will discover a new force between some new particles that you could imagine supporting some new kind of much smaller star. But if so, then general relativity asserts that your imagination is wrong.

One of the most amazing results from general relativity is found in an equation called the Tollman–Oppenheimer–Volkoff (TOV) equation, which describes the battle between gravity and interparticle forces in a star. The details of this equation are beyond the scope of this book, but the result is not: When space is curved beyond a certain amount, there is no force capable of halting a star's total collapse, *not even a force we haven't discovered yet!* You can get a feeling for how such a claim is possible by remembering that, in general relativity, *energy* as well as mass causes spacetime to curve. So if you try to invent a stronger force to halt the star's collapse, that force contributes to the energy, curving space even more and requiring a stronger force to resist collapse, and so on. The TOV equation expresses this property of general relativity.

The collapsed star that results from this extreme space curvature is called a **black hole**. Most stars that start their lives more massive than the Sun by a factor of about 30 times will ultimately become black holes. Black holes are popularly defined as "objects whose gravity is so strong that not even light can escape." You'll be ready for a better definition by the end of this section, but you can already start to see what the popular definition means: In a black hole, space is so tightly curved that even a photon's geodesic never emerges. If not even a photon—traveling at the speed of light—can escape, then you can appreciate why the TOV equation denies *any* means of avoiding collapse. In this sense, "black hole" is a great name: It's a hole because things get stuck in it, and it's black because those things include photons (light).

The blackness of black holes makes them hard to find in space. They're also very small—generally only a few kilometers in diameter—which steps up the detection challenge yet another notch. Even though astronomers currently have no way of directly seeing a distant, tiny, black object like this against the black backdrop of space, they don't have to give up altogether: They look for indirect evidence instead.

In practice, astronomers look for matter orbiting near a black hole. This matter can be either (a) one or more stars orbiting a dark, unseen object or (b) hot gas spiraling into the unseen object. Either way, the "smoking gun" is the size and speed of orbiting stars or gas they see around the hole, which indicates the strength of its gravity. This evidence is very much like some of the evidence for dark matter from Section 4.3, in that astronomers rely on the visible orbiting objects to tell them about the invisible object they orbit around. In the case of black holes, these orbits are very small, so there isn't room for much inside the orbit, apart from the black hole itself. The high speed of the gas spiraling inward toward the black hole means that collisions among gas particles generate enough heat to produce x-rays; aside from black holes, few other compact sources in the universe are that hot. Many of these sources believed to be black holes have been found by the Earth-orbiting *Chandra* x-ray telescope.

The popular science press often reports the "discovery of a new black hole"; perhaps you've read such a story. These claims are probably true, but you should bear in mind that what the discoverers actually *observed* is some matter which appears to be *near* a black hole, and not the black hole itself. Figure 6.12 shows, and its caption describes, the probable detection of a black hole (as do the next several

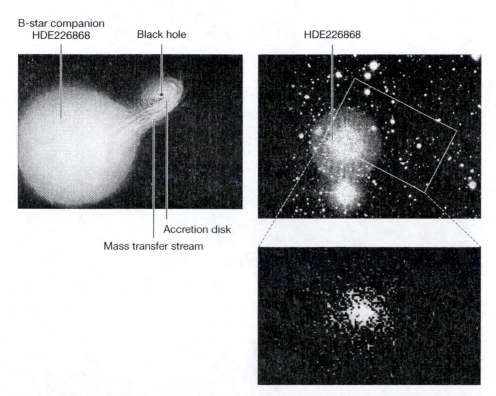

FIGURE 6.12 Black Hole Detections   Since the black hole cannot be seen directly, astronomers look for binary systems in which one star is a black hole, accreting material from its larger partner and emitting x-ray light in the process. The left panel is an artist's conception of this process for a binary system. The very bright star in the right top panel is thought to be the large companion of the black hole Cygnus X-1, responsible for the x-ray emission shown (in false color) in the right bottom panel.

figures); you can examine some observations of candidates for black holes in greater depth in Further Exploration Questions 4 and 5 at the end of this chapter.

Most black holes are "stellar mass black holes," which means that their mass is comparable to that of a star. Astronomers understand these holes to have been born from supernova explosions of larger stars (more on this in Chapter 9). But there are two other types of black holes out there. **Supermassive black holes** (SMBHs) have been (indirectly) observed at the centers of many galaxies. Astronomers now believe them to be a standard component of a large spiral or elliptical galaxy. These holes are millions or billions of times the Sun's mass, way too massive to have formed in the explosion of a single star, although they may have started out that way and then rapidly grown from there. Most, or maybe all, galaxies like ours probably have an SMBH at their core. Even so, SMBHs remain very difficult to detect, because their influence is minimal on galaxies, whose total masses are typically huge—around a trillion solar masses each.

Recently discovered **intermediate-mass black holes** (IMBHs)—possibly an adolescent stage between youth (stellar mass) and old age (supermassive)—occupy the mass range from hundreds of solar masses to tens of thousands. (See Color Figures 31, 32, and 33 in Color Gallery 1, page 15, for SMBH and IMBH discoveries.) Figure 6.14 shows several IMBHs near the core of our own Milky Way Galaxy, where they are probably en route to merging with the SMBH at the center; such mergers are almost certainly part—but not all—of the standard growth process that builds SMBHs.

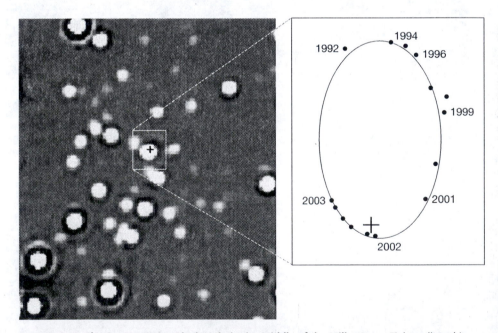

FIGURE 6.13 The Supermassive Black Hole in the Middle of the Milky Way   Tight stellar orbits near the galactic center strongly imply that an SMBH must be their gravitational anchor. No other object would plausibly have such a strong effect on nearby orbits and yet occupy so little space; the left frame is only about 0.1 light-year across. The spots are individual stars, and the cross is the location of the SMBH. More detailed tracking of one of these stars (inset, right) reveals that it moves in a very close elliptical orbit around the central black hole, again denoted with a cross. (◆ Also in Color Gallery 1, page 15.)

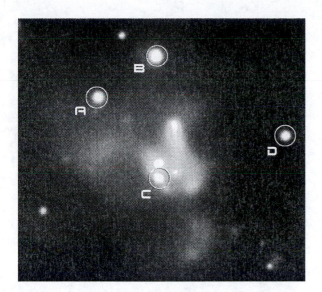

FIGURE 6.14 IMBHs at the Center of the Milky Way An image of the galactic center, surveyed by the *Chandra* orbiting observatory. Circled and labeled regions of bright x-ray emission are probably IMBHs caught in the process of spiraling their way toward the SMBH residing at the center (less than 3 light-years away from each IMBH). The IMBHs are expected to eventually merge into the SMBH.

We'll have more to say about SMBHs throughout the rest of the book, including the next section. But for now, let us address whether SMBHs might help solve the problem of identifying the dark matter in galaxies. You saw in the last section that black holes and other MACHOs account for a little of the "missing" dark matter known to inhabit galaxies due to their light curves. (Recall Section 4.3.) Wouldn't a supermassive black hole contribute even more?

The answer is, definitely not! At first glance, SMBHs seem like great candidates for dark matter because we already know that some black holes fit the bill as MACHOs and the supermassive variety have much more mass. But they have nowhere near enough mass. One SMBH weighs in at a few million to a few billion solar masses. But the mass of a galaxy—mostly dark matter—is typically a few *trillion* solar masses. This means that an SMBH is a significant gravity source, *but only near a galaxy's core*. Beyond the core, the rest of the galaxy— stars, dark matter, gas, etc.—is much more massive and therefore has much more gravitational influence than the SMBH itself. SMBHs just aren't "SM" enough to amount to anything significant in the dark matter world.

Relativistic Behaviors

Black holes offer extreme examples of curved space (strong gravity), which makes them ideal for elucidating the core concepts of general relativity, concepts that play an important role in the modern understanding of our universe. General relativity specifies that *matter and energy dictate the curvature of space, and that curvature influences all motion through the curved space*. (If you ever find your grasp on this chapter's inherently tricky material waning, then this last sentence is the place to reground yourself.) With that point in mind, we can understand a black hole as a place where an extremely compact region of matter or energy—like a collapsed star—has created a dramatic, localized warp in the shape of space.

One consequence of such a warping is a black hole's **light-capture radius**. At some distance from a black hole, a passing light beam will be deflected into the black hole, where its photons will remain. The effect is similar to the Earth's effect on the trajectory of a passing asteroid. A black hole is strong enough to pull in a

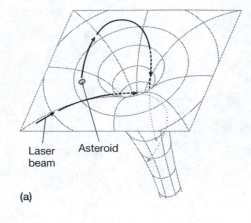

(a)

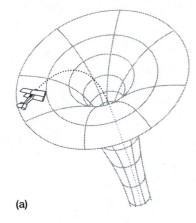

(a)

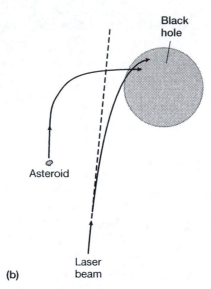

(b)

FIGURE 6.15 Capture by a Black Hole   Both the asteroid and the light ray are shown being captured by a black hole (in two-dimensional space). Frame (a) shows the rubber-sheet reasoning; frame (b) shows the actual light-capture radius for the light beam and the asteroid, as seen by two-dimensional astronomers living in the rubber sheet. The slow-moving asteroid is easily captured by the gravity of the black hole, even far away from it. But the fast-moving light will be captured only if it passes very close to the hole.

Top view

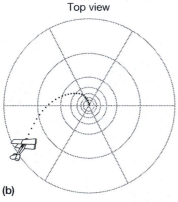

(b)

FIGURE 6.16 Gravitational Time Dilation   (a) A spread of bullets entering a black hole undergoes a time dilation from our point of view outside the hole. Bullets are shown in their locations at 1-second time intervals, as if captured in a time-lapse photo illuminated by a strobe light. In the top view (b), you can see the pileup of bullets approaching, but never quite reaching, the center. Evidently, they appear to slow down.

photon of light, whereas the Earth couldn't capture anything moving that fast. (See Figure 6.15, which compares the capture of asteroids and light rays.)

Another black hole phenomenon derives from its warped space*time*. Imagine a machine gun firing a long stream of bullets into the hole. (See Figure 6.16. This may not be as unlikely as it sounds. In the 1980s, one of the tabloids, whose hard-hitting journalism can be found at any supermarket checkout line, reported a cover story titled "World War II Fighter Planes Sighted Near Mars." One of those planes might fire a spread of bullets into a black hole.)

While the bullets are still far from the hole, they're in flat space. Then they reach the curved region around the hole, the "slope" of which increases toward the center.

Seen from the top view in frame (b) of the figure, the bullets appear to bunch up as they approach the center. This bunching of the bullets occurs because they are following their natural motion in a region of spacetime that is now curved. From this simple thought experiment, we gather that the bullets arrive at the hole with a faster pulse rate than they had when they emerged from the barrel of the gun. That is, if they're fired as "bang—bang—bang—bang" from the barrel, they arrive as "bangbangbang-bang" at the black hole. The reverse is also true: If bullets are fired from near a black hole, but aimed away from it, they emerge from the gun as "bang—bang—bang—bang" and end up far from the hole as "bang——bang——bang——bang."

Of course, there's nothing particularly special about machine gun fire in this example. The bullets could be pages shooting out of a photocopier, drops of fuel steadily leaking from one of the WWII fighter planes, or —and here's the point— anything at all that happens with a particular timing. General relativity tells us that it's not just the spacing of the bullets; it's actually a *change in the flow of time*. If we throw a ticking clock into a black hole, its ticks will slow down as they emerge from the hole, so we'll actually see the clock tick more slowly. There's nothing special about a clock either! You could throw a live parrot into the black hole and watch its beak squawk in slow motion, too: "Caw!——I'm——a———pretty——bird!" Time flows more slowly and people age more slowly near a black hole, compared with the rest of us here on Earth. The effect is called **gravitational time dilation**. It's the *comparison* that matters: From the point of view of an observer falling in alongside the clock or parrot, the ticks or squawks are not slowed down at all.

Let's suppose that we're talking about photons of light, instead of bullets or chattering parrots. If you point a flashlight toward a black hole, the photons arrive in more rapid succession than when they left the flashlight. That means two things. First, because photons are particles, the faster arrival rate means that the light is brighter—there are more photons per second. Second, because the photons are also waves (recall Section 2.4), the faster arrival *frequency* means a different *color*. After all, the increase in frequency is due to a change in local timekeeping, so it affects anything sensitive to timing, including the rate at which light waves repeat. Someone near the hole will see a flashlight beam arriving that looks blue. If that person replies by aiming his own (yellow) flashlight outward, it will be seen far away as red.

Thus, the bending of spacetime creates a **gravitational redshift** (and blueshift). This effect has even been measured in the curved spacetime caused by the Earth. In fact, if you've ever used a GPS receiver to locate your position, you've made use of the effect of gravitational time dilation. The GPS receiver works by timing how long it takes for signals from different satellites orbiting Earth to reach your receiver. It computes your position on the basis of the very precise timing of the signals, which would be off by a little bit without correcting for this effect of general relativity.

The Event Horizon

As you can see from any of the rubber-sheet diagrams in this section, the inward curvature of a black hole keeps getting steeper and steeper the closer in you go. You might wonder just how close you can go before it's too steep to get out again. To make the question specific, suppose that you have a spacecraft that is somehow capable of reaching the speed of light. (According to what scientists understand today, this isn't possible, not even in theory. But it *is* theoretically possible to get very, very close to light speed, so let's just pretend that you can get all the way there.) At light speed, you have the maximum possible ability to escape the black hole.

Now, suppose also that you're hovering at a safe distance from the black hole and an alien ship pulls up alongside you, wanting to play chicken. Well, you're not about to let some space aliens think they're braver than you are, but you also don't want to die in a black hole. It's all about knowing just how close you can get. That boundary, between being trapped and being able to escape when moving at the speed of light, is called the **event horizon**—an imaginary enclosure around a black hole.

Why is the boundary called an "event horizon"? The answer comes from Einstein's relativity theory, which tells us (among the many things it tells us) that it's impossible to transmit information faster than light. Radio transmissions, for example, travel at exactly light speed. As you just sit and watch television, the sequence of events in your TV show is communicated to you by light signals leaving the TV screen and traveling to your eyeballs at the speed of light. But if the TV were inside a black hole's event horizon, then its light signals couldn't escape. Those of us outside the event horizon would never see the show.

Following that logic to its completion, no event that takes place inside a black hole's event horizon will ever be known by those of us on the outside. Why? Because the fastest that information can be communicated is at the speed of light, and that's not fast enough to escape the hole. They say that "what happens in Las Vegas stays in Las Vegas"—well, that applies to black holes as well. On Earth, the horizon is an imaginary line beyond which you can't see, caused by the shape of the Earth. For a black hole, the event horizon also marks the edge of what you can't see, because the light inside it is trapped there. In this sense, a black hole is a region of space that's disconnected from our universe. Once you enter a black hole (whether you are a person or a photon), you're cut off, and no one on the outside will ever hear from you again.

An event horizon is not a real surface; there's nothing there to stand on. Inside the event horizon, the curvature is so great that there is no rocket imaginable that could prevent you from falling to the very center of the black hole. Physicists call that center point a *singularity*, and the rules of general relativity tell us that anything inside the event horizon, including light, must go there to stay.

Figure 6.17 shows some geodesics near the event horizon. Photon geodesics (solid lines) also involve gravitational red- and blueshifting. Matter geodesics (dashed arrows) describe how slower-than-light spacecraft would travel. As usual, all of this is determined by the curvature in the region.

One geodesic is of particular interest. Recall that the circular orbit of a satellite around the Earth is the result of the moderate curvature induced by the Earth (as in Figure 6.3). Near a black hole, where the curvature is extreme, it is possible to have a photon traveling on a circular geodesic; that is, a beam of light can orbit the black hole, and it "thinks" it's going straight. This represents the best possible way to see the back of your head in order to style your hair perfectly. Light leaving the back of your head is bent around the event horizon along this circular geodesic and ends up entering your eyes. But don't get your hopes up: Household black holes, for both hair styling and garbage disposal would be dangerous and expensive, and you'd probably have to get a special government license to operate them.

We are now nearing the end of our discussion of black holes, where, as promised, you know enough to absorb a better definition of them. When a star collapses far enough inward, it forms a black hole. Our new definition is this: *A black hole exists when a region of matter and energy (such as a star) is compressed into a region smaller than its own event horizon.*

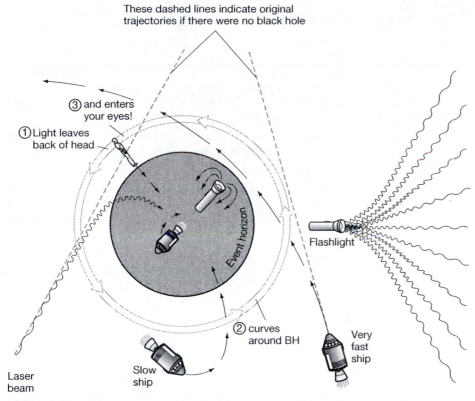

FIGURE 6.17 The Event Horizon   Take a moment to see what geodesic is followed in each case shown. Note gravitational red- and blueshifts, drawn here as wavelengths changing in flight.

The size of a black hole's event horizon is usually described by its **Schwarzschild radius** (the actual radius of the sphere), named after Karl Schwarzschild, who first computed it with general relativity. He obtained a fairly simple formula for the radius:

$$r_{\text{Sch}} = 2\frac{GM}{c^2} \ .$$
(eq. 6.1)

That is,

The size (radius) of a black hole = mass of the black hole
× a bunch of constants: $2G/c^2$.

For a black hole with the mass of the Sun, the Schwarzschild radius is only about 3 km. (The Earth's r_{Sch} would be only a few centimeters!) For a supermassive black hole of a few million solar masses, like the one at the center of the Milky Way, the radius is about 0.1 AU. The biggest known supermassive black holes in galaxy cores are about 1000 times that big, or 100 AU. That's a region only a few times bigger than the orbit of Neptune around our Sun enclosing the mass of *several billion stars*. In other words, in order to make a billion-solar-mass object capable of trapping light, that matter must all be compressed into a volume the size of a solar system—a volume that's usually occupied by only one star (or, more often, none!). This again is the general nature of a black hole: The mass–energy is crammed into a very small space—big M, little r_{Sch}.

EXERCISE 6.4 Little Green Men

Imagine two cities inhabited by little green men. (Other cities are inhabited by women, too, but in these particular cities it's just men.) One city is on its home planet, Greenland. The other city is on a space station, in a tight orbit around a nearby black hole. Both cities manufacture high-power telescopes that let the little green men in one city watch the little green men in the other city. The two cities are struggling for peaceful coexistence, and since you're studying the spacetime effects of their environment, they have consulted you on the issue.

(a) Consider social issues. What sort of skin color prejudice might arise between the two cities? Are there any other sorts of differences the two cities' inhabitants might notice in one another, based on watching each other move about by foot or by car?

(b) Consider communications. The two worlds have two-way radio transmission technology, so little green men can talk on cellular telephones across space. How should each city tune the frequency of its radio antenna to hear the other city? What will they sound like to one another when they speak on this interplanetary telephone? What do their radio stations sound like on opposite planets? For example, suppose that the orbiting city listens to its own radio station, KLGM, at a frequency of 98.5 megahertz (MHz), but this station is just static back on Greenland.

6.6 Quasars

At this point, we are equipped to interpret the light from another class of astronomical object, called a quasar. On the night sky, **quasars** are point-sized light sources, like stars, except that they have extraordinarily large redshifts. The name "quasar" was originally coined as shorthand for "quasi-stellar *radio* sources": pointlike objects discovered by their very bright radio emission. A related term, "quasi-stellar object" (QSO) was invented to describe similar *optical* light sources; one is shown in Figure 6.18. QSOs and quasars seemed substantially different because of the different frequencies at which they were discovered, but the physical source is probably very similar. So today, even professional astronomers often knowingly misuse the terminology by lumping everything into the category "quasar." If they can get away with it, then so can you, unless your instructor is an active researcher in this field, in which case you might want to learn the distinction for the sake of your grade.

According to the Hubble law we learned in Section 5.1, quasars must be extremely distant objects because of their large redshifts. Usually any distant object bright enough to see would be *extended*—taking up a large volume of space, like a galaxy or a cluster of galaxies. But quasars appear as tiny points. They are evidently very distant, very small, and—in order to be visible from very far away—incredibly bright.

Without general relativity, this combination of "small and bright" would present a problem, because it's hard to explain the existence of an ultrabright source without a *large* region from which to draw its energy. Stars, for example, operate by nuclear energy, but even a large galaxy full of stars is dimmer than a quasar. General relativity gives us the explanation. We need a highly compact and highly efficient source of energy, and that source is a black hole.

An isolated black hole is genuinely black, in the sense that it eats light and doesn't emit it. But a black hole surrounded by infalling gas is very different. The hole itself is the same—"eating" and not shining—but the surrounding gas is greatly energized by the hole. The hole doesn't shine, but the gas near the hole shines more efficiently than any other known source.

FIGURE 6.18 An Example of a Quasar Indicated by an arrow, the quasar is the most luminous and most distant object in this image. (Other objects appear brighter only because they're much closer.)

Accretion Power

Gas located near a black hole spirals its way inward due to the tremendous curvature of space (tremendous gravity) in that region. In so doing, it forms into the shape of a disk, for the same reasons discussed in Section 4.2 on the formation of disk galaxies. But unlike a disk galaxy, where things are more spread out and whose stars almost never come close enough to one another to interact, the black hole is compact, so the gas clumps interact frequently.

The process works in the following way: The black hole supplies all the energy with its gravity. This causes the gas to rush in. When that happens, there is a trade-off as gravitational energy is converted to kinetic energy and the infalling gas accelerates to high speeds. The gas, now arranged into a disk, undergoes many collisions—atom with atom, cloud with cloud—which preserve the high speeds, but randomize their directions. While the overall gas flow is still a spiral inward toward the black hole, a close-up view would show the gas particles bouncing around in a fast, random, and violent way. We saw long ago (Section 3.1) that fast-moving particles imply hot temperatures and that hot material glows as a bright blackbody. That's why we see bright emission from the vicinity of a black hole.

To summarize the process, we can say that the ultimate energy source is the gravity of the black hole, which generates the high speeds that in turn generate the blackbody emission. This entire process of gas spiraling into a black hole is called **accretion**, and the process takes the form of an **accretion disk** (a disk of material surrounding an object with a strong gravitational field). One can also say that the gas "accretes onto" the black hole.

Black holes are dynamic in other ways. Typically, an accretion disk around a black hole undergoes magnetic interactions as well. The interplay between the magnetic energy and the high-speed circling motion has been shown to produce jets—rapid, narrow, and ongoing blasts of matter away from the vicinity of the hole in the polar directions with respect to the accretion disk; these jets are depicted in Figure 6.19, which shows photographs in multiple frequencies, along with an artist's depiction of the situation. We now know of many such systems, some with jets pointing right at us, and others tilted away to show their disks instead.

Now pause a moment to appreciate how well the black hole explains quasar properties, without the need to introduce any new assumptions beyond the direct and necessary consequences of some well-known gas dynamics. General relativity explains the presence of the compact gravity source, needed to construct a compact light source. Gas speeds up as it follows a geodesic inward. The disk forms, the particles interact, and the temperature rises.

If you understand this process, then you understand *a lot*, because it pulls together many concepts from this chapter and earlier: point sources and angular sizes from Chapter 2; blackbodies and particle behavior from Chapter 3; disk structure from Chapter 4; the Hubble law from Chapter 5; and curved space and black holes from Chapter 6.

There is one more detail to fill in: A normal stellar-mass black hole isn't powerful enough to account for the bright, distant quasars we observe. We need a supermassive black hole instead. Major nearby galaxies, including our own, harbor ample evidence for SMBHs at their cores, so we speculate that quasars are the bright, active cores of distant galaxies. In other words, when we see a quasar, we're seeing light from a distant galaxy: We see the bright SMBH core region, while the rest of the galaxy is usually too faint to detect.

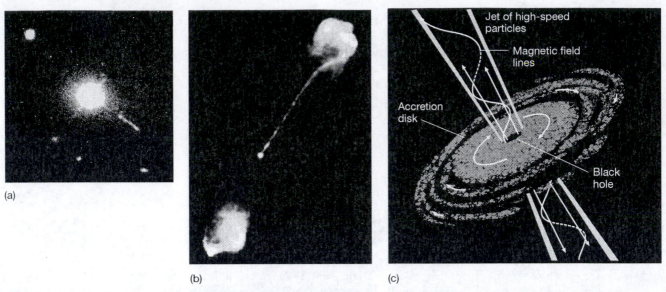

FIGURE 6.19 Accretion Disks With Jets (a) Quasar 3C 273, emitting a jet from its lower right. (b) Quasar 3C 175 as seen through a radio telescope. Consider the scale: The jets stretch for about a million light-years, while the galaxy quasar resides in is at least several times smaller than the jet and the black hole itself is at least a thousand times smaller than *one* light-year! (c) An artist's conception of how such jets are formed by quasars, with magnetic forces redirecting disk material into the jets.

In modern cosmology, quasars are very useful sources. Because they're so bright, they can be seen from a great distance, and cosmology is all about the quest to understand the universe on such vast scales:

- Quasars are important sources in redshift surveys (Section 5.2). These surveys can pick up quasars in addition to visible galaxies, and they allow researchers to obtain large-scale galaxy survey information, such as information about walls and voids, at greater distances.

- In Chapter 8, we'll encounter quasars again, where they'll afford us an important insight into how light is affected by intergalactic material during long journeys. (This insight will help us understand the light from the even more distant microwave background we encountered in Chapter 5).

- In Chapter 10, we'll see that quasar light is very useful in estimating the cosmic abundance of heavy hydrogen, because the quasar light travels through so much of the universe on its long journey here. (The resulting information about heavy hydrogen in turn will yield important evidence about the very early universe and will help solve the mystery of the abundances of the light elements, from Chapter 5.)

Right now, we'll use quasars for one more purpose before we close out the chapter: as a means for measuring great distances. This is an important task in cosmology, as you know from Chapter 3, because distance measurements are always needed, but are not always easy to make.

Supplemental Topic: Gravitational-Lensing Time Delays

Equipped with a knowledge of quasars and with an understanding of gravitational lensing from Section 6.4, you have earned the *absolute* distance indicator to be discussed next. Unlike the *relative* indicators on the distance ladder in Section 3.3, an absolute indicator does not need to be calibrated by other techniques, which means that it has fewer types of uncertainty.

Figure 6.20 shows the effect of gravitational lensing on a quasar. Distant quasars can be multiply imaged by a massive galaxy in our line of sight, so that you can see several images of the *same* quasar. This is the setup we need in order to obtain distance measurements. The figure also shows the arrangement that produces doubly lensed quasar images. This arrangement explains the multiple imaging, since the light from the same quasar approaches the observer from four different directions, resulting in four different spots on the sky. This type of arrangement is particularly useful if one light path is longer than the others. Many such systems have been identified.

Now, suppose that something about the source quasar changes suddenly. For quasars, this happens quite often, and you can probably imagine why if you think about how they produce their light. The gas flow into the central SMBH determines the quasar's light production; if the gas flow is disturbed (a little more gas here, a little less there), then the light will temporarily brighten and/or dim.

The distance indicator arises when we observe this glitch (the quasar light brightening or dimming) in two or more lensed images *at different times*. Because the light paths are unequal, the same glitch from the same quasar typically shows up several *years* apart in the two images. Obviously, to obtain these measurements, astronomers have to monitor the same quasar over a long time. The light travel times for the two images we see are affected by the differing path lengths *and* the amount of *gravitational time dilation* along each path (Section 6.5) through the lens's gravitational field. Although this sounds complicated, you've already studied all the relevant physics. The theoretical reasoning and the associated math are well established.

If researchers can figure out the structure and the distance to the object doing the lensing—typically a galaxy or a cluster—then they can calculate the difference in light path lengths and, finally, the distance to the source quasar. Such calculations are often possible, because (a) we may be able to see the lens object directly and (b) the manner in which it warps the background quasar light tells us about how its mass is physically arranged in space. (This, by the way, is an excellent way to study dark matter as well, by mapping its spatial distribution in the lens object.) Things also get easier if there are more than two lensed quasar images.

Gravitational-lensing time-delay systems combine a number of exotic studies, including the gravitational bending of space and time, the variability of quasar light with time, the spatial distribution of dark matter, and light travel across cosmic distances.

FIGURE 6.20 Einstein Cross   At the left is the image formed of the distant quasar, split into four images by the foreground galaxy. At the right is an artist's conception of the lensing geometry that produces this image.

Over time, we should expect more and more interesting scientific information from these studies. As a distance tool, such systems measure the largest scales we are capable of measuring today.

Einstein's Legacy of Unification

Nature does not always show us directly how the physical world works. In any given experiment, the details can obscure the more fundamental science we seek to uncover. Consider an example from the beginning of the chapter: Drop two different objects side by side to see how the Earth's gravity affects them. It matters tremendously which two objects you choose, because if you choose a bowling ball and a feather, then the bowling ball will win the race and lead you to incorrectly conclude that gravity makes more massive objects fall faster. In this case, the effect of air resistance is a powerful distraction from the nature of gravity, which was what you originally sought to investigate.

So pernicious was the mask of air resistance, and that of our instinct gleaned from the experience of lifting heavy objects, that it took our species until 1585 to actually discover the "true" nature of gravity, which was then improved by Newton in 1687 and subsumed by Einstein in 1916. But we're not done yet: Physicists today agree that general relativity isn't the final word and that it, too, will eventually be incorporated into a grander understanding called *quantum gravity*—an attempt to unify the theories of general relativity and quantum mechanics. And even then, how will we know if we've got the theory of quantum gravity right?

We might seek some comfort in the scientific approach by looking for a general pattern that accompanies our successes. When Newton formulated his theory of gravity (which was an extremely important scientific advance, even if it was eventually replaced with something better), it provided a single construct for understanding multiple different phenomena in nature. In particular, it linked the properties of planetary orbits to those of objects under Earth's gravity—objects we can experiment with in a laboratory. Planets orbiting the Sun share the same basic physics as a thrown rock. We might take comfort in the fact that major advances in science often *unify* such formerly disparate phenomena as these.

In the epic saga of scientific unification, Albert Einstein is the protagonist. In his theory of special relativity, he united the formerly unrelated concepts of space and time by determining how they would be experienced at different velocities. When he generalized this theory to account for acceleration, it became known as general relativity—the union of space, time, and gravity, all expressed with geometry. And as we'll see in Chapter 9, Einstein's unification story goes further: With the most famous equation of them all, $E = mc^2$, he showed that matter (m) and energy (E) are just different manifestations of the same essence.

Matter and energy, spacetime and gravity—those are some of the biggest, but there are other brilliant unification stories, too. Some have already been written, and some have yet to occur. In the human enterprise of science, it has become clear that our greatest progress is that which merges avenues of past research onto the same highway. You could argue that your own route to wisdom and success has the same merging structure, which is an excellent reason for studying cosmology.

Quick Review

In this chapter, you've learned that gravity can be interpreted as the bending of space and time in the presence of matter and energy, rather than as a force. By characterizing gravity in this new way, general relativity has been able to explain observations that Newtonian gravity cannot explain. The insights into space and time provided by general relativity will play an important role in understanding the overall nature of the universe.

Try the following crossword puzzle to test your knowledge of key terms and concepts from Chapter 6:

ACROSS

5 Principle linking the effects of gravity and acceleration
6 Shortest path between two points
11 Acronym for the type of black hole often found at the centers of galaxies
12 When gas spirals onto an object with strong gravity, like a black hole
13 Gravitational ____ are ripples in space-time curvature traveling away from some source
14 How photons lose energy as they fly away from a strong gravity source
17 Change in velocity
18 Type of curvature in which a triangle's interior angles add up to more than 180 degrees
19 "Normal" geometry, applies to a flat universe
20 Most natural state of motion in general relativity

DOWN

1 Gravitational microlensing is a good way to look for this type of dark matter.
2 Inventor of general relativity theory
3 Gravitational bending of light that changes the appearance of a distant light source
4 Type of curvature in which parallel lines diverge
7 Radius of a black hole

8 General relativity depicts gravity as a result of ____, rather than as a force.
9 Tiny, distant, and extremely bright light source
10 Considered to be the edge of a black hole
15 Useful wavelength range for observing black holes
16 Resistance to changing a state of motion

Further Exploration

(M = significant mathematical content; I = integrative—builds on information from more than one chapter; P = project idea; R = reflection; D = suitable for class discussion)

1. (R) Describe what a geodesic between two points looks like in a flat space. Draw two points on your paper, and connect them with a geodesic. Now repeat the same task, but this time bend the paper into a tube shape before you try to connect the dots. Is your new geodesic different? Why or why not? (You can unroll the paper to compare it with the flat page.)

2. (RD) Use the expanded concept of weightlessness in general relativity to explain why astronauts feel weightless inside their orbiting spacecraft. (*Hint:* It is not because they're far from a gravity source! In fact, the acceleration due to gravity at the international space station's altitude is about 8.7 m/s^2, which is not all that different from 9.8 m/s^2 on the ground! The question therefore becomes, How does being in orbit prevent moving toward Earth at 8.7 m/s^2?)

3. (R) Answer the following questions about geodesics on Earth:
   (a) Are any, or all, latitude lines great circles? Show your reasoning by sketching flight paths between a few pairs of cities located at the same latitude. Choose one pair of cities, both of which are on the equator.
   (b) If you extend the line for a flight path in both directions, what happens?
   (c) Starting right where you are, if you could walk any distance on Earth without turning (or drowning), which directions would send you on a great circle?
   (d) Imagine that you could turn off gravity and stick a metal marble on the surface of a smooth magnetic globe. Think about what happens if you flick the marble in some random direction. How is its motion captured by the concept of a geodesic, the shortest possible path in a curved space? Would the same thought experiment work on a magnetic saddle? What if you flick two metal marbles at the same time and in the same direction on a magnetic globe or a magnetic saddle?

4. (P) The existence of black holes has been a topic of debate since 1939, when Robert Oppenheimer and Hartland Snyder published a paper arguing theoretically (using Einstein's general relativity) that black holes should exist as the collapsed final state of very massive stars. That same year, Einstein independently published a paper arguing that such objects could not exist. Beginning with the detection of Cygnus X-1 in 1970, a growing body of circumstantial evidence convinced most astronomers that stellar-mass black holes really do exist.

Look up the evidence for stellar-mass black holes, and write a page summarizing your view on the issue and the evidence supporting your view. A good place to start your research is "Unmasking Black Holes," by Jean-Pierre Lasota, *Scientific American*, May 1999, pp. 40–47.

5. (P) As a follow-up to the previous question, what about supermassive and intermediate-mass black holes? How might we confirm that they really are black holes? On the basis of current observations, could they be something else? Write a page summarizing the evidence for SMBHs and IMBHs, including your view of how convincing the evidence is. Here are some suggested places to start, although you can find your own starting points with a Web search:
   • For SMBHs, Laura Ferrarese and David Merritt, "Supermassive Black Holes," *Physics World*, June 2002, pp. 41–46.
   • For IMBHs, http://www.astrobio.net/news/article276.html, from *Astrobiology* magazine.

6. (MR) Is the surface of a cylinder a curved space or a flat space? What about the flared end of a trumpet? Justify your answers with the properties of curved spaces discussed in Section 6.1.

7. (MRI) Consider two SMBHs orbiting one another.
   (a) Using Newtonian gravity and/or Kepler's laws, argue that two isolated SMBHs will never collide or even change orbits. You may assume that one hole is much more massive than the other if that helps.
   (b) Using general relativity instead, explain how the binary SMBH system will lose energy over time, starting with the continual reshaping of spacetime in its vicinity. Since this energy loss leads to less gravitational binding energy (careful: remember that E_g is a negative number!), show that the black holes will eventually merge.

8. (RI) Redraw or photocopy Figure 4.25. Use the principles of general relativity from this chapter to sketch a few photon geodesics across the region shown, with different starting and ending points.

9. (R) Why do MACHO observations by gravitational lensing produce only a brief brightening of the source object, rather than creating a visible, lasting Einstein Cross?

10. (R) Apart from MACHO lensing searches, how can the phenomenon of gravitational lensing inform us about locations and amounts of invisible dark matter?

11. (R) Consider the Einstein cross in Figure 6.20. Under what conditions would you see a whole ring in your image, rather than just four points along it, as in the cross?

12. (MR) Use the surface of a sphere, such as a globe, as a sample geometry to demonstrate that the circumference of a circle is less than $2\pi r$ in a positively curved space. (The reverse is true in a negatively curved space; in that case, the circumference is greater than $2\pi r$.)

13. (R) If a black hole is curved such that photon geodesics never leave the hole (it's called "black" for that very reason), then why do we suspect that black holes serve as the light *sources* for quasars, which are incredibly bright?

14. (R) Use black holes to argue that a redshift could be *faked* by the shape of spacetime and that the redshifted source need not be traveling away from you, as the Doppler effect interpretation would have us believe. (This problem is a warm-up for the next chapter.)

15. (RD) If the Galaxy were littered with lots and lots of black holes, how would we know? Would we know? Try to think of several different ways to answer.

16. (R) According to the theory of general relativity, why do you fall back to Earth after you jump?

17. (R) The following questions are designed to help you understand general relativity with your body, heart, and soul:

    (a) If gravity is not a force, and your sense of weight is based on the hard Earth's ability to keep you from your natural state of free fall, then why do your arms fall back to your sides whenever you stop holding them in some other position?

    (b) If gravity is not a force, then how do you explain the uncomfortable feeling you get when you're falling (or when you dream that you're falling)? Some describe this feeling as "my heart jumped into my throat."

    (c) If gravity is not a force, then how might your soul drift upward (if you've been good!) when you die? For this question, you may assume that you have a soul and that it's lighter than air, like a helium balloon.

18. (MI) It's surprisingly easy to obtain the radius of a black hole's event horizon with a Newtonian formulation. (Normally, one shouldn't put too much faith in an approach that excludes general relativity, but it happens to provide the correct result in this case.) Recall from Section 4.1, Equation 4.5, that the total energy of a system of two bodies in space is the sum of the kinetic (motional) energy and the (negative) gravitational binding energy. Rewrite that equation so that r refers to the radius of an event horizon—the precise boundary between where light can escape and where it cannot. This means doing two things. First, recognize that the velocity in question is the speed of light, $v = c$. Second, since we're interested in the border line between bound (trapped, negative energy) and unbound (escaping, positive energy), we seek the case where $E_{\text{total}} = 0$. Proceeding in this way, derive the Schwarzschild radius, Equation 6.1.

CHAPTER 7

An Expanding Universe

> *"At first glance, all of this sounds like medieval mystics discussing the music of the spheres, angels on the head of a pin, or some similar early approach to cosmology. Is it just a mathematical game we are playing, is it just semantics, or is it reality?"*
>
> —Leon Lederman and David Schramm

This chapter is really a continuation of the last one, which armed us with Einstein's theory. General relativity describes the nature of space, time, and gravity. In the last chapter, we applied general relativity to localized phenomena, such as gravitational lenses and black holes. All that remains is to apply the theory to the universe itself.

What follows might feel a bit strange. How do you describe a "spacetime?" How can you write down some math and come away with knowledge of the universe? Perhaps it seems self-aggrandizing to develop "the universe equation" and then quote its results—but that's exactly what we're about to do. The universe is a *system*—like a star, or a software package, or a living cell. Just because it's a big system doesn't stop us from learning its behavior.

7.1 Spacetime and the Cosmological Horizon

Inside a black hole's event horizon is a region whose events are disconnected from the rest of the universe (Section 6.5). The term "horizon" indicates a fundamental barrier between what we can observe (outside the hole) and what we can't (inside). However, it is not the only fundamental barrier of this sort in the universe: There's also a *cosmological* horizon. To understand what the cosmological horizon is, we have to mix time more explicitly into our thinking.

To specify a location, one must give three coordinates: length, width, and height; x, y, and z; or latitude, longitude, and altitude. If you go to 40° north latitude, 75° west longitude, and 30 meters above sea level, then you'll be in Philadelphia, but you won't witness the Continental Congress signing the Declaration of Independence. For that, you have to move four dimensionally and specify the event's four coordinates by

194

adding time: July 4, 1776. Locations have three coordinates, and **events** have four. We can think of all activity in the universe as a collection of events that exist within a four-dimensional coordinate landscape called **spacetime**.

The specific choice of event coordinates you use depends on your frame of reference. For example, you might measure distance from your residence, using your watch to tell time, while another observer might measure distances from Tokyo and use Tokyo time instead. He might even be one of those people who sets his watch five minutes fast in order to avoid being late. The point is that anyone, from his or her *own* perspective, can pinpoint any event. By contrast, sharing knowledge about that event with someone in a different reference frame is trickier. We've already seen that space and time "flow" differently for different observers, like the little green men living near a black hole in Exercise 6.4, so it's useful to know how to translate a description of events from one reference frame to another.

One way to communicate about events is with a graph called a **spacetime diagram**, which is similar to a standard map you're more familiar with, except that it also includes *time* as a coordinate. Just as you can locate a specific point in space as a dot on the map (e.g., Philadelphia at 40° north, 75° west), a spacetime diagram represents a certain place *and time* as a dot (e.g., Philadelphia in 1776 is represented by a different dot than the one that represents Philadelphia in 2035).

A complete spacetime diagram would need to be four dimensional (to include all three directions of space and one of time), which is a little hard to draw. So in practice we never draw the whole thing. But you're familiar with this kind of limitation, too: A typical map just gives latitude and longitude, leaving out elevation. (This is an omission that becomes apparent if you're using the map to plan a walk between the north rim and the south rim of the Grand Canyon, which might be only about 10 miles apart on a map, but the down-and-up dimension makes a big difference!) In a typical spacetime diagram, time appears on the *y*-axis and distance along some spatial direction is plotted on the *x*-axis. As an example, the vertical line in Figure 7.1 describes the thrilling life of a large rock that remains stationary relative to us. The rock just sits there while the years go by, so its path in the figure proceeds forward in time (up–down), but doesn't change in space, or location (left–right).

In astronomy, it's sometimes convenient to use years as units of time and light-years for distance. In that case, light, or anything else that travels at light speed, follows a path on the spacetime diagram for which the change in physical distance is equal to the change in time. That's because a photon travels 1 light-year in 1 year (by definition), so we have 1 = 1. On a spacetime diagram, a photon traces out a line with a slope of 1 (see Figure 7.1), because its rise-over-run is 1-over-1, or 1/1, which equals 1. If the light changes direction by reflecting against a mirror, then its slope will be $1/(-1) = -1$. A piece of matter, like a penguin, moving more slowly than light, would have a *steeper* slope, since it would cover less distance than light would in some fixed span of time (less run per rise).

The lines shown in the figures, and any lines on any spacetime diagram, are called **worldlines**. So far, you've seen the worldlines of a rock and a photon; Figure 7.2 is a spacetime diagram of the Earth and the Sun. What would the worldlines of photons look like in this diagram? Would their slopes be steeper than the Earth's up-spiraling line? Would they spiral as well? What *possible* slopes would exist on this diagram for the worldline of a spaceship leaving the Earth? This is valuable stuff to think about; it's worth your time to try to answer these questions for yourself before going on.

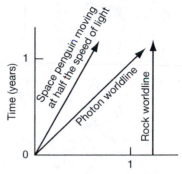

FIGURE 7.1 Spacetime Diagram for Light and Matter   How far does light travel in one year? According to the photon's worldline, angled at 45° on a spacetime diagram, it travels a distance (horizontal axis) of one light-year in a time (vertical axis) of one year. The worldline of matter, however, is steeper, because matter travels less distance (horizontal axis) in some amount of time than light travels. On a spacetime diagram, steeper means slower. The position of a rock remains fixed (with respect to our location) as the years roll by, so its worldline is a vertical line.

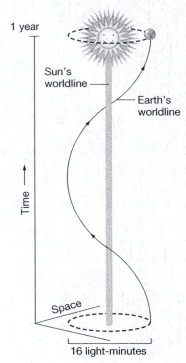

1 year

Sun's worldline

Earth's worldline

Time →

Space

16 light-minutes

FIGURE 7.2 Spacetime Diagram for the Earth Moving Around the Sun Here, we plot two spatial dimensions (both horizontal) along with the time dimension (vertical). The Earth completes one orbit every year, returning to the same place (with respect to the horizontal axes), but at a later time (vertically higher). That is, the endpoint (top) of one orbit is directly above the starting point (bottom).

For the remainder of this section, we apply the unified spacetime concept to *a* universe—but not our own universe. We'll look at a *static universe* first, because it's easier to think about. In the next section, we'll upgrade to an *expanding universe*, like the one in which we actually live.

In a static universe, space never changes: A stationary galaxy 2 billion light-years away will always be 2 billion light-years away. We will also suppose that our hypothetical universe has a *finite age*. That means that it hasn't always existed.

Let's say that our pretend universe is about 14 billion years old, as our actual universe appears to be. Imagine a galaxy 14 billion light-years away from us, shining light in every direction, including toward us. When this universe popped into existence, complete with full-grown galaxies (hypothetical universes can do that), light from that galaxy began its journey, and those photons are just now arriving here. Another galaxy, located 15 billion light-years away, also began sending photons our way 14 billion years ago, but since 15 billion years haven't elapsed yet, those first photons haven't arrived here yet, so we can't see that galaxy.

Any source whose light we can't yet see lies beyond our **cosmological horizon** (or "cosmic" horizon, sometimes)—the boundary between what we can and can't see. The distance from here to our cosmological horizon defines the size of the **observable universe**. Be it our hypothetical static universe or any other, the observable universe is the only region that we can see, no matter how good our telescopes are, because only light from within the observable universe has had enough time to get here. As such, the observable universe is a spherical region with the Earth at its center, 14 billion light-years in radius for a static universe. The horizon is a boundary in *space*, but it originates from a boundary in *time*, illustrating again the close relationship between space and time.

Warning: Very often, when magazine and newspaper articles talk about "the universe," they are referring only to the *observable* universe. These sources frequently say things like "when the universe was *x* seconds old, it was only the size of *y*." So far, however, the data allow the possibility that the universe is infinite in size, in which case the size of *y* had better be infinite, too (unless it started out finite and then somehow became infinite later on). But the *observable* universe is always finite, provided that its age is finite, so the observable part would have been small in the early universe. When the hypothetical static universe was seven years old, for instance, the observable universe was seven light-years in radius. As the universe got older, its horizon grew, although nothing got any bigger physically. A billion years into the future, the horizon would be 15 billion light-years away in every direction. (Can you articulate why?)

For the case of our own *expanding* universe, the situation is more complicated. The distance to the horizon in light-years is not just the age of the universe in years. Instead, the relationship between horizon size and age depends on the details of how the expansion proceeds over time. You are now ready to examine this more realistic case; it's time to get acquainted with your physical universe.

Exercise 7.1 Consider Olbers's Paradox Again

Refresh your memory on Olbers's paradox from Section 5.6, and consider how the notion of a cosmological horizon might affect it. How are the paradox and the cosmic horizon related? Under what circumstances would the horizon concept help resolve the paradox? Construct a spacetime diagram to illustrate your reasoning, and distinguish clearly between those sources we can see and those we cannot. Also, distinguish clearly between "here" and "there" and between "then" and "now" in your diagram.

7.2 Cosmic Expansion

At the end of Section 6.3, we pointed out that general relativity implies a profound conceptual shift in our understanding of what space is. One important consequence that we'll examine later in this section is that it's very natural in general relativity for space to expand and contract, like a rubber sheet stretching or shrinking.

At this point, it's fair to point out that the expanding universe is a difficult concept to work with, even for professional scientists. Many people casually throw around the words "expanding universe" without knowing what it means or why it's happening. They view the whole idea as some obscure megagenius concept that *they would never understand even if they tried*, so they have to just accept it. But we don't want *you* to feel that way! It's true that this subject matter is complicated and that understanding every detail generally requires many years of dedicated study. However, you can get a satisfying *sense* of the expanding universe without all that effort. That's the purpose of this chapter, particularly this section.

An expanding space is like a photocopy enlargement: Two different marks on the page don't get up and walk away from each other across the paper, but they end up farther apart after the photocopy anyway. A ruler would count off more inches in between the marks after the copy. One difference between a photocopy enlargement and some scene from an expanding universe is that objects in the universe generally *can* get up and "walk across the page," too. In the universe, when the distance between two galaxies gets larger, it could be because (a) they're actually moving away from each other, passing other galaxies as they travel, or (b) they're sitting as still as they can, and it's space that's expanding.

Recall what we already know from general relativity: A particle moves on a geodesic trajectory, which is the closest thing to a straight line that's possible in curved space. The new concept to add to your library of cosmology knowledge is this: Particles continually get farther away from each other, which is the closest thing to sitting still that's possible in an expanding space. A "particle" in this sense can be anything—a photon, a galaxy, a penguin—whatever.

Now, it is possible to *prevent* the distance between particles from growing. This can be accomplished with rockets or explosions, or with something more subtle. For example, the trees in a forest aren't expanding away from each other, even though they inhabit an expanding universe. They remain in place because they're holding onto the ground with their roots and the ground is holding itself together, using interactions between molecules in the soil and using the Earth's gravity. In other words, there are stronger influences on a tree than cosmic expansion. It's similar to the way a wooden bridge pier doesn't get carried downstream by a river: The pier is bolted in place, but if it weren't, then it would be carried along with the current.

Gravitational curvature keeps the Moon from flying away from Earth in the expansion. Earth curves space, and that curvature has a stronger effect *near Earth* than the expansion does. Although the space around the Earth would expand as the universe expands, the curvature that keeps the Moon here continually "updates itself" to keep its range of influence the same, keeping the Moon where it is. After all, that curvature is caused by the Earth's mass and energy, which are not changing significantly over time. Put another way, the expansion changes space throughout the universe, but masses like the Earth have a local effect, controlling just the space around them.

The Scale Factor

For now, let's proceed by considering large and predominantly empty regions of space, where the expansion is the dominant process. In such places, the effect of the expansion is quite simple: It is a continuous *scaling up* of distances in space, like a photocopy enlargement. The mathematical construct that cosmologists use to characterize this expansion is called the **scale factor** of the universe, usually denoted with a lowercase *a*, sometimes written as *a(t)* to remind us that the scale of the universe changes as a function of time. Amazingly, this one number *a* describes the bulk expansion of the entire universe, with distances growing in the same way everywhere in space and along every possible direction. If the scale factor doubles, then the distance between two distant galaxies doubles. If the scale factor decreases, which might happen someday, but probably won't (Chapter 11), then the contraction is equally simple: All distances shrink by the same amount, everywhere and in every direction.

Here's how you use the **scale factor** *a(t)*: The distance between two stationary objects—usually two galaxies—can be computed at any time, past or future, if we know that distance today. The distance at any time is just the distance today, multiplied by the scale factor for that time:

$$\text{Distance at time } t = \text{Distance today} \times a(t) \ . \qquad \text{(eq. 7.1)}$$

Mathematically speaking, if we denote distances with the *x*-coordinate (why not?), then

$$x(t) = x_{\text{today}} \times a(t) \ . \qquad \text{(eq. 7.2)}$$

Cosmologists have chosen to represent the present time with a zero, which means that we write

$$x(t) = x_0 \times a(t) \ . \qquad \text{(eq. 7.3)}$$

We also choose to set $a = 1$ for the present time, when $t = 14$ billion years (an estimate of the age of our real universe). So it follows that, in our expanding universe, *a* used to be a fraction smaller than 1 and will be bigger than 1 in the future.

In Figure 7.3, the scale factor goes from $a = 1$ (today) to $a = 2$ (in the far future), showing how galaxies that are "trying" to sit still end up at double their current separation. For this calculation, you don't ever need that actual time in the future; that is, you don't need to know exactly when the universe doubles in size. It might be tempting to assume that *a* becomes 2 when the age of the universe doubles from 14 to 28 billion years old, but this is incorrect. The relationship between *a* and *t* is more complicated, but that doesn't prevent us from *using a* for scaling, as in the figure.

The Einstein and Friedmann Equations

As we mentioned at the start of this section, the expansion is not an extra feature of the universe, beyond those predicted by general relativity. It is *because of* general relativity that the universe expands. Matter and energy tell spacetime how to curve, and curved spacetime tells matter how to move. We've examined this premise many

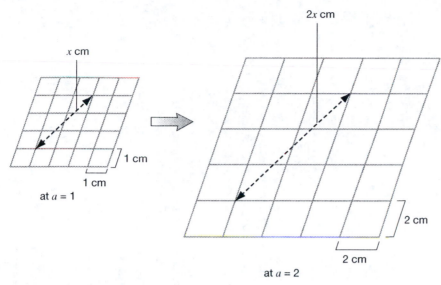

FIGURE 7.3 An Expanding Space   The growth of the scale factor means that all the relative positions remain the same, but all distances get bigger by the changing scale factor. In the figure, the scale factor doubles, so every distance doubles as well.

times now, but so far we've looked only at cases where some individual object, like a black hole, causes *local* curvature. Now let's apply the idea to the entire universe all at once. Within the context of general relativity, the matter and energy content *of the entire universe* curves the overall fabric of spacetime *everywhere*, leading to some particular geometry, and expansion, for the universe overall.

General relativity describes the universe mathematically with the **Einstein equation**. The details are complicated (too complicated even for most undergraduate physics majors), but the insights we need can be seen best by looking at the equation conceptually:

> Very complicated math describing the curvature of spacetime = Very complicated math describing the distribution of matter and energy in spacetime   (eq. 7.4)

The curvature of spacetime is what determines the geodesics along which particles travel, meaning that it defines the gravity at every location. The right side of Equation 7.4 describes the matter and energy content of space. So the Einstein equation links the universe's contents to its curvature; it quantifies what we've been talking about since the last chapter: The distribution of matter and energy tells spacetime how to curve, and the curvature of spacetime tells matter how to move.

Here's the new part: Cosmologists can try to solve the Einstein equation to figure out the overall curvature of spacetime. We'll soon see that this boils down to solving for (drum roll, please...) the scale factor $a(t)$, which will turn out to be very enlightening. It's possible to solve the Einstein equation if one asserts that the universe is *homogeneous* (looks the same from all locations) and *isotropic* (looks the same in all directions) on large enough scales. (See Figure 7.4.) Taken together, these two properties are known as the **cosmological principle**, which says that, on large scales, all places in the universe are alike. Today there is good direct

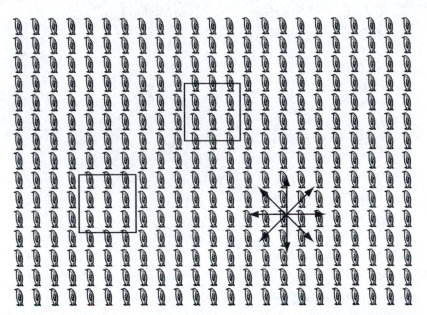

FIGURE 7.4 Homogeneous Pattern This penguin pattern is *homogeneous* on scales larger than the size of a penguin: The boxes shown all have about the same contents, no matter where they are located. But the pattern is *not isotropic*: From any location, the view looks different in different directions (such as the directions indicated by the arrows). In a universe that satisfies the cosmological principle (on large scales), the distribution of material would be both homogeneous and isotropic everywhere.

evidence for both homogeneity and isotropy from galaxy surveys; you might revisit Section 5.2 to remind yourself about this important evidence. Effectively, these two assertions mean that wherever you go in the universe, and whatever direction you look, you'll see galaxies, galaxies, and more galaxies, off into the distance. So as long as you are looking at properties averaged over a large enough volume (a few hundred million light-years will do), the universe looks roughly the same everywhere.

It's similar to the density of air in a room. If you go to very small scales, the situation varies dramatically as you change location or look in a different direction. One place might be inside the dense nucleus of an atom, while another is empty space. So the density is quite different in different locations. But if you average over a volume the size of a shoe box, the density of air is very nearly the same everywhere in the room: A shoe-box-sized region at one end of the room has basically the same arrangement of air as one at the other end. Similarly, the universe is very uniform as long as the size of the things you're looking at is bigger than a giant's shoe box that's a few hundred million light-years across.

With homogeneity and isotropy, any changes in space are the same in all directions, and the equation for "how spacetime bends" as a result of "how matter and energy are distributed" takes on a much simpler form. In that case, the Einstein equation can be boiled down to a simpler expression, called **the Friedmann equation**. As we did with the more general Einstein equation, we'll write the Friedmann equation in conceptual form, with the left side describing the curvature of spacetime, while the right side describes the distribution of matter and energy in the universe. But this time we can be more specific about what the terms on each side are, because there are only a few to consider:

| Slightly complicated math describing how the scale factor (a) changes with time + the curvature of space | = | Slightly complicated math describing the average density of matter and energy + the average density of dark energy   (eq. 7.5) |

Asserting that the large-scale universe contains no special places and no special directions allows us to capture all the dynamics of spacetime in terms of a single parameter: the **scale factor** a. It also allows us to express the distribution of matter and energy as just an average density—how much stuff there is, typically, in a certain volume of space. (Just ignore the "dark energy" density for now; we'll return to that a little later.) In fact, the Friedmann equation can be viewed as a recipe for calculating how the scale factor changes with time, depending on the density of matter and energy. In other words, this equation is the most compact way to see how everything affects everything else in the universe. (We'll discuss this point in much more detail when we consider the fate of the universe in Chapter 11.)

Here's where we are right now:

- General relativity, whose predictions (like gravitational lensing and black holes) have been repeatedly validated by observations, is mathematically expressed by the Einstein equation.

- When the Einstein equation is mixed together with two observations—that the universe is homogeneous and isotropic—it can be manipulated until it becomes the Friedmann equation.

- The Friedmann equation includes the scale factor $a(t)$, which depends on *time*. Therefore, solving for $a(t)$ tells us something about how the universe *changes* over time. This is where the insight we've been promising from general relativity reveals itself. According to general relativity, it's natural for a to change with time. If a is increasing with time, space is expanding. If a is decreasing with time, space is contracting. A static universe would be a special case in which the other parts of the Friedmann equation exactly balance out so that the scale factor remains constant in time. (There is an analogous feature in Newtonian cosmology—without general relativity—that makes it difficult for galaxies to remain static: The gravity force is always attracting them together. You're familiar with this effect from simply throwing a ball in the air while Earth's gravity acts on it: It's easy to make the ball rise (analogous to an expansion of space) or fall (analogous to a contraction of space), but very difficult to make it hover motionless in midair. The situation is similar in general relativity, but now it is *spacetime itself* that cannot remain static.)

First let's take a closer look at the Friedmann equation, to better understand what each part means. The left side expresses the structure and dynamics of spacetime:

- *How the scale factor changes with time.* This is the *time* part of the spacetime curvature. It tells us how the size scale of the universe changes. That is, it tells us how fast the universe expands or contracts.

- *The curvature of space.* This is the concept we talked about in Section 6.1. The curvature of space can be positive (like the surface of a sphere), negative (like a saddle), or zero (flat). But here it's a *global* curvature, describing the shape of space everywhere (rather than a *local* curvature, caused by some nearby object like a star or a black hole).

The right side of the Friedmann equation speaks of matter and energy—things that affect the dynamics of spacetime:

- *The average density of normal matter and energy.* This is how empty or full the universe is, on average. To compute it, you take a huge region of the universe (an imaginary box 100 million light-years on a side, for example), add up all the matter and energy in it, and divide by the volume. More density means more stuff in the same box, for any such large box anywhere in the universe. Density is often referred to by the symbol ρ (Greek lowercase "rho").
- *The average density of dark energy.* A measure of the amount of "dark energy" in the universe. Dark energy is a special type of energy that tends to make space expand faster; we'll address this topic next. Dark energy is often represented by the symbol Λ (Greek uppercase "lambda").

The last two points deserve some attention. The presence of matter and energy tends to curve space positively; that is, both matter and energy gravitate, pulling things inward, towards themselves. This inward pull should slow down the expansion. **Dark energy** does the opposite; it is an unknown agent that speeds up the expansion. What's interesting right now is the reason the Λ term appears in the equation at all.

Today we've actually *measured* the expansion of the universe and determined that dark energy is needed to produce the observed rates. But before those measurements, Einstein still referred to Λ, which he called the **cosmological constant** (i.e., just some attribute of the universe, like the speed of light or the charge of an electron). In fact, that was before scientists even knew about the expansion at all. There was no particular reason to stick the cosmological constant in the Friedmann equation, except that Einstein himself expected a *static* universe! After all, why would space be expanding? He realized that without the extra Λ term, the equations did not readily yield a static universe. The scale factor a would be changing if normal matter and energy were the only things in the universe.

You can see all this for yourself in the Friedmann equation. Suppose that the universe is *flat* (as it appears to be in daily life and as our best cosmological measurements indicate today), so that the curvature of space is zero. Then any nonzero amount of matter or energy—and you personally *are* a region of nonzero matter and energy in the universe—would require a nonzero expansion rate. Otherwise, the equation would say that zero on the left side of the equation equals nonzero on the right side, and then the equation wouldn't be true: Nothing doesn't equal something.

Einstein introduced the term we now associate with dark energy so that there would be something to oppose the contracting effect of the normal matter and energy, ρ, in order that the end result would fit a static universe. (In fact, he also needed the space curvature term in his model.) The normal matter and energy would tend to contract the universe, while the dark energy would tend to expand it, and the result would come out perfectly balanced in between. The two energies would somehow tie, so that the expansion could be zero, too. That was the idea.

Aleksandr Friedmann knew of Einstein's work on a general relativity–based static universe (and the related work of another researcher named Willem de Sitter). But he realized that a much more natural set of universe models was possible (with the Friedmann equation) if you allow for a *nonstatic* universe. He reported on his nonstatic universe as follows:

> "The goal of this Notice is...the proof of the possibility of a world whose space curvature is constant with respect to three coordinates that serve as space coordinates and dependent on the time, i.e., on the fourth—the time coordinate....The assumptions on which we base our considerations...coincide with Einstein's and de Sitter's assumptions [including the existence of matter in the universe]. . . ."
>
> A. Friedmann, "On the Curvature of Space"

Friedmann goes on to discuss a few possible solutions to his equations. In one of these solutions, he indicates the need for an *expanding* universe:

> "It follows that [the scale factor] is an increasing function of *t* [time]."

Einstein initially thought that Friedmann's universe looked "suspicious." But then Friedmann wrote to Einstein, to explain his calculations in detail, and Einstein publicly admitted that he was wrong and that "Mr. Friedmann's results are correct and shed new light."

The importance of all this is as follows: The observed expansion is *not* a strange feature of the universe, as it may have originally seemed. Rather, it is *expected*, given our understanding of spacetime and gravity from general relativity. The Friedmann equation predicts a nonzero expansion; that is, *the scale factor (a) increases with time*.

Aleksandr Friedmann *(1888–1925) was a Russian scientist who developed not only the equation that bears his name, but, along with it, a class of models of the expanding universe. Together with other researchers, he constructed another, general equation for a general relativistic universe, thereby providing a mathematical basis for modern cosmology.*

As we mentioned earlier, even in the Newtonian picture of gravity (before general relativity was proposed) something like an expansion or a contraction is a reasonable expectation. The only way you could explain a universe of galaxies holding still would be to make up an "antigravity force." The cosmological constant in general relativity is analogous to this kind of antigravity force in the Newtonian picture of gravity: a force that just balances the pull of gravity.

Under the general relativity framework, in order to counteract the predicted expansion, you'd have to do what Einstein did and make up a completely unexpected "cosmological constant," assigning it the precise numerical value needed to balance out the contracting effect of normal matter and energy. This is a pretty contrived thing to do, and Einstein later called it the "biggest blunder" of his life.

Ironically, Einstein was closer to being right the first time: Modern observations show that the universe *does* contain dark energy, even though we don't need it to "fix" the Friedmann equation any more. That's why many astronomers, physicists, and cosmologists were surprised when the discovery of dark energy was announced in 1998. Today there are multiple lines of evidence for it, and the presence of dark energy seems to be much more helpful than harmful to our understanding of the universe—apart from the one nagging detail that we don't technically know what it *is*. We'll have more to say about dark energy later in this chapter, and more again in Chapter 11.

To summarize, one equation alone manages to characterize the expanding universe, and it does so with a single parameter *a*, which operates the same way in all directions. The Friedmann equation demonstrates how general relativity helps us understand the universe-wide properties of space and time. We can now use the tools we've been building in this chapter to understand the observations of distant galaxies—just keep reading.

7.3 Recession and Redshift

Galaxy Recession

In Chapter 5, we noted many observations of the cosmos as clues to its nature and its history. One of these clues is the recession of galaxies: All but some of the closest galaxies have redshifted spectra. This indicates that they are probably becoming more distant from us, because when light is Doppler shifted toward the red, it means that the source is moving away from the observer. The cosmic mystery is this: Why should all these galaxies be moving *away* and not toward, and why are they moving away *from us*? What makes us so special?

A more natural initial expectation is that galaxies should be moving every which way, with half coming toward us (or toward any other point in space) and half going away. Near our own Galaxy, that expectation turns out to be true, but outside of our local galactic neighborhood, all galaxies appear to be moving away from us. Can the notion of expanding space make sense of this observation? Edwin Hubble, in his famous paper on the redshifts of galaxies (Section 5.1), thought so:

> "The outstanding feature, however, is the possibility that the velocity–distance relation may represent...the general curvature of space."
> —Edwin Hubble, "A Relation Between Distance and Radial Velocity Among Extra-Galactic Nebulae"

Let's tackle the distant galaxies first—the ones that are all redshifting, without exception. If space is expanding, then distances are growing, including the distance between us and these galaxies. Therefore, the expansion totally explains the recession of distant galaxies, provided that they're receding at the predicted speed, which is a point we'll return to momentarily.

How can we explain the motion of the nearby galaxies, which don't particularly recede? Simple! Just as your own body isn't expanding, because there are molecular forces holding it together (or if your body *is* expanding, you can't blame it on the scale factor of the universe!), a local system of galaxies is more strongly affected by gravitational forces between galaxies than by the universal expansion.

The example we used before is the Moon, because it's not being carried away from the Earth by the cosmic expansion. Similarly, local galaxies are kept local by the curvature created by other local galaxies, such as our own Milky Way. Only far away, where the Milky Way and its neighbors have little influence, will the cosmic expansion be the predominant effect.

You might wonder, if the universe appears to be homogenously filled with galaxies, why there would ever be a galaxy that's sufficiently isolated from its neighbors to be carried away by the expansion. And you would be correct. There appear to be galaxies everywhere, and those which are close enough to clump together (like our Local Group) can resist expanding away from one another. The resulting *clumps*, however, become isolated, and the expansion separates the clumps. Imagine a person as a clump of organs (brain, heart, etc.), the way a group or cluster is a clump of galaxies. Then even if two star-crossed lovers are dragged away from each other by a cruel and divisive society, the organ clumps remain intact: Juliet's liver gets no farther from her spleen.

There's more to come in this story about clumping (Chapters 10 and 12). For the remainder of this chapter, however, you might find it more convenient to forget about the clumps and just picture a universe full of individual, distantly isolated galaxies, free to move with the expansion. That's not cheating; it's just choosing to pay attention only to what's most relevant.

Now let's quantify the recession velocity that the expansion should cause. Recall from Section 5.2 that the recession of galaxies is systematic: More distant galaxies recede faster than those close to us, and they do so according to a simple linear equation called the Hubble law. Remember that it says

$$v = H_0 d \; , \tag{eq. 7.6}$$

so that the recession velocity v of a galaxy increases when its distance d from us increases. That increase is calculable from the Hubble law equation: You just multiply the distance by a constant number called the Hubble constant, H_0. The question is, Would the recession of galaxies yield this same relationship if it were caused by the expanding universe?

Four dots labeled A, B, C, and D, representing galaxies in space, are drawn on a rubber band in Figure 7.5. (If dots bore you, feel free to imagine little ants instead; just tell them not to move around.) These "galaxies" start out evenly spaced 1 cm apart, but then the rubber band is stretched to simulate the expansion, and they are 2 cm apart 1 second later. You can, of course, physically perform this experiment yourself; if anyone asks what you're doing, you can respond, "I'm investigating the nature of the universe." Try to keep a straight face.

Once the band is stretched, not only are your dots farther apart, but the farther apart they started, the farther they are carried by the expansion. This is a crucial point to grasp in understanding the universe, so follow these numbers closely.

Galaxy A watched as galaxy B moved from 1 to 2 cm away, during 1 second. The inhabitants of galaxy A conclude that B is *receding* (moving away) at a velocity of 1 cm/s. Galaxy C went from 2 to 4 cm away in 1 second and therefore receded from A at 2 cm/s. (Make sure that you believe this; see the figure or use your own rubber band.) Galaxy D moved at 3 cm/s. (Verify this for yourself.) Notice that distance and velocity are proportional: From A, C is *twice* as distant as B and

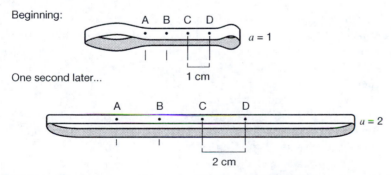

FIGURE 7.5 Expansion and the Hubble Law   When the rubber-band universe is stretched, all the dot-galaxies separate by the same scale factor. But when you check the rates at which *every* dot-galaxy moves away from *every* other one, the Hubble law naturally emerges!

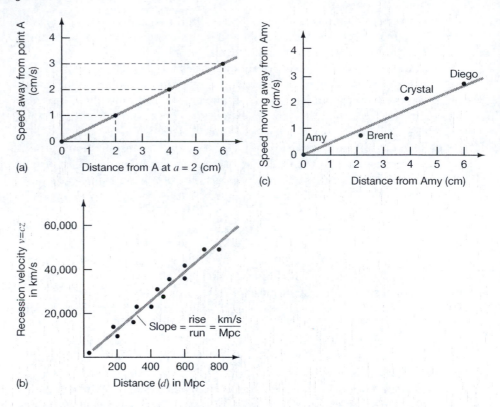

FIGURE 7.6 Hubble Diagrams (a) Hubble diagram for rubber-band universe of galaxies (drawn from the perspective of someone on the dot representing galaxy A, but it would look similar from other perspectives). (b) Hubble diagram from Chapter 5. Remember that the galaxies in (b) are from all over the sky, not just along a one-dimensional line. (c) Same as (a), but with ants that can move around a little, instead of fixed dots. If galaxies move around a little, too, that would help explain the scatter in the real Hubble diagram in frame (b).

moves *twice* as fast as B. This means that, on a graph such as Figure 7.6(a), all the galaxy-dots lie on a common straight line. The pattern is the same as on a Hubble diagram for real galaxies; you can check this in frame (b) of the figure.

Thus, we have demonstrated what we had hoped to demonstrate: that a homogenous and isotropic expanding universe perfectly reproduces the Hubble law that we actually observe in our night sky. Furthermore, it doesn't matter which point you choose as your "observation station" for the law to work. Go ahead and think outside the box by abandoning alphabetical order. From the perspective of point B (not A), point A was 1 cm away, ends up 2 cm away, and therefore moved away at 1 cm/s. Point C did the exact same thing in the opposite direction: 1 cm away, then 2 cm away, receding at 1 cm/s. But double your distance by going to point D, and you get double the velocity: 2 cm/s.

This feature of the expansion is very important. It doesn't matter where you live—you could live off to the left on point A or somewhere in the middle on point B—and in every direction you look, you see points moving away from you. This provides us with a nice answer to one of our previous mysteries: Why are we so special that everything moves away from us? Evidently, we're not! Everything moves away from everything else, and Exercise 7.2 is your chance to see this important insight for yourself.

In our rubber-band analysis, we assumed (until now) that none of the galaxy-dots were moving of their own accord; they were just sitting still in their expanding, one-dimensional rubber universe. Real galaxies, however, do move through space, because the curvature caused by other galaxies and galaxy clusters affects them, so they move accordingly on their local geodesics. The rate of such local motions *not* caused by the cosmic expansion is called a galaxy's *peculiar velocity*. (Motion caused by the cosmic expansion is called a galaxy's *recession velocity*.)

To make our rubber-band example more realistic, we should give the dots peculiar velocities; so, instead of dots, imagine that they are ants, free to crawl around. Give the ants creative names, like *Amy*, *Brent*, *Crystal*, and *Diego*, and let them walk along the rubber band while you pull on it. The result? The Hubble law still holds, but now there's some scatter. (See frame (c) of Figure 7.6 to learn how this scatter improves the way the data are captured by the expansion model.)

The Hubble law holds only for distant galaxies, and the ant scatter offers a reason for that as well. The closest ant moves only 1 cm/s due to the expansion. But an ant might move faster than that anyway, just walking around. That is, its peculiar velocity could exceed its recession velocity. If the nearby ant were a nearby galaxy, it might not appear to obey the Hubble law: It might have been moving toward us with a peculiar velocity faster than the Hubble expansion makes it move away. But now look at the most distant ant (Diego). From Amy's perspective, Diego receded at about 3 cm/s, which is probably faster than his peculiar velocity. Some other ant, much farther down the rubber band (say, Zelda) recedes at 25 cm/s. Give or take a peculiar velocity in the vicinity of 1 cm/s, Zelda obeys the Hubble law. Similarly, distant galaxies have recessional velocities that far exceed their peculiar velocities. The Hubble diagram still has some scatter, but you can easily make out the linear velocity–distance relationship, and its slope is still H_0.

Now, throughout this analysis, we just arbitrarily let the galaxy-ant-dots move apart such that two neighboring dots went from 1 to 2 cm apart in 1 second. But we could have chosen 2 seconds or even 50. This rate is an example of what's meant by the first term in the Friedmann equation (Equation 7.5): "how the scale factor changes with time." If the rubber band were to have stretched in 2 seconds instead of 1, then the expansion would have taken place more slowly. All the velocities would then be half as large, so the slope on the Hubble diagram would be less steep, also by half. (Verify this!) The point is, the Hubble constant *is* the present-day expansion rate. This rate plays directly into the Friedmann equation and therefore is related to the other quantities in the equation: the spatial curvature and the density of matter, energy, and dark energy in the universe.

The Redshift

At this point, we've used general relativity and the isotropic expansion of space to neatly and completely account for one of the most important observations ever made about the universe: the recession of the galaxies. We've been able to explain why galaxies recede. We've also been able to explain why nearby galaxies don't necessarily recede. And we've been able to explain the Hubble law—the observed relationship between distance and recession velocity whereby more distant galaxies consistently show a faster recessional velocity than closer ones.

Or have we?

One seemingly minor glitch demands our attention first. Consider how we *measure* the recessional velocities of galaxies: We measure a galaxy's redshift, not its velocity. To extract a velocity from the redshift, we must *interpret* the reddening of a galaxy's light by *assuming* that the colors have reddened due to the Doppler effect. In that case, the reddening is caused by the source galaxy *moving away from us while it emits its light*. (It's worth looking back to Section 2.4 to remind yourself why a moving source generates Doppler-shifted light.)

But is this really true? After all, if the expansion of space is causing the recession, then galaxies aren't actually *moving* away from us. Apart from any peculiar velocity a distant galaxy may have (which generally produces much smaller shifts, for which red and blue are equally common), the galaxy is *just sitting there*. We are left to contemplate a potential flaw in our logic: Under the expansion model, *should there even be a Doppler shift at all?*

Don't panic. You don't need to abandon everything in this chapter that you've worked hard to understand. You don't need to give up and pursue some easier science, like chemistry. Instead, you need to make sure that you've applied the rules of general relativity systematically to every particle involved.

Light from distant galaxies travels to us through expanding space. Space grows uniformly, enlarging all lengths (except those fixed by some local force) according to the scale factor *a*. The only such length that we forgot to scale up is the *wavelength* of the light.

As a photon of light travels through space, its electromagnetic properties oscillate as a wave. (Recall Section 2.1.) Light of a particular color (frequency), traveling at the fixed speed *c*, traverses a specific distance in the time it takes to oscillate one full wave—up and down and back again. That distance expands as the spacetime fabric expands under it, causing the photon's wavelength to grow. (Think of a wave drawn on a piece of graph paper. If you enlarge the image with a photocopier, the features of the wave are still at the same places on the grid, so the wavelength of the wave you drew grows along with the grid.) A photon's wavelength defines its color and its energy. Longer waves mean redder colors and less energy. As the universe expands, its light energy gets redder and weaker, always.

We arrive, then, at the resolution of this once-threatening glitch: Although the Doppler effect, in its usual sense, does not produce cosmological redshifts (i.e., those caused by the Hubble expansion), the expansion of photon wavelengths does. The more distant a galaxy is, the more time its light must spend traveling through space to reach us. And the more time a photon travels through expanding space, the longer its wavelength becomes. As a result, the most distant galaxies show the greatest redshifts, exactly matching our observations and reproducing the Hubble law. From this new perspective, it makes more sense to plot the Hubble diagram with the directly measured redshift (rather than the interpreted velocity) versus distance. In this way, we can link our modern understanding of space, time, and gravity—this whole chapter—directly with global properties of the universe that have actually been measured.

Exercise 7.2 The Hubble Law and the Expansion of the Universe

Draw 15 or 20 galaxies on a piece of paper. Make them somewhat distinctive so that you can tell different galaxies apart. Next, use a photocopier to make a 25% enlargement. Try to get all the galaxies from the original page on the enlargement as well. For this exercise, you'll need two copies of the original size and two copies of the enlarged size.

(a) Assume that the enlarged photocopy shows the distribution of galaxies today and that the original drawing shows the distribution at some time in the past. Figure out some way to determine the scale factor a for the past state, assuming, as usual, that $a = 1$ today.

(b) Pick a galaxy somewhere near the center of your drawing, and circle it in whatever color says "home" to you. That one will be the Milky Way. Now pick five other galaxies at random, circle them in a different color, and use them to check the Hubble law. To accomplish this, measure the distance between each of these five galaxies and the Milky Way with a ruler, both before and after the photocopy enlargement. Then, assuming that 1 second elapsed from "before" (the original) to "now" (the enlargement), compute the observed recessional velocities of the five galaxies. Make a Hubble diagram of these data, using distances from the enlarged copy.

(c) Repeat part (a), but this time choose a different "home" galaxy, one far from the center.

(d) Compare your results from parts (a) and (b). How can you apply these results, from a uniform-expansion experiment, to the universe? What does applying them in that manner mean to you? Does it say anything about the total size and shape of the universe? About whether the universe is infinite or not? About our place in it? Do your conclusions from this simple expansion experiment leave any wiggle room for interpreting your answers in some other way?

Exercise 7.3   Cosmological Redshift

In this exercise, you'll use your photocopy enlargement from Exercise 7.2 to predict the Hubble law without ever computing velocities. Instead, you'll use only distance and redshift, because those are the measurable quantities. Start with the scale factor you computed for the past state (your original drawing of galaxies, before the photocopy enlargement).

(a) Set up a ratio of scale factors to determine the growth of a photon's wavelength in a light beam coming from a particular galaxy to us. If the original wavelength when the photon was emitted was $\lambda_{emitted}$, what is the wavelength of the photon as it arrives here today?

(b) Defining the redshift as before, namely, $z = (\lambda_{received} - \lambda_{emitted})/\lambda_{emitted}$, calculate the value of z.

(c) Show that, in general, $1 + z = a_0/a(t) = 1/a(t)$.

(d) Repeat parts (a) and (b) for the other four galaxies you circled in Exercise 7.2. You may use your result from part (c).

(e) Make a new Hubble diagram, plotting redshift (z) on the vertical axis and today's distance on the horizontal axis.

(f) How is the information content in this Hubble diagram different from that in the previous one from Exercise 7.2?

7.4 Dark Energy

The Hubble constant H_0 is an important parameter of the expanding universe. If this were a more math-intensive cosmology book, we'd also be making a big deal of another expansion-related quantity: the *deceleration parameter* q_0. In a universe containing nothing but matter and light—which, at first glance, ours appears to be—the expansion would be expected to decelerate, or slow down. After all, both matter and light are affected by gravitational attraction, according to Einstein. So while the expansion moves galaxies apart, gravity tries to pull them back together. Who wins? That's not so simple, but since gravity is *opposing* the expansion, the expansion should at least be slowing down. That's how the logic goes—hence the name *de*celeration parameter.

But we should also keep in mind something the Friedmann equation taught us: We are permitted to introduce a dark energy term Λ that could be zero or nonzero. That we are permitted to do this tells us something. It tells us that even though we don't immediately *see* any dark energy around—in our garages or backyards—our theory of spacetime might be "expecting" it, so we shouldn't be alarmed if it turns up somewhere.

The effect of dark energy in the equations is to *help* the expansion. That is, it does the opposite of what matter does: It tends to *speed up* the expansion. So *now* who wins? To answer this question, scientists must conduct an experiment designed to measure the (presumed) deceleration. Such an experiment is difficult, but not impossible, to carry out. The difficulty lies in the fact that any measurable change in the expansion rate would need to be observed over very long time scales, which means that we would need to observe very, very distant sources. And as you know, with large distances come large uncertainties in the distance ladder.

Still, if we pick a single distance measurement technique and use it consistently for many different measurements, we can hope that any error in the technique would affect all the sources in the same way. In other words, we should still be able to tease out the information we're looking for, although we should certainly try to double-check the results with some other, independent type of measurement. We'll return to that task later in the book; here we want to show you the most direct evidence.

The technique that pays off is using type Ia supernovae as standard candles. (We saw this method in Chapter 3.) Here's the idea:

- You measure the distance to the supernova with the standard-candle technique, instead of plugging the supernova's redshift into the Hubble law.
- You still measure the redshift, but since you already know the distance, you can use the redshift to characterize the expansion of the universe. This property follows directly from our investigation of redshifts in the last section: If photon wavelengths expand in the expanding space, then the redshift directly measures how much the universe has expanded between the time the supernova went off and the time its light arrives here.

This technique is a clever reversal of the usual practice for very distant objects. Usually, we measure redshifts, *assume* that the expansion of the universe obeys the Hubble law (since we have experimentally verified this for less distant sources), and thereby acquire distances. But in the present case, we already know the distance, so with the measured redshift, we can actually measure the amount of expansion.

At this point, all we need is a set of calculated models to tell us how much expansion we should expect to measure for various values of the deceleration parameter. In other words, we have to decide what kind of model would produce the data we see: Do the data agree with how an expanding, but decelerating, universe would behave? Does the universe expand at a perfectly steady rate? Or does it actually *accelerate*, indicating the action of dark energy?

The results are graphed in Figure 7.7, along with the model lines we just described, so you can compare for yourself. The uncertainties in distance are somewhat large and the model lines are close together, making it hard to read the results by eyeball, so cosmologists use more reliable statistical methods instead. But you can still see the result by eye if you look closely. The majority of dots lie *below* the constant Hubble-expansion line, along the *accelerating*-universe line!

What does this measurement mean? It means that the expansion is speeding up, not slowing down. It means that dark energy *does* exist; it's opposing the universe-contracting effect of gravitating matter, and it's winning. It means that the *dec*eleration parameter is actually a negative number; we should have named it the *ac*celeration parameter! (Of course, it's too much trouble to actually rename the parameter and correct it wherever it's written, so we just accept the "negative deceleration," but we still feel a little silly every time we refer to it.)

Perhaps above all else, these supernova results mean that we're applying general relativity correctly to characterize the universe. We used relativity to derive the Friedmann equation, and it had this allowance for dark energy built into it. If we had just ignored the dark energy term and simply declared it to be zero without any real justification, that would be cheating. It would mean that we had invoked general relativity in a biased way—in order to see what we wanted to see. The fact that we now accept the reality of dark energy, despite its being something we can't see or explain (yet), at least confirms that we're being true to the scientific method. (This will all feel a little better in Chapters 11 and 12, where we'll show you other cosmological data sets that support the existence of dark energy as well.)

So what is dark energy? The name itself suggests our ignorance. In fact, it's probably more of a placeholder than a name. Someday, if we discover that it's really caused by alien green-blaster rays, we'll call it that instead. For now, the name "dark energy" reminds us that although we've measured its effect, we still don't know the cause.

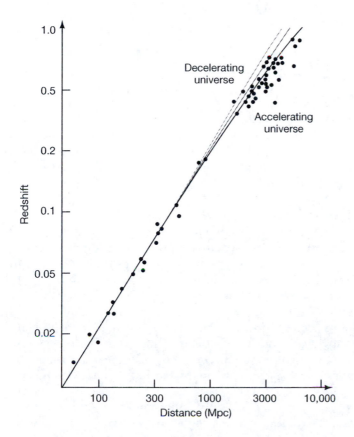

FIGURE 7.7 Evidence for Dark Energy Data points indicate measured redshifts and distances (from Type Ia supernovae standard candles) for a collection of galaxies. Solid lines show the predicted relationships for decelerating, constant, and accelerating expansion models. Thus, the data suggest that our universe is accelerating in its expansion. (◆ Also in Color Gallery 2, page 27.)

It could be something *in* the universe, like an undiscovered particle that performs an antigravity interaction. (But it's *not* the same particle as dark matter. Dark matter particles (WIMPs) behave as gravitating matter—exactly the opposite of dark energy. The names sound similar, but they're not directly related, at least not in any known way. You can't multiply dark matter by c^2 and get dark energy!) Or it could be a property of spacetime, a possibility we'll explore further in Chapter 11. There are lots of good ideas, but few are well constrained by current data.

One interesting possibility is that dark energy could be what physicists refer to as *vacuum energy*. Among many interesting discoveries of quantum theory is that empty space—or vacuum—is not really empty. Even if you remove every single particle from a region, there is always a background level of particle activity; new particles called "virtual particles" just appear and disappear, over and over again, on their own. Vacuum energy is the energy of these particles that somehow always exist, even in empty space. The reason that vacuum energy, which has been observed in experiments, is such an intriguing candidate for dark energy is that we already know that it exists and we already happen to know that it would act like dark energy, rather than normal inward-gravitating energy.

If vacuum energy provides such a reasonable explanation, then why do we hesitate to claim that the dark energy mystery has been solved? Because all attempts to match up the expected amount of vacuum energy with the needed amount of dark energy have failed miserably so far. So, without putting all of our faith in this connection, let's just ask what it would mean *if* vacuum energy or something like it ends up working as dark energy. There are two key points here, related to the fact that vacuum energy is tied to otherwise empty space:

POINT #1: As the universe expands, more space means more vacuum energy. In this sense, the density of vacuum energy is always the same, because if the amount of space in a region doubles, the amount of vacuum energy doubles along with it. This is exactly what Einstein meant when he called it the "cosmological *constant*" (Section 7.2). Now, what's the importance of this constant behavior? It means that as space expands, there is more and more dark energy in the observable universe: The density is always the same, but the total amount goes up. So dark energy is "winning" against gravitational attraction now, and the amount of dark energy compared with the amount of matter will increase in the future. We should therefore expect the cosmic acceleration to continue forever.

POINT #2: If dark energy is vacuum energy, then it's a property of space itself. The fabric of the universe has a physical reality and behavior of its own. Together with the expansion of space, we also have the (nonzero!) energy of empty space. The importance of empty space having nonzero energy is this: We can't think of space as being just an empty stage upon which the things *in* the universe act. Instead, space is a dynamic and evolving entity itself; it has a "life" of its own.

So, is dark energy just vacuum energy? We don't know yet. It might even be vacuum energy *and* something else; it seems likely that vacuum energy should at least have some significant impact on the expansion of the universe. For now ("for now" in this book *and* "for now" in this world!), you have to accept a little mystery. Dark energy, like dark matter, is a well-measured feature of our universe that we just have to admit we don't fully understand yet. After all, this is the land of cosmology, where everything is possible, but nothing comes easy!

7.5 Consequences of the Expansion

That the universe expands like a photocopier enlargement *sounds* simple. That the expansion accelerates by dark energy doesn't sound too much worse. But the immediate consequences of these simple-sounding premises are surprisingly complicated.

The observable universe, as we saw in Section 7.1, is the region whose light has had enough time to reach us. We saw that in a hypothetical static (nonexpanding) universe, approximately 14 billion years of age, the observable universe is a spherical region 14 billion light-years in radius, with us at the center. As beings who live in that static universe, we can't know anything about what transpires beyond that distance, because even at the speed of light, the information couldn't have reached us yet.

But now throw in the expansion, and the situation changes. Imagine a light source that *used to be* near us when the scale factor was smaller, but is now very far away. If that source emitted light at the very beginning, and some of that light is just reaching us now, then its current distance is at our horizon, at the edge of the observable universe. How far away is that?

During the universe's 14-billion-year history, the expansion has widened the distance between the source and us, so it is now *farther* than 14 billion light-years away. To put it another way, the light from this source didn't have to travel the full present distance, because the source used to be closer to us. So the source can be much more than 14 billion light-years away *now*, and used to be much less than 14 billion light-years away long ago, even though its light reached us in 14 billion years. Exactly how far away the horizon lies depends on the details of how much matter and dark energy inhabit the universe, since these entities influence the expansion rate. In our universe of both matter and dark energy, the cosmological horizon turns out to be 46 billion light-years distant; that's the size of the observable universe. This is depicted in Figure 7.8.

How quickly is that source galaxy's distance from Earth increasing? The distance is growing at a rate that is faster than the speed of light! Of course, the source itself is not moving *through* space that fast, so it's not really a speed at all; the expansion *of* space is to blame. That's why there isn't a problem: Matter isn't permitted to win a race against light *through* space, but space itself can do whatever it "wants." Since the Hubble law tells us that farther away means faster recession, there must be a distance at which the recession rate is faster than the rate at which light travels. That distance is well within our horizon.

Now imagine a light source at a great distance away, but still within our horizon. For definiteness, we'll suppose that the source is a quasar, being fueled by gas accreting onto a supermassive black hole. Suppose that the gas approaches the black hole in sporadic blobs, rather than in a smooth stream, and that the quasar light "hiccups" as a result. One little pulse of light is emitted from the quasar over some particular period, but the expansion has *two* effects on the traveling photons. One effect is the redshift, which you already know about, stretching out the wavelength of each photon. But the expansion also stretches out the distance *between* the photons, which means that they're *farther apart* when they arrive here. In other words, we observe a *longer duration* pulse than was actually emitted. (See Figure 7.8.) We count all the photons that were emitted toward us, but now they're redder *and* it takes longer to collect them. So, what if we witness an explosion and want to measure how long it lasts? Is it actually that simple? No! We also need to know the redshift of the light emitted by the explosion, so we know how much the expanding universe has artificially increased its duration.

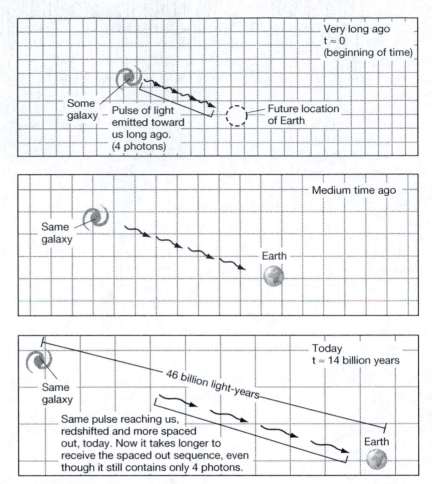

FIGURE 7.8 Light in an Expanding Universe   When a photon from a distant galaxy first begins its journey toward Earth (which had not yet formed), it has only a fairly short distance to get there. But by the time it has traveled part of the way to Earth, the distance has increased due to the expansion. And by the time it has traveled part of this increased distance, the distance has increased again. Thus, it takes longer to complete the journey than it would in a static space. In addition, the expansion of the universe increases both the wavelength of the light and the duration of an emitted pulse.

Now what about dark energy? Dark energy is causing the universe to accelerate. To understand the effect of acceleration, imagine first the case of a *deceleration*. If the expansion were slowing down, it would eventually (and perhaps temporarily) approach a static universe, in which the horizon grows in a simple fashion: Another year of age equals another light-year of horizon distance. In that case, more and more objects come into view as time goes by.

That's for a slowdown, but in the case of a speedup, the reverse is true: The universe does not become static; instead, the expansion pulls objects away from us faster and faster, so that our horizon will never catch up. In a universe controlled by dark energy, there exists another, more permanent sort of horizon. Light emitted today from beyond this limit will *never* arrive here. Our horizon continues to grow, but does so too slowly to keep up with the accelerating expansion. Photons from sources beyond the limit are being emitted toward us and will chase us forever without ever catching us. This phenomenon adds an interesting aspect to the horizon idea: The observable universe is the set of spacetime events whose light is observable to us *today*, but due to dark energy, some distant events won't be visible to us *ever*. If there are bloodthirsty aliens beyond that distance, we can rest easy knowing that they will never be able to eat us or even send us a threatening message!

We end this chapter on an even grander note: Is the universe finite or infinite? Oddly, considering how much the expansion seems to affect every other conceivable question, the fact that space is expanding doesn't determine whether the universe is finite or infinite. The scale factor will grow—increasing distances between galaxies, causing photons to be redshifted, and continually changing the size of our horizon—whether the universe is infinite or not. But since the observable part of the universe today is 46 billion light-years in radius, we can confidently say that even if the universe isn't infinite, it is, at the very least, really, really big.

What Does the Universe Expand Into?

This question *seems* direct and necessary. When people gesture with their arms to indicate the expanding universe—as if holding a growing basketball—their universe expands *into the room*, or at least into the surrounding air. All the analogies everyone uses have the same feature: Balloons, rubber bands, and raisin bread all appear to expand into a larger space. But universes aren't always as easy to imagine as balloons, and in some ways they might be very different.

The basic problem with the universe expanding into something else lies in the nature of spacetime. We've seen in this chapter and the last one that space and time are dynamic features of the universe, so, in order for the universe to expand into some larger space, that larger space would need to be *another universe*. After all, what does "space" even mean outside of the universe? The whole concept of size—lengths, areas, volumes—requires a physical space in which to measure that size. If you try to imagine watching some universe expanding from the outside, then aren't you putting yourself in a physical space outside of our universe? And if so, doesn't that mean you're in another universe that surrounds ours? And what if that "outer universe" is expanding, too? It might obey a Friedmann equation of its own, in which case, what does *it* expand into? A third universe?

In the beginning of this chapter, we identified the structure of the universe in terms of spacetime *events*: sets of four coordinates for everything that happens, with three for the spatial dimensions and one for time. By this approach, the universe is a *list of events*, like "August 14, 2009: lightning bolt strikes the Earth at 37° north latitude, 108° west longitude, and 3350 meters altitude." But this type of thing would make a very long list, so perhaps we can imagine a DVD collection instead, labeled "The Universe, Volumes 1 through 16." The act of you reading this page takes place in Volume 9.

In the master DVD collection of the universe, there would need to be lists of numbers for spacetime coordinates. From one disc to the next, all these numbers get bigger because of the cosmic expansion. At one point on disc 11, the scale factor is 4.8; later, on disc 15, it's 6988.2, a bigger number that makes the disc fill up a little sooner. *But the disc doesn't get any bigger*. It doesn't physically expand into anything. It's a disc space that's growing, not a physical one—gigabytes, not light-years.

The key to understanding all of this is a radical change in perspective. Is the universe a collection of things, like x-rays, atoms, and killer whales, or is it a collection of *information*? If you take the spacetime event picture to its logical limit, the universe begins to look more like the latter. Of course, our immediate daily perception tells us the former, but our immediate daily perception also tells us that the world is flat and that the flow of time actually speeds up when we're having fun.

All of this is highly speculative, and if you don't like it, you don't have to believe it, and there will probably never be a scientist on television telling you you're wrong. The scientific method requires us to obtain knowledge through experiments, and there appears to be no way to conduct experiments "outside" of our universe. Our bodies and our instruments are made of particles such as electrons that live *in* our universe, and they might be wholly incapable of probing anything beyond it.

But don't despair: Nothing can stop the march of cosmological theory. Once we've fleshed out the big bang theory, we'll look at some extravagant, but vaguely promising, ideas, like extra spatial dimensions beyond our three, universes that cycle over time, and even a "multiverse."

Quick Review

In this chapter, you've seen how your new understanding of gravity (from the perspective of general relativity) can be applied to the universe as a whole. The average distribution of matter and energy tells spacetime how to bend—which includes expanding and contracting with time. General relativity predicts that the universe is expanding, and this expansion naturally explains the observed redshift–distance relation for galaxies that is summarized in the Hubble law. However, new observations indicate that the expansion is not constant in time; rather, it is currently accelerating due to the influence of an unknown agent called dark energy.

Try the following crossword puzzle to test your knowledge of key terms and concepts from Chapter 7:

ACROSS

3 Four-dimensional construct in which we live

7 A general term for something happening at a particular place and a particular time

9 A universe that's neither expanding nor contracting

10 A light source that's observed from a great distance to help us study dark energy

12 General relativistic equation for a homogenous, isotropic universe

16 Motion of galaxies away from us (or an economic slowdown)

17 Our cosmological horizon is 46 _____ light-years away in every direction.

18 The cosmological _____ is a limit to how far across the universe we can observe.

19 Everything inside our cosmological horizon is part of the _____ universe.

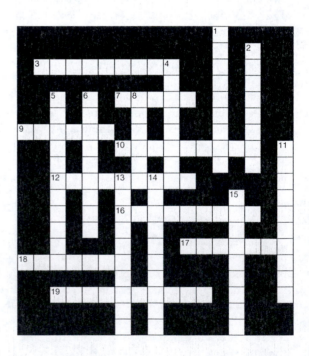

DOWN

1 Continuously increasing scale factor
2 The component of a galaxy's velocity that's not associated with the expansion of the universe
4 General equation for general relativity
5 A number that describes the separation distance for all distant pairs of galaxies in the universe
6 Homogeneity and isotropy together constitute the cosmological _____.
8 Quantum mechanics tells us there's always energy present, even in a complete _____.

11 The cosmological redshift is caused by photons' expanding _____.
13 Unknown agent that tends to speed up the cosmic expansion
14 The rate of expansion of the universe is not constant; it has been observed to _____.
15 An object's path on a spacetime diagram

Further Exploration

(M = mathematical content; I = integrative—builds on information from more than one chapter; P = project idea; R = reflection; D = suitable for class discussion)

1. (MR) What is the smallest physically allowed slope of a particle's worldline, drawn in a spacetime diagram with distance in light-years and time in years? Explain.

2. (RI) Explain how the cause of galaxy redshifts differs from the Doppler shift you learned about in Chapter 2.

3. (RI) (a) How big is your "personal horizon"—the region from which information concerning events that occurred since your birth could have reached you by now? For now, you may assume that the universe is static. What objects are physically inside your personal horizon? (You might want to look back to Section 3.3 for help.) (b) If the universe is expanding, would it mean that there's more stuff or less stuff in your personal horizon than you indicated in part (a)? Why?

4. (M) In the past, when $a(t) = 0.5$, how many times smaller than today's sizes were

   • the length of a distance across space (a line)?
   • the area of a square?
   • the volume of a cube?

   (You may assume that all of these examples change when the scale factor changes.)

5. (MR) (a) Consider the cube from the previous question, and suppose that somehow it contains nothing but a gas of matter (atoms, but not photons). In what way(s) is this cube of matter different at $a = 0.5$ than it is now? (b) What if the cube contains only photons instead?

6. (PRD) Assume that the effect of dark energy is caused by some so-far undiscovered particles that repel (rather than attract) one another through their form of gravity. They

behave as a cosmological *constant*. Now, create a representation of the universe on graph paper or some similar paper. Make "before" and "after" pictures, using $a = 1$ and $a = 2$. Include matter, photons, and dark energy particles in your universe. Think carefully about how each of these constituents behaves differently in the expansion of the universe; the different behaviors should be immediately evident in your drawing.

7. (PRD) On the basis of the previous question, what will the universe look like in the far future?

8. (RD) How are dark matter and dark energy similar? How are they different?

9. (M) What redshift z is associated with a photon that travels through space while the scale factor a grows from (a) 1 to 2? (b) 1 to 3? (c) 2 to 4? (d) 0.1 to 1, today? How big can cosmological redshifts get?

10. (RD) Invent a situation in which a photon undergoes a combination of Doppler, gravitational, and cosmological effects, causing a very large redshift when the photon ends its journey by arriving at the Earth.

11. (RI) Figure 7.8 and the discussion surrounding it make a testable prediction. By the expansion of space, the redshift of a photon beam should be accompanied by a stretching in the sequence of photons, too, so that fewer photons arrive per second (the beam is of less intensity) and the signal takes more time to arrive than it took to send. Suppose you wish to test this prediction against a different model in which photons lose energy for some reason other than the expansion of space—a model that would not predict the stretching beam. To perform such a test, we need a source whose emission lasts for a known amount of time. That way, we can compare the known time with the measured time for a distant source and see if

the measured time is longer (by the appropriate amount). Identify a few viable sources for this experiment. They must have known durations of emission, and they must be visible at great distances across the universe.

12. (M) Using the Hubble constant $H_0 = 71$ (km/s)/Mpc, calculate the recession speed of a galaxy located 50 Mpc away. At this speed, by what percentage will the galaxy's distance change over the course of one day? one year? one century? Note that 1 Mpc $= 3.09 \times 10^{19}$ km.

13. (MR) In Figure 7.6, the expansion rate for the rubber band is 0.5 (cm/s)/cm, while the expansion rate for galaxies is the Hubble constant $H_0 = 71$ (km/s)/Mpc. Convert Mpc into km (using the conversion factor 1 Mpc $= 3.09 \times 10^{19}$ km) to express the Hubble constant in (km/s)/km. Argue that you now have the same units for both expansions (i.e., (cm/s)/cm $=$ (km/s)/km), and compare the expansion rate of the universe with that of the rubber band.

Photons and Electrons

> *"What, exactly, is electricity?...we know from our junior high school science training that electricity is actually a fast-moving herd of electrons, which are tiny one-celled animals that can survive in almost any environment except inside a double-A battery, where they die within minutes."*
>
> —*Dave Barry*

A good cosmological theory must satisfy many requirements to win our acceptance. It must consistently connect a diverse set of physical laws to every possible system in the universe. It must explain the history of history, dating back to a time long before people, and long before Earth. It must be testable in many different, independent ways. Unless the theory meets all these requirements, how could anyone believe that we really understand something so grand and so ancient as the entire universe?

Perhaps the most compelling line of cosmological evidence derives from the all-sky cosmic microwave background (CMB) radiation that you read about in Chapter 5; for that reason, the CMB is critically important in developing a good cosmological theory. Here's the bad news: As with most worthwhile endeavors, understanding the cosmic microwave message takes some hard work. The message comes from photons (light—recall Section 3.1) whose most recent interaction was with electrons, and that interaction took place more than 13 billion years ago. Extracting meaning from these particles is like deciding whether to trust politicians: You have to learn a lot about them, because only then can you correctly interpret what they're saying.

There is some unexpected good news, however. Once you understand photons and electrons, you begin to appreciate the CMB the way you do any other great beauty in nature—a pristine forest or a cloud-piercing sunset over water. In fact, you can think of this chapter as an art appreciation lesson, for a scene painted by particles rather than people and on sky rather than canvas. (You might look ahead to the modern all-sky map of the CMB, shown in Color Figure 48 (a) in Color Gallery 2, page 22; doesn't it have a sort of pointillistic, modern-art quality to it?)

But let's return to our initial thought: If someone were to tell you a story and claim that it explains the entire universe, would you accept it? What evidence would be needed to convince *you* that the history of the universe has largely been figured out? Would you have to actually *see* that history somehow? It just so happens that the universe provides an ongoing microwave movie, so that no matter where in the universe you live, and no matter in which direction you look, you can actually watch its history unfolding.

8.1 Blackbody Radiation

The cosmic microwave background is a *blackbody* source. We encountered this term first when we discussed starlight in Chapter 3 and then again when we investigated the CMB in Chapter 5. Now that we're ready to look at the CMB in more detail, we need a deeper understanding of blackbody radiation.

We can start by recalling the general nature of blackbody radiation from Section 3.1:

- Blackbody radiation is the natural glow of a hot object. An object that's "red hot" (like the burner of a stove) or "white hot" (like the filament of an incandescent light bulb) shines as a blackbody.

- Blackbody radiation is characterized by a unique *spectrum*. In other words, it is distinctive because it follows a very precise recipe: A particular fraction of its energy shines at each frequency. If the source is visible to human eyes, then its distinctive spectrum boils down to an exact prescription for how much light of each color to mix together.

- The hotter the glowing object, the *brighter* it glows. Not only does it glow brighter at *some* frequency (color), but it also glows brighter at *every* frequency (color). If the temperature of a red-hot coal goes up, then it will emit more red light than before, and more yellow light, and more blue light, and more ultraviolet light, etc.

- The hotter the glowing object, the higher is the peak frequency at which most of the light is emitted. More casually stated, the hotter the object, the *bluer* is its light: More of the light will be composed of high-energy frequencies, which often tend *toward* the blue end of the electromagnetic spectrum. A red-hot coal that goes up in temperature will emit more orange light than red and will therefore appear orange. Note that "bluer" is a somewhat inaccurate term: If the light was already blue, then it will go to higher energy—in which case we'll still *say* "bluer," but we'll *mean* "higher energy" or "more violet, ultraviolet, x-ray," etc.

For our investigation of cosmology, we need to define the physical environment inside a blackbody. Be warned, however, that the analysis which follows will make sense only if you can imagine a material object as it really is: an assembly of tiny particles separated by large empty spaces. Shrink yourself down until you can go inside a solid object, and wander among its atoms; then resume reading. Remember to shrink this book, too, so you can bring it with you.

An object produces a blackbody spectrum when two conditions are met:

CONDITION #1: At any location within the object, there exists a single *temperature*.

This condition may seem like a no-brainer; after all, what object doesn't have a temperature? But physically, it's significant. It means that all the particles (atoms, say) in the object have interacted with each other enough to "share" their energy. For example, imagine that the molecules inside some hot object act like billiard balls. (If you're not in the mood to imagine, then just look over Figure 8.1.) When the first player breaks, by hitting the cue ball into the triangle of solid and striped balls, the energy imparted to the cue ball gets spread out among all the others on the table. This creates a common characteristic energy—like an average energy—for all the billiard balls. Even if you start with the cue ball holding all the energy, after a series of collisions every ball shares in that energy. The balls don't share the energy *equally*, in that different billiard balls end up moving at different speeds, but they don't end up moving at *wildly* different speeds from one another either. That's why it's fair to speak of a "characteristic" energy that's described by the overall temperature.

Inside a hot object with a single temperature, all the particles "bump into" each other—interacting by some mechanism—often enough to spread out all their energy. The result is an object whose temperature is constant all over its glowing surface. The constant temperature is caused by a rapid and continuous series of interactions among particles in which their energy is shared. Remember from Chapter 3 that this is the physical meaning of *temperature:* a number that characterizes the average energy of all the particles inside an object.

The second condition a blackbody satisfies is

CONDITION #2: The object is *opaque.*

This condition means that inside a hot object, such as a solid brick or a gaseous star, a photon of light will "bump into" lots of other particles before it reaches the edge of the object and escapes. How opaque a material is—its *opacity*—is a measure of how many times the photon will typically bump into other particles before it emerges. It's somewhat like trying to escape from a crowded room in an emergency: You start running, but keep bouncing off of other people until you emerge at the edge of the crowd. A photon—a particle of light—exists inside a blackbody object, and no matter where it's aimed, it interacts with many other particles before it gets out. Each interaction causes it to zigzag and change speed as it works its way outward. This means that when the photon leaves the object, its properties have been affected by conditions throughout the whole object, not just by the conditions that were present where it originated. You can visualize all this in a number of ways. (Recall, for example, Figure 3.4.)

Both of the preceding conditions depend on the *communication* of energy within the blackbody. There is an ongoing energy exchange among particles of matter (we know this because of the object's single temperature) and between particles of matter and photons (because the object is opaque). Inside a glowing blackbody like the Sun, for example, the energy of one electron is shared with neighboring electrons by way of interactions: Electrons get near each other, push apart, and thereby exchange some energy. On average, the result is a common range of energies throughout the object. In this way, the blackbody spectrum is regulated: Interactions average out the energy, and this averaging always generates the same mixture of wavelengths—the same type of glow.

We already know that hotter means both brighter and bluer; the blackbody spectrum displayed in Color Figure 16 in Color Gallery 1, page 8, shows why. Raising a blackbody's temperature shifts its blackbody curve upward (toward greater brightness) and sideways (toward greater energy, higher frequency, and

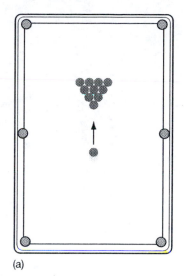

(a)

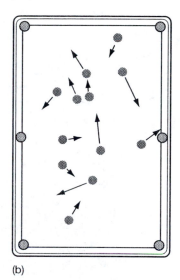

(b)

FIGURE 8.1 The Particle-level Meaning of Temperature Just before the cue ball breaks, a system of billiard balls is not at a single temperature, because the cue ball has all the energy and the numbered balls have none. (b) Soon after the break, all the balls share the cue ball's initial energy and the system represents a single temperature while the balls are all in motion.

bluer color). Notice that an increased temperature causes more light output at every frequency, but also shifts the *peak frequency* sideways (toward a higher frequency). A mathematical relationship called the *Wien law* relates a blackbody's surface temperature to the peak frequency of the light it emits. Wien's law states that

$$T = \frac{0.0029}{\lambda_{\text{peak}}}, \quad \text{or equivalently,} \quad \lambda_{\text{peak}} = \frac{0.0029}{T}. \qquad \text{(eq. 8.1)}$$

The number 0.0029 that appears in the formula carries the units "meter kelvins," so each of the two versions of this formula in Equation 8.1 requires you to use surface temperatures measured in kelvins and wavelengths measured in meters. For example, if we plug in the Sun's surface temperature, 5800 K, we find that the peak wavelength is $0.0029/5800 = 5 \times 10^{-7}$ meters, or 500 nanometers, which is somewhere in the middle of the range of wavelengths of visible light, as expected.

If we count all the photons leaving the surface of the blackbody each second, we can express the energy emerging from the surface as a flux (F), or a luminosity per square meter of the emitter's surface. That is, more area means more luminosity, which is consistent with what we expect. For example, a tiny speck of hot metal glows intensely, but doesn't give off enough total light (luminosity) to help you see very well in the dark. By contrast, a whole wall covered with those specks at the same temperature would light up the room. (If your memory of flux and luminosity is a little rusty, you may want to review the discussion of the inverse-square law in Section 3.2.) The total blackbody flux can be computed from the *Stefan–Boltzmann law*, or

$$F = \sigma T^4, \qquad \text{(eq. 8.2)}$$

where σ is just a number, called the *Stefan–Boltzmann constant*, equal to 5.67×10^{-8} in SI units. In Equation 8.2, temperature is raised to the fourth power, meaning that if you double a star's temperature, you increase its light flux by $2^4 = 2 \times 2 \times 2 \times 2 = 16$ times! Thus, the study of blackbodies provides a powerful result: The light output depends very strongly on the temperature. In other words, change the temperature a little, and you change the output flux a lot.

It's often more useful to write Equation 8.2, the Stefan–Boltzmann law, in terms of luminosity, to get rid of the "per square meter" dependence. To do that, we just multiply by the emitting area (similar to what we did in Equation 3.2). Since most astronomical emitters are spherical (e.g., stars) we can usually use the surface area of a sphere, which is $4\pi R^2$. Then a blackbody's luminosity is

$$L = 4\pi R^2 \sigma T^4, \qquad \text{(eq. 8.3)}$$

or

$$\text{luminosity} = \text{some constants} \times (\text{radius})^2 \times (\text{temperature})^4.$$

So add this equation to your library of blackbody knowledge. You know how the curve looks, where it peaks, and what it means; and now you have its total luminosity.

EXERCISE 8.1 Toward a New Disrespect for Red-Hot Objects

On weather maps, red lines are warm fronts and blue lines are cold fronts. This type of color coding, with red indicating hot, occurs often. But it doesn't show the real connection between temperature and color. For each of the following examples, try to use your understanding of light and blackbodies to construct a better intuition for yourself.

(a) A "red dwarf" star is smaller and dimmer than the Sun. A "red giant" star is much bigger and (therefore) much brighter than the Sun. Both are actually red, in that their blackbody curves peak at a red frequency. Which star is hotter? That is, which has a higher surface temperature?

(b) The Sun, when viewed from space, is approximately white in appearance. What *is* a "white-hot" object like the Sun? Answer in terms of its blackbody curve. (*Hint:* Think about the composition of white light. To help you get started, realize that there is no such thing as a white photon! We'll see more on this in Exercise 8.2.) Is the Sun's surface hotter or colder than the surface of the red stars from part (a)?

(c) Suppose that a particular star's blackbody curve peaks in the infrared part of the electromagnetic spectrum, so that most of its starlight emerges at an energy weaker than that of red light. What would this star look like to your eyes? Most of its light would be invisible to you, but would you see *any* of its light?

(d) Many stars in the night sky are visibly blue to the naked eye. Where does a blue star's blackbody curve peak? (Note that the Sun's blackbody curve peaks around green, even though the Sun doesn't look green.)

(e) The core of a candle flame is it's hottest point; the temperature diminishes as you move outward. Sketch a candle flame the way it would appear on a weather map, with red designating hot and blue designating cold.

8.2 Photons and Bound Electrons

Not all light sources shine like blackbodies. For example, fluorescent lamps, lasers, and LEDs all shine quite brightly without a particularly hot source (i.e., no filament and no flame). In Chapter 2, we noted that certain atoms emit (and absorb) **line radiation**—light of a few specific wavelengths only.

Line radiation can be complicated, but from it, we learn a tremendous amount about our universe, including things that once seemed impossible to know. "There are some things of which the human eye must remain forever in ignorance," said French philosopher Auguste Comte in 1835, "for example the chemical constitution of heavenly bodies." Take the Sun, for instance: Even if we travel to it, how could we ever get close enough to perform experiments? It's too hot! But in 1859, it was discovered that absorption lines form distinctive patterns from each chemical element. In the early 1900s, atomic physics finally taught us what those lines were saying. As it turns out, and contrary to Comte's assessment, they were saying a lot.

The Orbits of Planets and Particles

A *bound electron* is an electron that inhabits an atom. Because the atom's nucleus is charged positively and the electron is charged negatively, the two attract each other. That attraction *binds* the electron to the atom, because the electron doesn't have enough energy to escape the pull of the nucleus. The situation is similar to that of the Earth and the Sun: The Earth (electron) is bound to the Sun (nucleus) because of their gravitational (electrical) attraction. If the Earth (electron) had more energy, it could fly farther away from the Sun (nucleus) and maybe even leave the solar system (atom) altogether. An electron that escapes its home atom like this is called a *free electron*—a particle that will require our attention in the next section. Of course, by this analogy, a *free Earth* would be completely undesirable and would have nothing to do with democracy.

Figure 8.2 shows the analogy between planets and electrons. In frame (a), we see a sequence of scenes: planets orbiting the Sun; Earth being struck by a large asteroid and thus gaining enough energy to move to a higher orbit (one with a bigger semimajor axis); and Earth being struck again, so that it has enough energy to fly away completely. Frame (b) shows the same sequence for an electron in an atom: electrons bound within the atom; an electron being struck by a photon of light and

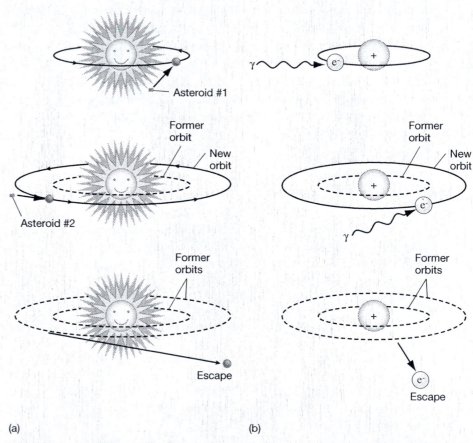

(a)   (b)

FIGURE 8.2 The Orbits of Planets and Electrons   (a) A planet in a circular orbit can gain energy from another body, such as an asteroid that collides with it. The collision moves the planet to an orbit with a larger radius or, if the asteroid has enough energy, could give the planet enough speed (v_{esc}) to escape from its sun. (b) In an atom, an electron can often be modeled with a circular orbit as well. A photon's electromagnetic energy could raise the electron to an orbit with a larger radius or enable it to escape from its atom.

thus gaining enough energy to move to a higher position; and the electron being hit again, so that it gains enough energy to break free of the atom.

There are, however, two major differences between the planet and the electron in this context. Both differences arise from the laws of **quantum mechanics**, a theory that describes physics on small scales, including atoms and particles. But don't let the size of its subject matter diminish the importance of quantum mechanics in your mind. Quantum mechanics is one of only two great theories (the other is general relativity) that accurately describe our physical universe today. Right now, we'll just need some of its results to point out the differences between the bound planet and the bound electron.

DIFFERENCE #1: The planet moves in an *orbit*—either a circle or an ellipse—along which it travels in a recurring, predictable way. The planet does not stray from the orbit, nor does it change speeds from one trip around the Sun to the next. It is systematically regulated as we described in the last chapter: The Sun curves the space around it, and the planet follows a geodesic path through the curved space.

The electron moves in an **orbital**. The motion does not follow a circle or indeed any particular path, such as a geodesic. One of the truly great discoveries of quantum physics is this: The electron's behavior at any given moment is partially *random*. As you saw in Section 3.1, an atom's "electron cloud" only vaguely specifies the electron's motion and location, and different orbitals have "clouds" of different shapes and sizes. In fact, the only thing we can say about an electron's location is the *probability* that we'll find it in a particular region.

But our inability to precisely pinpoint the electron doesn't change the fact that it takes a certain amount of energy to get an electron from one orbital to another, just like getting a planet from one orbit to another. In fact, even though we know better, it is often extremely useful to *model* bound electrons as if they move in definite orbits like planets, because then we can *visualize* an electron's energy. This modeling turns out to be a valuable thing to do, because of

DIFFERENCE #2: Oddly, bound electrons are very picky about how much energy they want to have (or perhaps it's the atoms that are picky about how much energy they *allow* an electron to have). Planets are not like that: An asteroid could come along and whack a planet arbitrarily hard. The planet doesn't care; it'll just take whatever energy it's given and spend it all, moving to a larger orbit. That new orbit could be any ellipse, 1 inch larger, 1 light-year larger, or anything in between. But a bound electron has a fixed set of possible orbitals, and it *absolutely will not reside in between*.

This concept is purely quantum mechanical; there is nothing like it in classical, nonquantum physics. In fact, that the electron can have only certain specific energies is properly expressed by saying that the electron's energy is **quantized**. Because the concept is so foreign to our everyday lives, it's worth our time to consider an analogy.

Consider a staircase versus a ramp. When you ascend a ramp, you can stop anywhere, at any height. And since you can always move a tiny bit higher or lower, there are an infinite number of heights at which you could stop. But when you ascend a staircase, you can stop only on a step, and not in between steps. In this sense, a staircase quantizes your height: If there are 13 steps, then you can have 13 possible heights. A planet's orbit is like your height on a ramp; an electron's orbital is like the stairs.

The Quantum Atom

In 1913, Danish physicist Niels Bohr invented a model for the simplest atom, hydrogen, to accommodate the electron's quantized energy. He did this by assuming that the single electron orbits the single-proton nucleus of hydrogen in a perfectly circular orbit, but that such orbits are quantized. That is, the electron can't remain in just any orbit; it can be in *this* orbit, or *that* orbit, but nothing in between. This situation is shown in Figure 8.3. A low-energy electron occupies the innermost orbit. With more energy, the electron could pull away from the electrically attractive, positively charged nucleus, to occupy a more distant orbit. But an electron can't occupy just any orbit at all. It's like the difference between an essay exam (on which a student–electron can write any possible answer), and a multiple-choice exam (on which the student–electron must choose among options that already exist). Bohr's electron faces the analogue of a multiple-choice exam.

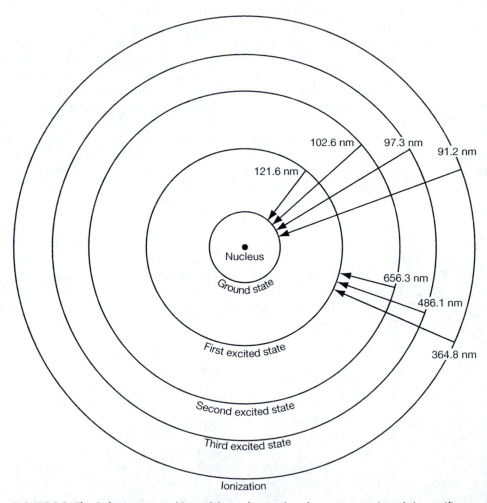

FIGURE 8.3 The Bohr Atom   In this model, we observe that electrons may exist only in specific, quantized, circular orbits. Which circle determines how much energy. When the electron orbits in the innermost circle, the atom is in its "ground state." Subsequent circles, as you move away from the nucleus, are "excited states." The energies associated with a few possible transitions from one state to another are indicated by arrows and wavelengths, corresponding to photons with the same energy. (Radii are not drawn to scale here but a larger radius indicates greater energy.)

This model is called the **Bohr atom**. Of course, you already know that it's wrong: Electrons have complicated *orbitals*, not circular orbits, and a collection of electrons in orbitals forms an atom's *electron cloud*. But scientists still use the Bohr model, and textbooks still teach it. It's not just because of tradition, or to show respect for famous dead people like Niels Bohr. It's because the Bohr model is really good! Even a wrong model can be useful, provided that you use it only to study certain questions, for which its flaws don't affect the answers.

The success of the Bohr atom is very telling about the nature of science. As a student of science, you have the authority to invent whatever mental images are useful to you. They don't have to be true; they only have to be *true enough*. By this measure, the Bohr atom is a great success. We just have to agree that we'll listen only when the model talks about the electron's energy (for which the model is measurably correct), and not when it talks about where electrons actually are (for which it is wrong).

Refer to Figure 8.4 as we explain how the model works. Each atom has a central nucleus, where the electrons never go. Surrounding the nucleus is a succession of concentric circular orbits, as drawn in the figure. In each orbit, an electron can have only a specific energy. Just as you must consume more of your own muscle energy to climb six steps than five, an electron must consume more energy to get to higher orbits, farther from the nucleus. When you descend stairs, the opposite happens: You release energy. (You tend to notice this energy more when you stumble and fall down the stairs, rather than just walk.) When an electron descends to a lower orbit, it releases energy, too.

The real difference between stairs and atoms lies in the type of energy involved. The way electrons (usually) absorb and release energy is based on their electrical nature: They absorb or emit electromagnetic energy. (Recall Section 2.4.) That is, they absorb or emit *photons* (light, from Section 3.1). An electron in a particular

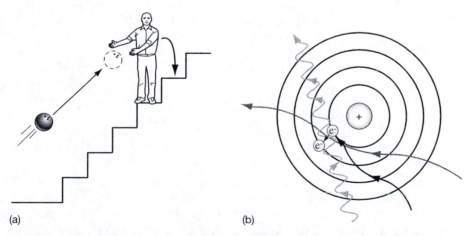

(a)   (b)

FIGURE 8.4 Quantum Energy States   (a) On stairs, height means energy, since you can hurt yourself more by falling from a greater height. The staircase quantizes that energy, since you can't hover in between steps. If a fast-moving bowling ball is used to increase your energy, then it must be thrown exactly fast enough to bump you up a whole step (or more than one), because if its energy is sufficient only to bump you up a half step, or two-and-a-half steps, etc., then you'll trip and fall. (b) An atom is the same way: You can shine light of many different colors (energies) on it, but the electron will absorb just a few specific colors—those which carry just the right amount of energy to move it up to another orbit. Absorbing any other color would be like moving up a fraction of a stair step.

orbit—or *energy level*—will always have exactly the same amount of energy, according to the Bohr model. To move to a higher level, it must acquire extra energy, which it accomplishes by *absorbing* that energy from a photon. This process of moving an electron to a higher orbit is called *exciting* the electron, and it can happen as a result of shining energetic light on its atom.

Neatly, electrons perform the opposite process as well. Excited electrons can descend to a lower energy level by *emitting* a photon—shining a little blip of light. The higher, excited energy gets split: Part of it goes to the electron's new, lower energy level, and the rest goes into the photon, which flies away and never returns. An electron in an atom can take energy from incoming light and give it away again with outgoing light.

The real beauty of quantization, and of the Bohr model, lies in the specific amount of energy. Since that energy is carried by a photon, an "amount of energy" really means a *color*. If an atom were like a ramp, instead of a staircase, then its electron could absorb *any* photon, which would carry the electron up to whatever height the photon's energy enabled it to do. A weak photon (red, perhaps) would carry an electron up a little ways, and a stronger one (blue, perhaps, or ultraviolet) would send the electron much farther upward.

But with the quantized staircase, the electron can't absorb just any photon. Only certain photons will do. The electron needs a photon that has *exactly* the amount of energy (color) required to *just barely* make it to a higher energy level. A weaker photon does the electron no good, since the Bohr atom allows only electrons in certain preexisting orbits, which come with exact energy specifications.

Interestingly, a photon with extra energy does no good either. The photon must have exactly the amount of energy needed to get the electron to another energy level and no more. More energy would again put the electron in between levels. Figure 8.4 illustrates this state of affairs with an analogous process for people climbing stairs, using incoming and outgoing bowling balls for energy instead of photons. (If you're one of those people who hasn't personally encountered a high-speed bowling ball while climbing stairs, then just pretend, won't you?) The process is similar, in that an incoming bowling ball at the right speed will send you up a stair or two, but if its speed is in between, you'll end up off balance, instead of squarely on a particular step. Because of the latter possibility, you might choose not to catch a bowling ball with the wrong speed; similarly, an electron "chooses" not to intercept photons with the wrong color.

This is the nature of the bound electron. Since it has only certain available energy states, it interacts only with certain photons. All other photons are allowed to pass through the atom.

We need to add only one more detail in order to appreciate the value of the preceding discussion: The Bohr atom is quite successful for hydrogen, the simplest element, but more complex models are needed for more complex atoms. It's not worth getting into those models here: They're complicated and not terribly rewarding. And it doesn't matter anyway, because they all follow the same basic premise as the Bohr hydrogen atom: When electrons move, only particular photons are involved. Because every atom is different, every atom has a unique set of colors (or frequencies, since the light could be nonvisible) that it interacts with. In other words, every atom and every molecule has a spectral fingerprint. Light generally includes such fingerprints, hidden in its spectrum of colors, that tell us what elements the light source is made of.

Consider streetlamps, which are not blackbodies, but rather consist of electrically excited sodium atoms. When these atoms *de-excite*, or descend to lower energy levels, they emit only those colors which correspond to the exact energy gaps between electron energy levels in sodium atoms. Sodium's most prominent gaps correspond to yellow photons. The element has two very frequent electron jumps, and each is a highly precise, slightly different shade of yellow. See Color Figure 18 in Color Gallery 1 on page 8, which shows a sodium streetlamp and what its light looks like when broken into its component colors (with a prism, for example).

The sodium lamp's illumination is an example of *emission:* Atoms are excited, and we see their light when they de-excite, because that process emits (yellow) photons. The opposite process, atomic *absorption,* wherein unexcited electrons go up in energy level by absorbing passing photons, might seem more troubling. You might ask yourself, "Why would there *ever* just *happen* to be a photon passing by with *exactly* the right amount of energy for some bound electron to jump up a level?" But you should rephrase the question and ask yourself instead, "Is there a light source that has *every* type of photon in it—like a photon "buffet line"—so that an atom can just take the one it needs?" There is such a source: a blackbody.

The Sun emits blackbody radiation, since it has a surface temperature and is opaque inside. (Review Section 8.1.) But the Sun's *atmosphere* (outer gas layer) contains lots of different atoms, each of which absorbs certain colors from the blackbody sunlight. The solar spectrum, then, when broken into all of its wavelengths, looks like a rainbow, with certain precise colors removed. (No atom can remove "orange," but many atoms remove their own narrow *shades* of orange.) Take a moment to appreciate all the detail hidden in sunlight; the solar spectrum is shown in Color Figure 19 in Color Gallery 1, page 9. All of those lines are the complicated result of a single idea: the quantum interaction of photons and bound electrons.

You can see, in the high-resolution breakdown of sunlight, that the Sun's atmosphere contains thousands of *absorption lines*, from thousands of possible bound electron transitions. Some of these are from "normal" atoms. Some are from ions, which (recall from Section 3.1) would be normal atoms, except that some of their electrons are missing, which happens at high temperatures. At lower temperatures, particular lines could come from molecules—chemical combinations of atoms. In molecules, atoms share electrons in complicated ways, but the physics is always the same: The interaction with light is based on specific, quantized energies. Thus, we have a comprehensive catalog of *spectral lines*—fingerprints—of all the atoms, ions, and molecules we encounter. In this way, we can usually figure out what we're looking at, even from far away.

The study of spectral lines, called *spectroscopy,* has a number of powerful benefits for the study of objects in space:

- We can recognize spectral patterns in different sources, such as stars or galaxies, and identify the elements they contain and in what relative amounts they exist.

- We can use that recognition to identify objects such as stars used for distance measurements by main sequence fitting. (See Chapter 3.)

- We can determine, based on how spectral lines can sometimes be altered, the environment of those elements, including important properties like surface gravity, temperature, and pressure.

And perhaps its most important benefit in cosmology:

- We can use spectral lines to determine cosmological redshifts. (Recall Chapter 5.) For example, we look at features in a galaxy's spectrum—namely, lines from particular elements that we recognize—in order to tell that they have shifted.

In the next section, we'll see how understanding photons and electrons reveals another cosmic message, from what turns out to be the most distant light source ever seen.

8.3 Photons and Free Electrons

Before we can ascribe meaning to the cosmic microwave glow, we need one more photon–electron interaction in our arsenal. What happens when photons encounter electrons that are running around loose, not attached to any nucleus and hence not restricted to fixed energy levels? (This is the case for electrons in hot ionized gases.) What happens when electrons are free to absorb *any* amount of energy from *any* photon?

Often, when people (and textbooks) talk about light, they describe light *waves*. They say, correctly, "Light is an electromagnetic wave." But think back to our discussion of this phenomenon in Chapter 2, and ask yourself, What does that really mean? It means that light is an electrical (and magnetic) disturbance, which gains and loses strength over and over again as it travels. It means that when a light wave arrives, electrical charges get pushed around. In other words, it means what we already know: that light interacts with electrons. Said differently, it means that light can move an electron, and conversely, an electron can disrupt a ray of light.

As we saw in the last section, electrons bound up in atoms can disrupt light only of *particular* frequencies: those corresponding to each "step" between energy levels. But free electrons can be pushed around by light of *any* frequency, because nothing restricts them from following the pushes and pulls of the light wave. Other charged particles (such as the positively charged protons) are also pushed around by light, but since electrons are nearly 2000 times less massive than protons, electrons are much easier to push around. So whenever light passes through a bunch of free electrons, the energy from the original light goes into jiggling the electrons, and the jiggling electrons in turn emit new light.

At this point, the particle view of light becomes valuable in combination with the wave picture. It's similar to the way we can *choose* to use part of the Bohr model, even though we know it's not the complete story; analogously, we can choose to view light in whichever way we find most useful for any given situation. When light interacts with things like electrons, it acts as a particle called a photon: a packet that transmits a definite amount of energy.

Physicists indicate photons with the lowercase Greek letter gamma, which is written γ. Electrons are represented with a lowercase e. So we can represent the interaction between a photon and an electron as

$$\gamma + e \ .$$

When people interact, say, by a handshake, you can describe the interaction as a physical contact between two solid surfaces (hands). But does that same model fit for photons and electrons? Some physics and chemistry textbooks do draw electrons as little solid balls, as in Figure 8.5(a).

But are electrons really solid? Can you "touch" an electron and feel its edge? This ambiguity leads some books to draw them as "fuzzy" balls instead, as in Figure 8.5(b).

Photons are even more distressing, since they don't need to be balls at all; they can be waves, rays, or wavy rays, as shown in Figure 8.5(c).

At this point, you might worry that we have *no* single compelling way to describe our particle-based world. If you recognize that deep down, most of your life is assembled from interactions between photons and electrons (see Exercise 8.2) then this page might threaten to shatter your confidence! But rather than turning off the lights in order to avoid photons forever, you might be better advised to fall back on something you know is safe:

> You can think of particles as "localized bundles of energy."

What they actually "feel like" up close is a whole other story—a story that we don't need here. In fact, that story may be a purely human illusion: Our bodies are large collections of millions of trillions of trillions of particles. Our perception of things having surfaces and textures is just that—a perception, based on an intricate blend of many particle-level phenomena.

(a)

(b)

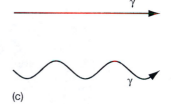

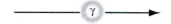

(c)

FIGURE 8.5 Electromagnetic Particles   How should you picture electrons and photons?

EXERCISE 8.2   Reality and Your Senses

Write a page or two connecting your human-level experience with the particle-level one:

(a) Ask yourself what it really means to *touch* something. Does your hand have a hard surface? What would an ultrapowerful microscope reveal about the particles in your hand if you cranked up the magnification high enough? Would you see an array of little fuzzy balls? What would this tunneling viewpoint tell you about the fundamental nature of your perception of touch? (Have you ever *actually touched* anything?)

(b) What about sight? If you choose to view light as photons—particles, or little "wavelets"— then what would the ultrapowerful microscope tell you about light entering your eye? What about the little fuzzy balls in your retina?

(c) Using the particle level as your frame of reference, compare your senses with those of a machine that could be built with present-day technology. Specifically, speculate on how different an electromechanical touch sensor, such as a laptop computer's touch pad, is from your own skin. How different is a camera from an eye?

(d) Comment on what all of this means to you personally. Can you think of your own body in terms of little fuzzy balls? Is the picture believable?

Let's suppose for simplicity that the interacting γ and e particles don't actually touch. (Indeed, on the basis of the preceding discussion, it's not even clear what it would *mean* for them to touch.) Instead, two particles enter some region, interact due to the oscillating electrical properties of the photon, and then two particles exit:

$$\gamma + e \rightarrow \gamma + e \ .$$

This type of interaction is called **scattering**. It means that the same type of particles exit as those which entered, namely, γ and e. Neither one was transformed into

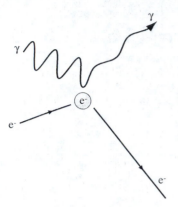

FIGURE 8.6 Photon–Electron Scattering The photon and electron interact, change direction, and exchange energy. You can tell that there has been an energy exchange in the picture because the photon's wavelength is different after the interaction than it was before.

something new, like a neutron, or an oxygen atom, or a pony. But—and this is important—it's *not* as though *nothing* happened. The interaction generally involves a *change of direction* and an *exchange of energy*. (See Figure 8.6.)

Now, for a photon, a gain or loss in energy means a gain or loss in frequency and color. So you can think of a γ–e interaction as a way to *change* light, by changing its direction and its color. Or, if you prefer, you can imagine that the electron *kills* the original photon and creates a different one. Either way, we are left with an inescapable conclusion: *A ray of light can't travel unchanged through a region with free electrons*. Whatever light comes out will not look the same as what went in the other side. (See Figure 8.7.)

In more common language, this means that free electrons are *opaque*: You can't see through them, because photons can't get through unchanged. Free electrons are not the only way to produce such opacity: Complicated materials like wood are opaque, too, without having any free electrons. But among simple, transparent atomic gases in space (e.g. hydrogen and helium), ionizing them to generate free electrons is how they become opaque.

By way of contrast with the opacity of free electrons, consider bound electrons in the Bohr model of an atom, as we discussed in Section 8.2.

Bound electrons—those inside atoms, ions, and molecules—are relegated to specific energy levels only. We already saw what that means: Bound electrons can interact only with a very particular set of photons. Specifically, bound electrons can absorb photons only if they have the precise amount of energy (the precise color) that the electron needs to jump to a different, quantized energy level. So the description breaks down like this:

Free electrons are opaque. Light cannot travel unchanged through free electrons.

But

bound electrons are transparent (except to photons with very particular colors). Most photons can travel straight through a gas of normal atoms. Only a few very

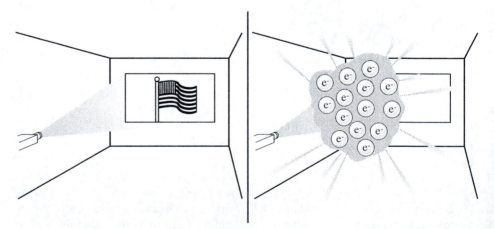

FIGURE 8.7 Effect of Free Electrons on Light In the left panel, the light streams freely from the projector to the screen, transmitting an accurate image. In the right panel, a cloud of electrons gets in the way, so the light reaching the screen last interacted with the electrons, not the projector, resulting in a smeared-out and unrecognizable image. Because of the scattering, the smear of light ends up all over the room, not just on the screen. Fortunately for movie audiences, hot plasmas with free electrons are much more common in stars than in theaters.

specific shades of a few very specific colors are hindered by bound electrons. All other photons—the vast majority—don't encounter any obstruction at all. For example, compare Figure 8.7 with Figure 8.4(b).

The cosmic microwave background radiation, which arrives here on Earth from every direction in the sky and appears to pervade the entire universe, appears to have a history involving *both* types of electrons. We can't see anything *through* the CMB; its source appears to be totally opaque. But today, the microwaves are unhindered as they travel through space. That is, the universe is transparent to them, which means that the photons are still out there, freely traveling every which way, including toward our planet. Because nothing is obstructing them, they are able to travel here from long ago and far away, and they provide us with an image of the universe's past.

Supplemental Topic: the Sunyaev–Zel'dovich Effect

Before we move on to study the CMB in more detail, we can take advantage of our knowledge of photon–electron scattering and cosmic microwaves in order to augment the astronomical distance ladder from Chapter 3. This topic is something of a detour, and you can skip this subsection and still follow the main "story" of the chapter. But the subject is worth thinking about because it beautifully ties together a collection of seemingly disparate topics that we've studied up until now. Its logic follows from a series of principles that we've already seen:

ALREADY SEEN PRINCIPLE #1: Rich galaxy clusters are pervaded by a hot ionized gas called the *intracluster medium* (from Section 4.3). Since the gas is ionized, it is flooded with free electrons.

ALREADY SEEN PRINCIPLE #2: These galaxy clusters are located in the universe, which is all we need to conclude that they're also flooded with CMB photons. There's the setup: photons and free electrons.

ALREADY SEEN PRINCIPLE #3: Because the intracluster gas is hot, its electrons are highly energetic. By comparison, microwave photons have low frequency and low energy. Therefore, when the photons and electrons scatter, $\gamma + e \rightarrow \gamma + e$, the energy exchange preferentially *up-scatters* the photons (up to higher energy) and *down-scatters* the electrons (down to lower energy). That is, energy is transferred from electrons to photons. Up-scattering is shown in Figure 8.8.

In 1969, Soviet physicists Rashid Sunyaev and Yakov Zel'dovich predicted that we could observe this effect upon looking at CMB light coming from the direction of a rich galaxy cluster (and *only* light emanating directly from such a

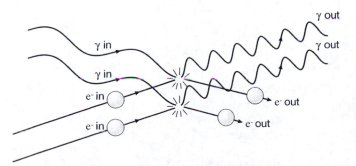

FIGURE 8.8 Up-scattering of CMB Photons   CMB light, which appears to flood the universe uniformly, is "up-scattered" when it passes through the middle of a galaxy cluster whose intracluster medium (ICM) is filled with high-energy free electrons. The free electrons have more energy than the CMB photons, so interactions (such as collisions) between a photon and an electron tend to increase the photon's energy at the expense of the electron's energy. This effect can be seen in the figure: Wavelengths of photons shorten by absorbing electrons' kinetic energy (and hence speed).

FIGURE 8.9 The Sunyaev–Zel'dovich Effect   This picture is an artistic blend of the two data sources needed to study the SZ effect. The array of radio/microwave ground telescopes pictured at the bottom is the Caltech radio millimeter interferometer, used to image the CMB. The contour lines in the middle of the picture show where these instruments find a deficit of CMB photons that are otherwise "supposed to" be there, but have been up-scattered out of the detector's frequency range. The orbiting Chandra x-ray telescope (top) is used to image the x-ray light (bright spot, middle) from a galaxy cluster's intracluster medium (ICM), whose electrons are responsible for the up-scattering. Notice that the x-ray hot spot coincides perfectly with the microwave contours.

particular direction). We should see an *up-scattered* blackbody, with an unusual excess of higher-than-peak-frequency photons and a corresponding reduction in photons below the peak frequency. This **SZ effect** has in fact been observed. It becomes useful as a distance indicator when we consider the strength of the effect in a given cluster, because that strength depends directly on the number of free electrons in the cluster. The more free electrons, the higher is the probability that a particular CMB photon will be up-scattered. Therefore, the strength of the observed SZ effect informs researchers about the number of electrons involved. All of this leads to a distance measurement to the cluster when we figure in the following already seen principles 4 and 5:

ALREADY SEEN PRINCIPLE #4: The intracluster medium is observable in x-ray light. (Recall Section 4.3.) The luminosity depends on the number of electrons and on the gas temperature, both of which are known: You get the number of electrons from the SZ effect, and you get the temperature from the x-ray spectrum. The combination of microwave and x-ray data that we're talking about here is illustrated in Figure 8.9.

Finally, we have

ALREADY SEEN PRINCIPLE #5: Now we know the x-ray luminosity, so we can measure the x-ray flux arriving here and solve for the distance with the inverse-square law (from Section 3.2).

Estimating distances via the SZ effect may seem like a complicated chain of logic, and it certainly is an advanced cosmology topic, but it follows directly from the five principles listed. The SZ distance measurement technique isn't particularly useful all by itself; it suffers from fairly large experimental uncertainties, mostly because clusters aren't perfectly spherical, so the ionized gas tends to occupy an unpredictable spatial arrangement. Another limitation of the method is that it works only for rich galaxy clusters with hot x-ray-emitting gas, and not for ordinary galaxies or other sources.

The value of SZ distances is twofold: (1) The technique works for very distant sources, comparable to the type Ia supernova standard-candle technique. These are the sources with the greatest experimental uncertainties by other methods, so having a second measurement method to confirm distances is very useful. (2) The measurements provide an *absolute* distance indicator, as opposed to a *relative* one. They therefore do not need to be calibrated by more local measurements, as the distances obtained by most techniques do. (Recall Section 3.3.) For example, if Cepheid variable star measurements turn out to be all wrong for some reason, then supernova Ia light-curve distances will be flawed, too, but SZ distances will be unaffected.

8.4 The Cosmic Microwave Background

The last three sections involved quite a bit of small-scale physics. Now, together with what you know of the cosmic microwave background already, your knowledge of photons and electrons qualifies you to interpret the CMB light.

We begin with the observed CMB itself. From Chapter 5, we know that

- It appears to be *cosmic:* It is everywhere in our sky, coming from every direction in space. (The cosmic nature of the CMB is corroborated by the SZ effect from the previous section, which tells us that the CMB light has passed through distant galaxy clusters.)

- It is *uniform:* It looks remarkably the same in every direction
- It is a *blackbody:* Its spectrum is that of a blackbody source with a single temperature of 2.725 degrees above absolute zero everywhere.
- And it is the *best* blackbody known: Unlike the spectra of stars, the CMB's blackbody spectrum is not contaminated by absorption lines.

As the low-energy microwave photons fly through space today, they encounter no obstacles (or at least, exceedingly few of the photons encounter obstacles). We see CMB light from everywhere in the sky, coming through galaxies and galaxy clusters, near and far. Apparently, these photons arrive from very far away, and almost none of them interact with anything during their journey. That is to say, the universe is *transparent* to them and has been for a long time.

A Scattered Sky

Based on the physics from the previous sections, we can understand the transparency of space in terms of electrons. The CMB photons evidently have not encountered any *free electrons* for a long time. They may encounter any number of bound electrons because photons don't interact with them.

Now consider that these photons correspond to a remarkably perfect blackbody. From Section 8.1, we know what a blackbody source must be like:

- It must be unified by a single temperature—which is what we see directly in the CMB spectrum today.
- It must be opaque—meaning that photons couldn't travel very far without interacting with something and thereby exchanging energy with the medium they originated from.

In other words, long ago and far away (before the Galactic Empire, the Galactic Senate, or the Jedi), photons were constantly interacting. We can again understand this behavior in terms of our study of bound and free electrons: CMB photons must have emerged from a medium flooded with *free electrons*, which served to make the source opaque. This event is known as *last scattering:* the last time the photons interacted with matter (electrons), before the universe became transparent.

The imperfections in non-CMB blackbodies—hot stars, hot coals, hot people—are caused by the effects of quantized energies in bound electrons. Evidently, the CMB's source was devoid of these flaws (i.e., devoid of atoms).

We have determined, then, the necessary history of the CMB:

- It started in an opaque gas with free electrons (completely ionized, so there were no atoms).
- Then it traveled through a transparent gas of neutral atoms.

All of this is pretty compelling; it's hard to imagine any other scientifically viable explanation for the *observed* CMB, which is a distant source with a perfectly uniform origin. (Although if you're not convinced about this scattering story yet, hang on until Section 8.5, where it gets even better.) Now we turn our attention to how the cosmic microwaves might maintain their perfect blackbody spectrum as they travel across the universe and through many atoms. Should there be absorption lines? Should the cosmic expansion affect the blackbody spectrum?

An Unblemished Blackbody Sky

Let's start by considering a point source. The brightest and most distant point sources are quasars. Recall from Section 6.6 that quasars are starlike points in the sky, except that their power source appears to be much more energetic—presumably a supermassive black hole. So distant are these quasars, that by the time their light arrives here for us to see, it has traveled through much of the observable universe. En route, it passes through many objects in space, such as galaxies and gas clouds, which are full of atoms. Perhaps we can use quasar spectra, then, to calibrate our expectations about the effect of this gas on traveling CMB light.

Recall from Chapter 7 that our universe appears to behave as an expanding space and that light traveling through this medium stretches in wavelength due to the cosmological redshift. Quasar light is no exception. Light from a quasar is continuously redshifted as it travels through some gas-filled object, then out into emptier space, then through another object, and so on, until it arrives here. Each object's hydrogen atoms absorb hydrogen line frequencies and thereby add dips in the quasar's light spectrum. But since all the wavelengths expand (the colors redden) from one object to the next, the spectrum gets loaded up with many absorption lines (essentially all from the *same* electron transition in hydrogen), but all at *different* wavelengths, due to the ever-growing redshift. (See Figure 8.10.)

How could the CMB, which arrives from an even greater distance than the quasar light, have avoided absorption lines? How could it avoid atomic and molec-

FIGURE 8.10 Absorption of Light from a Quasar   Quasars are bright, distant light sources whose light travels through many different intergalactic gas clouds along the way before arriving here. With each cloud, more light is absorbed, always at the same frequency, determined by the hydrogen atoms. But since the light is continually redshifting as it travels, these absorption events show up in many different parts of the spectrum we observe today.

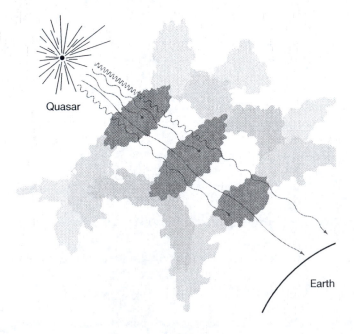

ular absorptions when it traveled through all the same gas that absorbed quasar light?

For one thing, the CMB temperature is very low, and has been for much of its journey, so its photons don't carry enough energy to cause the kind of hydrogen absorptions we see in quasar spectra. Those particular lines correspond to temperatures around 100,000 kelvins; the CMB, over its flight time, has never been nearly that hot. So absorptions in the CMB would have to come from less energetic electron transitions caused by far less abundant atoms and molecules. Second, there are around a *billion* CMB photons per hydrogen atom out there, so any absorption lines might be difficult to detect. That is, on average, there are a billion untouched CMB photons for every time one is absorbed. Still, what if there are some low-energy absorptions that could accumulate over the very long time the CMB light has been traveling through gas systems? Should they produce lines that we might be able to detect someday in the CMB spectrum?

An interesting and instructive response comes not from physics, but rather from geometry. Recall that the CMB is everywhere, traveling in every direction. It is illuminating to contemplate this geometry in general, since the cosmic nature of the CMB is so important to our study of the universe. Let's begin by supposing (incorrectly) that CMB light is absorbed often, all over the universe. Then the atoms that absorb the light eventually reemit it, also all over the universe.

From every direction, CMB light rays headed here would get absorbed and reradiated away, toward some other random direction. Those rays would never get here, so we should see absorption lines—dark gaps—in the CMB spectrum. But other CMB rays, which weren't aimed toward us to begin with, would also be absorbed, and when the electrons that absorbed them returned to their previous energy levels, that light would also be reemitted in some random direction. The consequence is this: On average, an absorption that removes light from our line of sight would be countered by some *other* absorption that reemits light *back into our line of sight*. Emission balances absorption.

Notice that this effect wouldn't hold for a point source like a quasar. Since all its rays originate from the same place, most of the rays would scatter *away* from our line of sight, and essentially none would scatter into it. The geometric effect for the *cosmic* microwave background would sharply mitigate the observable effects from any bound electron interference its photons might encounter.

Now, what about the cosmic expansion? The CMB today has the spectrum of a 2.725 K source. But its photons have been traveling across space for a long time, and in that time they have been stretched by the cosmological redshift (just like any other photons in space). Does it make sense for a blackbody spectrum to be redshifted and yet remain a blackbody spectrum (with the appropriate shape for its temperature)?

The answer turns out to be yes: The shape of the distribution of blackbody radiation remains intact when that radiation is exposed to a uniform redshift. The math behind this claim is quite complicated, but the reasoning isn't too bad; work through Exercise 8.3 and see for yourself. The resulting spectrum is still that of a blackbody, but one of lower temperature. Therefore, astronomers can use measurements today to calculate the conditions of the past, at the time of last scattering. Evidently, the CMB was formed within a hot blackbody and has since been redshifted to look like a cold one.

EXERCISE 8.3 Redshifting a Blackbody

To better understand why our measurements of the CMB today inform us about the universe in the past, consider the two ingredients necessary for the cosmological redshift to take a hot blackbody and turn it into a cold one:

(a) We would need to replace hotter ("bluer") photons with cooler ("redder") photons; that is, we would need to shift the blackbody curve's peak to the left. Refresh your understanding of the cosmological redshift by writing a quick explanation for how this happens.

(b) We would also need to reduce the overall intensity at *every* energy, including microwave energy; that is, we would need to drag the blackbody curve downward. How might this come about? To begin linking important physical concepts together, try to answer this question on the basis of general relativity from Chapter 7. That is to say, assume that general relativity provides the *reason* behind the redshift, in the form of expanding spacetime, and use that reason to speculate on how the total light intensity would diminish, in addition to shifting the color.

A Spotted Sky

We noted in Chapter 5 that the CMB is almost perfectly *isotropic*—that it looks the same in every direction on the sky. We're now ready to quantify the level at which imperfections in this CMB isotropy begin to show up. If we look with high enough precision at the cosmic microwaves, we see that there are minute temperature variations—**anisotropies**—from one direction on the sky to another. Figure 8.11 is a map of the CMB anisotropy as observed by the WMAP ("Wilkinson Microwave Anisotropy Probe") satellite in 2003. This is the same microwave sky as in Figure 5.6 (and Color Gallery 2, page 22), except that here the contrast is greatly exaggerated in order to display the tiny fluctuations. Red and blue spots on this map differ in temperature by only a few *thousandths* of a percent!

Both the remarkable isotropy and the magnitude and size on the sky of the slight anisotropies are worth appreciating in the context of your study of photons and electrons.

FIGURE 8.11 CMB Anisotropy   The variations in color (shades of gray here) indicate the temperature, just as in the maps of the sky and the Earth in Chapter 5. Here the variations shown are only some hundred thousandths of a degree. (◆ Also in Color Gallery 2, page 22.)

ISOTROPY: The CMB is the same temperature to within ±0.003% in every direction on the sky, which is remarkable. To see why it is so remarkable, observe that if you point your arms in two totally opposite directions on the sky, then the source of one direction's microwave photons is very far from that of the other source. Somehow, evidently, both locations were perfect blackbodies with the same temperature prior to last scattering. In both places, the free electrons moved just as fast and interacted with photons just as often. Then, somehow, in both places, the free electrons vanished.

The isotropy that resulted remains an important clue to the history of the universe. Why did the universe have such a precisely uniform distribution of matter (electrons) in the past? After all, it certainly isn't that way now: All the matter is concentrated in objects like galaxies, separated by millions of light-years of practically nothing. As we assemble a cosmological theory to explain our universe, keep in mind that it will have to account for this striking change over time.

ANISOTROPY: When we pay attention to the 0.003% level (which is less than 0.0001 K, since 0.003% of 2.725 K is 0.0003×2.725 K $= 0.00008$ K), we see microwave spots—anisotropies—as in Figure 8.11. Some are brighter and hotter than average (but only a little!), and some are dimmer and cooler, but still nearly perfect blackbodies. In terms of the particles that create the blackbody spectrum, we can assert that hot spots on the sky come from distant regions of space with higher-than-average temperature—that is, *faster particles*.

It's useful to ask whether there's a pattern to the CMB spots. Were patches of faster moving particles littered around randomly, or were they arranged in some more regular way? Do all the hot spots usually imply the same amount of excess speed? Are certain speeds more common than others? Are small hot spots more common than large ones?

It's difficult to answer these questions just by inspecting Figure 8.11 by eye. Fortunately, it's possible to perform a statistical analysis on the spot pattern with a computer. It turns out that there is a correlation among the different sizes and types of spots. Hot and cold spots occur equally often, and spot sizes are generally distributed in the same way everywhere. If you look in one direction, you'll see approximately the same number of hot spots of size X, and cold spots of size Y, as in any different direction.

What does the statistical analysis tell us about the hotness and coldness of the spots? First, the excess speed is not the same in every hot spot: Some spots are hotter than others. That much you can see directly in the WMAP figure. But although these particle speeds are not all the same, they do follow a distinctive pattern. We find that the temperature variations, hot or cold, are *most pronounced in spots of certain sizes*. For example, the biggest departures from 2.725 K (which are still less than 0.0001 K) occur in spots that occupy about 1 degree of angle on the sky; bigger and smaller spots tend to be closer to the average temperature. A few other spot sizes are favored like this, too, and the pattern is far from simple. We'll have more to say about the pattern in Chapter 12; at the moment, the key point to recognize is that the spots are not purely random and their complicated pattern will need a natural explanation.

Whatever sort of disturbance causes the anisotropy, it must maintain the blackbody nature of the source, while at the same time boosting its temperature only by a tiny percent. Hot spots on our best local blackbody, the Sun, are caused by changes in its gas medium: localized variations in pressure and density. But even if some similarly mundane effect generated the CMB's anisotropies, it still must have some very unusual properties:

- Its effect is remarkably weak (0.003 %).
- Its effect is remarkably uniform: The hottest spots are just as hot everywhere on the sky. Even the *an*isotropy shows remarkably isotropy!
- The spot pattern isn't random; some spot sizes are more prevalent than others. There must be a reason for this.

EXERCISE 8.4 Isotropic Anisotropies

Compare the WMAP view of the microwave sky (below, left) with the with the satellite picture of downtown Chicago (below, right). In each image, look for the pattern shown in the inset (a particular hot spot in the microwave map on the left and Daley plaza on the right). Once you find each (allow six to eight weeks), comment on what your search tells you about the microwave anisotropies. In particular, think about how every segment of the image you scan is unique, yet they all manage to look the same anyway. In what ways is the CMB random, and in what ways is it not?

8.5 Polarization

We now investigate one final, and cosmologically important, piece of photon–electron physics.

Polarization is a property of light, like intensity or wavelength. It's often observed in light that has bounced off of something, and the most common example in everyday life is *glare*. You can buy polarized sunglasses to reduce glare; these glasses are particularly useful if you do some glare-intensive activity like boating or skiing, or just driving west in the late afternoon. (See Figure 8.12.) Polarized sunglasses block the glare-related photons, but let others pass through. How do the glasses know the difference?

The answer has to do with the polarization of light. You first read in Chapter 2 that light can be viewed as an *electromagnetic wave*, which involves an oscillating electrical disturbance, getting stronger and weaker, stronger and weaker, and so on. That disturbance is known as an **electric field**. (There's also an oscillating magnetic disturbance, called a magnetic field, but we'll focus on the electrical one here.)

In Section 8.3, we made the point that the electromagnetic properties of light enable it to push and pull on electrically charged particles, such as free electrons. But that's not the complete story, because electric fields have a *direction* to them, as indicated in Figure 8.13. Because of this directedness, a particular light wave (or photon) can push electrons only left and right, or up and down, *but not both*. Now we can officially define **polarization**: it's the property of a light wave that describes

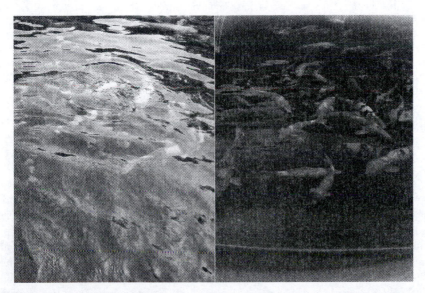

FIGURE 8.12 Polarization and Glare   Photographs of fish in a pond without (left) and with (right) a polarizing filter in front of the camera (◆ also in Color Gallery 2, page 23). The filter blocks light that has reflected off the surface of the pond, making it easier to see the light coming from the fish below the surface.

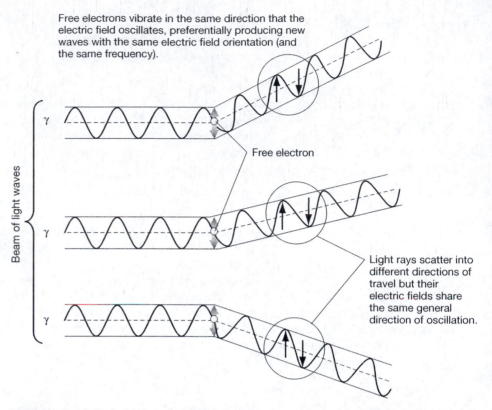

Free electrons vibrate in the same direction that the electric field oscillates, preferentially producing new waves with the same electric field orientation (and the same frequency).

Free electron

Light rays scatter into different directions of travel but their electric fields share the same general direction of oscillation.

Beam of light waves

FIGURE 8.13 The Polarization of Light After Interacting with Free Electrons   A beam of light waves, all polarized in the same up–down direction, shines on free electrons. The electrons respond by jiggling up and down, since the light's electric field oscillates in those directions only. Then, since jiggling electrons cause light waves (recall Chapter 2), these outgoing light waves share the same vertical polarization from the same up-and-down jiggling electrons, even though they scatter into different directions. Note that vertically polarized light is not equally free to scatter in all possible directions. Scattering is largely confined toward directions that lie in the horizontal plane, perpendicular to the up–down jiggling of the electrons.

the direction along which its oscillating electric field points (and along which it will jiggle a charged particle like an electron). For example, light that pushes electrons up and down is referred to *as vertically polarized*, while light that pushes electrons left and right is called *horizontally polarized*.

As you have done many times in this chapter, imagine a light ray shining on a free electron. The photon's oscillating electric field shakes the electron (up and down, say), there is some energy exchange, and the light beam deflects into a direction that preserves this electric field orientation (up and down). This is essentially how photon–electron scattering works in detail, although this description is somewhat simplified in order to convey the most relevant aspects for our interest in CMB cosmology. The incoming and outgoing light rays tend to share the same electric field polarization, because both rays are linked to the motion of the same electron. You could say that the incoming ray *causes* the motion of the electron and the outgoing ray *is caused by* the motion of the electron.

All of this is much easier to *see* than to read about, so direct your attention to Figure 8.13. There you'll see that when the light deflects off the electron, its deflection will be constrained to a particular range of directions (specifically, those that lie roughly within a plane perpendicular to the polarization of the photon). This constraint is produced by the combination of two facts: (1) The polarization (electric field orientation) of the incoming photon tends to be the same as the polarization of the outgoing photon. (2) The polarization of a photon is always perpendicular to the direction the photon is traveling. As a result, light with a particular polarization is not free to scatter into just any direction. Figure 8.13 on the previous page illustrates that vertically polarized light is constrained to preferentially scatter in directions that lie within a horizontal plane. Similarly, horizontally polarized light is contrained to scatter roughly within a vertical plane, and so on for light with other electric field orientations.

The key insight from the previous paragraph is that light of particular polarizations scatters preferentially into particular directions. But most of the time, a beam of light contains many photons with lots of different, random polarizations. Some photons have their electric fields running up–down, others left–right, and still others at every possible angle in between. This sort of blend is called *unpolarized* light, meaning light that contains photons of all polarizations. When a beam of unpolarized light scatters off of a group of free electrons, particular polarizations will scatter in particular directions, according to what we learned in Figure 8.13. This means that *scattering has the effect of separating out unpolarized light*, sending different polarizations in different directions. (See Figure 8.14(a).) Since we view the scattering from one particular direction, this means that much of the light that makes it to our viewpoint *is polarized the same way*. The constraint of "scattering from one particular direction toward another particular direction" has the important consequence of polarizing formerly unpolarized light.

This, by the way, is the miracle behind polarized sunglasses (Figure 8.14(b)). If unpolarized light traveling in a particular direction creates a glare by deflecting off a surface, then the light headed for your eyes becomes polarized in a particular direction by the deflection. The glasses simply employ a filter which eliminates photons polarized in that one direction, so that the remaining light that passes through to your eyes does not include the glare. For the sake of these sunglasses, it is convenient that glare is usually caused by light deflecting off a *horizontal* surface, so the direction of polarization is the same every time. If you tilt your head, ear-to-shoulder (as if you're eating a taco), then your polarized sunglasses will no longer be of much help. (Fortunately, most taco eating takes place indoors.)

We need just one more example before we're ready to understand the polarization of the CMB. Imagine that light comes *from* a variety of directions and scatters off of free eletrons *into* a particular direction, such as toward you. This situation is drawn in Figure 8.14(c). This scattered beam that you see is unpolarized because it's made up of a variety of photons with electric fields oscillating in all different orientations. So the only way for electron scattering to cause unpolarized light to become polarized is if *both* the incoming and outgoing directions are constrained. It's not sufficient that light scatter toward a particular observer; it has to scatter *from* a particular direction as well (as we saw in Figure 8.14(a)).

Now that we have some understanding of what polarization is and how it can be produced by scattering, we will use this understanding to gain insight into the origin of the CMB. If the CMB is truly of cosmic origin, traveling in all places and

FIGURE 8.14 Polarization of Light due to Electron Scattering   (a) If a beam of unpolarized light comes from a particular direction, as from a single light source, then those rays which scatter into a particular direction will share a common polarization. In this way, unpolarized light can become polarized through electron scattering. That is, when a beam of unpolarized light *scatters* toward us, it becomes polarized. (b) Polarizing sunglasses (and camera filters) are oriented to weed out rays polarized by bouncing off a flat horizontal surface (such as a road or a lake). (c) Unpolarized light originating from many different directions remains unpolarized after scattering. Multiple rays that get scattered into a particular direction, toward a single observer, remain unpolarized.

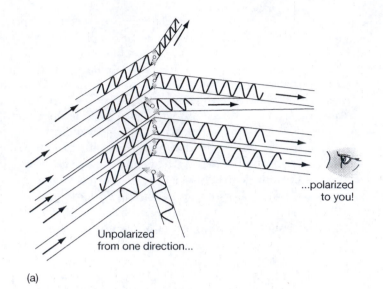

...polarized to you!

Unpolarized from one direction...

(a)

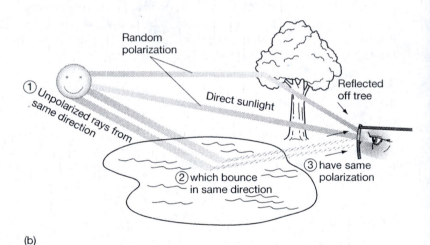

Random polarization

Direct sunlight

Reflected off tree

① Unpolarized rays from same direction

② which bounce in same direction

③ have same polarization

(b)

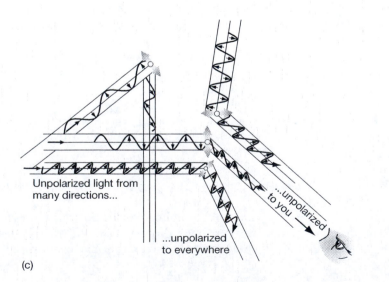

Unpolarized light from many directions...

...unpolarized to everywhere

...unpolarized to you

(c)

in all directions after it last scattered off a blackbody field of free electrons, then there is a necessary consequence: *The CMB should be polarized* in a specific way.

EXERCISE 8.5 Polarization Experiments for Freedom-Challenged Individuals

Imagine a prisoner behind bars. The prison librarian wheels a cart of books and magazines around, cell to cell, offering reading material to the inmates. To hand the prisoner a book, the librarian must turn it sideways to pass it through the bars. But what if the librarian were to save time by standing in the middle of the cell block and just throwing books randomly, with random orientations (some toppling end over end, some spin stabilized like a Frisbee, etc.) with respect to the cell bars? Construct as complete a model as you can with this idea, to explain polarization phenomena by analogy. Give some examples, such as what happens to a book that hits the bars but still goes through.

To see how the CMB polarization works, we can look at some particular hot spots in the anisotropy field. Since these spots are measured to be perfect black-bodies just like the rest of the CMB, but slightly hotter than the surrounding regions, they will have an excess of light radiating between them. We can think of think of the excess photons as traveling along particular channels between hot spots, like cars traveling along particular highways between cities. These channels, where the excess photons run back and forth *in a particular direction* (along the channel), provide an environment fit for creating polarization. If we have some knowledge that constrains the direction that light was traveling before and after it scattered, then we can figure out the direction of its polarization. In the present discussion of CMB polarization, we're halfway there, because we know the direction the light was traveling before it scattered.

We earthlings (particularly cosmologist earthlings) are concerned with the CMB rays that happen to scatter toward us; those are the only rays we get to see. Thus, our observations also constrain the direction after scattering. In other words, the situation with CMB hot spots is like that in Figure 8.14(a) or (b), in which both the initial direction (between hot spots) and final direction (toward Earth) are constrained. In Figure 8.15, we see how this should appear in the microwave background radiation. Light rays traveling between distant hot spots and that deflect toward us ought to have electric fields oriented in a particular direction (indicated with double-headed arrow in the figure).

In reality, the hot and cold spots on the CMB sky are more complicated than just "spots." They're splotches, wavy lines, speckles, and so on, so the polarization they produce is more complicated than Figure 8.15 indicates. The first detection of CMB polarization was made by a team of researchers led by John Carlstrom at the University of Chicago, using the Degree Angular Scale Interferometer (DASI) experiment that they designed and built. They published the following results, in which they highlighted the scientific significance of observing the CMB polarization, in addition to the CMB temperature fluctuations (anisotropy), which had already been observed:

> "Within the context of this [standard cosmological] model, recent measurements of the temperature fluctuations have led to profound conclusions about the origin, evolution and composition of the Universe. Using the measured temperature fluctuations, the theoretical framework predicts the level of polarization of the CMB with essentially no parameters. Therefore, a measurement

*Among many CMB pioneers are **David Todd Wilkinson** (1935–2002) and **John Carlstrom** (born 1957). Wilkinson, of Princeton University, was a founder of both the COBE and WMAP satellite-based CMB experiments. (The WMAP satellite CMB experiment was named in his honor: the Wilkinson Microwave Anisotropy Probe.) Carlstrom, of the University of Chicago, pioneered the DASI experiment, which made the first-ever detection of the CMB polarization. Carlstrom also performs leading cosmological research involving the SZ effect.*

of the polarization is a critical test of the theory and thus the validity of the cosmological parameters derived from the CMB measurements. Here we report the detection of polarization of the CMB with the Degree Angular Scale Interferometer (DASI). The polarization is detected with high confidence, and its level and spatial distribution are in excellent agreement with the predictions of the standard theory."

—J. M. Kovac, E. M. Leitch, C. Pyrke, J. E. Carlstrom, N. W. Halverson, and W. L. Holzapfel, "Detection of polarization in the cosmic microwave background using DASI"

In other words, within the context of our current theory of the universe (coming in Chapter 10), cosmologists can predict the CMB polarization pattern. The first measured data matched that predicted pattern perfectly. In the sky map from DASI (Figure 8.16), dashlike lines indicate the direction of the electric field (the polarization) for light rays headed toward us.

The detection of CMB polarization is a wonderful achievement in cosmology. It confirms, beyond a reasonable doubt, that the CMB must have been formed by light scattering against free electrons. Without the polarization evidence, you might be tempted to say that a uniform blackbody glow could be caused by some other source with a uniform temperature. You'd still have to explain the precision in the blackbody spectrum and in its distribution of spots, but maybe you can imagine somehow doing that. The measurement of polarization, then, and therefore of free-electron scattering, is important; it tells us specifically that the universe used to be opaque with free electrons and became transparent with neutral atoms. This means that the universe is *evolving*—changing over time—and doing so in a specific way that needs to be justified by a cosmological theory. And it means that

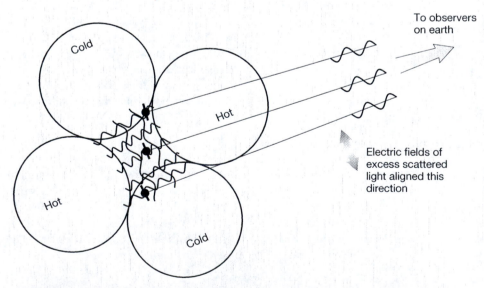

FIGURE 8.15 CMB Polarization   Light reaching us from the vicinity of, for example, CMB hot spots, should be polarized in a particular way. The light traveling among hot spots exceeds that traveling along cold spots, because hotter blackbodies are brighter. In order for some of this excess light traveling along a hot-spot "channel" to scatter in our direction, the scattering electron would have to jiggle in a particular direction also. In this way, CMB light that we observe from various directions on the sky should generally have a particular polarization pattern.

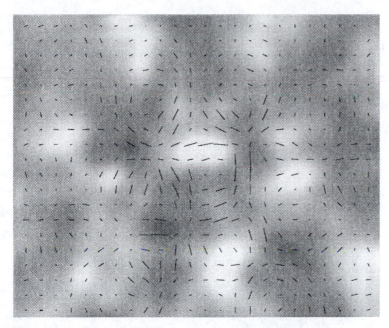

FIGURE 8.16 Measured CMB Polarization   Dashes indicate the strength and direction of polarization of light coming from the direction on the microwave sky where each dash is located. The overall pattern is a complex blend of polarization patterns from a variety of distant hot and cold spots. (◆ Also in Color Gallery 2, page 23.)

the CMB is worth our study; it is an image of the last interaction of distant, cosmic light—*an image of the history of our universe.*

Perhaps most of all, the polarization of CMB light means that we actually know what we're talking about. Without it, our understanding of photons and electrons, applied to redshifted light from deep space, might seem iffy. We can't directly travel to the light source and measure its properties in person, so it helps to have the polarization. It's a nitpicky detail that perfectly matches our expectations, so we can be confident that we're using the right physics to explain the light we see.

In fact, CMB polarization has become an area of dedicated scientific study, with variations in the polarization *pattern*—polarization anisotropies—being measured in great detail all over the sky. The work involves an intricate statistical analysis, but the results fit the predictions for electron-scattered, uniform-blackbody light. We've gotten good at studying the CMB—which is important, because now we can use it to become "good at" knowing the universe (starting in Chapter 10).

There is one more polarization detail that we haven't mentioned yet. It turns out that we measure *two* separate levels of polarization in the CMB. One is on small scales, corresponding to the spots in the microwave sky map; that's the one we've been talking about throughout this section. But the other spans larger scales, ranging over many spots. In Chapter 10, we'll see that the larger polarization patterns were imprinted more recently, requiring an *additional* round of free-electron scattering long after the original one, affecting some CMB photons already in flight. For now, we'll leave this as a minor mystery—another observation that we'll have to explain with a cosmological theory of the history and evolution of our universe.

Simplicity and Complexity in Two Particles

The world of particle physics has many inhabitants. Atomic nuclei are made of *quarks*, and quarks interact with each other with the help of *gluons*. In Chapter 9, you'll be introduced to particles called *neutrinos*, which often interact with help from *W* and *Z bosons*. You're already familiar with *gravitons*. Dark matter also must be made of something; perhaps that's another particle (a WIMP), or several other particles. Existing physical theories postulate the existence of other new particles, too, and there could always be more out there that we have yet to imagine.

But this chapter, whose observational motivation comes from the most far-reaching and pervasive radiation source in the universe, is based only on photons and electrons. In physics circles, the study of photons and electrons is a field all its own, called quantum electrodynamics. And if the name sounds complicated, that's because the science is. Still, without committing ourselves to that level of study, there are some interesting things to notice about photons and electrons in our world.

When charged particles such as electrons interact with each other, the process usually involves photons, and only photons. This wasn't expressly stated in the chapter, but it ought to sound reasonable, since photons transmit electrical disturbances. They might as well be the ones to carry electrical disturbances from one electron to another. Perhaps the most common electron interaction in our lives is that between atoms and between molecules. Chemical bonds are based on the spatial arrangement of electron charges. This means that, apart from the mere *existence* of nuclei, photon–electron interactions are responsible for the material world as we experience it. They allow the oxygen atom to bond to itself in O_2, the gas we breathe. They cause liquids and solids to be liquid and solid. They transmit energy from place to place, including to and from human skin, allowing us to feel things like hot and cold. And behind the scenes, they orchestrate all the other interactions we witness: pressure, condensation, magnetism, elasticity, etc.

Consider sight. Photons enter your eye and interact with the molecules in the rod and cone cells of your retina, via their electrons. A signal is transmitted electrically to your brain. The brain is an electrochemical processor, with neurons acting like silicon chips. Even if you see something only with your mind's eye, rather than with an actual eye, the action is solely that of photons and electrons the whole time.

That's not the story for the entire universe. The universe employs nuclear power and dramatic extremes of gravitational power. Its motions are dominated on large scales by dark matter, not photon-based electrical interactions. But as it turns out, the world that human beings experience every day, apart from simple gravity, boils down to the behavior of only two particles—particles with a rich portfolio of properties and interactions.

Perhaps therein lies a useful model. Do you choose to understand the universe in terms of all the various *arrangements* of particles—galaxies and stars, planets and people—or do you assemble your knowledge from only a few particles studied in great depth? How far can you go with a truly comprehensive understanding of just a few core concepts, like money, fitness, or courtesy? (After all, computers seem to accomplish a lot without knowing anything beyond ones and zeros.)

Quick Review

In this chapter, you've learned how to extract more information about matter by looking at the light it emits. In particular, you've learned that the cosmic microwave background radiation must have been emitted from an opaque gas of free electrons and has since traveled nearly unimpeded through the transparent universe of mostly neutral atoms. By looking at the details of temperature and polarization of the CMB light coming from different regions of the sky, it's even possible to discern the properties of the electron gas from which the CMB was last scattered. This information will provide important clues for our overall theory of the history of the universe.

Try the following crossword puzzle to test your knowledge of key terms and concepts from Chapter 8.

ACROSS

3 An electron with too little energy to escape from its atom

5 This type of radiation is the natural glow of an object due to its own temperature.

9 Mathematical relationship between temperature and peak wavelength for a blackbody

10 When a photon's energy is exactly used up in order for an electron to move from a lower energy level to a higher one

11 Abbreviated name for the frequency boost in CMB light passing through a galaxy cluster

13 A number that characterizes the average energy of the particles in an object

14 An electron in an atom moves within one of these

15 Type of radiation occurring only at certain specific colors

16 Describes an electron not bound to an atom

18 A star's luminosity depends on its _____ and its temperature.

19 The property of the CMB that describes the spots of slightly hotter- or cooler-than-average temperature

20 Increasing the temperature of a blackbody makes its spectrum both _____ and "bluer."

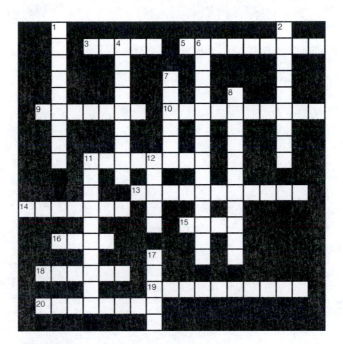

DOWN

1 Word that means only specific electron energy levels are allowed in a Bohr atom

2 A beam of light that's scattered from a particular direction to another particular direction by free electrons becomes _____.

4 Interactions between low-energy photons and high-energy electrons tend to _____ the photons.

6 Took place just before the CMB photons began traveling freely across the universe

7 A dense medium of free electrons is _____ to light.

8 Medium that allows photons to pass through unaltered

11 Two-particle interaction in which the same two particles leave as those which entered

12 Polarization refers to the alignment of the electric _____ for a beam of light.

17 Polarized light caused by reflection off a road's or lake's surface, for example

Further Exploration

(M = mathematical content; I = integrative—builds on information from more than one chapter; P = project idea; R = reflection; D = suitable for class discussion)

1. (RD) If you were to view people as particles, which team sports might be characterized by a single temperature on the playing field? Which of these sports are hottest and which are coldest? (Figure 8.1 and the surrounding discussion might help.)

2. (MRI) When a normal star (like the Sun) reaches "old age," its outer layers swell outward, so that the star grows in radius by about 50 times. Due to changes inside the star, the luminosity will increase by about 100 times. At this point, it is called a *red giant* star. The "giant" part comes from the increase in radius.
   (a) Use Equation 8.3 and your knowledge of blackbodies to demonstrate that red giants are actually red. (It is sufficient to show that they are red*der* than a white-looking star like the Sun.) Be quantitative in your answer.
   (b) Because the star's outer temperature decreases, fewer of its hydrogen atoms will be ionized. Explain why in terms of the Bohr model. Also, explain why "fewer" hydrogen atoms will be ionized, instead of "none." That is, why isn't there a hard cutoff?
   (c) The red giant's outer layers can keep on growing, ever outward, until they just dribble off into space. (This creates what's called a *planetary nebula*, which astute students of astronomical naming conventions might correctly suspect has nothing to do with planets.) List *two* reasons the growing gas layers are becoming increasingly transparent.

3. (RDP) Invent another analogy for the Bohr hydrogen atom, instead of the staircase analogy used in the chapter. Try to make it accurate in detail; for example, a staircase isn't quite accurate, because all the stairs are the same height, unlike the hydrogen energy levels in Figure 8.3.

4. (RI) This question concerns the interplay of the Bohr model and the Doppler effect. Consider a glowing cloud of hydrogen gas, emitting only line radiation.
   (a) If the cloud is moving toward you, what lines will you observe? Since these lines are not the same as the hydrogen lines you would measure in a laboratory (which would not be moving toward you!), how could you determine whether they even *are* hydrogen lines, rather than lines from some other element?
   (b) Imagine instead that the cloud is sitting still with respect to you, but that it's hot, and therefore all the emitting atoms are zooming around within the cloud, so that some of them are coming toward you, others are moving away, and others are moving sideways with respect to you. Now what would you expect to measure?

5. (RI) In Chapter 4, you read about the WHIM, or "warm–hot intergalactic medium," as a small constituent of dark matter. Assume that the WHIM is made entirely of hydrogen, and use the Bohr model to justify the existence of a "warm–hot" temperature range in which hydrogen could behave as dark matter in terms of its being detectable via photons. (*Note:* At very high temperatures, hotter than "warm–hot," ionized gas emits strong x-ray radiation. Why? Because ionized matter is full of free electrons, which are forced to accelerate when they pass near the positively charged ions, and that causes them to radiate. But this radiation is hard to detect at lower temperatures.)

6. (R) Radiative *recombination* is a process in which ionized matter (plasma) cools and becomes neutral again. The formerly free electrons *recombine* with the ions to make neutral atoms. Would you expect a source of this sort to produce *line emission* or *continuum emission* (without lines, like what a blackbody produces)?

7. (R) (a) Summarize the SZ effect, including any concepts you need along the way. (b) Draw a picture showing SZ-up-scattered CMB photons heading for Earth. Show where they up-scattered (what's located there?), and indicate where the photons came from before up-scattering.

8. (RD) Now that you've studied some of the physics behind the CMB, discuss the constraints on a theory that seeks to explain its origin. In particular, what must the source of the CMB be like in order to satisfy the observed overall isotropy, detailed anisotropy, and nearly perfect blackbody radiation? Your answer should include some mention of the history of electrons in the universe.

9. (R) What does an anisotropy spot tell you about the region that emitted its CMB light?

10. (RD) Would there be polarization in the CMB, caused by its last scattering, if (a) it had no anisotropy at all, or (b) it had a different average temperature, or (c) its blackbody spectrum was less perfect? Briefly explain your answer for each part.

11. (R) Recall from the very end of Section 8.5 that the CMB contains hints of larger scale polarization patterns than those discussed in the text so far. Speculate about what might cause them.

12. (M) It can be useful to have ballpark figures handy for what temperature of blackbody radiates at what peak color. Using Figure 2.20, or something like it, calculate approximate temperature ranges (in kelvins) for blackbodies of each of the following "colors": radio, microwave, infrared, red, green, blue-violet, ultraviolet, x-ray, and gamma ray.

13. (MI) You radiate essentially as an infrared-peaked blackbody.
    (a) What is your peak wavelength? (Notice that you'll need your body temperature in kelvins to figure this out!)
    (b) What's your overall flux? If you use Equation 8.2, your flux will come out in watts (like a light bulb) per square meter of your skin.

    (c) Suppose that the palm of your hand is similar in area to a 10 cm by 10 cm square. That's 0.01 meter by 0.01 meter, or 0.0001 square meter. Estimate how many watts of (mostly infrared) energy your hand puts out. Is this a flux or a luminosity?

14. (RPD) Sometimes astronomical photographs, including some telescope images of galaxies and nebulae, are displayed in "false color," meaning that those colors are different from what your eye would see. But is the human eye the best instrument available to determine what we consider "true" color? With a quick Internet search, find out how the human eye sees color and then write a short essay to address the following issues:
    (a) Which colors, if any, are we most sensitive to?
    (b) Which colors are the "primary colors of light"? Is their primariness a feature of the colors themselves, or is it just a feature of the human eye?
    (c) When you see yellow light, does it necessarily mean that yellow photons are entering your eye?
    (d) Think about this, and then offer your opinion: The Sun's blackbody curve is peaked in the "visible light" range of wavelengths, and the human eye is sensitive to essentially the same range. Is this coincidence interesting? Is it surprising?
    (e) Here's another opinion question: Is a true-color image superior to a false-color one?

CHAPTER

9

The Nuclear Realm

"To make an apple pie from scratch, you must first invent the universe."

—*Carl Sagan*

The next chapter, Chapter 10, pulls together the observations and concepts we've encountered so far into a single, coherent cosmological theory. This would be difficult to do from scratch, but we've already assembled most of the needed pieces. We have studied the theory of general relativity, to understand the expansion, the horizon, and the redshift. We have examined blackbodies, atoms, and scattering, to understand the microwave background. Just one major physical topic—nuclear interactions—remains to be investigated before we can understand the cosmic uniformity in the abundances of the elements.

One part of this story takes place in the bellies of stars; the other takes place very early in cosmic history. Both settings have temperatures beyond what can be reproduced for very long in a laboratory, and both settings are opaque, so they cannot be observed directly with telescopes. Yet neither of these limitations actually prevents us from discovering what happened. Nor do they relegate our studies to the purely theoretical: The nuclear reactions involved occur at rates that are measured in laboratories on Earth, and the cosmic reactions leave behind traces of material that can be observed in the universe.

As you read this chapter, try to appreciate the remarkable scientific achievements discussed here, such as understanding the center of the Sun, measuring invisible particles, and explaining the evolution of entire galaxies in terms of the behavior of tiny atomic nuclei. The domains of the unimaginably tiny and the unimaginably vast may sound distant and impersonal; but events in these unfamiliar domains are responsible for the emergence of everything you can see or touch around you. Every object in your personal experience—including your planet, your possessions, and every part of your body—shares the common history described in this chapter.

9.1 Energy

Although we'll be using it to help us understand distant astronomical objects, the modern formalized concept of energy emerged as a way to organize and refine the description of some very common and familiar experiences with the world. For example, you get tired when you climb a long flight of stairs. You get hungry if you exercise for a long time without eating. A lamp won't shine if it isn't plugged into a power outlet. A car stops moving if it runs out of gas. You speed up as your bike coasts down a hill, but you slow down again when you coast back up the next hill. All these experiences have the common feature that a particular change in the state of one thing is accompanied by corresponding changes in the states of other things. You can coast from the bottom of the hill to the top, but not without slowing down. You can move from the bottom of the stairs to the top, but not without feeling a little more tired. You can make your car move, but not without using up gasoline in the process. These observations point toward a familiar principle: "There is no free lunch," or "You can't get something for nothing." The lesson is that any change you *want* to make requires some sort of capacity to cause that change, and this capacity must be taken away from something else in the world, and given to the thing you're trying to change.

Let's try to sharpen this idea a little further by zeroing in on the notion that there is "something" that gets passed along or transferred as the "capacity to make things happen" moves from one part of the world to another. Imagine the chain of events that starts inside the Sun, where hydrogen (somehow) gives up "something" that is passed along to the light that streams toward the Earth. Some of the light is absorbed by plants, transferring this "something" from the light to the plants and enabling them to grow. Then we eat the plants (or we eat the animals that eat the plants) and gain the ability to do important things, like study cosmology.

So far, this idea is just a rough, qualitative statement. In order to understand that we are talking about energy, and to see the full power of the concept, we need to assign specific numerical values to this thing that gets passed from one part of the world to another. Once we know how to calculate these numerical values in different situations, we'll see that we can make a much more sweeping statement than our previous observation that in order to change one thing about the world, there must be a corresponding change in something else. We'll find that the changes obey a precise mathematical relationship.

Forms of Energy

Energy is the capacity to rearrange some part of the universe in certain ways. (That may not seem like a satisfying definition, but we challenge you to find or invent a better one!) Energy comes in many forms and carries the SI units of "joules." Officially, a joule ("J") is the amount of energy needed to maintain a force of 1 newton over a distance of 1 meter (as when pushing a box).

There are specific rules for calculating the amount of energy to assign to a certain object or situation. For example, if the part of the universe we're looking at consists of just a single particle of mass m moving at speed v, then m and v are the only two relevant properties of the system. Recall from Chapter 4 that the energy physicists assign in this case depends only on those two properties:

$$E_k = \frac{1}{2}mv^2 .$$

(eq. 9.1)

Another previous example is a system which stores gravitational energy (such as that between the Earth and Moon, also from Chapter 4)

$$E_g = -G\frac{mM}{r} \ ,$$

(eq. 9.2)

since a gravitating system is characterized only by masses m and M, and their separation distance r.

Another specific formula is that for thermal energy (introduced in Chapter 3),

$$E = mC_p\Delta T = \text{mass} \times \text{"specific heat capacity"} \times \text{temperature change} \ .$$ (eq. 9.3)

This relationship was a very important discovery in the history of science, because it showed that thermal energy can be interchanged with other, more useful forms of energy. Thus, it opened the way for understanding that energy is something that is never lost. For example, slamming on the brakes doesn't *kill* your car's kinetic energy; it merely *transforms* it into heat.

There are many other forms of energy: Light carries energy, electrical systems and chemical bonds store energy, etc. The central point is that the energy is quantified with a specific number associated with each arrangement of a system, and there is a well-defined procedure for computing energy in terms of the properties or qualities of that system, such as mass and temperature. This concept of an energy that we can compute is just a quantitative refinement of the general concept we discussed earlier: "something" that gets passed along from one part of the world to another and that represents the capacity to make things happen.

Conservation of Energy

The quantitative measure of energy allows us to formulate one of the deepest and most general laws of physics: the **law of conservation of energy**, which is also known as the **first law of thermodynamics**. It's a way of summarizing all of the experiences we've been describing. It says that energy cannot be created or destroyed. Energy can be transformed into different forms, but if you add up all the different forms, the total amount of energy never changes.

Another way of stating this law is that the total amount of energy is *conserved*. You can take energy from one place and move it somewhere else (the way sunlight travels through space) or change it from one form into another (as when the energy in sunlight heats up some water). When you add up all the energy at the beginning (in sunlight, in space) and at the end (in the hot water), making sure that none of it slipped away unnoticed (by sunlight reflecting off the water's surface, maybe), then the total amount always stays the same. One useful way to summarize this idea is to say that the *energy lost in one place or one form always equals the energy gained in another place or another form*.

The situation is similar to financial accounting for a complex corporation, where money stands in place of energy. The company may have many different bank accounts and many different channels through which money goes in and out (payments for supplies, purchases by customers, salaries paid to employees, bribes to government officials, private jet trips to islands in the South Pacific for the

CEO, etc.). But if you keep careful account of everything, no matter how many transfers occur, no money ever disappears; someone always has it. The same is true for energy: Anytime it looks as if some has been lost, it turns out that there is another "account" somewhere that we didn't know about before. Once we factor that account in, we find that no energy has really been lost.

This is the real benefit of the concept of energy: Once you know *how much* energy is available, you no longer need to know exactly *how* it operates; now you just have a question of *quantity*. Something is possible as long as there's *enough* energy. In this way, energy allows us to organize our understanding of the myriad processes and transformations that occur in the universe. Everything from bacteria, to plants, to cities, to stars, to galaxies can be described in terms of energy. The flow of energy drives the formation of all the incredible structure that we see on many scales throughout the universe.

Energy Requirements for Stars

In order to investigate energy use in astronomical objects like stars, let's add one more energy-related term to our vocabulary: **Power** is how fast energy is transferred from one form to another or from one place to another. In common language, power may be referred to as the rate of energy *used*, by an electronic device, for example. The SI units of power are watts (W), and one watt equals one joule of energy transferred (or used) per second. When we refer to such a rate of *radiant* (light) energy, we refer to that type of power as *luminosity* (which we introduced in Chapter 3). So luminosity, being a type of power, is measured in watts.

When light is quantified as a *flux* (also from Chapter 3), it is the rate of energy transfer per unit time *and* per unit area. Its units are therefore joules per second per square meter, and since W = J/s, flux is quoted in units of W/m^2. The flux of energy pouring onto the Earth in the form of sunlight (ignoring reflection back into space) is about 1400 watts per square meter (1400 W/m^2); this is called the **solar constant**. Because 1400 W is roughly the power needed to operate a microwave oven, the Sun could power one microwave oven on each square meter of land it directly illuminates. If you average the solar flux over the Earth's surface around the globe (since not all of the Earth's surface faces the Sun all at once), you get about 350 W—roughly the power drawn by a television—for each square meter of the surface of the Earth.

Now, the luminosity of the Sun is 3.9×10^{26} W. This energy is somehow released from inside the Sun and radiates out into space in all directions. (Recall the discussion of the inverse-square law in Chapter 3.) What source of energy could provide such a tremendous power output, lasting over the 4.6 billion-year present age of the Sun? We know from the law of conservation of energy that something must be giving up this amount of energy per second.

Astronomers puzzled over this phenomenon for a long time. The puzzle is sharpened by comparing the power output with the mass of the Sun. The Sun has a mass of 2.0×10^{30} kg, so, on average, each kilogram of the Sun is pouring out energy at a rate of

$$\frac{3.9 \times 10^{26} \text{ W}}{2.0 \times 10^{30} \text{ kg}} = 2.0 \times 10^{-4} \text{ W/kg} \qquad \text{(eq. 9.4)}$$

and has been doing so for billions of years without running out of fuel. Watts are joules per second, so energy released at 2.0×10^{-4} joule per second for 4.6 billion years (1.5×10^{17} seconds) adds up to about 30 trillion joules:

$$(2.0 \times 10^{-4} \text{ J/s kg}) \times (1.5 \times 10^{17} \text{ s}) = 2.9 \times 10^{13} \text{ J/kg} . \qquad \text{(eq. 9.5)}$$

What source of energy can release over 30 trillion joules from each kilogram of matter? Burning 1 kg of gasoline provides only about 50 million J, so the Sun must be powered by something nearly a *million* times better! This simple comparison provides strong evidence that the Sun, though fiery in appearance, is not powered by *burning* anything—at least not in the usual sense of the word "burn." Burning, or combustion, despite being one of the most energetic chemical reactions, doesn't even come close to being able to power the Sun.

EXERCISE 9.1   How Do We Know the Luminosity of the Sun?

Recall the standard-candle method for measuring distances in Chapter 3. There, we assumed that we knew the intrinsic luminosity of the object, and we measured the flux of light received on Earth, in order to calculate the distance to the object. But we can also use the same relationship in a different way, to calculate the luminosity of a star whose distance we know, given its apparent brightness (the flux of energy we receive). Calculate the luminosity (or "wattage") of the Sun in this way, using the flux that arrives here at the Earth, 1400 W/m², and the measured distance to the Sun, 1 AU $= 1.5 \times 10^{11}$ m.

9.2 Nuclear Interactions

Nuclear physics is the secret behind the power of the Sun and other stars. So before we go any further, let's refresh our knowledge of nuclei from Section 5.4 and elsewhere:

- All the matter we're familiar with, in all its variety and complexity, is made up of only 92 natural building blocks: chemical elements listed in the periodic table. In different combinations, these elements are the ingredients that make up our bodies, this book, the air, etc. (If you're paying close attention, you might complain that there are more than 92 elements in the periodic table. However, atomic numbers 93 and higher are generally human-made elements that we don't find occurring naturally on Earth.)

- The elements are atoms, which are built from just three building blocks: the proton, neutron, and electron. The variety in the periodic table arises because elements are made up of different combinations of protons, neutrons, and electrons.

- Nuclei, consisting of protons and neutrons, are only a few *femto*meters (10^{-15} m) across; by contrast, the surrounding electron cloud is generally about 100,000 times larger.

- A single element, defined by its atomic number (number of protons), can exist in several different isotopes—that is, with different numbers of neutrons.

By convention, to identify the isotope, we write the total number of protons and neutrons (i.e., the mass number) as a superscript before the chemical symbol. For example, the periodic table lists helium as atomic number 2. (Helium has two protons). Therefore, ^{3}He must have one neutron in addition to its two protons, and ^{4}He must have two neutrons.

The power source of the stars comes ultimately from the binding energy of the **strong nuclear force**, the natural force that holds the nuclei of atoms together. This is a force of attraction among certain particles, including protons and neutrons. It has *no effect* on certain other particles, such as photons and electrons. Because of the strength of the strong nuclear force, or equivalently, because of its large binding energy, nuclei are bound together tightly.

As the name "strong nuclear force," or sometimes just "strong force," implies, this force is much stronger than other natural forces, provided that we restrict our attention to the close-range interactions of particles inside a nucleus. For example, the electric force causes the positively charged protons in the oxygen nucleus to repel one another, but (fortunately) this electric repulsion is overwhelmed by the strong force's attraction. Atomic nuclei are therefore stable. However, the strong force diminishes much more rapidly with distance than the electric force does. Beyond about 10^{-14} m, the strong force becomes negligibly weak, and electrical repulsion takes over. We know that this must be true, because if it were not the case, then nuclei of atoms could grow much larger than they are.

Nuclear processes involve changes in the composition of a nucleus. Because these changes involve the strong force, we can correctly expect that the energies released by nuclear processes are much greater than the energies of chemical processes, which involve changes only to the arrangement of electrons on the outskirts of atoms (more than ten thousand times beyond the range of the strong force). The most important nuclear process in cosmology is **nuclear fusion,** a process that combines, or fuses, two nuclei into a more massive one and releases energy in the process. That is,

$$\text{nucleus 1} + \text{nucleus 2} \rightarrow \text{heavy nucleus} + \text{energy} . \qquad \text{(eq. 9.6)}$$

Now, we know what you're thinking. You're thinking, "That can't be right; Equation 9.6 violates energy conservation! All the same ingredients are on both sides, so this reaction produces free energy from nowhere." And *now* you're thinking, "Woohoo! The world's energy problems are over!"

Well, you're totally right about needing to conserve energy, and we're glad you brought it up: There's no free lunch and no free nuclear energy. The energy on the right side of Equation 9.6 must come from somewhere. And it does: It comes from the change in (strong) nuclear binding energy brought about by *rearranging* the protons and neutrons. That's the idea, but we need to make it more concrete.

In Chapter 4, we encountered gravitational binding energy. We saw that more tightly bound systems have *more negative* binding energies. This is because more tightly bound systems are deeper "in the hole" as far as energy is concerned. (The "hole" can be metaphorical or literal; just think of a squirrel falling down a hole.) The more negative the binding energy between two objects (like the squirrel and the Earth), the more energy you must give them in order to separate them. (For example, the squirrel has to use some of its own energy in order to climb out of the hole. Once it's out, the squirrel is still gravitationally bound to the Earth, but it's not *as* bound, because it would now take less energy to lift the squirrel to some height above the surface of the Earth than it would if you started lifting it at the bottom of the hole.)

The same is true for nuclear (or electrical, or whatever) binding energy. The particles in the two nuclei on the left side of Equation 9.6 are less tightly bound than when they are part of the single nucleus on the right side. As the particles rearrange into the more tightly bound single nucleus, the extra binding energy is released as kinetic energy (just as when an object falls to the ground on Earth, it speeds up as it loses gravitational binding energy and converts it into kinetic energy).

This conversion of binding energy into kinetic energy can be made even more tangible through Albert Einstein's famous insight that *mass* is also a form of energy. It turns out that the total mass of the two nuclei on the left side of Equation 9.6 is greater than the mass of the single nucleus they combine to form. Apparently, decreasing the binding energy (i.e., becoming more tightly bound) also means *losing mass*!

The more binding energy (i.e., less negative binding energy) a nucleus has, the more mass it has. This is ordinary mass, which can be directly measured in the usual ways. For example, if we're talking about helium nuclei, we could put a little vial of helium on a scale, or otherwise measure its mass on the basis of how much force is needed to move the vial around. So even if the nuclei on both sides of the reaction contain exactly the same ingredients (protons and neutrons), the right side is *lighter*. This idea takes some getting used to: Mass is not only based on the "amount of stuff," but also the nuclear *arrangement* of that stuff. When two protons and two neutrons are arranged into a helium nucleus, the measured mass is less than it was for the same four particles before they were assembled into a helium nucleus. If you never noticed this before, it's only because you've never performed nuclear reactions and accurately weighed the reactants and products.

So we can express things in this way: The energy of fusion comes from what was previously stored in (part of) the mass of the two light nuclei. The mass of the resulting heavy nucleus is slightly less than the sum of the masses of the two lighter nuclei. Of course, the mass of the pair of light nuclei wasn't much to begin with, so how much energy could it even be worth? As it happens, it's worth a lot. The formula that tells you its worth is this:

Energy released (in the form of heat and light) = mass lost (converted to heat and light) × speed of light, squared,

or, more succinctly,

$$E = mc^2 .$$ (eq. 9.7)

The speed of light, remember, is $c = 3 \times 10^8$ meters/second, and $c^2 = 9 \times 10^{16}$ m²/s². The quantity c^2 shows up in this equation because the equation is derived from Einstein's theory of special relativity, which involves the speed of light. The nuclei are not moving at that speed; it's just a huge number, so when you multiply a tiny mass loss by this huge number, you get a substantial amount of energy. If you fuse a large supply of nuclei in this way, the energy output can be enormous. Mass is just another form of energy, but it's a very concentrated form, as Equation 9.7 reveals.

In order for this fusion process to take place, the two light nuclei have to get very close together—within the short range of the strong force, so that it can bind all the participating protons and neutrons together. The problem with this scenario is that all nuclei have *positive* electrical charges, which repel each other as the nuclei approach. It becomes a question of distance: Will the nuclei get close enough to fuse, or will electrical repulsion keep them out of range?

The electrical force gets stronger and stronger as the charged nuclei approach each other. Therefore, in order to overcome the repulsion, the nuclei must move toward each other with great speed. It's like trying to ride your bicycle up a steep hill, knowing that it's downhill on the other side. If you are going too slowly, you'll stop and roll back down again (repulsion). But if you build up enough speed, you can

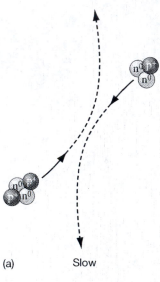

(a)   Slow

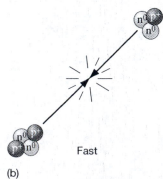

(b)   Fast

FIGURE 9.1 Fusion Two positive charges repel (a), so it requires high speed (b) to get them close enough to fuse into a new nucleus and release energy.

FIGURE 9.2 A Gravitational Analogue of Nuclear Fusion   If the bicyclist (incoming nucleus) is moving fast enough to make it over the hill (electrostatic repulsion), then it coasts downhill (pulled by the strong nuclear force) to a stable state in the valley on the right (fused with another nucleus).

Successful
fusion

make it over the top and the hill can no longer repel you. (See Figures 9.1 and 9.2.) One way to get the energy necessary for fusion is to heat the nuclei to a very high temperature, greater than about 10 million kelvins (10^7 K), so that all the nuclei are moving around at very high speeds. This process is known as **thermonuclear fusion**; the "thermo" part refers to the high temperature. Where is it hot enough for thermonuclear fusion? We'll look at one such site next.

9.3 Thermonuclear Fusion in Stars

In the 1920s, British astronomer Arthur Eddington and others argued that the gravitational energy within stars that are forming heats their cores to more than 10 million degrees. As the particles that compose the Sun were originally pulled inward due to their mutual gravitational attraction, they sped up and collided with one another, making the core hot. Thus, thermonuclear fusion is *expected* in stellar cores, and it explains how the Sun has shone so brightly for 4.6 billion years. (See "Nuclear Power in the Sun" below.)

Nuclear Power in the Sun

The primary source of energy in the Sun is a sequence of fusion reactions whose net result is to turn four hydrogen-1 (^{1}H) nuclei (i.e., four protons) into one helium-4 (^{4}He) nucleus. From $E = mc^2$, we can calculate the fusion energy released by the loss of mass. The combined mass of four protons is 6.6943×10^{-27} kg, and the mass of a helium-4 nucleus is 6.6466×10^{-27} kg, so the mass lost in the fusion reaction is

$$m_{\text{lost}} = (6.6943 \times 10^{-27} \text{ kg}) - (6.6466 \times 10^{-27} \text{ kg}) = 0.0480 \times 10^{-27} \text{ kg} .$$
(eq. 9.8)

Multiplying by the square of the speed of light, we find that the energy released by the fusion of four protons is

$$E = (0.0480 \times 10^{-27} \text{ kg}) \times (3.0 \times 10^8 \text{ m/s})^2 = 4.3 \times 10^{-12} \text{ joules} .$$ (eq. 9.9)

That amount may not sound like much, but it comes from the fusion of only one helium nucleus. You may recall from chemistry that there are "Avogadro's number," or 6.02×10^{23}, hydrogen atoms per gram; therefore, there are 6.02×10^{26} hydrogen atoms per *kilo*gram. With four hydrogen nuclei going into one helium nucleus, this adds up to

$$E \text{ (per kg)} = (4.3 \times 10^{-12} \text{ J}) \times \frac{6.02 \times 10^{26}}{4 \text{ fusions per kg}} = 6.5 \times 10^{14} \text{ J/kg} . \text{(eq. 9.10)}$$

That's the amount of energy released for each kilogram of hydrogen fused. Notice that it's more than the required amount we calculated earlier in Equation 9.5. Now, at the Sun's current luminosity of 3.9×10^{26} watts, where 1 watt equals 1 joule per second, the Sun fuses

$$\frac{3.9 \times 10^{26} \text{ J/s}}{6.5 \times 10^{14} \text{ J/kg}} = 6.0 \times 10^{11} \text{ kg/s} . \qquad \text{(eq. 9.11)}$$

The Sun evidently fuses 600 billion kilograms of hydrogen every second. (That's more than the mass of *the entire human race*, every second!) Spread out over the Sun's 4.6 billion-year age, with 3.15×10^7 seconds in a year, the total mass of hydrogen fused by the Sun so far is

$$(6.0 \times 10^{11} \text{ kg/s}) \times (4.6 \times 10^9 \text{ yrs}) \times (3.15 \times 10^7 \text{ s/yr}) = 8.7 \times 10^{28} \text{ kg} .$$
$$\text{(eq. 9.12)}$$

And since the mass of the Sun is 2.0×10^{30} kg, the total mass of hydrogen fused by the Sun so far amounts to only about 4% of the Sun's mass. Therefore, thermonuclear fusion does the trick: After 4.6 billion years, only 4% of the Sun's mass has been spent as nuclear fuel.

What does it take to make thermonuclear fusion happen inside the Sun and other stars? It is only with tremendous technological difficulty that thermonuclear fusion can be sustained under controlled laboratory conditions on Earth. Yet the recipe is surprisingly simple: You need (a) a sufficient concentration of nuclear fuel, in the form of light nuclei, and (b) a sufficient temperature so that the nuclei can overcome their electrostatic repulsion. That's all, and both conditions are generally met inside stellar cores.

In the Sun, as in most stars, the nuclear fuel is hydrogen. At the temperatures in question, it is always totally ionized, which licenses us to ignore the electrons—or at least, to deal with them separately. Instead, we'll focus on the **baryons**—a class of particles of which the proton and neutron are the only members found in normal matter. (It is possible to *construct* other baryons briefly, in the laboratory, but these other baryons have no particular importance to our study of cosmology. Therefore, you may think of baryons as essentially meaning "protons and neutrons.") Baryons are affected by the strong nuclear force and are therefore capable of nuclear fusion. In the current context, baryons are the components of nuclei, which explains why nuclei are capable of nuclear fusion.

A number of nuclear fusion reactions are involved in energy production in the Sun. The most important are collectively called the **PP chain**, which is short for the "proton–proton chain reaction." A chain reaction is a sequence of nuclear reactions in which the products of one reaction become the ingredients for the next reaction. The first reaction in the PP chain begins with the simplest and most abundant ingredients: hydrogen-1 nuclei. Remember, ^{1}H is just a proton, so when two of them combine as follows, the "proton–proton" chain is initiated:

$$^1\text{H} + {}^1\text{H} \rightarrow {}^2\text{H} + \text{e}^+ + \nu . \qquad \text{(eq. 9.13)}$$

Let's look at the products of this reaction. Hydrogen-2, also known as deuterium, or ^{2}H, is a proton bound together with a neutron. (Why wouldn't *two protons* bound together be called ^{2}H? *Hint:* what makes one element different from any other, in general?) Right off the bat, then, the PP chain reaction delivers unwanted complexity: Why is one of the protons replaced with a neutron? And what are these extra e^+ and ν particles? All of these questions are interesting and important enough to deserve their own section. We'll give them that, in Section 9.5. For the time being, however, just file them away as something to come back to. They don't actually play much of a role in the rest of the chain reaction sequence; it's just bad luck that the first reaction happens to be the most complicated.

Now that we've made deuterium, the chain reaction can build upon it:

$$^1\text{H} + {}^2\text{H} \rightarrow {}^3\text{He} + \gamma \ . \qquad \text{(eq. 9.14)}$$

As the chain reaction continues, successive fusions combine the products into larger and larger nuclei. In the reaction given by Equation 9.14, the result is helium-3 (containing 2 p and 1 n), a relatively rare isotope of helium.

Far more common in nature is helium-4 (2 p and 2 n), which is produced in the Sun by this next reaction in the chain:

$$^3\text{He} + {}^3\text{He} \rightarrow {}^4\text{He} + {}^1\text{H} + {}^1\text{H} \ . \qquad \text{(eq. 9.15)}$$

In this third reaction, helium-4 is produced only when the first two reactions, Equations 9.13 and 9.14, have each run *twice*. The two helium-3 nuclei produced in this way are fused together, but instead of a six-baryon nucleus emerging, two of the baryons (both protons) are released as by-products. If you count the protons used in the three reactions, you'll find that

- two protons are used in the first reaction (one gets converted to a neutron), Equation 9.13, and
- one more proton is added in the second reaction, Equation 9.14, so that's three so far, and
- reactions 1 and 2 have to run twice to make two ^{3}He nuclei, so now we've used six protons, but
- reaction 3, Equation 9.15, returns two protons, so we've really consumed only four.

That is, the PP1 chain reaction is a process for merging four hydrogen-1 nuclei into one helium-4 nucleus. Helium-4 is one of the most tightly bound nuclei, which makes it extremely stable. This means that once helium-4 is formed, it takes a lot of energy to change it. In the Sun, then, the chain reaction is finished with the construction of helium-4. Figure 9.3 shows all three reactions that got us here.

This process is the basic idea behind the power source of stars. We can think about it in the big-picture view of energy transformations: The process of a star shining amounts to taking energy frozen in the form of mass and releasing it by fusion into thermal energy and radiant energy. The light emitted by the star then shines out into space. In the case of the Sun, some of its light hits the Earth and drives processes such as weather and life. So our existence on Earth is supported by the Sun's permanently losing about 4 billion kg of mass every second to make 3.9×10^{26} joules of light. That's about half a kilogram of mass lost for *every person on Earth*, during each

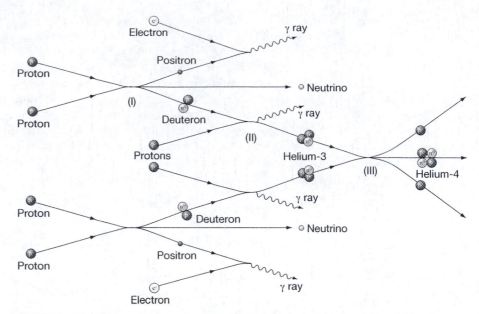

FIGURE 9.3 Solar Fusion   Fusion processes in the Sun convert hydrogen nuclei into helium by a three-step chain of reactions (from left to right, three sets of collisions) and release energy that supports the Sun against its own gravity.

second the Sun shines. (The "mass lost" we're talking about is the amount of mass converted to heat and light, not the larger mass which is the amount of hydrogen converted into helium.)

Note that the fusion of hydrogen into helium is only one specific example of nuclear fusion that releases energy. In principle, *any* transformation that converts light nuclei into a heavier nucleus can release energy and help fuel a star, provided that the total mass of the pieces is greater than the mass of the resulting nucleus. However, higher temperatures are needed to fuse heavier nuclei (which have more protons than lighter nuclei), since there is a greater electrical repulsion that must be overcome to get them to fuse.

As an example, consider the next possible step after the fusion of hydrogen into helium. Being very stable and having a charge of +2 (owing to its two protons), helium won't fuse into anything else unless the temperature rises by nearly a factor of 10. When it does fuse, it does so like this

$$^4\text{He} + {}^4\text{He} + {}^4\text{He} \rightarrow {}^{12}\text{C} + \gamma + \gamma \ , \qquad \text{(eq. 9.16)}$$

and then sometimes like this

$$^4\text{He} + {}^{12}\text{C} \rightarrow {}^{16}\text{O} + \gamma \ . \qquad \text{(eq. 9.17)}$$

You're probably wondering why *three* helium nuclei are needed in Equation 9.16, rather than just two. The reason, which turns out to be very important to cosmology, as you'll see at the end of this chapter, is simply that a blend of two helium nuclei happens to be radioactively unstable: The nuclei won't "stick" together.

Only if you slam a third one in there before the other two fall apart will they all stick together. Carbon-12 and oxygen-16, once formed by this process, are quite stable.

As we'll see in the next section, some stars give away most of the elements they build. They send this newly constructed material off into space, causing the composition of galaxies to evolve over time.

EXERCISE 9.2   Understanding the Strong Nuclear Interaction

Nuclear reactions, like other natural processes, obey certain rules. These rules take the form of *conservation laws*, in which some quantity is *conserved* (doesn't change) during the reaction. Two of these rules, not specific to nuclear processes, but obeyed by them, are:

- Conservation of energy
- Conservation of electrical charge

Nuclear reactions in particular obey two more conservation rules as well. One of them will wait until Section 9.5, but the other one is

- *Conservation of baryon number*, which says that the total number of baryons on the left side of the reaction equals the total number of baryons on the right. Notice that it's not conservation of protons or of neutrons, but rather the sum of both. If a reaction involves only protons on the left and only neutrons on the right, then baryon conservation is satisfied as long as there's the same number of each.

Use the preceding laws to answer the following questions:

(a) Show that charge and baryon number are conserved in each of the three reactions in the PP chain. (*Note:* The e^+ particle has a $+1$ charge, exactly the same as that of a proton; the ν particle has zero charge. Neither e^+ nor ν is a baryon; you'll see what they actually are in Section 9.5.)

(b) Does the first reaction in the PP chain conserve the number of protons or of neutrons?

(c) All three reactions in the PP chain help to power the Sun. Take one in particular, say, Equation 9.14. In this reaction, is there more *kinetic and radiant* energy (combined) on the left side or on the right? (*Hint:* Don't forget about the mc^2 energy that all matter has!)

(d) In nuclear **fission** reactions, lighter nuclei are formed by breaking apart heavier ones. (This is the opposite of fusion.) In the following example of this process, in which a neutron breaks up a nucleus of weapons-grade uranium, what number should x be? $n + {}^{235}U \rightarrow {}^{93}Rb + {}^{x}Cs + 2n$.

9.4 Heavy Elements and Stellar Genetics

An important by-product of stars being powered by nuclear energy is that they manufacture a variety of elements that would not otherwise exist. We've already seen how the Sun can convert hydrogen into helium. Other stars transform a variety of lighter elements into heavy ones via nuclear fusion and ultimately build up most of the elements we see in the periodic table and in our own bodies.

Think about where the building blocks come from that make up the paper in your book, the plastic in your pen, and the metal in your watch. What is the origin of the calcium in your bones and the iron in your blood? The history of the universe should be able to provide the origin of every element—the source of the *whatever* in your *whatever*.

It turns out that nearly all of the chemical elements in our world were formed by stars. So in order to understand where we come from, we need to learn a little more about the lives of stars and the stages during which different elements are formed.

Overview We'll start by sketching the basic picture, and then we'll fill in the details. Inside a galaxy, clouds of gas occupy some interstellar space (the space between the stars). Normally, the gravity of a gas cloud is too weak to trigger any significant change on its own. But when some external agent comes along to compress it—perhaps the passage of a density wave associated with a galactic spiral arm, or a shock wave from a supernova explosion—the cloud begins to collapse under its own weight, as its constituent particles are drawn together by gravity. Astronomers say that a star is "born" when some ball of collapsing gas reaches a threshold core temperature and pressure, initiating nuclear fusion.

A new star begins as mostly hydrogen and helium (recall Section 5.4 on the observed abundances of the light elements) and starts turning hydrogen into more helium in its core. Some smaller stars will never go any further than the production of helium. Other, larger stars will eventually form heavier elements such as carbon, oxygen, silicon, and iron. Some stars keep all the elements they make, and others disperse them into the galaxy. Such a dispersal can be slow and gentle, or it can be explosive—and during that explosion, even heavier elements are forged, such as gold, platinum, lead, and uranium. The newly created elements enrich the galaxy. When new stars emerge from this enriched gas, they contain more and more of the metals and minerals needed to build planets.

Now we look at how all of this happens.

Birth and Youth Star formation begins when a cloud of gas and dust is compressed and becomes too dense to support its own weight. The cloud contracts under the force of its own gravity, and does so in a lumpy fashion, fragmenting into smaller clouds that will eventually become stars. As the material falls inward in each of these smaller clouds, it heats up (why? Think about energy conservation), increasing the temperature. More temperature implies faster moving particles, which push harder when they collide with one another and thereby create more *pressure*. The forming star exists in a battle between gravity and pressure (although other factors, such as magnetism and cloud rotation, play a part as well). Gravity relentlessly tries to pull material inward and shrink it as small as possible, while pressure fights to expand the material against gravity, just as air pressure in a balloon fights to expand the balloon. For a star, the battle is usually a tie, and the star neither grows nor shrinks, in which case it is said to be in **hydrostatic equilibrium**.

To halt the collapse, the star must be able to generate sufficient pressure. This requires sufficient thermal energy—sufficiently fast-moving gas particles—which happens as soon as the star begins nuclear fusion in its core. Prior to the onset of nuclear fusion, the star's only source of heat is the energy of the gravitational infall, so hydrostatic equilibrium is not achieved until the star produces its own nuclear energy to stop the infall. However, once the star does initiate fusion in its core, hydrostatic equilibrium is maintained for millions, billions, or even trillions of years, until a significant fraction of the hydrogen fuel is used up. We assign the term *protostar* to the pre-fusion phase and *main-sequence star* (like the Sun) to the phase with hydrogen fusion in the core. Figure 9.4 depicts the star birth process; Figure 9.5 shows telescope images of star formation, including stars surrounded by disks of debris that will likely evolve into systems of planets.

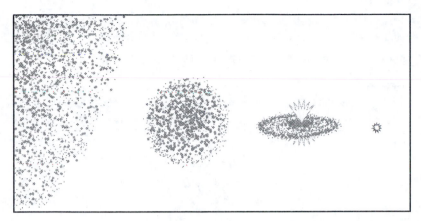

FIGURE 9.4 Star Formation   Evolution of an interstellar cloud as it compresses by its own gravity, flattens into a spinning disk, and spawns a star at the center, possibly with orbiting planets forming within the disk.

FIGURE 9.5 Infrared Images of Stars Forming in the Orion Region Dusty disks surrounding young stars are seen in these two remarkable visible-light "top-view" images, which happen to be aligned so that we view the disks from "above." (◆ Also in Color Gallery 1, page 10.)

Middle Age   Once the star reaches a stable balance point where it can release enough energy by fusing hydrogen into helium, it stays in this state for most of its active life. Now, which periods of a star's existence get termed "life" is quite subjective; however, people usually mean the periods in which the star is engaged in some kind of nuclear fusion. The hydrogen fusion phase lasts as long as the star has enough hydrogen remaining in its core to fuse.

The smallest main-sequence stars have the least gravitational pull inward and therefore the least gas pressure pushing outward, since hydrostatic equilibrium requires that the two be in balance. Relatively low pressure means relatively low temperature, which in turn means relatively low nuclear reaction rates. (Why? Think kinetic energy.) Therefore, the star is relatively cool—which makes it visibly red in color—and it uses up its hydrogen fuel very slowly. These *red dwarf* stars (which we encountered earlier in Section 4.3) are only about one-tenth of the Sun's diameter and start their lives with much less than the Sun's mass of hydrogen fuel. Even so, they fuse it so slowly that they live longer than any other star—up to tens of trillions of years, which is much longer than the present age of the universe.

The largest main-sequence stars have just the opposite properties: tremendous gravity balanced by tremendous pressure and temperature. With higher core temperatures come faster moving nuclei and thus a faster rate of fusion. More fusion creates more luminosity, and a higher surface temperature creates a visibly blue appearance. Such *blue supergiant* stars can be tens or hundreds of times the Sun's diameter, and they start their lives with much more hydrogen. But they use it up so quickly that they are the shortest-lived stars, coming and going in as little as a few million years. Figure 9.6 compares the sizes of various stars; also refer to Color Figure 5 in Color Gallery 1, page 3, and Color Figures 11 and 12 in Color Gallery 1, page 6, to see the color distinction.

EXERCISE 9.3   One Star, Two Star, Red Star, Blue Star

Answer the following questions about the colors of stars, as they would appear to the naked eye:

(a) The typical surface temperature of a red dwarf star is about 3000 K. Thus, this type of star emits mostly *infrared* light. Use a blackbody curve to show why such a star would appear red instead of some other color or instead of being completely invisible.

(b) Typical blue supergiant surface temperatures are around 30,000 K. These stars emit primarily ultraviolet light. Use a blackbody curve to show why they appear blue in the night sky.

(c) Most stars we can see in the night sky look approximately white (as does the Sun when viewed from above our atmosphere). Such stars have intermediate temperatures—the Sun's is 5800 K, for example—whose blackbody peaks fall in the middle of the visible part of the electromagnetic spectrum. Draw a blackbody curve to show why these stars look white (which is how your brain interprets a blend of other colors) instead of green (where the Sun's spectrum actually peaks).

Old Age The real cosmological beauty of *stellar evolution*—that is, the lives of stars—takes place at the end, when a star begins to run out of hydrogen fuel in its core. The star can no longer generate enough nuclear energy to supply the heat and pressure needed to resist gravity, and hydrostatic equilibrium is broken: The star's core begins to collapse inward. The infall is powered by gravity, which means that the gravitational energy must go somewhere. It becomes thermal, which means, oddly, that when the nuclear power is turned off, the core of the star actually heats up as a result. Equally oddly, this heat causes most of the star to expand, even while the core contracts. The star swells up, and its surface cools down (while its core heats up!). At this point, the star is called a *red giant*. When the core gets hot enough, nuclear power can resume, fusing helium into carbon and some oxygen in accordance with Equations 9.16 and 9.17. Meanwhile, although hydrogen

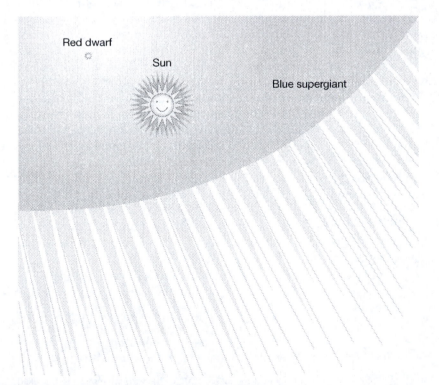

FIGURE 9.6 Relative Sizes of Main-sequence Stars Shown here are the smallest main-sequence stars (red dwarfs), medium-sized main-sequence stars (like our Sun), and the largest main-sequence stars (blue supergiants), approximately to scale.

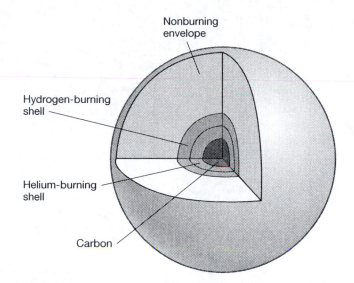

Nonburning
envelope

Hydrogen-burning
shell

Helium-burning
shell

Carbon

FIGURE 9.7 Structure of a Low-mass Star During Old Age Nuclear fusion takes place only where there is available fuel. The core has already fused all of its nuclear fuel, so fusion can only take place in **shells** surrounding the core. Each fusion reaction produces a heavier nucleus than the ones it was fused from, so this heavier nucleus sinks toward the center.

fusion is extinguished in the *core*, there is still hydrogen left on the core's outskirts, and a *shell* of hydrogen fusion continues to operate there. (See Figure 9.7.) As you can imagine, with multiple fusion processes going on in different places in the star, things get complicated at this point.

At this point, a split occurs between stars less than about 10 solar masses and those greater than 10 solar masses (summarized in Figure 9.8):

- "Low-mass stars" no longer undergo nuclear fusion in their cores, which never get hot enough to form anything beyond carbon and oxygen (and, in some cases, not even beyond helium). With no more nuclear power, gravity contracts the core once more, releasing gravitational energy as heat, just as it did in the red giant phase. This time, however, the rest of the star swells up even more, and now it's called a *red supergiant*. Over time, its outer layers (mostly H and He, plus some C, N, and O) blow off into space, creating a cloud of gas called a *planetary nebula*. (This type of nebula has absolutely nothing to do with planets; the name derives from the fact that such nebulae sometimes *look like* planets through a small telescope.) What's left behind is an Earth-sized carbon core called a *white dwarf*. Figure 9.9 shows some spectacular telescope images of white dwarfs surrounded by the gas they previously expelled.

- Once a white dwarf is formed, it remains forever, provided that other stars leave it alone. Otherwise, if another star deposits more gas onto the white dwarf (which may happen if the white dwarf is part of a binary system—two stars orbiting close together), then the mass of the extra gas gravitationally squeezes the star. The squeeze spawns such a sudden flash of nuclear fusion that the star explodes as a *type Ia supernova*, the standard candle we examined back in Section 3.2.

- "High-mass stars" require greater pressure to support their greater weight, so their cores need to be hotter—hot enough to fuse heavier elements after hydrogen becomes helium and helium becomes carbon. Figure 9.10 shows the complicated core-and-shell structure of these stars, which fuse heavier nuclei beyond carbon and oxygen to neon, magnesium, silicon, and iron. In lesser

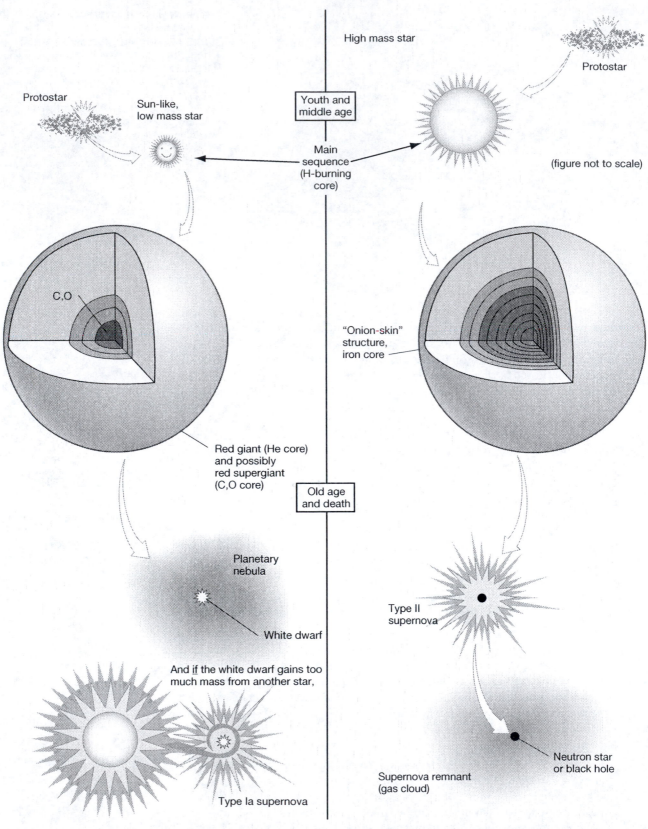

High mass star

Protostar

Youth and middle age

Protostar

Sun-like, low mass star

Main sequence (H-burning core)

(figure not to scale)

C,O

"Onion-skin" structure, iron core

Red giant (He core) and possibly red supergiant (C,O core)

Old age and death

Planetary nebula

White dwarf

Type II supernova

And *if* the white dwarf gains too much mass from another star,

Type Ia supernova

Supernova remnant (gas cloud)

Neutron star or black hole

FIGURE 9.8 Stellar Evolution   The left side summarizes the evolution of a low-mass star, such as the Sun; this summary is valid for a star that's less than about 10 solar masses. The evolution of more massive stars is summarized on the right side. (See the text for details.)

quantities, they produce many other elements, such as sulfur, calcium, phosphorus, etc.—almost everything below iron (atomic number 26) on the periodic table. At the center, a core of iron forms and becomes unstable, triggering a series of events that lead the star to explode.

In a high-mass star, the core uses one element for nuclear fuel and then runs out of that element, relegating its fusion to a shell. The core contracts by gravity, since there's no nuclear power there to resist the gravitational contraction. The core heats up from the release of gravitational energy and becomes hot enough to run the *next* nuclear reaction on whatever elements it has available. That is, the core quits fusion until it contracts enough to restart with a different fuel. We left off earlier with carbon in the core. Carbon fuses into magnesium as follows:

$$^{12}C + {}^{12}C \rightarrow {}^{24}Mg + \gamma \ . \tag{eq. 9.18}$$

Oxygen fuses into neon via the reaction

$$^{16}O + {}^{4}He \rightarrow {}^{20}Ne + \gamma \ , \tag{eq. 9.19}$$

and into sulfur via

$$^{16}O + {}^{16}O \rightarrow {}^{32}S + \gamma \ . \tag{eq. 9.20}$$

Processes like these lead to heavier and heavier nuclei. The elements that result most often from these reactions tend to be the ones that have *even* atomic numbers,

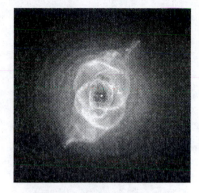
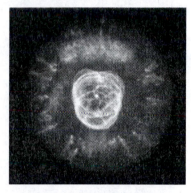

FIGURE 9.9 Planetary Nebulae and White Dwarfs   Notice the white dwarf at the center of each planetary nebula. These nebulae are made from material expelled by dying low-mass stars, including hydrogen, helium, carbon, nitrogen, and oxygen. These are some of the most abundant elements in the universe and, apart from helium (as we'll see in Chapter 13), the most abundant ingredients in living organisms as well. (◆Also in Color Gallery 1, page 10, with additional planetary nebulae shown there too.)

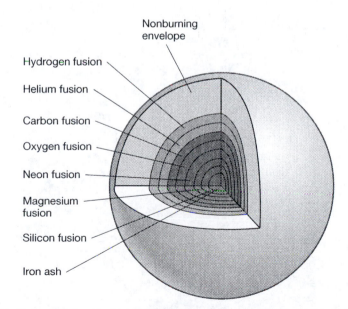

Hydrogen fusion
Helium fusion
Carbon fusion
Oxygen fusion
Neon fusion
Magnesium fusion
Silicon fusion
Iron ash

Nonburning envelope

FIGURE 9.10 High-mass Star During Old Age   This is called the "onion-skin" model, in which the fusion of various elements takes place in a series of concentric shells, with different elements being fused in different shells. A core of nonfusing iron is formed at the center.

and atomic masses in multiples of four. This is because it's easy for elements to build up by squeezing in more and more ^{4}He nuclei, as in Equation 9.19. In addition to products from the reactions already shown, we get substantial production of silicon, argon, calcium, titanium, chromium, and, ultimately, iron. Odd atomic-numbered nuclei are generally produced in lesser quantities by reactions such as the following ones for nitrogen (found in air), sodium (found in salt), and phosphorus (found in DNA):

$$^{17}\text{O} + {}^{1}\text{H} \rightarrow {}^{14}\text{N} + {}^{4}\text{He} \ , \tag{eq. 9.21}$$

$$^{12}\text{C} + {}^{12}\text{C} \rightarrow {}^{23}\text{Na} + {}^{1}\text{H} \ , \tag{eq. 9.22}$$

$$^{16}\text{O} + {}^{16}\text{O} \rightarrow {}^{31}\text{P} + {}^{1}\text{H} \ . \tag{eq. 9.23}$$

Two details remain for our contemplation. First, how does the universe manufacture elements heavier than iron? Second, how does it disperse them into space? It turns out that both questions are answered by the same event.

Death The "death" phase of a high-mass star is, as promised, a supernova. It is triggered in the iron core of the star (refer back to Figure 9.10), which has reached a point of no return.

Figure 9.11 graphs an important property of nuclear physics. It shows that the isotope iron-56 (^{56}Fe) has the lowest binding energy per baryon of any stable element. If the baryons in an iron nucleus had more energy, they would be more free and less bound. That means that iron-56 is the most "content" nuclear arrangement, if you like. Lighter nuclei can release energy by fusing into heavier ones, and heavier ones can release energy by breaking apart into smaller pieces. But iron-56 is the cutoff point. Once a star's core is made of iron, it can't use nuclear fusion to generate the heat and pressure it needs. Fusion of elements heavier than iron would *consume* energy (and create mass), which would make the core less capable of supporting the star's weight.

Here's the point: Once the star's core is iron, no nuclear fuel remains to save the star from its own gravity. The core must collapse in on itself. The star will crush down until the core itself—which is about double the mass of the Sun—is like one giant

FIGURE 9.11 The Stability of Iron This graph displays the nuclear binding energy "felt" by each proton or neutron inside a nucleus (vertical axis) versus the number of baryons in each nucleus (horizontal axis). The lowest energy state, at the bottom of the curve shown, is the most stable, because it has the most (negative) binding energy. Iron is therefore the most tightly bound nucleus. In terms of nuclear reactions, every other nucleus "wants" to become iron, either by the fusion of lighter nuclei or by the fission of heavier nuclei.

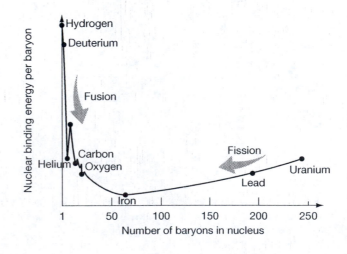

atomic nucleus, because all of its baryons will be tightly packed right next to each other. The density of this nuclear matter is enormous—about a thousand trillion (10^{15}) times that of water—such that a pair of dice made from this stuff would weigh more than all the people on Earth combined. At this density the core stabilizes, unless the original star was more than about 30 solar masses, in which case the collapse can't be halted even by nuclear forces and results in a *black hole* instead (Section 6.5).

During the collapse, the iron nuclei disintegrate from the heat (and 10 million years of nuclear fusion, spent building up iron in the first place, are undone just like that!). Because it is photons that deliver the heat energy to the iron nuclei, the process is called *photodisintegration*. The collapsing core is made of elementary particles now. As gravity squishes everything together, an *electron capture* reaction occurs:

$$p^+ + e^- \rightarrow n + \nu . \qquad \text{(eq. 9.24)}$$

(Don't act so surprised! Admit it: secretly, you always suspected that you could think of a neutron as a proton and an electron squeezed together, so that the plus and minus combine to become neutral.) The extra "ν" particles fly away (more on these in the next section, as promised), and the resulting object is made almost entirely of neutrons. This former stellar core is now a **neutron star**. Despite being more massive than the Sun, it's only the size of a small island, about 30 km in diameter. Color Figure 23 in Color Gallery 1, page 11 shows a remarkable x-ray telescope image in which the tiny neutron star is actually visible.

Meanwhile, another 20 or so solar masses worth of material from the original star remains outside the core. When the core finally stops collapsing, it rebounds and blasts the rest of the star off into space in a massive explosion. (See Question 10 at the end of the chapter for more on how that happens.) This **type II supernova** is so energetic that when Chinese astronomers documented one in the year 1054, it was bright enough that it could be clearly seen *in the daytime*, despite occurring nearly 6000 light-years away. The debris from that explosion is still visible and still expanding; it's known as the Crab Nebula (Figure 9.12). (You might remember from Chapter 3 that the labeling of supernovae—type Ia, type II, etc.—is based on distinctive features in the observed spectrum of light from the supernova.)

The explosion carries with it a fleet of high-speed neutrons produced during electron capture (Equation 9.24). Since free neutrons are electrically neutral, there is no electrical repulsion to prevent them from fusing with existing nuclei. In this way, the nuclei can grow rapidly. Some neutrons are able to transform into protons once inside a nucleus, by a process something like running Equation 9.24 in reverse; we'll examine this process further in the next section. All the heavy elements, such as gold and platinum, are formed *during* the explosion by this neutron-adding mechanism.

So the expanding debris from the supernova carries with it elements formed at two different times:

- *during the star's life* and prior to the collapse of the core; throughout this period, the star manufactured elements up to ^{56}Fe.

- *during the star's death*, which is to say during the supernova itself; throughout this period, the star manufactured elements beyond iron on the periodic table.

Stellar Genetics   We now have a comprehensive story detailing the origin of most elements. Hydrogen and helium, along with traces of lithium, appear to be ancient,

FIGURE 9.12 The Crab Nebula This nebula is the remains of a relatively nearby and relatively recent supernova in our region of the Galaxy a little less than a thousand years ago (♦ Also in Color Gallery 1, page 11.)

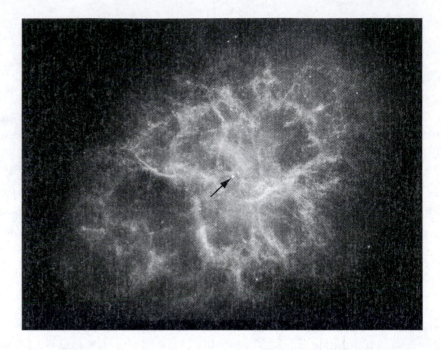

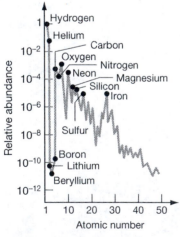

FIGURE 9.13 Abundances of Elements When you look back at how stars produce and distribute new elements into space, does it seem reasonable that stars produced this abundance pattern? Compare with Figures 9.7 and 9.10, for example.

existing in the oldest stars we can find. (Recall Section 5.4.) Helium is also produced in stars. Carbon, nitrogen, and oxygen are produced by both low- and high-mass stars. In low-mass stars, these elements are blown off into space just before the stars turn into white dwarfs. In high-mass stars, elements from carbon through iron are produced inside the stars and dispersed throughout galaxies by supernova explosions. The elements above iron in the periodic table are less common because they form only while the explosions are actually happening, and not during the previous 10 million years of the star's life.

We repeat here a figure from Chapter 5, so that you may now see the source of the abundances of the elements in the universe. Verify for yourself, in Figure 9.13, that these elements could reasonably have formed in the abundances shown, by the mechanisms just described. (Of course, there are models and equations that predict more exact amounts, and it's impressive that they match up with what's observed spectroscopically, but you can see the general trend for yourself even without the exact numbers.) You might notice that iron seems particularly abundant, considering that most of it disintegrates during the collapse of the core of a type II supernova. You are quite right; in fact, the quantity of iron and a few elements near iron in the periodic table is enhanced because they (and they alone) are also produced by type Ia supernovae. These supernovae, from white dwarfs that get too massive, are nuclear blasts that rapidly and repeatedly fuse all the white dwarf's carbon and oxygen nuclei, until they wind up as iron.

Low-mass stars take a long time to live and die, and when they do die, they keep most of the elements they made during their lives. But high-mass stars live and die quickly by cosmological standards—millions of years, not billions—and when they die, they give the elements they made back to the galaxy. With every generation of high-mass stars, a galaxy's gases are increasingly enriched with heavier and heavier elements. Each new star born from some galactic gas cloud contains the elements that a former star made. "Parent" stars therefore affect "daughter" stars. (This imperfect biological analogy is sometimes called *stellar genetics*).

There is an ebb and flow of energy. As heavy elements are assembled, the mass of the material world diminishes slightly (Figure 9.11 and Equation 9.7). But the distribution of those elements into space requires an explosion, triggered by the destruction of iron nuclei in a collapsing stellar core. This explosion restores some of that lost mass at the expense of heat energy from the core. The gravitational energy, too, which was spent by pulling the star together in the first place and heating up its core until hydrogen could fuse for the first time, is partially restored by the explosion. Now most of the mass of the original star is expanded again and thereby resupplied with the gravitational energy needed to form new stars. And what might trigger this new burst of star formation? *Another* supernova, whose shock wave compresses enriched interstellar gas clouds until their gravity can do the rest.

Over time, the composition of the universe is no longer just the hydrogen, helium, and traces of lithium that we observe in the oldest stars. If the Sun had formed around the same time as the birth of the universe, it would have contained no carbon, no nitrogen, no oxygen, etc. But it formed 9 billion years later, after a thousand generations of massive stars had lived, died, and scattered their chemically enriched ashes. About 2% of our Sun is therefore made from these heavier elements. So when our solar system formed, it had all the materials needed to make water, rocks, and metals. Heavy and abundant iron sank to the core of the molten Earth as it formed, while a crust of oxygen and silicon hardened above. We'll see in Chapter 13 that the materials that make up our biosphere as well, including human beings, are those produced in abundance by stars.

You were born at a time when the accumulated knowledge of a thousand generations of people made it possible for you to study the history of the universe. Your world was born at a time when the accumulated elements of a thousand generations of stars made it possible for to you to exist at all.

EXERCISE 9.4   History of Common Objects

Pick a favorite object and do some research on the Internet or in a chemistry book to find at least two of the chemical elements that the object is made out of. Trace the history of those elements as far back into the past as you can. In other words, where have those atoms been during their time on Earth and before?

9.5 Exotic Particles

Since experiment after experiment confirms the law of conservation of energy, the scientific community has long regarded it with the highest confidence. So how should we react when an apparent counterexample arrives?

In 1930, physicist Wolfgang Pauli wrestled with this very problem. The specific case involved was *beta decay*, a process in which an element spits out a high-energy electron. This process is a common variety of natural *radioactivity* in which the electron does not come from the atom's electron cloud, but rather is somehow emitted by the nucleus (which never contained any electrons!). After the decay, the atom's chemical properties are different: The atom has changed its position in the periodic table. For example, when radioactive carbon-14 decays,

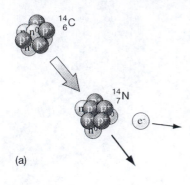

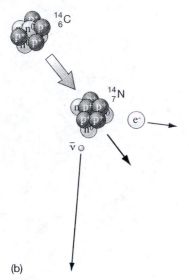

FIGURE 9.14 Beta Decay   When, for example, a carbon-14 nucleus decays radioactively, it emits not only the high-energy electron that we can detect relatively easily, but also a neutrino, which is essentially invisible.

an electron is emitted and what's left behind is no longer carbon; it's nitrogen-14. So it *seems* like the following is happening:

$$^{14}_{6}\text{C} \rightarrow {}^{14}_{7}\text{N} + \text{e}^- .\qquad \text{(eq. 9.25)}$$

Since the mass number is 14 both before and after the decay, the total number of baryons (protons plus neutrons) is evidently unchanged. But the atomic number has gone up from 6 to 7. To accommodate both of these facts, it must be that a neutron in the nucleus has somehow *turned into* a proton and spit out an electron in the process. That is, one of the neutrons in the carbon-14 nucleus *seems* to have done this:

$$\text{n} \rightarrow \text{p} + \text{e}^- .\qquad \text{(eq. 9.26)}$$

The rest of the baryons appear to have remained unaltered. This reaction helps to explain how an electron comes from a place with no electrons: It's as though it was somehow "hidden inside" a neutron. (We caught a glimpse of this idea earlier, in Equation 9.24.)

But why do these reactions only "seem" to be the case? Pauli's conundrum was this: There appeared to be more energy in the original samples (^{14}C in our example) than in the products of the decay (^{14}N + e$^-$ in our example). Figure 9.14(a) depicts this conundrum, and Figure 9.14(b) shows Pauli's resolution. In a letter he wrote to his colleague Enrico Fermi, Pauli proposed that an invisible particle flies off during the decay and carries with it the missing energy. In other words, given the choice between believing the apparent experimental results and believing in the conservation of energy, he chose the conservation of energy.

Fermi named the hypothetical particle **neutrino**—Italian for "little neutral one," and it wasn't actually detected for another 26 years. The neutrino exactly fixes the energy drain problem, just as Pauli predicted, so the preceding incorrect reactions correctly become

$$^{14}_{6}\text{C} \rightarrow {}^{14}_{7}\text{N} + \text{e}^- + \bar{\nu}\qquad \text{(eq. 9.27)}$$

and

$$\text{n} \rightarrow \text{p} + \text{e}^- + \bar{\nu}\qquad \text{(eq. 9.28)}$$

where the symbol ν indicates the neutrino. (We'll explain why there's a bar over it momentarily.)

A similar process, called "inverse beta decay," takes place in the PP1 chain reaction in the Sun (Equations 9.13, 9.14, and 9.15). When two hydrogen-1 atoms fuse into one hydrogen-2 atom (deuterium), a proton turns into a neutron. How does it do this? By an *inverse* beta process

$$\text{p} \rightarrow \text{n} + \text{e}^+ + \nu ,\qquad \text{(eq. 9.29)}$$

which is evidently what's going on behind the scenes in the first reaction of the PP1 chain:

$$^1\text{H} + {}^1\text{H} \rightarrow {}^2\text{H} + \text{e}^+ + \nu .\qquad \text{(eq. 9.30)}$$

In these reactions, we have introduced **antimatter,** which, as the name implies, has some properties that are the opposite of normal matter. A normal electron has a little minus sign, e⁻, and a normal neutrino is written simply as ν. Their *anti*particles, however, are e⁺ and $\bar{\nu}$. In each case, the matter particle and its antimatter counterpart are *almost* identical. The antielectron, or **positron**, for example, has the exact same mass as the electron, but carries a positive electrical charge instead of a negative one. The antineutrino has the same mass as the neutrino, and neither particle carries an electrical charge; they differ only by a quantity that particle physicists call "lepton number." We need not concern ourselves with this level of detail right now; you can investigate it in Exercise 9.5. For now, feel free to imagine the difference however you like: Neutrinos could be orange and antineutrinos purple, or neutrinos could zip around humming rock music while antineutrinos prefer jazz, etc.

Neither positrons nor neutrinos have much of a future in the Sun, for very different reasons. Consider positrons first. In this chapter, we've concerned ourselves largely with atomic nuclei and we've ignored the surrounding electron cloud. The atoms in the hot core of the Sun are ionized anyway, so their electrons just fly around freely. But what you've seen in sci-fi films is true: When matter and antimatter collide, there's a burst of energy. In this case, each positron produced by the PP1 chain (Equation 9.30) quickly collides with one of the electrons flying around, and the two **annihilate**—that's the official term for matter and antimatter particles wiping each other out and converting their masses into other forms of energy. The burst of energy produced takes the form of two gamma-ray (high-energy) photons:

$$e^- + e^+ \rightarrow \gamma + \gamma \ . \tag{eq. 9.31}$$

These two photons serve as an additional source of heat for the Sun's core. In this way, the Sun produces antimatter and then eliminates it as well. But even though positrons exist only briefly in the Sun, and their presence can only be inferred, they are well studied in laboratories on Earth. Physicists can make antimatter, in the form of individual particles, and even atoms of antihydrogen (an antielectron in an orbital around an antiproton). Perhaps some pocket of the universe contains antistars made from antihydrogen, with circling antiplanets and anticivilizations of antipeople. If so, you could safely introduce yourself to them over a radio transmission, but we wouldn't recommend a handshake.

EXERCISE 9.5 Understanding the Weak Nuclear Interaction

In addition to the rules for nuclear reactions that we've seen so far—conservation of energy, momentum, charge, and baryon number—there is another one to learn. The *conservation of lepton number* enables us to keep track of **leptons**: electrons and neutrinos. Electrons (e⁻) and neutrinos (ν) are each assigned a lepton number of +1; their antimatter partners—positrons (e⁺) and antineutrinos $\bar{\nu}$—each have a lepton number of –1. Reactions always conserve lepton number, so that the same lepton number count exists on both sides of a nuclear reaction.

(a) Show that Equations 9.25 and 9.26 both violate the conservation of lepton number and therefore can never take place. Then verify that Equations 9.27 and 9.28 fix the problem.

(b) Come up with another way to fix Equation 9.26, other than what's done in Equation 9.28. Remember not to violate conservation of charge either.

(c) Look at Equation 9.31. At particularly high energies, this e⁻–e⁺ collision could produce two neutrinos instead of two photons. Write out this alternative annihilation reaction so that it obeys all the rules. Neutrinos have much less mass than electrons, so how is energy conserved? How was it conserved with photons as reaction products?

(d) List all of the allowed reactions that interchange protons and neutrons. Equations 9.28 and 9.29 are two of them; there are eight total, but you can reduce the number to four by letting your reaction arrows point in both directions.

Neutrino Astronomy

Neutrinos are not destroyed in the Sun the way positrons are; rather, they *leave* the Sun as soon as they're created. We called the neutrino "invisible" earlier, which just means that it fails to interact with the eyeball's retina the way a photon does. But neutrinos turn out to be invisible in other ways, too. They interact via the **weak nuclear force**, which is essentially just another way for particles to affect each other—gravitationally, electrically, weakly, etc. This force is named "weak" because it really is weak in comparison to the other nuclear force, the strong force. (Sometimes physicists are very creative in naming things. This time they were not.)

Consider this: an x-ray, which is a photon and therefore interacts by the electromagnetic force, can pass through about an inch of solid lead. A neutrino, however, can fly through a *light-year* of solid lead without interacting with anything. That's how weak the weak force is: It's not zero, but it's close. In principle, any force is capable of stopping a particle's progress through a medium of other particles, such as lead; the incoming particle pushes or pulls on one or more particles of lead until it loses all of its energy. But the weak force is effective only at extremely close ranges, such that a neutrino is exceedingly unlikely to get that close to the particles of lead. Weak interactions become likely only when a particle travels though a lot of material (a light-year of lead, say), just like winning a slot machine jackpot becomes likely only if you play a lot.

The Sun is nowhere near a light-year across, nor is it as dense as lead, on average. So a neutrino produced by the PP1 reactions in the core has no trouble flying unhindered through the Sun, emerging from it, and never returning. In other words, the Sun is shining neutrinos, as well as shining ordinary light. Antineutrinos are also *weakly interacting* particles (via the weak nuclear force); therefore, once they are produced (by other mechanisms), they easily escape the Sun as well.

Weakly interacting particles are of some importance to cosmology. Ideally, we'd like to study them in a laboratory setting; this turns out to be quite difficult, though not totally impossible. To see why, imagine a neutrino and a photon, both heading toward your hand (Figure 9.15). Then ask what each "sees" as it approaches. The (electromagnetic) photon, for example, "sees" a fleet of atoms in your hand that harbor (electrically charged) electrons. In other words, the photon sees an impenetrable layer of charges that it has no choice but to interact with. The photon either will give up its energy to those electrons, warming up your hand, or will reflect back.

By contrast, the neutrino has no electromagnetic properties—only weak nuclear properties. It therefore "sees" only a collection of tiny particles with huge expanses of empty space between them. Without the threat of an electromagnetic interaction, an electron is just a little dot. Now you understand why a neutrino can penetrate a light-year of lead: To the neutrino, it's like a light-year of empty space!

This state of affairs presents quite the observational challenge: How do you detect a particle that goes right through all your instruments? (Imagine a neutrino entering an optical telescope and just laughing at the mirror that is supposed to reflect it!) The only hope is to put a *lot* of atoms in its way and hope that, once in a great while, one of them will perfectly intercept the neutrino's path. For this

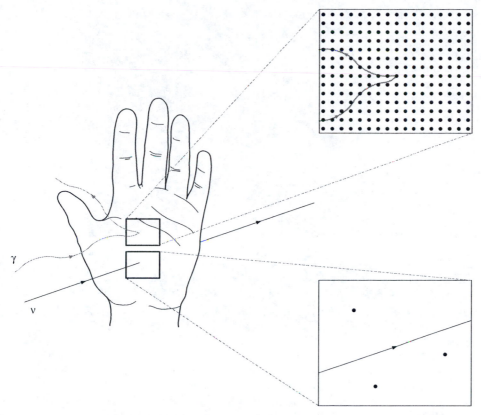

FIGURE 9.15 What Neutrinos "See"   When a photon (an electromagnetic wave) approaches your hand, it "sees" a fleet of electrical particles to interact with. When a neutrino (which does not interact electromagnetically) approaches your hand, it "sees" almost perfectly empty space, since it doesn't notice the electrical charges of the particles in your hand.

reason, *neutrino telescopes*, such as those shown in Figure 9.16, involve enormous quantities of matter. When these instruments detect enough neutrinos from a source, it is possible to construct a false-color "neutrino image" of the source, just as detecting photons of visible light with an ordinary telescope allows us to make an optical image. Figure 9.17 is just such an image of the Sun, from its "neutrino glow."

In general, neutrinos are produced in dense, high-energy regions, such as the core of the Sun and of supernovae, from which other particles are unable to emerge directly. Therefore, neutrino astronomy, using neutrino telescopes, lets us "see" into regions that leave no other direct residue. Recently, we've been able to "see" the core of the Earth from the neutrinos produced by its natural radioactivity. We'll learn in Chapter 10 that another one of these hidden places in which neutrinos play a key role is the early universe itself.

Weakly Interacting Massive Particles

Setting neutrino cosmology aside for the moment, there is another cosmological context for weakly interacting particles, a context introduced in Section 4.3. Based on their rotation speeds, galaxies typically appear to have at least 10 times more matter than we can directly observe. A little of this *dark matter* has been identified as hot gas, or dense objects called MACHOs. (Recall Section 6.4.) But most of the

FIGURE 9.16 Neutrino Telescope Unlike normal photon telescopes for collecting light, neutrino telescopes need to be made from huge quantities of matter, in the desperate hope that a few of its many atoms will interact with at least a few of the incoming neutrinos. One example is the Super Kamiokande neutrino observatory in Japan, in which neutrinos that interact within a vast tank of water are detected by photomultiplier tubes (like very sensitive cameras) lining the walls of the tank. (◆ Also in Color Gallery 1, page 12.)

FIGURE 9.17 The Sun in "Neutrino Light." False colors are assigned to place white pixels where many neutrinos were observed and red pixels where fewer were observed. (◆ Also in Color Gallery 1, page 12.)

dark matter has proved to be much more difficult to detect or explain, which would make sense if it were made of weakly interacting particles.

The majority of the dark matter might take the form of weakly interacting massive particles, or **WIMPs**. It's the "M" that makes them differ from neutrinos. A halo-sized swarm (recall that a galaxy's dark halo is about 10 times the size of the visible galaxy) of such particles would be massive enough to exert the gravitational influence observed on a galaxy's stars. And because their only nongravitational interaction is the weak nuclear force, they would pass through normal matter as neutrinos do and thus be extremely difficult to detect. This is essentially the *definition* of dark matter: material that generates a great deal of gravity without any other observable interactions, such as the emission or obstruction of light. We can imagine a dark matter halo, then, as a collection of WIMPs. You might wonder whether *neutrinos* could be the dark matter. But neutrinos could make up only a tiny fraction of the dark matter. Although their masses are not yet *precisely* measured, they are by far the lightest particles known, apart from totally massless ones like photons, and we can confidently assert that neutrinos account for less than 2% of the mass of dark matter in the universe. (WIMPs and neutrinos are depicted in Figure 9.18.) But although neutrinos are not a major constituent of dark matter, they at least confirm the *principle* of **exotic dark matter**—that some of the dark matter is not made from normal atomic particles: protons, neutrons, and electrons. The well-documented existence of neutrinos makes it plausible that some *other* weakly interacting particle could also exist, and that particle could match what we know about dark matter. The scientific community just hasn't detected it yet.

In practice, the nonexotic variety of dark matter is usually called **baryonic dark matter** or sometimes **normal dark matter**: dark matter that's made of protons, neutrons, and electrons. (Equating the terms "baryonic" and "normal" like this is partially incorrect, because electrons are part of "normal matter," but they are not baryons. However, no real ambiguity results, because the electrical attraction

FIGURE 9.18 Weakly Interacting Particles　What (a) a neutrino and (b) a WIMP would actually look like to the human eye.

between electrons and (baryonic) nuclei ensures that they're always found close to one another. And it's safe to ignore the electrons' contribution to dark matter anyway, since baryons are nearly 2000 times more massive than electrons.)

There are, in fact, several independent and compelling lines of evidence for the existence of *exotic* dark matter. You'll read about them in Chapters 10 and 12. But although that discussion will have to wait until those chapters, the results are worth knowing right now: *About 85% of all the matter in the universe is exotic dark matter*. WIMPs are the leading contender.

Of course, cosmologists have to at least *try* to detect them. (If you think that the cosmology community is going to shy away from this detection effort—just because WIMPs are, by definition, virtually impossible to detect—then you really haven't been paying close enough attention throughout this book!) The most direct method is to build some sort of detector. If the WIMP idea is correct, then our solar system continually passes through the swarm of WIMP particles that make up the Milky Way's halo. In Figure 9.19, we zoom in on our local neighborhood within the Galaxy. Since our planet flies through a field of WIMPs, which, like neutrinos, are weakly interacting, they must be passing *through* our planet (and our bodies!) all the time. We could build a detector to sense them, although it's harder than detecting solar neutrinos, because we encounter fewer WIMPs.

So far, these detection efforts have produced inconclusive results. One collaboration claims to have detected a type of WIMP that another collaboration claims to have effectively ruled out! Often, an experiment may not be able to directly observe the phenomenon in question, but can still usefully eliminate possibilities. At this point, it is probably accurate to characterize the state of affairs in that light: The dark matter particle, if it exists, has a possible *range* of values (e.g., for its mass) attributed to it. Researchers are slowly narrowing those ranges, zeroing in on the particle's true nature.

Another avenue for finding WIMPs is less direct. Researchers postulate that WIMPs and anti-WIMPs should *both* exist, in equal amounts. (No, MACHOs are *not* the same thing as anti-WIMPs!) Electrically interacting antiparticles such as positrons can't exist in any appreciable quantity today, because they would have annihilated with normal matter by now. But weakly interacting particles don't

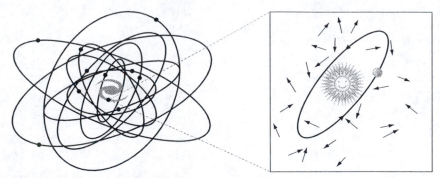

FIGURE 9.19 Our World, Moving through a Halo of WIMPs The Galaxy inhabits the center of a halo that is likely to be a huge collection of swarming WIMP dark matter particles. Each WIMP moves along an orbit around the Galaxy and halo, bound by gravity. Since our solar system orbits the galactic center as it moves through the halo of WIMPs, there are always WIMPs right here on (and in) Earth, where we might be able to detect them.

interact strongly enough to annihilate very often. So we look toward the dense core of some galaxy's dark matter halo and watch for the little gamma-ray flashes from WIMP–anti-WIMP annihilations (similar to that indicated in Equation 9.29). This method also gives us some insights into the nature of dark matter particles, but has not yet made any firm detections.

As you can see, the study of particle dark matter is a difficult work in progress. However, the idea of particle dark matter is sufficiently well grounded to justify the uphill effort. And as it turns out, the dark matter problem is not the only reason to hypothesize new weakly interacting particles.

In particle physics, a theory called *supersymmetry* postulates the existence of a new *super*partner particle for every known particle. Photon particles should have "photino" partners, for example, and electrons have "selectrons." (The "High Energy Physics New-Particle-Naming Institute" is not accepting any more job applications. Once the squarks were named (really), there just weren't enough sparticles (really) left to be named!) Most such supersymmetric particles should have decayed away by now, but one, called the "neutralino," would be stable. Some fairly general calculations reveal that neutralinos have the right mass and interaction strength, and should exist in sufficient quantities, to account for the missing dark matter.

Now, the theory of supersymmetry is substantially beyond the scope of this book. The relevant part, however, is this: Standard particle physics has a couple of minor glitches that appear to require new particles. The leading amendment to the standard model is supersymmetry, which specifically identifies a particle that ought to behave like dark matter. But—and here's our point—supersymmetry was *not* invented to solve dark matter; it was invented to solve the glitches in particle physics, and it just *happens* to solve dark matter, too. Exotic new science is made more believable when it can be justified from multiple, independent approaches.

9.6 Elemental Abundances

We've learned that stars transform the material from which they are born, changing its composition over time and returning this processed material back into the interstellar medium, where it becomes the raw material for the next generation of stars. The exact composition of material after stellar processing depends on the composition of the material that went into the star. And if stars are steadily changing the composition of the material in the universe, then it makes sense to ask whether there was a time *before* stars existed, and if so, what the initial composition of the *first* stars was.

Recall from Chapter 5 that we have observational information about the composition of matter in the universe. Hydrogen is by far the most abundant element, making up about 74% of baryonic matter in the universe by mass, followed by helium at 24%. All the other elements combined make up only the remaining 2%. Recall also that we see consistent baseline amounts of hydrogen (the majority), helium (23%), and lithium (two hundred millionths of a percent) in all stars, no matter how long ago they formed. These consistent numbers show up only for these three lightest elements (in six particular isotopes, as described in Chapter 5). Therefore, it's quite natural to suppose that the same baseline quantities also went into the very first stars.

Starting with the assumption that the preceding figures were the baseline abundances built into the first stars, astronomers can model the chemical changes from one generation of stars to the next, over a variety of stellar masses and lifetimes, along the lines of the stellar genetics of Section 9.4. Figures that match the total abundances measured today can indeed be arrived at in this way. The same cannot be said for most other initial compositions; starting with all hydrogen, for example, does not work. So, with one initial success in hand, we should proceed to seek other, independent evidence to support the idea that the observed baseline abundances were in the primordial gas before there were any stars. It would feel more satisfying if—with nothing more than our understanding of nuclear processes from this chapter—we could derive a solid, theoretical reason to *expect* the first three elements, but no others.

We will endeavor to develop this theoretical justification, starting with some highlights from our studies so far:

- The uniformity of the baseline abundances in every location in space suggests that a single event, probably before the birth of the first stars, produced these abundances everywhere. The only known way to create these elements' nuclei is by nuclear reactions, so we'll imagine some sort of cosmic nuclear burn—a flash or a fizzle.

- Some stars are hot enough to fuse elements heavier than helium, starting with the reaction given in Equation 9.16, in which three ^{4}He nuclei simultaneously fuse into ^{12}C. In cooler stars, this process never takes place.

- The absence of beryllium (Be, atomic number 4) and boron (B, atomic number 5) from the baseline abundance list is particularly conspicuous, since the present-day reactions that produce lithium also produce beryllium and boron. (It might be worth revisiting Section 5.4 to brush up on this material.)

- Not all combinations of protons and neutrons are stable; some are radioactive and decay into other elements, as in Equation 9.27.

With these ideas in mind, we'll proceed with *pure speculation* about the early universe. Our task is to make plausible the creation of H, He, and Li, but nothing else. In the next two chapters, we'll see timelines and justifications for the events of the early universe; but right now, we're just asking what nuclear physics can tell us.

We can begin with the ingredients: protons and neutrons. Never mind how they first appeared in the universe; they exist *now* (inside of nuclei), so they must have appeared *somehow*. Since we have no particular reason to start with any one arrangement of these baryons over any other arrangement, we'll just build a universe littered with p and n particles. You might think that this sounds awfully naive—we're talking about astrophysics and cosmology, after all—but there doesn't seem to be much else we can do. Let's just try it and hope it gets us somewhere.

The general approach we're taking here, with some differences, was first pursued in a famous paper published in 1948. Doctoral student Ralph Alpher and his thesis advisor George Gamow studied the possibility of an episode of nuclear reactions taking place very early in the history of the universe. For the sake of comedy, Gamow decided to include the name of his friend and well-respected colleague, Hans Bethe, on the paper. That way, the authors, "Alpher, Bethe, and Gamow," would sound like the first three letters of the Greek alphabet, "alpha, beta, and gamma." (Go ahead and ask the physicist in your life if he or she thinks this is funny; we promise you, the answer will be yes.) As it turned out, Bethe's interest was piqued, and he did go on to research this topic further.

Alpher and his collaborators presented the idea of nuclear fusion in the early universe as follows:

> "As pointed out by one of us, various nuclear species must have originated not as the result of an equilibrium corresponding to a certain temperature and density, but rather as a consequence of a continuous building-up process arrested by rapid expansion and cooling of the primordial matter. . . .The radiative capture of the still remaining neutrons by the newly formed protons must have led to the first formation of deuterium nuclei, and the subsequent neutron captures resulted in the building up of heavier and heavier nuclei."
>
> —Ralph A. Alpher, Hans Bethe, and George Gamow, "The Origin of Chemical Elements"

The paper actually suggested that *all* the elements could be formed in this way, even though we know now that only hydrogen, helium, and traces of lithium existed early on. Still, the approach outlined in the paper is valid for these lightest elements; the other elements would generally have to be made by stars. The reaction that seems likely to get the element-making ball rolling is

$$p + n \rightarrow {}^{2}H + \gamma .$$ (eq. 9.32)

(The rules of nuclear physics require the photon on the right side, and that is why Alpher referred to it as a "*radiative* capture" of a neutron onto a proton. But the photon doesn't affect our immediate interest in tracking baryons, so feel free to ignore it.) This reaction produces deuterium (^{2}H), although not the same way as the Sun does (Equation 9.13). The Sun's method starts with two protons instead of one proton and one neutron, because there are no free neutrons in the Sun.

Then what? Deuterium usually reacts as follows to make two different three-particle nuclei:

$$ {}^{2}H + {}^{2}H \rightarrow {}^{3}He + n \quad \text{or} \quad {}^{2}H + {}^{2}H \rightarrow {}^{3}H + p .$$ (eq. 9.33)

So our hypothetical early universe is now populated with p, n, ^{2}H, ^{3}H, and ^{3}He. With more ingredients, a greater variety of reactions becomes possible. Many

*Coauthors of the famous "alphabet paper" are **Ralph Alpher** (born 1921), **Hans Bethe** (1906–2005), and **George Gamow** (1904–1968). ("Alpher, Bethe, and Gamow" was meant to sound playful, like "alpha, beta, and gamma," the equivalent of "ABC" in the Greek alphabet.) Their paper paved the way for one of the great successes in modern cosmology: an account of the abundances of the light elements. Alpher is known mainly for his discoveries in two major cosmology papers: the alphabet paper (see accompanying text) and another work in which he and Gamow predicted the existence of the CMB. Bethe is considered one of the foremost contributors to astrophysics, particularly in the field of stellar nucleosynthesis. Gamow's work ranged from stars and cosmology all the way to DNA genetics; he is generally recognized as a chief developer of the big bang theory.*

of these subsequent reactions synthesize helium-4, sometimes in obvious ways, as in

$$^3\text{He} + \text{n} \rightarrow {}^4\text{He} + \gamma \quad \text{and} \quad {}^3\text{H} + \text{p} \rightarrow {}^4\text{He} + \gamma \ , \quad \text{(eq. 9.34)}$$

but more often in less obvious ways, such as

$$^2\text{H} + {}^2\text{H} \rightarrow {}^4\text{He} + \gamma \quad \text{and} \quad {}^2\text{H} + {}^3\text{H} \ (\text{or} \ {}^3\text{He}) \ \rightarrow {}^4\text{He} + \text{n} \ (\text{or p}) \ . \quad \text{(eq. 9.35)}$$

Lithium-7 is produced by two reaction paths:

$$^4\text{He} + {}^3\text{H} \rightarrow {}^7\text{Li} + \gamma \quad \text{(eq. 9.36)}$$

and

$$^4\text{He} + {}^3\text{He} \rightarrow {}^7\text{Be} + \gamma \ , \quad \text{followed by} \quad {}^7\text{Be} + \text{e}^- \rightarrow {}^7\text{Li} + \nu \ . \quad \text{(eq. 9.37)}$$

Equation 9.36 is quite direct. Equation 9.37 looks less direct. Note that because ^{7}Be has 4p and 3n, whereas ^{7}Li has 3p and 4n, the conversion from ^{7}Be to ^{7}Li requires a proton to transform into a neutron, via a weak nuclear process. (Pause a moment to answer for yourself how the electron capture process shown accomplishes that.)

Remember that helium-4 is a very tightly bound nucleus. Its great stability implies that once it's formed, it's unlikely ever to be destroyed. Equations 9.36 and 9.37 are examples of reactions that destroy ^{4}He, so they must be very rare, and we should expect very little lithium-7 to be produced by them. The production of lithium-6 is much rarer still, because the primary reaction that makes it, namely,

$$^4\text{He} + {}^2\text{H} \rightarrow {}^6\text{Li} + \gamma \ , \quad \text{(eq. 9.38)}$$

is a much less probable way to use up hydrogen-2 (deuterium) than Equations 9.33 and 9.35—besides which, both lithium-7 and lithium-6 can be destroyed in favor of returning to helium-4, as follows:

$$^7\text{Li} + \text{p} \rightarrow {}^4\text{He} + {}^4\text{He} \quad \text{and} \quad {}^6\text{Li} + \text{p} \rightarrow {}^3\text{He} + {}^4\text{He} \ . \quad \text{(eq. 9.39)}$$

So far, so good: We've produced helium and a little lithium, even with our naive starting point! But to be consistent with the actual abundances of the light elements, we also need to *avoid* producing anything else. Beryllium and boron are the next two elements after lithium in the periodic table. The next element after those, carbon, is also a possibility, since it *is* constructed from helium in stars. Let's see if we can fuse any of these elements.

If we try to construct any nucleus with five particles in it—^{5}X—where "X" might mean "He," or "Li," or "Be," etc., we hit a snag. Although you could imagine some appropriate reactions, like

$$^4\text{He} + \text{p} \ (\text{or n}) \rightarrow {}^5\text{X} \quad \text{or} \quad {}^3\text{H} \ (\text{or} \ {}^3\text{He}) + {}^2\text{H} \rightarrow {}^5\text{X} \ , \quad \text{(eq. 9.40)}$$

none would be successful, because *there is no stable nucleus with exactly five baryons.* The components of the left reaction, for example, won't stick together. And we already know that the reaction on the right makes ^{4}He instead of ^{5}X, as we saw in Equation 9.35; at best, the ^{5}X is a brief, intermediate step.

We've successfully constructed nuclei with masses 6 and 7 (^{6}Li and ^{7}Li); what about mass number 8? Here we encounter the same snag as before. *There is no stable nucleus with exactly eight baryons.* Even though you could make one quite easily, via the reaction

$$^4\text{He} + {}^4\text{He} \rightarrow {}^8\text{Be} + \gamma \ , \qquad \text{(eq. 9.41)}$$

it won't last, because beryllium-8 is radioactively unstable and immediately decays right back into the two helium-4 nuclei it came from:

$$^8\text{Be} \rightarrow {}^4\text{He} + {}^4\text{He} \ . \qquad \text{(eq. 9.42)}$$

The only remaining ways to build a larger nucleus are as follows:

- Fuse something with lithium? . . . But lithium is already incredibly rare, and the reactive species that could combine with lithium (^{2}H, ^{3}H, and ^{3}He) are much more likely to get used up in other ways, as in Equations 9.34 through 9.37. The most common reaction for ^{7}Li to undergo is to react with a proton, but that results in two ^{4}He nuclei; that is, it rearranges the lithium into *smaller* nuclei, not bigger ones (Equation 9.39).

- Fuse three helium-4 nuclei together simultaneously to make carbon? . . . This does occur in some stars, as we saw earlier in the chapter, which means that under the right temperature and density conditions, it is definitely possible. However, stars have an advantage that the primordial gas we started with lacks. When a star's core runs out of hydrogen fuel and therefore can't fuse hydrogen in order to support its weight, its gravity causes the star's core to compress. The compression increases the temperature and density until they are sufficient to form carbon-12. But in the absence of a localized gravity source like a star, our nuclear process would need another, similarly prolonged period of high temperature and density in order to continue. Evidently, the early universe didn't provide the right conditions, since we don't observe carbon-12 in the primordial gas.

So, starting with nothing but some mixture of baryons, we have identified a sequence of natural nuclear processes that automatically produces H, He, Li, and nothing else—just as we set out to do. However, let's keep in mind some caveats and demand that our overarching theory address each of them in the next chapter.

CAVEAT #1: In this section, we just *invented* a period of nuclear activity, because we believe that there must have been one in order to create the baseline quantities of elements that made up the first stars. But a theory of cosmology should provide a good *reason* for the onset of nuclear reactions.

CAVEAT #2: Since carbon-12 does not have a significant baseline quantity, the early universe's nuclear episode could never have been too hot for very long; otherwise, ^{12}C would have been produced along with H, He, and Li. To put it another way, in order for our reasoning in this section to be true, there couldn't have been too much of a nuclear "flash"; at best, it could have been a flash that fizzled out. A successful theory will need to explain why it never got hot enough for long enough to make carbon.

CAVEAT #3: We found a way to make H, He, and Li exclusively, but we haven't tried to make them in the proper *quantities*. The quantities produced strongly depend on the initial mix of protons and neutrons. Our theory will need to provide the correct initial ratio of protons to neutrons. (You're already familiar with a mechanism that can accomplish this: Weak nuclear interactions such as Equations 9.28 and 9.29 could turn protons into neutrons and vice versa, until the right mixture is obtained.)

CAVEAT #4: We considered these nuclear processes in an environment without photons and other particles that might interfere. We also failed to include the expansion of spacetime, which would continuously cause any photons that were present to be redshifted. Redshifting reduces each photon's energy and the overall temperature, affecting which nuclear reactions are able to proceed and how quickly they go. A good theory will need to include these things, in such a way that doesn't ruin the successes we've already had in this section.

So without knowing the *actual* nuclear history of the universe, we still managed to identify some key features of that history, based on the nuclear physics in this chapter. We did the same for the CMB in the previous chapter, identifying necessary elements of its history based on the physics of photons and electrons.

The Ingredients of Planets and People

The material world, apart from exotic dark matter, is constructed from atoms of different elements, all of which are manufactured naturally somewhere in the universe. The manufacturers are:

- *the early universe*, which made hydrogen, helium, and small traces of lithium.
- *low-mass stars*, which make carbon, nitrogen, and oxygen.
- *high-mass stars*, which make everything between carbon (6 protons) and iron (26 protons), inclusive.
- *type Ia supernovae*, explosions of white dwarf stars, which produce iron and the elements near iron in the periodic table.
- *type II supernovae*, explosions of massive stars, which make all the naturally occurring elements above iron in the periodic table, up to uranium (92 protons). We see the elements above uranium when they are deliberately constructed by people.

There's also one manufacturer we haven't yet mentioned:

- *interstellar collisions* between nuclei in space will sometimes break apart a larger nucleus into smaller pieces. In this way, relatively light and abundant nuclei such as carbon, nitrogen, and oxygen can be fragmented. Lithium, beryllium, and boron (with three, four, and five protons, respectively) are formed predominantly in this way. (*Most* of the lithium in the universe formed this way, but the *earliest* lithium originated as described in Section 9.6, before the first stars, and therefore before carbon, nitrogen, or oxygen, even existed.)

Thus, the gold in your jewelry was forged not in the fires of Mordor, but in a type II supernova whose brightness rivaled a hundred billion stars combined. Gold and other heavy nuclei were assembled by repeatedly assimilating uncharged neutrons into existing nuclei—the neutrons having been produced in abundance by electron capture reactions just before the explosion. From some of the planets in the galaxy where all this was happening, the explosion would have been visible in the daytime sky.

When you eat your potato salad, sushi, or samosa, you're bringing into your body a collection of nuclei with a distant and ancient history. They have lived in hundreds of different stars (at least) before they arrived here on Earth. But they're permanent residents now, bound by the Earth's gravity.

Maybe there's a mosquito on your arm, drinking your blood, which contains iron. Probably each of the 30 neutrons inside every iron atom has spent most of its life bound inside a helium nucleus. When the helium nuclei found their way into the core of a star, their neutrons had to sit around watching for at least a million years before they finally got the chance to join a carbon or oxygen nucleus, where they could get close to other neutrons. Eventually, the small star they inhabited got too massive to support its own weight and exploded as a type Ia supernova. The fusion that ensued gave the neutrons the opportunity to experience the camaraderie of the most tightly bound nuclear community out there: an iron nucleus. It's an experience they will continue to enjoy somewhere, probably forever, even after you heartlessly squish the mosquito.

Each second, 10 trillion neutrinos enter your head. They were produced by nuclear power in the core of the Sun, and they carry enough energy to kill you. But instead, the 10 trillion neutrinos leave your head without even slowing down. The energy needed to make each neutrino previously resided in hydrogen nuclei, some of which have remained completely unchanged since our universe began, until just over 8 minutes ago, when they underwent the first reaction in the PP chain in the Sun (just over 8 light-minutes away from Earth). The hydrogen nuclei gave up a little of their mass as they merged into deuterium, so that a neutrino might pass through your head from space. Meanwhile, other neutrinos and antineutrinos streak through your head from below, produced by natural radioactivity (beta decay) in the dirt beneath you. The uncountable trillions cross paths inside your brain without affecting you or each other in any way.

Your diet soft drink may contain just one calorie, but it probably also contains one WIMP. That particle has been around since a time long before the existence of stars, galaxies, or material objects of any kind, mindlessly following its changing geodesic trajectory over cosmic time. It wandered into your Galaxy and then into your beverage. It leaves your soft-drink can right after it enters, but then another one shows up. It doesn't care that you're drinking diet cola instead of regular, and the M in WIMP stands for "massive," which might explain why it's so hard to lose weight (just kidding).

Quick Review

In this chapter, you've formalized your understanding of energy and used that concept to learn about the nuclear power source of the stars. Stars shine by converting mass energy into light energy. In the process, they change the composition of the universe. Over time, stars have built up all the naturally occurring elements in the periodic table, beginning with just a few of the simplest. Other nuclear reactions, taking place long before the first stars were born, were able to construct only hydrogen, helium, and traces of lithium, which explains why we see baseline quantities of just those elements in the universe.

Try the following crossword puzzle to test your knowledge of key terms and concepts from Chapter 9:

ACROSS

1 Stable, four-baryon nucleus produced by many different nuclear reactions in the early universe

8 Formed in the early universe by fusing a proton and neutron together

9 Power source for main-sequence stars

11 Weakly interacting particle with very little mass

12 One joule of energy transferred every second

13 A star "being born," in the stage before hydrogen fusion begins in its core

17 Produced by a type II supernova of an extremely massive star (more than about 30 solar masses)

21 Nuclear force that makes protons and neutrons bind together to make atomic nuclei

22 Small main-sequence star with low surface temperature

24 The element with the most tightly bound nucleus

25 Small, ultra-dense relic star that usually remains after the death of a high-mass star

26 The amount of energy produced by a nuclear reaction is proportional to the amount of _____ lost.

27 Nuclear process in which two lighter nuclei merge into one heavier nucleus

28 Conserved property of some object or system, measured in joules

29 Explosion signifying the death of a high-mass star (or a white dwarf that becomes too massive)

30 Power source for the cores of red giant stars

DOWN

1 Type of equilibrium achieved by stars: a balance between inward gravity and outward pressure

2 Power in the form of light

3 Large star with low surface temperature

4 Nuclear reaction sequence by which the Sun converts H to He

5 In order for two nuclei to fuse, they must approach one another fast enough to overcome their mutual _____ repulsion.

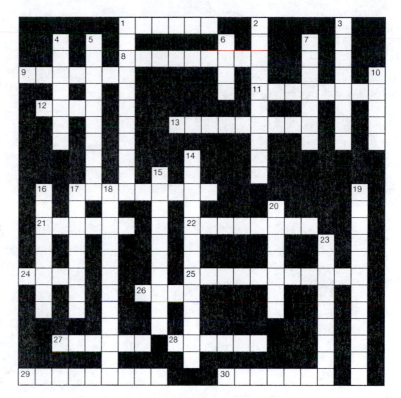

6 Weakly interacting massive particle that might account for most of the dark matter in the universe

7 When protons undergo electron _____, neutrons and neutrinos are produced.

10 Rate of energy transfer or consumption

14 Flux of sunlight arriving at Earth

15 A low-mass star after its "death"

16 Antimatter partner of the electron

17 Type of dark matter made mostly from protons and neutrons (by mass)

18 Burst of energy caused by matter colliding with antimatter

19 Type of physical law governing properties like energy or electrical charge that remain constant in nature

20 Class of particles, including protons and neutrons, that are affected by the strong nuclear force

23 Radioactive process in which a nucleus emits a high-energy electron

Further Exploration

(M = mathematical content; I = integrative—builds on information from more than one chapter; P = project idea; R = reflection; D = suitable for class discussion)

1. (R) What is needed to make natural thermonuclear reactions occur somewhere in the universe? Is any otherwise sufficient concentration of nuclear fuel enough? Are there many other requirements?

2. (M) Verify that the Sun gains no net charge during the PP chain reaction. Make sure that you count *all* the relevant charged particles: nuclei, electrons, and positrons.

3. (M) How many neutrinos are produced every time the Sun makes one ^{4}He nucleus by the PP chain reaction? How many photons and how many positrons are produced?

4. (RDI) Read "Solving the Solar Neutrino Problem" in the April 2003 issue of *Scientific American*. In no more than one page, summarize the nature of this problem, how long it was a problem, and when and how it was solved. Did it turn out to be a problem with our understanding of the Sun or our understanding of neutrinos?

5. (RDI) Read the March 2003 issue of *Scientific American* article titled "The Search for Dark Matter," and write a one-page summary on the techniques and technologies employed in this search.

6. (M) In "Nuclear Power in the Sun" in Section 9.3, we calculated the mass used by the Sun for nuclear fusion each second. How much of that mass used per second is converted into energy, and how much is converted into helium? Express both answers in kilograms and as percentages. Then verify that the Sun can stand to *lose* mass at this rate for 4.6 billion years without substantially diminishing in mass during that time.

7. (MR) In Equation 9.37, two reactions are listed that produce lithium-7.
   (a) Why does the first reaction, as stated, produce ^{7}Be instead of ^{7}Li?
   (b) Why does the second reaction involve a neutrino on the right side?
   (c) What conservation law(s) would be violated if you were to replace the neutrino in the second reaction with an electron? a positron? a hydrogen-1 nucleus? an antineutrino? a puppy?

8. (M) Let's figure out the surface gravity of white dwarfs and neutron stars. Surface gravity is the acceleration of an object as it falls into the star's surface; on Earth, it's 9.8 m/s^2, or equivalently, 32.2 ft/s^2. You can compute surface gravity by dividing Newton's gravity force, Equation 4.2, by the mass of the falling object (m), along the lines of Figure 4.6. The result will be in units of m/s^2.

   (a) How many times bigger is the surface gravity on a white dwarf than on Earth? Assume that the white dwarf has the Sun's mass (2×10^{30} kg) and the Earth's radius (6.37×10^6 m).
   (b) What about the surface gravity on a neutron star, with double the Sun's mass and about a 15 km radius?
   (c) Argue that in either case the surface gravity is tremendous and therefore a tremendous internal pressure is needed to prevent the star from totally collapsing. (Why? There's a great two-word answer!) Argue further that since both the white dwarf and the neutron star might just sit there forever, continuously cooling down, ordinary gas pressure is *not* sufficient to prevent the collapse. Thus, a new form of pressure must (and does) arise somehow; it is called *degeneracy pressure*.

9. (DP) Refer back to Figure 9.10. What would such a picture look like for the Earth? Search the Internet to help you make your drawing. Compare (a) the phase of matter (solid, liquid, gas, or plasma) with (b) the physical activity going on between the star figure (Figure 9.10) and your Earth figure.

10. (M) In this problem, you'll investigate how neutrinos travel through a neutron star, which will help explain why type II supernova explosions occur. One way to characterize how weakly neutrinos interact with matter is by their *cross section*, $\sigma = 10^{-48}$ m^2. In other words, when normal matter encounters a neutrino, it "sees" a circle of area $\pi r^2 = 10^{-48}$ m^2. (You can imagine the neutrino as a sphere with the same radius.)

    (a) Verify that the quantity $m_n/\rho\sigma$, where ρ is the density of the neutron star (in kg/m^3) and m_n is the mass of a neutron (1.67×10^{-27} kg), carries the units of distance (meters). This distance is roughly how far a neutrino of cross section σ can travel through an "ocean of neutrons" of density ρ before it hits something.
    (b) Compute the distance from part (a), using a forming neutron star's enormous density of about 10^{17} kg/m^3. Recompute this distance just for fun, using the density of ordinary rock, about 3000 kg/m^3. Your two answers will *not* be close to one another: You should be able to conveniently express your answer for a neutron star in kilometers and for a rock in light-years! (Recall that 1 light-year = 9.46×10^{15} m.)
    (c) Compare your answer for how far a neutrino travels inside a neutron star with the typical radius of a neutron star, about 15 km. Think about this and about Equation 9.24, and write an explanation of how electron capture reactions could create a "neutrino bomb" to help a type II supernova happen whenever

a neutron star forms. (Note that as soon as a neutrino hits something, that something will hit other somethings, each of which will hit more somethings, and so on.)

11. (MR) When heavy elements such as gold form in the universe, how does it happen, and do the nuclear reactions involved produce energy or consume energy?

12. (M) To what extent is sunlight able to heat up the oceans? Let's investigate for each square meter of the oceans' surface. The specific heat capacity of water is 4186 J/kg, and the average depth of the oceans is about 4 km. You may use a globally averaged solar constant of 350 W/m² and a water density of 1000 kg/m³.
    (a) About how long would it take for sunlight to raise the temperature of the ocean by 1 degree Celsius? Use your answer to comment on why coastal regions tend to experience mild seasons.
    (b) The calculation in part (a) is quite rough; what assumptions went into it, and what did we neglect to account for?

13. (M) The city of Chicago covers an area of 230 km² (square kilometers), or 2.3×10^8 m².
    (a) With a maximum of 1400 W/m² of sunlight, could solar power accommodate the electricity needs of Chicago's 3 million residents? Assume that each resident consumes a total of 7000 W, on average (electricity and fuel combined). What would it take to convert Chicago to solar power?
    (b) The total annual human energy use worldwide is about 5×10^{20} J. What percentage of global energy use corresponds to a single large city like Chicago? How much power is used by U.S. citizens, on average, compared with the same quantity for an average person anywhere in the world? (To answer this last part, find an estimate of the current worldwide population.)

14. (M) In engineering, the efficiency of an energy conversion device is defined as the energy or power of the desired form that it produces, divided by the energy or power it receives. For example, typical solar panels are 15% efficient, meaning that 15% of the incident radiant energy gets converted into electricity; the rest turns into heat. Typical incandescent light bulbs are about 5% efficient, converting only 5% of the electrical energy they consume into light.

Imagine reading a book indoors. The book's dimensions are 20 cm by 30 cm (0.2 m by 0.3 m); let's use a solar panel of that same size, outdoors, exposed to direct sunlight. The solar panel sends electricity to an incandescent lamp indoors. Is the resulting wattage enough to read indoors by? (Note that if you just read outdoors by daylight, the wattage is more than adequate.)

15. (R) Why can't you make a neutrino telescope out of lenses and mirrors, like an optical telescope?

16. (RD) Comment on why Figure 9.11 looks the way it does. For example, why is the "happiest" nucleus that of iron, somewhere in the middle of the periodic table, rather than being either the lightest or the heaviest nucleus?

17. (ID) Match up what you learned in this chapter with what you read in Chapter 5 about lithium, beryllium, and boron. What is answered now, and what still needs explaining? A few paragraphs should suffice here.

18. (IDP) Look up what the world is made of: What elements are in the Earth's crust and its interior? (You can use the primary source of all human wisdom, google.com.) Then indicate where each of these elements comes from. Finally, recall from Section 9.4 that, in the periodic table, even-numbered elements between oxygen and iron tend to be more abundant than odd-numbered ones. Does this appear true from your Internet search? Does it appear true of the objects you see around you in your daily life? Give several examples.

19. (R) What does nuclear binding energy have to do with the mass of a particular nucleus? What would happen if you put a ^{4}He nucleus and a ^{12}C nucleus on one side of a balance and an ^{16}O nucleus on the other? How does nuclear binding energy help to explain what happens?

20. (RI) In a few sentences, indicate the ways in which nuclear physics is related to cosmology.

21. (RI) How might an *expanding* universe affect the story presented in Section 9.6 of the creation of the light elements? List a few of these effects, with a sentence or two explaining each.

22. (RI) Why would white dwarfs, neutron stars, and black holes make perfect MACHOs? Answer in terms of what a MACHO is, what it's made of, and how we detect them. (See Sections 4.3 and 6.4.)

CHAPTER

10

The Big Bang Theory

"The big bang theory describes how our universe is evolving, not how it began."

—*P. J. E. Peebles, cosmologist*

"The evolution of the world may be compared to a display of fireworks that has just ended: some few red wisps, ashes and smoke. Standing on a cooled cinder, we see the slow fading of suns, and we try to recall the vanished brilliance of the origin of the worlds."

—*Georges Lemaitre, one of the pioneers of the big bang theory*

We now have in hand the key observations from physics and astronomy that inform us about our universe. We also have the physical concepts needed for interpreting the observations. Our next task is to weave all of these observations into a coherent and consistent picture of what is going on. This is not an easy task. Any theory that survives must explain a host of disparate observations. Here is a summary of what we'll need to explain:

- *Recession of galaxies.* Beyond our local galactic neighborhood, all galaxies are getting farther away from us.

- *Redshift–distance relation (Hubble law) for galaxies.* The physical distance between our Galaxy and other galaxies is increasing at a rate proportional to that distance. On average, this relation governs the recession of galaxies uniformly in all directions. Galaxy survey data suggest that we are not in a unique position, so we suspect that galaxies receding according to the Hubble law could be observed from planets in distant galaxies, too.

- *Large-scale structure.* We don't see a center or an edge to the universe. We see galaxies everywhere we've been able to look, arranged in a pattern of walls and voids.

- *The cosmic microwave background (CMB).* Our world is being bombarded by microwaves from every direction in space. We need a theory to account for many details of the distribution of microwaves: the perfect blackbody spectrum, the temperature, the extraordinary uniformity, the specific anisotropy pattern, and the specific polarization pattern.

- *Abundances of the light elements.* Throughout the universe, we see mostly hydrogen and helium with traces of heavier elements. Deuterium, helium, and lithium have specific, universal baseline abundances. Any theory we adopt must explain where these elements came from and why they formed in their observed quantities.

- *Astronomical ages.* We don't see arbitrarily old objects. Everything is younger than about 12 to 13 billion years.

- *The darkness of the night sky (Olbers's paradox).* This observation puts limits on either the spatial or temporal extent of the universe—or both.

You may have heard of the big bang theory. It provides a single coherent interpretation of all the cosmological observations we have discussed. Many refinements to this theory are made to accommodate new data as it is acquired, but the basic framework has been in place for more than 70 years. Some of the more recent cosmological measurements provide spectacular supporting evidence for the big bang. We'll present this evidence in Section 10.3 and add greater detail to the picture in Chapters 11 and 12.

The *words* "big bang" have entered popular culture, but a corresponding understanding has not grown as quickly. To some extent, this lesser understanding can be attributed to the theory itself: Its scale is vast, its features involve exotic physics, and its measurements are complicated. But one of the biggest obstacles to understanding the theory is just its name.

The name "big bang" calls to mind an explosion sending debris out into fixed, empty space. But that's not what the theory actually says. Rather, it describes a universe in which *space itself* is expanding, carrying everything along with it. A more appropriate name might be the "expanding space theory," but big bang is a catchy name, and we're stuck with it.

10.1 Overview of the Theory

In this first section, we'll explore three aspects of the big bang theory. We will summarize the nature of the expansion, consider what the universe was like in the distant past, and see what the theory has to say about the age of the universe. Then, in the rest of the chapter, we'll be able to address a particular misconception about the big bang theory, summarize the variety of evidence for the theory, and ask whether or not the theory leaves us any wiggle room for changes to it.

The Expansion

The observation of galactic redshifts beginning in the 1920s was the main motivation for developing the big bang theory, because it's very difficult to explain these observations in terms of independent motions through space of galaxies that are billions of light-years away from each other. How could they all have communicated their plan to one another? How could they have coordinated their motions such that they all recede from the Milky Way, in every possible direction, and at speeds carefully prescribed by Hubble's law?

It is much simpler to explain the redshifts, and the redshift–distance relation, if the galaxies are embedded in a space that is uniformly expanding. Then we can explain *all* of the galactic motions with just *one* basic premise: expanding space. To put an expanding universe in proper context, recall the following points from Chapters 6 and 7:

- The observational evidence in the latter half of Chapter 6 leads us to conclude that general relativity is an accurate theory of space and time.
- The theory of general relativity provides a context in which the notion of expanding space seems more natural than static space. Expanding space is part of the solution to the Einstein equation, for a homogenous and isotropic universe.
- From Section 7.3, the redshifts of galaxies, together with Hubble's redshift–distance relation, can be explained by an expanding universe. The redshift is the actual growth in a photon's wavelength as it travels through the expanding space. The expansion rate (today) is expressed with a single number called the Hubble constant, H_0.
- The galaxies themselves don't expand; rather, the space between them does. Internal forces within galaxies (and, for that matter, within stars, planets, and people) overcome the expansion.

So, accompanying the direct evidence from galactic redshifts, there is a framework in place to characterize an expanding space. Keep this notion of an expansion based on general relativity in mind as you read on, because it leads to some very interesting consequences.

The Universe Was Hotter and Denser in the Past

The universe is currently expanding at a specific rate (H_0), so if we go back in time, galaxies must have been closer together. Extending this reasoning back further, we see that there should have been a time when all of the material in the universe was packed in very tightly. That is, the *density* of matter (and light) would have been very high everywhere.

Photon wavelengths were shorter in the past, too, which means that the photons had a higher frequency and were more energetic. When these photons interacted with other particles, such as free electrons before the formation of the CMB, the photons exchanged energy with the electrons, which pushed and pulled on nuclei, too. So the high-energy photons of the past really mean high-energy *everything* in the past. Put another way, the universe used to be *hotter*. Temperature is a measure of the average energy of a collection of moving particles, so the temperature used to be higher. The further back in time you go, the shorter and bluer the photons were, and therefore the hotter everything was. Figure 10.1 displays the distinction between the universe's past and present.

So the upshot is this: The big bang describes a universe that was increasingly hotter and denser in the past, all the way back to the beginning of time. The universe has been cooling and thinning out ever since, a direct consequence of expanding space.

The Universe Has a Finite Age

Imagine watching the history of the universe in rewind mode. The expansion becomes a contraction, and galaxies rush toward each other. Eventually, the matter from different galaxies will overlap, and if the rewind continues, even

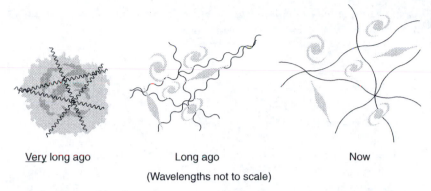

Very long ago   Long ago   Now

(Wavelengths not to scale)

FIGURE 10.1 The Big Bang Beginning: Hot and Dense   Today's cold universe of vast empty spaces evolved from an early state of extremely high density and temperature.

subatomic particles get squished together until they overlap one another. If we think of this ultrahigh-density state as time zero, then the universe has a finite age.

Look at it another way: We can measure the average distance between galaxies today, and we can measure the expansion rate, too. Combining the two pieces of information, we can estimate how much time has gone by since the distance between galaxies was zero. (It's like "How long does it take to drive 50 miles at 50 mph?") But the galaxies (or particles, or whatever) can't get any closer together than zero separation, so you can run the expansion backwards only so far. Evidently, there was a beginning. (Actually, it *could* instead be that the expansion used to be slower and slower as you go back in time, so that if you watch the universe rewind, it never quite gets to the beginning. But the opposite turns out to be true: The expansion was very fast in the beginning, and you'd have to rewind in super slow motion to watch it. We'll examine this history in Section 11.2.)

The observed distance between galaxies today is known. If the expansion rate is fast, then it wouldn't have taken much time for galaxies to reach their current separation. If the expansion rate is slow, then it must have taken a long time for galaxies to get where they are today. This reasoning leads us to realize that *the universe's age is inversely proportional to the expansion rate*:

$$t \sim \frac{1}{H_0} \ .$$

(eq. 10.1)

So a faster expansion rate (bigger H_0) corresponds to a younger age today (smaller t). Depending on the details of the matter and energy content of the universe (Chapter 11), the actual age could be greater or less than indicated by this simple relation, but it ought to be accurate to within ±50% or so.

The Hubble constant is a special number that blends the universe's expansion rate with its current size scale. Recall, therefore, that its units are a bit strange. H_0 is the slope of a graph with velocity in km/s (on the y-axis) and distance in Mpc (on the x-axis), so H_0 has units of (km/s)/Mpc. To get the age of the universe, we must convert Mpc to km in order to cancel the distance units (km over km), so that we're left with only time units. For reference, 1 Mpc = 3.09×10^{19} km.

As an example, let's calculate the age of the universe if $H_0 = 71$ (km/s)/Mpc, which happens to be a recently measured value:

$$t \sim \frac{1}{H_0} = \frac{1}{71 \dfrac{\text{km/s}}{\text{Mpc}}} = \frac{1}{71 \dfrac{\text{km/s}}{3.09 \times 10^{19}\,\text{km}}} = \frac{1}{2.26 \times 10^{-18}\,\dfrac{1}{\text{s}}}$$

$$= 4.41 \times 10^{17}\,\text{s} = 1.4 \times 10^{10}\,\text{years} . \qquad \text{(eq. 10.2)}$$

The result is about 14 billion years, which agrees nicely with the astronomical ages presented in Section 5.5. This age estimate follows directly from the presupposition of expanding space.

EXERCISE 10.1   Age Calculations

Is a universe with an expansion rate of $H_0 = 77$ (km/s)/Mpc older or younger than a universe with an expansion rate of $H_0 = 63$ (km/s) Mpc? (These values are approximate upper and lower limits on the Hubble constant as currently measured by the HST key project.) Perform the appropriate calculation, and then use words to justify your calculated results.

10.2 Expansion, Not Explosion

Remember from the introduction to this chapter that the name "big bang" is misleading and contributes to the common misconception of the big bang as a conventional explosion. The big bang is *not* an explosion (in which stuff flies *through* space), but rather is an expansion *of* space. An explosion happens *somewhere*, but an expansion happens *everywhere*. Figure 10.2 contrasts expanding space with an explosion into space.

In the big bang, the expansion occurred (and is still occurring) uniformly, with all of space participating in the same way. It is very important to understand this aspect of the big bang, both for accuracy and for the way it connects you to the universe. The big bang did not just happen out there somewhere, far in the distance, but *right here* on the patio (or wherever *you* like to read about cosmology). This region of space where you are sitting was extremely dense and glowing hot about 14 billion years ago.

To clarify the difference between an expansion and an explosion, let's investigate the possibility that the big bang *was* like a conventional explosion that sprayed matter into fixed, empty space. As we explore the details, we'll see why the explosion theory struggles to explain observations that fit naturally within the expansion theory.

In an explosion, different bits of matter (shrapnel) fly outward from the center of the explosion at different speeds; some bits just happen to get a stronger kick than others. Depending on the details of the explosion, we might expect that some bits will move faster than others. The faster bits overtake the slower ones, and after a time, the slower bits remain near the center while the faster ones are far from it.

This result is at least qualitatively consistent with the observations summed up by the Hubble law: Galaxies more distant from us recede faster. This is so whether we are located near the center of the explosion or far from the center. In the latter case, galaxies in one direction are farther from the source of the explosion and hence are moving faster than we are. Galaxies in the opposite direction are closer

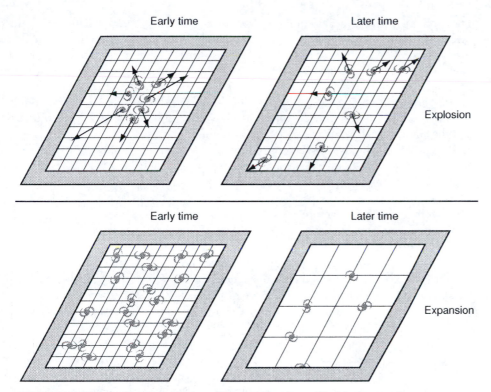

FIGURE 10.2 Contrasting an Explosion with an Expansion   All of the images show a two-dimensional universe viewed through a picture frame of fixed physical size. In the top panel, an explosion sends galaxies flying outward from a high-density center at different velocities into the fixed grid of empty space. In the bottom panel, an expanding space (represented by the grid lines) carries galaxies along with it, but they stay attached to the same points of space (same grid lines). In this example of expansion, the space is uniformly filled with galaxies at both times.

to the explosion, and hence move slower than we do. What we measure is the *difference* in velocity between our Galaxy and another galaxy, so in any direction we would see galaxies becoming more distant from us. This means that there is at least a chance that the observed Hubble law is consistent with the explosion idea.

However, there are some tricky details to be explained. First, it's very hard to imagine an explosion mechanism that would produce the wide range of velocities corresponding to the redshifts we observe (velocities ranging from around 100 km/s up to essentially the speed of light!). Second, it seems likely that the velocities of the galaxies would be related to some observable property of the galaxies. For example, if the explosion "kicked" each galaxy with the same amount of energy, then less massive galaxies would be moving faster than more massive ones. In this case, we expect a definite correlation between the velocity and the mass of the galaxies. But no correlation between galaxy mass and velocity is observed. The Hubble law works well even when we plot data only for galaxies of similar masses, which would not be the case if the big bang were like an explosion.

As a final illustration that it's wrong to picture the big bang as an explosion, continue with the idea that an explosion pushes galaxies (or matter that later becomes galaxies) outward, with faster galaxies leaving slower ones behind. As the faster galaxies move into empty space in all three dimensions, they become spread out. They isolate themselves by moving into unoccupied space, too fast for others to catch up with them. It's similar to a marathon (Figure 10.3), which starts with a mob of runners, but later has a few leaders with more space around them, ahead of the crowd. The result is that the galaxies closest to the center of the explosion (i.e., the slowest-moving galaxies) should be the most densely packed together, while those on the outskirts should be more widely separated. Again, this conflicts with observations discussed in Section 5.2 indicating that the universe is highly uniform on large scales: The universe appears to have the same density and organization (walls and voids) everywhere, in every direction.

The conclusion is that an explosion would require incredibly contrived conditions in order to explain the symmetry we see in the distribution of galaxies. Now consider an *expansion* instead, which explains the symmetry very naturally. As demonstrated in Section 7.3, an expanding space accurately reproduces galactic redshifts and the Hubble law. Furthermore, an expansion explains these observations even if we're not at the center of the universe. (Recall Figure 7.5 and the surrounding discussion from Section 7.3.) By way of analogy, if you glue a bunch of galaxy pictures to the *surface* of a balloon and then blow up the balloon, every galaxy moves away from every other galaxy, and *none* is located at the center. So *whichever* balloon–galaxy you look at, *all* the other galaxies are moving away from it as the balloon expands.

Not only are we not the center of the universe, but the universe doesn't even need to *have* a center! For example, suppose that the universe extends forever in all directions. (We haven't *seen* any edges, so this is a reasonable possibility.) In that case, *there is no center*, because there's just as much stuff on your left as on your right no matter where you go, since there's always *infinitely* much stuff in *every* direction. (Taking the infinite-universe logic even further, we could point out that if you travel through space in *any* direction, then *eventually* you'll find a planet inhabited by a civilization whose authors *also overuse italics*!)

The main point is this: In the big bang theory, space is filled uniformly with matter and energy whose density decreases (in the same way everywhere) as time

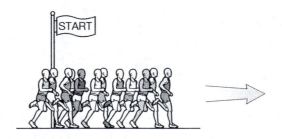

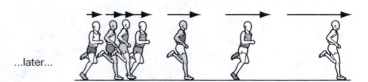

FIGURE 10.3 An Explosion Isn't Homogeneous   As a marathon progresses, the runners spread out in a manner similar to what happens in an explosion, because faster runners move farther and farther ahead. The result is thinning at the front and clumping farther back in the pack. By contrast, the distribution of galaxies in the universe is homogeneous on large scales.

passes. So once you've completed this book, and some skeptic seeks you out to ask, "Where in the universe was the big bang?" you can answer, "Everywhere!" And if the skeptic is not satisfied, just respond with an analogous question, like "Where in your body were you born?"

EXERCISE 10.2   Radial Marathon

Imagine a new kind of marathon, in which all the runners start simultaneously at the center of a huge circle, 26.2 miles in radius, and the perimeter of the circle is the finish line. The runners run outward, like spokes on a bike wheel. Draw two pictures, one at start of the race, and one an hour later, using about 25 runners in both pictures. Assume that some runners are fast and others are medium, slow, and very slow. (You can decide how many runners are in each speed category.) Show how the fastest runners break free of the pack, resulting in a circle with a dense clot of runners somewhere around the center, but with lower density near the finish line. This exercise demonstrates that an exploding universe would look different in different places, whereas the actual universe looks the same everywhere.

10.3 How the Big Bang Explains Our Observations

Now we look at how the big bang theory explains each of our key observations about the universe. These observations were discussed extensively in Chapter 5; Sections 5.1 through 5.6 directly match up with the subsections that follow here. As you read, pay attention to just how neatly the big bang accounts for each of the six major cosmic observables from Chapter 5. But also, keep in mind that some of the best evidence of all is the *combination* of many different kinds of observations, all of which support the same theory in concert. Having lots of evidence from a variety of approaches like this is very powerful, because it's extremely difficult to find any *one* theory to fit *all* the observations.

The Redshift–Distance Relation

The expanding-space interpretation of galactic redshifts explains quantitatively the observed redshift–distance relation, because it automatically produces the result that redshift is proportional to distance. We've already devoted much of the previous section and Chapter 7 to this topic. Here, we'll just summarize the big thoughts as they apply to the big bang:

- The observations reported by Edwin Hubble give us two important clues (Section 5.1): Nearly all galaxy spectra are redshifted, and the amount of this shift is proportional to the distance to the galaxy.
- The big bang theory easily accounts for Hubble's observations (Sections 7.2 and 7.3). The redshifts are caused by the expanding space, which stretches the wavelengths of photons in flight. The more distant a galaxy, the longer is the photon's flight to get here, and therefore the greater is the redshift.
- A Doppler shift interpretation (Section 5.1) could explain some of what Hubble saw if the big bang were an explosion. But the explosion hypothesis fails to match the observed uniform distribution of galaxies in deep space and other details pertaining to their masses and velocities (Section 10.2).

Galaxy Survey Data

Section 5.2 lists several informative discoveries brought about by collaborations of astronomers studying the large-scale distribution of galaxies in space. The big bang theory accounts for the following results:

- When observed on the largest scales, the visible matter in the universe—galaxies and clusters of galaxies—is arranged in a spongelike network through space.
- The network of galaxies and clusters is speckled with vast voids—empty regions of space—which are generally all about the same size.
- The pattern of walls and voids is similar everywhere in the universe, with no hint of a center, an edge, or any similar feature.

To understand the spongelike network of galaxies and clusters of galaxies, begin with the high density of the early universe, as postulated by the big bang theory. This high density implies that what used to be a uniform smear of matter particles has since evolved into the observed walls, sheets, and bubbles of galaxies and galaxy clusters. (How do we know that the universe was more uniform in the past? If the answer doesn't jump to mind, flip back to Section 8.4 to refresh your memory.)

Now simply add in the action of gravity, which always attracts chunks of matter toward each other. What's the result? A nearly smooth beginning gets clumpy over time. And since the gravitational force is strongest between nearby objects, it can't pull together things that are arbitrarily far away from each other with any appreciable strength. Thus, each clump will remain somewhat local: The whole universe can't all lump together in one spot, because that spot has a significant gravitational influence only over a local region around it. Astrophysicists have developed detailed computer models (Figure 10.4) of this clumping process, which

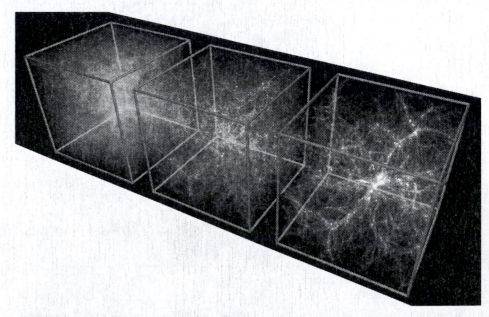

FIGURE 10.4 Simulation of Structure Formation   The cube is currently 100 Mpc on a side. The simulation shows how matter clumps under the action of gravity as time moves forward and redshift, z, decreases. (◆ Also in Color Gallery 2, page 25.)

we'll examine much more thoroughly in Section 12.3. The models accurately reproduce the observed arrangement of galaxies.

The formation of walls is the same process that fashions voids: As gravity pulls matter into clusters and walls, that matter must drain out of somewhere. So low-density regions, which have less gravity to hold them together, surrender their matter to neighboring high-density regions. This depletion of regions of space to form voids, like the buildup of galaxy walls that causes it, is well depicted in computer models which simulate the formation of the large-scale structures that populate the universe.

But what about the consistent size of voids, everywhere we can observe? This, too, is easily accounted for by the big bang, which says that the universe has a finite age. So it's just a question of "How much space can gravity empty out in 13.7 billion years?" The answer astrophysicists compute (again, with computer models) is a void that's about 100 million light-years across, which is what galaxy surveys find in the real universe. So because the same process (gravity) is depleting all the voids, and it has the same amount of time to do so everywhere (13.7 billion years), all the voids end up about the same size.

Knowing that the big bang invokes an expanding space rather than an explosion, we expect the pattern of walls and voids to be featureless, with no center or edge. If the beginning was uniform and relatively featureless, then the present large-scale arrangement of objects in space should likewise have no distinctive features anywhere. The expansion is happening in the same way everywhere because it's a property *of the spacetime* we inhabit, and not a property of objects *inside* the spacetime. Take a moment to convince yourself that this logic applies equally well whether the universe is infinite or not, provided that it's at least as large as the region around us that we can observe from here (i.e., our observable universe).

The Cosmic Microwave Background

So far, we've talked about the motion and arrangement of matter (galaxies) over a range of distances from here. But remember that we also observe a universal energy field of microwave light hitting us from every direction in space. The key observed features of the cosmic microwave background (CMB), from Section 5.3 and Chapter 8, are as follows:

- The CMB exists everywhere in the sky and does not change over the course of short time scales such as a human lifetime.

- The microwave intensity is almost perfectly uniform in every direction; variations in the intensity and temperature of the CMB are on the scale of a thousandth of a percent.

- These tiny variations from place to place contain a mathematically characterizable pattern of sizes and numbers across the sky. This pattern of spots or variations is known as anisotropy and turns out to be a fertile source of cosmological information.

- The CMB is by far the most perfect blackbody anywhere. Nowhere in nature or in human engineering is there a comparably precise thermal spectrum. That is, the spectrum from the glow of any other hot object, such as a star, contains imperfections that are not present in the CMB.

- The blackbody spectrum indicates a universal temperature of 2.725 degrees above absolute zero.

- The microwave light is *polarized* (like the late-afternoon glare on a road that polarized sunglasses can reduce). The pattern of polarization tells us that the last thing which CMB photons that we observe did, before they began traveling here, was scatter through a field of free electrons.

- There is also a second, fainter source of polarization, suggesting that a small percentage of the CMB photons scattered against free electrons *again* sometime after the original scattering.

The big bang theory explains the CMB as the glow left over from the hot early universe. In fact, the big bang explains *all* of the preceding features of the CMB in this context, as follows.

About 400,000 years after the big bang began, the temperature everywhere in the universe was about 3000 kelvins (remember, degrees Celsius above absolute zero). That was just barely hot enough for most electrons to avoid being bound inside atoms, given the conditions of the early universe.

So when the universe was hotter than about 3000 kelvins, there were free, unbound electrons. Slightly later, as the universe cooled down, *free* electrons basically went extinct by combining with nuclei to form atoms. This event is called **recombination**. But photons don't interact with neutral atoms the way they do with charged electrons, so there's a resulting process called **decoupling**, wherein the photons cease to interact with matter (Figure 10.5). Basically, it's like all your single friends getting married and settling down, so you don't see them any more. Your friends are the free electrons, their new spouses are the nuclei (to whom your friends are now bound), and you're the photon, cruising through life without any more interference from any of these people.

At 3000 kelvins, the universe was a thermal blackbody: Its photons acquired blackbody properties by interacting with the warm medium. These photons later became the CMB, so we should pause to appreciate two interesting details. First, the CMB has a blackbody spectrum because the universe *was* a blackbody, illuminated by its own warmth. Second, the CMB has a *perfect* blackbody spectrum

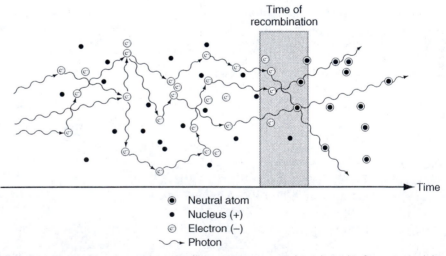

FIGURE 10.5 Recombination and Decoupling Time moves to the right in this figure. Toward the left, photons and free electrons collide regularly, which means that a photon can't get very far before it hits something. Then the universe cools and the free electrons recombine with nuclei to form neutral atoms, which don't interact with most photons. At present (toward the right), the photons are free to travel indefinitely in whatever direction they were headed when the free electrons vanished.

because the universe was very dense and very uniform, and there weren't any neutral atoms around yet to mess it up. That is, since the electrons weren't attached to atoms yet, they couldn't absorb photons in discrete wavelengths. (See Section 8.2 to refresh your memory on this point.) So the CMB photons started out with an unblemished blackbody spectrum.

When the universe cooled, the electrons joined nuclei to form normal neutral atoms, whereupon the universe stopped being opaque and became transparent. The photons could now travel without hitting a free electron. Most of those photons have done nothing except travel through space since then, and some of them happen to be arriving here on Earth today. This history leads to two more interesting consequences: (1) The last thing the CMB photons did before coming here was scatter with free electrons, so they *should* be polarized accordingly! (2) Since the photons have done nothing but travel through expanding space ever since the universe became transparent, they must have undergone a redshift. In Exercise 10.3, you'll verify that the redshift downgraded the original 3000 degree blackbody to a 2.725 degree blackbody. (Review Exercise 8.3 to brush up on this material.)

The best measurements today say that the CMB has redshifted by a factor of $z = 1089$. Recall from Exercise 7.3 that this redshift implies that wavelengths have grown by a factor of $1 + z = 1090$. (In other words, wavelengths today are 1090 times longer than when the CMB first formed.) So when the universe became transparent to CMB photons, they were 1090 times more energetic. Their 3000 K temperature corresponded to a blackbody peaked in the infrared, very close to red. Had you been there, wearing some kind of heat-shielding space suit, you'd have seen a red glow in every direction. (There were no galaxies, stars, or planets yet, so apart from enjoying the red glow, there wouldn't have been much to do.)

The big bang theory also explains the *second* polarization in the CMB, which occurred a few hundred million years later, after decoupling. (Cosmologists know that it occurred later because of the larger angular size of the polarization patterns compared with those from the first polarization. In Chapter 12, we'll see that the sizes of patterns observed in the CMB relate to how much time was available in the early universe to organize patterns of those sizes. Bigger patterns are imprinted at later times.) The polarization is the same sort as before, caused by free electrons. Although the free electrons went away when the CMB first formed, some of them came back later.

What could have reflooded the universe with free electrons? Recall from Chapter 8 what would be needed to create free electrons: The atoms of hydrogen and helium flooding the universe would need to be at least partly ionized, and this would be accomplished by high-energy photons—ultraviolet (UV) or better. Since the second CMB polarization appears to have taken place at around the same time all over the universe, we need a source that spontaneously flooded the universe with ultraviolet light. As luck would have it, we know of such a source: stars.

The first stars (ever!) probably produced mostly UV light. When these first stars "turned on," their UV photons *re*ionized the surrounding gas. (It was all ionized once before, prior to recombination.) This cosmic event is called **reionization**. The preceding era, before the first stars were born, is often called the *dark ages*. (Don't worry: Context clues usually make it pretty easy to tell whether someone is talking about cosmology or medieval Europe.) Before reionization, the universe had no light source whatsoever, apart from the CMB, which had redshifted out of the visual range shortly after it formed. Those ages were literally dark, until stars lit up for the first time.

Finally, the CMB anisotropy is also well explained by the big bang theory: The variations we see today were actual hot and cold spots when the universe was

400,000 years old. The temperature difference between spots carried with it a density difference—a lumpiness. The early universe was only *almost* perfectly smooth; tiny fluctuations of some kind must have permeated all of space, even when the universe was very young. Cosmologists have some knowledge of the cause of these fluctuations and understand which ones caused observable spots in the CMB. The explanation is far from simple, and a full treatment will have to wait until Chapter 12, where we discuss the origin of the spots and how they grow into the galaxy-filled universe we all know and love. Let us just advise you that the distribution of spots matches the big bang's predictions in great detail.

To seal the most important lesson from this subsection into your mind, consider this: Without the big bang theory, how easily could you come up with some way to explain *every* aspect of the CMB? How would you account for the uniformity across the sky? the tight blackbody spectrum and the temperature? the detailed anisotropies? both sources of polarization?

EXERCISE 10.3   Redshift of the CMB

From the temperature of the universe today and the measured redshift of the CMB, you can compute the temperature of the universe at the time the CMB decoupled. Temperature is related to particle energy, photon energy relates to frequency, and frequency depends inversely on wavelength. (We already saw this relationship between temperature and wavelength in Equation 8.1.) But wavelength is like any other length in the universe, so its growth (redshift) reflects that of other distances in space. It is then correct to say that the *ratio* of temperatures (then to now) equals the inverse ratio of wavelengths, and that ratio equals the ratio of scale factors (*a*, from Chapter 7):

$$\frac{T_{then}}{T_{now}} = \frac{\lambda_{observed,\ now}}{\lambda_{emitted,\ then}} = \frac{a_{now}}{a_{then}} = 1 + z \ . \tag{eq. 10.3}$$

Use the measured redshift of the CMB, $z = 1089$, and the measured CMB temperature today, $T_{now} = 2.725$ kelvins, to calculate the temperature at the time of decoupling. Verify that the universe glowed as an infrared-peaked (near red) blackbody back then by calculating the (former) blackbody peak wavelength and referencing Figure 2.20. Then verify that the present-day CMB blackbody is indeed peaked in the microwave band, and check to make sure that the ratio of peak wavelengths (now to then) is correct. How much bigger is the observable universe now than it was when the CMB formed?

EXERCISE 10.4   A Dark Expanse

Reionization didn't totally replace the original polarization from decoupling, which means that far fewer CMB photons scattered at reionization than at decoupling. Verify that this should be the case by showing that the universe was much more spread out when the second polarization was imprinted than it was for the first one, which implies that, for a typical CMB photon, there were fewer free electrons in its way the second time. To do this, calculate how many times larger length scales in the universe were at the end of the dark ages than they were at the beginning. In other words, compute *a* (reionization)/*a* (decoupling). You may take the redshift of decoupling to be $z = 20$, based on WMAP data.

Abundances of Elements

The big bang theory says that the universe was hotter and hotter as you go back in time. So there would have been a time, once in the distant past, when it was hot enough for thermonuclear fusion to occur naturally throughout the universe. That time was the first 20 minutes or so of the universe's existence, with fusion getting underway within the first few minutes. That fusion synthesized the light elements, and that's why the same base quantities of those isotopes exist everywhere we look. (Review Section 5.4.)

The abundances of the light elements can be explained by the big bang theory just that quickly—at least in concept. But, as with the CMB, the details are really impressive, too; so read on slowly, and verify to yourself at each step that things had to happen that way. Keep in mind that the physics of nuclear processes is well understood from tests that we can perform in terrestrial laboratories, giving us great confidence in the results.

Prior to this universal period of fusion, called **big bang nucleosynthesis (BBN)**, the temperature would have been too hot for nuclei to exist at all. We know this because the simplest fusion product is deuterium—a single proton fused to a single neutron—and deuterium breaks apart at high temperatures. The high-energy photons of the sort prevalent in the BBN era can split deuterium apart. As the universe expanded and the photons lost energy (reddened), a moment arrived when deuterium, once formed, remained stable. That's when the formation of nuclei first began. Like Baby Bear's porridge for Goldilocks, the temperature of the universe was "just right." (See Figure 10.6.)

Prior to the "just right" epoch, the universe contained only subatomic particles. Of these, protons and neutrons were the relevant participants in nucleosynthesis. The most important processes they underwent prior to BBN involved the weak nuclear interaction (Chapter 9), which can interchange protons and neutrons. This interchange is accomplished by a small set of two-way reactions that we explored back in Exercise 9.5:

$$\nu + n \longleftrightarrow p + e^- , \qquad \text{(eq. 10.4)}$$

$$e^+ + n \longleftrightarrow p + \bar{\nu} , \qquad \text{(eq. 10.5)}$$

$$n \longleftrightarrow p + e^- + \bar{\nu} . \qquad \text{(eq. 10.6)}$$

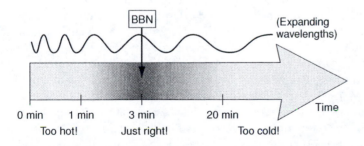

FIGURE 10.6 BBN Timeline   The arrow shown is a time line, indicating three phases surrounding the BBN process. First (left), during the first minute or so of the universe's existence, the temperature is so hot that once particles fuse, they immediately break apart again, due to a collision with an energetic particle. Then (middle), for the next several minutes, nucleosynthesis proceeds because the temperature has dropped to the point where newly synthesized nuclei can withstand collisions with (now less energetic) particles. Later on (right), the universe has cooled more due to redshifting, making it too cold for nuclei to fuse any further. Bringing two positively charged nuclei together requires more energy than is available, so BBN is over.

In each of the preceding reactions, there's a neutron on one side and a proton on the other. Neutrons turn into protons (plus other particles), and vice versa; this is well-established physics. But if you look up the mass of these two particles in the appendix of a physics book, the neutron turns out to be slightly heavier. That means its energy, calculated via $E = mc^2$, is greater than the proton's. Now, the average energy available to make each particle is determined by the surrounding temperature. So, as the universe expanded and cooled, there was a time after which it was too cold to make neutrons (by running Equations 10.4 through 10.6 from right to left), but it was still possible for neutrons to make protons (by running the reactions from left to right). So even if they started out in equal amounts, after a few seconds there were more protons than neutrons. This effect is calculable, and astrophysicists have determined that there were 7 protons for every 1 neutron when the fusion began. The 7-to-1 ratio determines the outcome of everything that happens next.

But why were there any neutrons at all? In Section 9.6, when we investigated how a BBN-type process could have worked, we blindly assumed that both protons and neutrons were present in abundance. If neutrons cost more energy, then why did the universe make any? The answer basically follows the same logic about temperature: In general, if it's hot enough to make a particle, and no particle physics rule prevents it, then that particle will be made. When the temperature drops too low, neutrons will stop appearing. But before that time, when the temperature was much higher, there was plenty of energy to go around. So the universe didn't care at all about wasting a tiny bit of extra energy to make a neutron instead of a proton. It's a bit like marbles rolling around on a board with divots in it—like a Chinese checkers board. If you keep shaking the board to give the marbles lots of energy (as baryons had before BBN), then they barely notice the divots, and can almost as easily be outside a divot as in one (just as the early universe can almost as easily make neutrons as protons). But if you stop shaking the board, the energy of the marbles drops, and they get stuck in the divots. They no longer have enough energy to escape. (No more neutrons are produced.) According to the big bang theory, then, the universe approached the BBN phase with an equal supply of protons and neutrons, but that 1-to-1 ratio changed to 7-to-1 over the first few minutes.

When it got cold enough for deuterium to be stable, protons and neutrons combined to form it. At this point, a sequence of nuclear reactions began, as we worked out earlier in Section 9.6. Depending on how precise you want to be, there are somewhere between about 12 and 70 or so relevant nuclear reactions to consider, and physicists have measured the reaction rates for them. That means that we can determine which elements were produced with great accuracy. Those produced include trace amounts of certain hydrogen, helium, and lithium isotopes. By far, however, the most common products turned out to be hydrogen-1 and helium-4.

Since normal hydrogen consists of one proton only, and helium-4 consists of two protons and two neutrons, we can extract the abundance of each directly, just from the ratio of protons to neutrons. For every neutron, there were 7 protons around. For every 2 neutrons, there were 14 protons. Both of these neutrons, and 2 of the 14 protons, would have been used to make one helium, and we can count the remaining 12 protons as 12 hydrogen atoms. That means that four-sixteenths, or one-fourth, of our particles (and thus one-fourth of the mass) were used up to make helium. The result is 75% hydrogen and 25% helium, which explains the observed 25% helium-4 abundance discussed in Section 5.4! (We actually said 23% to 25%, which is because the ratio wasn't *exactly* steady at 7-to-1; it's just easier to explain the math if we approximate things a little. The actual ratio was changing slightly, moment by moment, in accordance with Equations 10.4.)

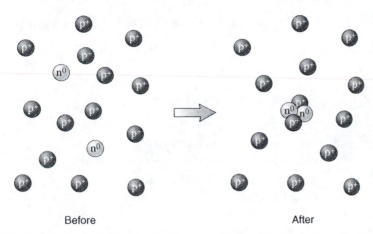

Before   After

FIGURE 10.7 Musical Chairs: Where Do the Neutrons Sit When the Music Stops?   The ratio of neutrons to protons determines the abundances of light elements that form. Since free neutrons are unstable, the only neutrons that survive the early universe are bound inside nuclei. When BBN occurs, it rearranges most of the protons and neutrons into hydrogen-1 and helium-4 nuclei, so the amount of helium-4 produced (relative to the amount of hydrogen-1) depends on the number of neutrons present initially (relative to the number of protons).

The process and its math are illustrated in Figure 10.7. As you look over the figure, make sure you appreciate how the initial number of neutrons present determined the resulting abundance pattern. This is a point where the big bang theory's predictive power allows us to test the science of the very first few minutes, with elemental abundances that are still observable today.

Why didn't the heavier elements form at this time? Part of the answer comes from our analysis of nucleosynthesis from Chapter 9. There are no stable five- or eight-particle nuclei, so you can't blend ^{1}H with ^{4}He, or ^{4}He with another ^{4}He, to make anything. Or at least, anything you make will fall apart again. But we know that some stars can make carbon-12 by fusing three helium-4 nuclei together at once; why didn't BBN do that?

This part of the answer comes from the big bang theory. Simply put, the universe expanded and cooled too quickly. Nucleosynthesis ended when it got too cold for any more fusion. Remember that nuclei are all positively charged because there are no negative charges in the nucleus. Since like charges repel, it takes energy to bring two nuclei close enough to fuse. Heavier elements require higher temperatures to overcome the electrostatic repulsion of their greater charges. But during BBN, expansion was rapidly dragging the temperature down.

So there was a window of opportunity for BBN. Early on, when it was still *hot* and dense enough for carbon formation via ^{4}He + ^{4}He + ^{4}He → ^{12}C, there was no ^{4}He available, because ^{4}He can't form until it's *cold* enough for its ingredients—^{2}H, ^{3}H, and ^{3}He—to exist. Later on, when ^{4}He was present, the universe was getting too cold to form carbon and the ^{4}He nuclei were too spread out—in each case because of the expansion. In other words, carbon formation requires a *sustained* condition of densely packed helium-4 at high temperatures. This is a condition that some stellar cores provide, but that BBN did not, because of the expansion of the universe.

As you can see, big bang nucleosynthesis is far from simple. We got remarkably far when we considered the relevant reactions in Chapter 9, but in order to calculate the outcome in detail, we need to include the effects of the expansion.

And the real glory of BBN is in the details. The full calculation involves

- The initial quantity of each type of particle, such as protons, neutrons, and photons

- The expansion rate, and even this *rate* was changing with time, but in a way that we can calculate with general relativity

- All the nuclear reaction rates, which depended on the temperature

- The temperature, which depended on *both* the expansion (via the photon redshift) and the nuclear reaction rates (which in turn depended on the temperature!)

- Careful tracking of all the nuclei and particles and where they ended up, such as tracking the (changing) rate at which neutrons vanished—either by the reactions described in Equations 10.4–10.6 or by incorporation into helium-4 nuclei, as happened to the vast majority of them.

These issues are a lot for anyone to stay on top of. That's okay; there's no shame in turning the task over to a computer. Once we've programmed in all the rules, our silicon-brained companion thanklessly chugs through all the math: the initial buildup of deuterium over the first few minutes, the flurry of reactions the presence of deuterium enabled, and the way those fusion reactions acted to raise the temperature—but then the expansion lowered the temperature, which lowered the reaction rates, which lowered the temperature even more, and separated the reacting particles, and gave free neutrons time to decay. . . .

One last complication remains before we can see the most powerful results. We introduce the **baryon-to-photon ratio** (symbolized with η, Greek lowercase "eta"), which compares the number of particles capable of fusion—the protons and neutrons—with the number of photon particles in the radiation field all around them. The baryon-to-photon ratio determines how far *all* the reactions could proceed, because more photons make it harder for deuterium to survive at the start. In other words, a larger η means fewer photons to break up deuterium, so BBN can get underway sooner. Once η is set, all the other BBN abundances can be calculated, including the small amounts of deuterium, hydrogen-3, helium-3, and lithium-7, as mentioned in Section 9.6.

The BBN-produced abundances calculated by computer are shown in Figure 10.8. Notice in this plot that there is a narrow range of values for η that correctly generates the measured abundances of *all* the BBN isotopes! (Hydrogen-3 is omitted from the plot, but only because it's hard to measure accurately enough for this purpose. However, measurements of hydrogen-3, rough as they are, do also agree with BBN predictions.) That such a common value of η exists shows us that the BBN calculation is valid.

The measured ratio is $\eta = 6.1 \times 10^{-10}$ baryon per photon, which you can see agrees with the predicted BBN abundances of all the elements, as shown in Figure 10.8. This means that we can trust BBN calculations and actual abundances to reliably measure the quantity of baryonic matter in the universe simply on the basis of η: You just multiply it by the quantity of (CMB) photons. The reason this is important is that if we already know the total abundance of matter in the universe (from orbital speeds or gravitational lensing), and, through BBN data, we discover the abundance of *baryonic* matter, then we can subtract to get the abundance of *exotic* dark matter (e.g., WIMPs). One of the great surprises of modern cosmology, that most of the matter in the universe is nonbaryonic (exotic), is determined in this way.

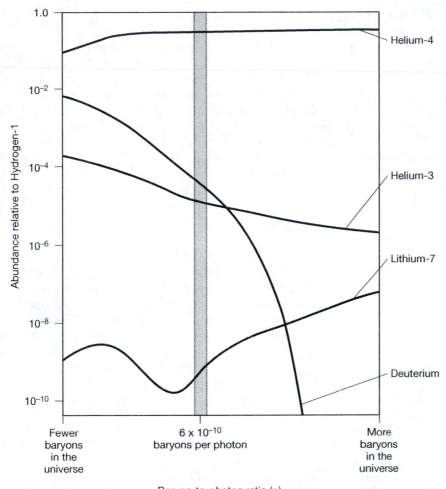

FIGURE 10.8 Elemental Abundances Predicted by Big Bang Nucleosynthesis   Lines show how the predicted abundance of each isotope varies with the baryon-to-photon ratio. The vertical stripe shows the range of baryon density that is allowed by other observations (Chapter 12). Therefore, lithium-7, for example, must be found within a range of abundance specified by just the *intersection* of the gray stripe with the bottom curve. (The abundance range is shown to the left of the intersection, on the vertical axis.) Remarkably, the observed abundances of *all* these isotopes fall within their allowed ranges. (◆ Also appears in Color Gallery 2, page 26.)

We'll see in Chapter 12 that the CMB provides another, totally independent way to measure the abundance of baryonic matter (and therefore that of exotic dark matter, too); the BBN and CMB figures agree with one another beautifully.

The BBN calculations shown in Figure 10.8 predict that BBN produced the following trace abundances:

- deuterium: 3 to 4 grams per 100,000 grams of normal hydrogen-1 nuclei
- helium-3: 1 to 2 grams per 100,000 grams of normal hydrogen-1 nuclei
- lithium-7: 2 to 4 grams per 10,000,000,000 grams of normal hydrogen-1 nuclei

Measuring the cosmic abundances of these isotopes is not simple, but it is possible. (We described some of the methods in Section 5.4.) The abundances we measure, from many different target objects, turn out to be remarkably consistent with the slim ranges listed above.

One important method of measuring primordial abundances relates nicely to some of our previous topics. Distant quasar light (Section 6.6) passing through intergalactic gas clouds (Section 8.4) is absorbed to reveal hydrogen absorption lines (Figure 8.10). There is a small, but measurable, distinction between absorption by hydrogen-1 and absorption by hydrogen-2 that allows for an accurate determination of the relative amount of each. Since these clouds are situated in the otherwise unoccupied space between galaxies, they could never have hosted any significant star formation. That means that the abundances in the clouds reflect big bang abundances, rather than stellar ones. The best deuterium measurements of this sort give a range of values narrowed down to between 2.40 and 3.22 grams per 100,000 grams of hydrogen-1.

Even the most ardent doubters of the big bang can't help but be impressed by numbers like these! They could perhaps argue that only *part* of the measured deuterium range overlaps that predicted, but considering the powers of 10 involved, that's splitting hairs! Dwelling on a discrepancy on the order of 0.1 gram, when viewed per 100,000 grams, is like winning a million dollars but complaining about the price of the lottery ticket. Similarly, the theory predicts a few grams of lithium-7 nuclei per 10 billion grams of hydrogen. The big bang theory is *that* specific, and yet it still matches observations of the real universe! The deuterium and helium-3 numbers are also impressive; each is a few parts in 100,000. And all the BBN isotopes—hydrogen-1, -2, and -3, helium-3 and -4, and lithium-7 —match the theory's high-precision predictions *simultaneously*.

In some ways, the BBN isotope that's produced in the smallest quantity, ^{7}Li, might be considered the most important. The way the big bang or any other scientific theory acquires its credibility is by making specific predictions that can be confirmed by experiment or observation, especially if those predictions can't easily be accounted for any other way. In Chapter 5, we noted that most of the lithium, beryllium, and boron in the universe today is formed by one mechanism: collisions involving heavier nuclei (such as ^{12}C or ^{16}O) in space. These nuclei break into smaller pieces, including Li, Be, and B. But in older and older stars (where there is less and less beryllium and boron because few of the needed collisions had happened yet), there *is* a fixed quantity of lithium-7: a few parts per 10 billion, just as BBN predicts. It's a tiny amount, but since BBN produced Li without Be and B, even though the universe has made no such distinction ever since, the ^{7}Li abundance has become a powerful test of the big bang theory.

Astronomical Ages

The ages of the oldest objects in the universe today top off at 12 or 13 billion years old, as we noted in Section 5.5. According to the big bang theory, this is because the universe is about that old. The most precise measurement of the age of the universe comes from observations of the CMB by the WMAP satellite experiment: 13.7 ± 0.2 billion years.

How do astrophysicists come up with the number 13.7 billion? Well, earlier in this chapter we calculated the age of the universe from the inverse of the Hubble constant and got a number around 14 billion years. So even a simple estimate like that gets us close. But the WMAP number is harder to derive. The idea goes like this: They can calculate how long it took the universe to expand to its current scale (e.g., the current separation between galaxies), but only if we know what the universe is made of. Matter gravitates, attracting everything toward everything else, and therefore tends to slow down the expansion. Dark energy (Section 7.4) has the opposite effect. So as long as we know how much of each exists, we can calculate the age of the universe.

As you might imagine, this manner of age computation—as well as many other CMB-based determinations of cosmological values—is quite complex in practice. We'll say more about precision cosmological measurements in Chapter 12.

Olbers's Paradox

Recall from Section 5.6 that the simple observation of a dark sky at night is something of a paradox. If the universe is infinite and filled with galaxies, then every line of sight into the night sky should eventually intersect a galaxy, so the night sky should be brightly lit in every direction.

The big bang theory offers two important insights related to Olbers's paradox:

INSIGHT #1: The universe has a finite age. Coupled with the finite speed at which light from distant galaxies travels toward us, the 13.7 billion-year age serves as a time limit: Light has 13.7 billion years to get here. If a galaxy is too far away, its light hasn't gotten here yet. And if the light hasn't reached here, then the galaxy can't brighten our night sky. This is the *observable universe* concept, from Chapter 7: We are isolated by our cosmological horizon.

INSIGHT #2: Even for some galaxies close enough for their light to have arrived here by now, the cosmological redshift due to expanding space has shifted that light out of the visual band. Most of the light emitted by distant galaxies starts out visible, but arrives here as infrared. So again, the light isn't *visibly* brightening the night sky. (Insight #1 would still solve the paradox even if our eyes could see infrared light.)

Notice how easily the big bang resolves the paradox. We don't need to make wild claims of what the universe is like beyond our cosmological horizon. We don't need an edge to the universe, or a limit on the number of galaxies. Even with a universe that looks everywhere like it does around us, as seen in galaxy survey data, the big bang still neatly answers the paradox.

10.4 Evaluating the Big Bang

All of the lines of evidence discussed in the previous section independently support the big bang theory. The big ones are historically called the "three pillars of the big bang":

- galactic redshifts according to the Hubble law
- the cosmic microwave background (CMB)
- big bang nucleosynthesis (BBN)

Toward the end of the 20th century, the abundances of the light element were the most compelling and precise evidence. Today, the CMB has become the best source of cosmological information. It provides us with a fleet of high-precision characteristics of the universe. Even so, much of our modern understanding of the cosmos comes from a blend of evidence from multiple sources. For example, to determine the breakdown of the universe in terms of normal matter, dark matter, and dark energy, we use information from many data sets, including the gravitational dynamics of galaxy clusters, light from distant supernovae, and the CMB. What's so remarkable about the big bang theory is the way in which its predictions match *all* of the *different* sources of measured cosmological data.

Some of the theory's supporting evidence comes from fairly direct reasoning, such as Olbers's paradox or reasoning about the ages of the oldest objects. Other evidence is derived from high-precision numbers, such as the CMB anisotropy or the cosmic abundance of deuterium, each of which involves a precision on the order of 1 part in 100,000, which would seem hard to reproduce under some different theory. The concordance of all of these approaches is particularly noteworthy when you consider how different the measurements are. The CMB, for example, reveals the universe as it was 400,000 years after it came into existence, whereas nucleosynthesis of the light elements happened in the first 20 minutes, and the Hubble law runs the show today. But the results still manage to fit together and support the same explanation. The big bang theory is *not* the wild speculation it is often mistaken for. Rather, it is a detailed, well-tested explanation of how our universe changed over cosmic time.

However, the scientific process is a comparative one, demanding that we always investigate other options. In so doing, we must remember that the strangeness of our universe is not something we've invented with our theories. Rather, it is revealed by the observations. (Return to the introduction to this chapter for a review of key observations that any viable theory of the universe must take into account.) The task of weaving together the detailed and sometimes surprising observations we now have is not an easy one. It may seem remarkable that *any* coherent theory can be developed to incorporate the diverse volume of detailed observations that have become available around the turn of the century, and this is what the big bang framework does. For this reason, it's highly unlikely for any alternative theory to survive without including the big bang's two core components: expanding space and the hot, dense early stage of the universe.

This book is written within the context of the big bang theory because that theory is the most solid scientific framework we have for building a sense of place in the universe. With that in mind, we have two long-winded remarks we'd like you to think about:

LONG-WINDED REMARK #1: The big bang theory itself is not a rigidly fixed and detailed description of *everything* about the universe. There are multiple possible scenarios for how the current expansion phase of our universe got started, how structures formed within that framework, and what will happen in the distant future. So there are alternatives for filling in the details of the big bang framework. (We will discuss some of these refinements and current cutting-edge ideas in Chapters 11 and 12.)

An excellent example of the flexibility of the big bang theory is the way it accommodates the discoveries of dark matter and dark energy. Dark matter, for instance, is not needed for the basic premise behind the big bang theory to work. An expanding universe that was initially hot and dense would be just fine with or without dark matter. However, as evidence for dark matter got stronger in the 1970s and 1980s, it became increasingly clear that the big bang theory would have to at least *allow* large amounts of dark matter to exist, whether or not it could explain *why* dark matter exists.

It's interesting that some of the first data sets forming the three pillars of the big bang provided no strong preference for or against dark matter. *Pillar 1:* The redshifts of the vast majority of the galaxies and the Hubble law are properties of the expansion, and the expansion history and present rate H_0 were not very well constrained yet. *Pillar 2:* Early measurements of the CMB, including even the COBE satellite measurements of the early 1990s, were unable to resolve small-scale anisotropies. Researchers at that time could determine that the CMB existed and that it had the right spectrum, uniformity, and temperature to confirm the big bang beginning, but this told them nothing about dark matter. *Pillar 3:* Measurements of the abundances of the light elements provided a constraint on the amount of baryonic matter in the universe, via the baryon-to-photon ratio. And although this would have been sufficient to see that large amounts of exotic dark matter were needed, the data on the light elements also depended on the poorly constrained details of the expansion history (because the expansion rate determined the rate at which temperature and density diminished, and temperature and density affected the nuclear reaction rates).

Today, all three of these approaches reveal the role of dark matter, due largely to newer and more accurate CMB measurements and computer models of large-scale structure formation, as in Figure 10.4 (more on all of this in Chapter 12). So at first, the big bang theory didn't particularly expect dark matter, but didn't forbid its existence either. Since then, as we have zeroed in on the detailed parameters of the theory with observational tests, we find that dark matter is, in fact, built right in. Current measurements indicate that the total matter budget of the universe breaks down as shown in Figure 10.9.

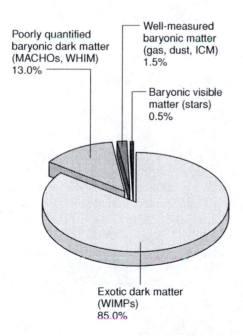

Poorly quantified baryonic dark matter (MACHOs, WHIM) 13.0%

Well-measured baryonic matter (gas, dust, ICM) 1.5%

Baryonic visible matter (stars) 0.5%

Exotic dark matter (WIMPs) 85.0%

FIGURE 10.9 Matter Breakdown   The pie chart shows the different components which contribute to the total matter content of our universe.[1]

---

[1]Return to Section 4.3 to refresh yourself on all the various DMAs (dark matter acronyms). (We're trying to get "DMA" recognized as a DMA.)

*Perhaps the most influential among the proponents of the steady-state cosmological theory was **Sir Fred Hoyle** (1915–2001), although he was not its inventor. (He was also fairly influential among supporters of the big bang theory, in that he coined the name "big bang.") Even though the theory he worked to justify (steady state) lost out to the big bang in the end, his work still had great virtue: Sometimes, hard work on the losing side can serve to inspire those on the winning side to find stronger evidence. In addition to Hoyle's work on cosmology, he is noted for important research regarding the nucleosynthesis of carbon in stars, via Equation 9.16.*

Dark energy is a similar story, one we've already told in Chapter 7. In that case, the equations of cosmology (particularly the Friedmann equation) permit the existence of an agent that acts to speed up the cosmic expansion. Mathematically, the value of that part of the equation could be zero. Therefore, the big bang theory doesn't require dark energy and has no immediate preference for or against its existence. Only in more recent data sets have scientists found that dark energy is needed to explain their findings. (Reassuringly, all of these data sets are satisfied by the same *amount* of dark energy.) In other words, despite the extreme sensitivity of the big bang theory to particular measurements—the primordial abundances of helium and lithium, the precise uniformity and polarization pattern in the CMB, etc.— the theory is flexible where it's needed.

LONG-WINDED REMARK #2: Although the vast majority of cosmologists agree that the core elements of the big bang are solid, there are people who disagree. These dissenting voices range from those with very little understanding of the modern scientific evidence and strong philosophical preconceptions against the notion of an evolving universe, to respected astrophysicists with an excellent understanding of the evidence and the history of scientific cosmology. The alternative theories this latter group supports range from young-universe models that are inconsistent with a wide range of well-documented observations to sophisticated theories that include the expansion of space and explicitly take into account many of the astronomical observations accounted for by the big bang theory.

The **steady-state theory** is a prominent example of a scientific theory that differs from the big bang theory. In a paper introducing the steady-state idea in 1948, the authors describe it as follows:

> "We shall pursue this possibility that the universe is in such a stable, self-perpetuating state, without making any assumptions regarding the particular features which lead to this stability. . . . Our course is therefore defined not only by the usual cosmological principle but by that extension of it which is obtained on assuming the universe to be not only homogenous but also unchanging on the large scale."
>
> —Hermann Bondi and Thomas Gold, "The Steady State Theory of the Expanding Universe."

Thus, the theory was designed to extend the cosmological principle to include time. That is, the steady-state theory proposes that, in addition to looking the same (on the large scale) in all locations and directions, the universe should look the same at all *times*. This notion is very different from the big bang concept, in which many major changes, such as nucleosynthesis or decoupling (or, for that matter, the beginning of the universe!), take place across the cosmos at different times. Throughout the 1950s, astronomers worked hard to find evidence that would distinguish between the two competing theories. For most of the astronomy community, that evidence arrived in the mid-1960s with the discovery of the CMB (Chapter 5).

However, there remains a modern version of the early steady-state cosmology: the quasi-steady-state theory proposed by astronomers Fred Hoyle, Geoffrey Burbidge, and Jayant V. Narlikar. This theory is a modification of their original steady-state theory and includes expanding space, just as in the big bang theory. As in the original steady-state theory, the quasi-steady-state theory allows for uniformity in time as well as uniformity in space. In order to maintain uniformity in time for an expanding universe, the quasi-steady-state theory proposes that matter is created at just the rate required to keep the average density constant when sampled over large intervals of time, so that there was never a beginning, or "time zero," and no early hot phase of the universe. In other words, they propose that the universe is expanding and has always been expanding. But because of their proposed mechanism for creating matter, the material in the universe would not be thinning out over time and therefore wouldn't have started out in a particularly dense state.

An important challenge for this theory is to account for the microwave background radiation, since there is no hot early phase to produce it in the manner of the big bang theory. The explanation the authors propose is that starlight absorbed by metallic "whiskers" scattered through extragalactic space may be re-emitted as microwaves, producing the observed nearly uniform background. The more detailed observations of the microwave background in the last few years make this explanation increasingly unlikely. For example, recent observations show evidence that the microwave background has been influenced by distant galaxies and so must originate from a very distant source.

It is important to keep an open mind about possible theories of our universe—a universe that is often full of surprises. Science is a dynamic and living enterprise in which even the most established concepts can be questioned. Some of these alternatives may hold pieces of the truth, reminding us of observations that challenge the big bang theory and help push forward progress in refining it. We'll see in Chapter 11 that some modern proposals for filling in details of the big bang framework contain concepts that were first introduced by alternative theories (including a variation on uniformity in time).

As we consider alternatives, it's also important to resist the temptation to look only at alternative explanations for *isolated* facts. As we've discussed earlier in this chapter, the concordance of many observations that all fit together is perhaps the strongest support for the big bang theory. Any competing theory must consistently account for all of these together in order to be equally viable.

"Just a Theory"

As someone who has studied cosmology, you may be approached by people who are interested in talking with you about the universe. Many of them will have heard of the big bang theory and will want to know more about it. Having just read this chapter, you are well versed in the evidence for the big bang, so you are probably well qualified to answer their questions.

Some people, however, actively doubt the theory. Their reasons can vary from person to person: Some have differing personal beliefs, and others just can't imagine how astronomers could possibly know what the very first few minutes were like. When conversation turns to the universe, these skeptics often point out that the big bang is "just a theory."

This reaction is not entirely surprising. The big bang is difficult to understand. It spans many fields of physical science: relativity; quantum mechanics; atomic, nuclear, and particle physics; thermodynamics; and cosmological physics, which has emerged as a field in its own right. And the big bang theory talks about things that happened both near and far, billions of years ago. It is hard for those without direct experience in the field to see how we can possibly know such things.

Surprisingly, though, such a reaction to many *other* scientific theories is uncommon. Why should other important theories—theories that can't be understood by mere intuition any more than the big bang can—be more readily accepted? For example, consider atomic theory, which says that everything in the world is composed mostly of empty space and secondarily of atoms, which are too tiny to see. Or consider the germ theory of disease, which says that illnesses occur not because some part of your body is broken, but rather because it is actively being attacked by microscopic life-forms, which actually mean you no harm. What is it that makes so many people preferentially assimilate these ideas over cosmological ones? Do people need to personally see an atom or a virus with an appropriately specialized microscope? Or is it just a question of *familiarity* with the evidence? How do people decide what they consider to be true? How do *you* decide?

In Section 1.2, at the beginning of this book, we made several points pertinent to the current discussion:

- A theory is a basic unit of scientific knowledge, much as a sentence is a basic unit of literary knowledge. In science, the word "theory" does not carry the speculative connotation it carries in nonscientific, everyday use. So, yes, the big bang is theory and not fact, but the same is true of electricity, buoyancy, acidity, and everything else in science.

- In science, evidence (but not proof!) comes by way of observational or experimental tests. So one needs to ask, "How many data sets support the big bang theory, and how good is their quality?"

- The strength of a scientific theory resides in its ability to be *falsified*. You've got a good theory if it could easily have been disproved by any of several measurements, but those measurements supported it instead. So when we measured the cosmic abundance of lithium, could the result have been a few parts per thousand, instead of a few parts per 10 billion? Or could the CMB have turned up *un*polarized, or not have turned up at all?

Regardless of how compelling the evidence seems, you can still answer people who tell you that the big bang is just a theory. You can tell them they're partly right: The big bang certainly *is* a theory. But it's not *just* a theory. In science, there's nothing "just" about a theory.

Quick Review

In this chapter, you've studied a theory that can piece together all the cosmological observations from Chapter 5. You learned how the expanding and cooling universe caused the BBN process to begin with a particular ratio of protons to neutrons, and how it caused the process to end with all the observed abundances of the light elements. You also learned why the universe, initially opaque, became transparent, thereby forming the CMB. Finally, you learned that the big bang theory simultaneously requires many different specific measured properties of the universe, while remaining flexible enough to accommodate the existence of dark matter and dark energy.

Try the following crossword puzzle to test your knowledge of key terms and concepts from Chapter 10.

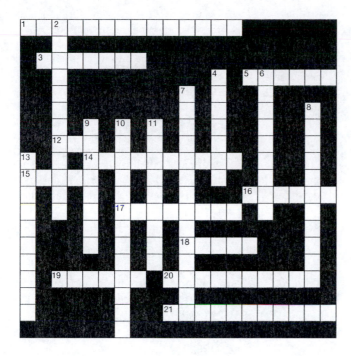

ACROSS

1 Important ratio whose value affected the abundances of all the elements formed by BBN

3 The pattern of galactic redshifts, the CMB, and the abundances of the light elements are together known as the three _____ of the big bang theory.

5 Where most of the free neutrons that existed prior to BBN ended up

12 Inverting the Hubble constant gives an estimate of the _____ of the universe.

14 The breaking free of photons from matter, thus forming the CMB

15 Regions of the universe in which most of the matter has been emptied out by gravity

16 The universe has evolved from a state that was both _____ and denser than it is today.

17 85% of the matter in the universe is not _____.

18 Responsible for the second round of CMB polarization

19 Even if the size of the universe is infinite, Olbers's paradox is resolved because the age of the universe is _____.

20 Last term in the Friedmann equation, which could be zero or nonzero, according to the big bang

21 An alternative cosmological theory

DOWN

2 Era in cosmic history when UV light from the first stars stripped electrons away from neutral atoms

4 Cosmological theory of the evolution of the universe

6 The big bang describes an expansion, not an _____.

7 Number characterizing the universe's rate of expansion

8 Aspect of the "cosmological principle" that would not be observed in a universe created by an explosion

9 Process responsible for reducing CMB photons' energy down into the microwave (radio) range

10 Cosmic event in which baryonic gas ceased to be ionized

11 BBN couldn't begin until the universe was cool enough for _____ to be stable.

13 Location of the big bang in our universe

Further Exploration

(M = mathematical content; I = integrative—builds on information from more than one chapter; P = project idea; R = reflection; D = suitable for class discussion)

1. (PD) What are some of the questions researchers are actively working on within the context of the big bang theory? To answer this question, you'll need to access a database of scientific articles on the subject. Go online to http://xxx.lanl.gov; choose "astrophysics" and then "search." This will bring you to a search form, on which you can enter keywords such as "big bang" into the "title" or "abstract" field, and call up articles that contain those keywords in the title or the abstract (summary) of the article. When you submit your search, the database will return a listing of articles. For any article, you can click on it to read its abstract. Find at least five relevant articles, and scan their abstracts to identify the questions the authors are investigating with their research. Make a list of these questions, and print out the five abstracts you find. If possible, compare your results with those of friends or classmates.

2. (RD) Where in space did the big bang occur? Explain as if you are talking to a friend who wants to know where in space they should look to see the location of the big bang.

3. (MDI) In Section 10.4, we listed the percentages of various types of matter in the universe. However, all matter is really a form of energy, because $E = mc^2$. Recompute the percentages given, but this time do it as a fraction of the *energy* in the universe. To do this, figure in dark energy, which has been measured to account for 73% of the universe. Essentially all of the remaining 27% is matter of some sort. What percentage of the universe's energy is made of atoms, as you are? What percentage is in a form that you can see with your eyes? Make a pie chart showing the percentages of the various forms of energy in the universe. The chart should correspond to Figure 10.9 showing the percentages of the various forms of matter in the universe.

4. (RI) Recombination took place when the temperature of the universe had dropped to around 3000 K, even though hydrogen gas becomes ionized in a laboratory at closer to 100,000 K. Use your knowledge of blackbodies to explain why the universe remained ionized so much below 100,000 K. (*Hint:* Consider the size of the baryon-to-photon ratio, as measured with BBN data.)

5. (RI) List all the evidence and/or observational methods you can think of for measuring the following quantities:
   (a) The amount of matter in the universe
   (b) The amount of baryonic matter in the universe
   (c) The amount of baryonic dark matter in the universe
   (d) The amount of exotic dark matter in the universe
   In some cases, you may need to compare results from different approaches you listed in (a), (b), (c), and (d); for example, you can measure the amount of exotic dark matter by subtracting the amount of baryonic matter from the amount of total matter. Your answers should be as comprehensive and as descriptive as possible! Expect to find *several* answers for each quantity listed.

6. (RDI) Imagine that the universe has a geometry similar to the surface of a donut. (Mathematically, such a surface is called a *torus*.) Envision a bunch of galaxies on the surface of the donut (denoted with chocolate sprinkles if you like), and pick one galaxy at random to be the Milky Way. (A little blob of red jelly might help to distinguish it.) Is this universe infinite or finite? Is it curved or flat (as defined in Chapter 6)? Does this universe have a center or an edge, and if so, where? If this were an expanding donut universe (stand back, and don't wear white!), where were all the chocolate sprinkles in the distant past? Where will they be in the future? What would a galaxy survey map look like in this universe? From the point of view of the Milky Way, is there any particular point that the universe is expanding away from or toward?

7. (R) List everything we learn from the various observed properties of the CMB.

8. (MRI) Consider the weak reactions that set the proton-to-neutron ratio for BBN and, in particular, the reaction in which the neutron decays into a proton, an electron, and an antineutrino.
   (a) How much total energy does the antineutrino have, assuming that none of the other particles has any kinetic energy (i.e., none is moving)?
   (b) What has become of those antineutrinos ever since the first minute or so?

9. (M) How many (a) nuclei and (b) grams of hydrogen-1 did BBN produce for each (a) nucleus and (b) gram of helium-4?

10. (RI) Within the universe's first few seconds, a large number of electrons and positrons annihilated: $e^+ + e^- \rightarrow \gamma + \gamma$. (a) What effect did this annihilation have on the timing of BBN, and why did it have that effect? (A change in timing causes a measurable change in the resulting abundances.)

11. (RI) The CMB is a collection of photons. What were these photons doing before decoupling? What have they been doing ever since decoupling?

12. (R) In a few sentences, summarize the reasons it's inaccurate to call the big bang an explosion.

13. (R) What are recombination and decoupling? How do these two distinct processes relate to each other?

14. (RI) Now that you've seen the big bang theory and some of its major predictions, what else is left? What *don't* you know about the universe yet? What would you *like* to

know? Write a few sentences summarizing your thoughts on these two questions.

15. (M) What if the value of the Hubble constant were 30 (km/s)/Mpc? Assuming, for simplicity, that the expansion rate was always constant, calculate the age of the universe in this case. Give your answer in years.

16. (MRI) Assume that the correct value of the Hubble constant is 71 (km/s)/Mpc. If you were not held together by any other forces, how fast would your head be moving away from your feet due to the Hubble expansion? Use your own height for the calculation. Give your answer in meters per second.

17. (RI) Look back at the list of observations at the beginning of this chapter. Which observation seems most difficult to explain without the big bang theory? Explain the reasoning for your choice.

18. (RDI) List two statements from your life that you consider "fact," and two statements that you consider "theory." Explain what differences in evidence led you to classify your statements as you did. Are there any such things as "facts," without any theoretical component whatsoever? Explain. What implications do your answers have for how we think about things, both scientific and nonscientific? For how we live?

19. (MRI) Refer to Figure 10.8 and explain why the abundance of deuterium is a more sensitive probe than that of helium-4 for learning about the baryon-to-photon ratio and therefore the baryon content of the universe.

20. (RI) Draw an annotated spacetime diagram that shows how the big bang theory explains Olbers's paradox.

21. (P) Research the steady-state theory of the universe, and write a short paper describing how it might explain each of the observations listed at the beginning of this chapter. One starting point for research is the book by Hoyle, Burbidge, and Narlikar, *A Different Approach to Cosmology: From a Static Universe Through the Big Bang Towards Reality* (Cambridge and New York: Cambridge University Press, 2000).

22. (RPD) In what ways is the big bang theory ingrained in our culture? Cite several examples from art, literature, film, television, etc., to support your answer. If you have trouble coming up with examples, you may want to start with an Internet search on "big bang."

23. (PDI) Do an Internet search for news stories and popular science articles that talk about the big bang. To what extent does it seem that the public is up on the theory, based on what's explained in the articles you find? What specific aspects does the public seem to misunderstand or need help with? Can you find any factually incorrect statements about the big bang theory in these articles? Look up at least five articles, and write a few paragraphs summarizing your responses to these questions.

11

History, Density, and Destiny

> *"This is the way the world ends*
> *This is the way the world ends*
> *This is the way the world ends*
> *Not with a bang but a whimper."*
>
> — *T. S. Eliot*

Often in daily life, we use mental models to fill in the gaps between the events we directly experience. For example, we don't have to observe each moment in the life of a friend to imagine some of the events in his day. If his teeth look clean and his breath doesn't smell too bad, we infer that he brushed his teeth this morning. Similarly, the big bang theory provides us with a framework that lets us fill in the gaps between those events we can directly measure, such as BBN and the last scattering of the CMB. Within the context of the big bang, we can re-create what must have happened at each stage and piece together our full cosmic history.

Of course, our predictions will be wrong in cases where our usual models don't apply. Similarly, our history of the universe is limited by the range of applicability of the laws of physics involved in constructing the big bang theory. In regimes where these laws are well tested and established, our history is most likely highly accurate. We must be careful, however, to remain aware of cases where we are trying to re-create events that occurred outside of these known regimes (at ultrahigh temperatures, for example).

In this chapter, we outline the major events in cosmic history within the framework of the big bang theory.

11.1 Density of the Universe

The defining feature of the big bang is that space is expanding as time goes on. This means that the density of the universe was once very high and is decreasing with time as space expands. So we'll begin by tracing how this general feature—the *density* of matter and energy—has changed during the nearly 14 billion-year history of the universe.

Introducing the Omegas

To organize our thinking about the evolution of density with time, we need to recall the structure of the Friedmann equation, which we expressed as follows in Chapter 7:

$$\begin{array}{l}\text{Some math describing} \\ \text{how the scale factor } (a) \\ \text{changes in time + the} \\ \text{curvature of space}\end{array} = \begin{array}{l}\text{Some math describing the} \\ \text{average density of normal} \\ \text{matter and energy + the} \\ \text{average density of dark energy}\end{array} \quad \text{(eq. 11.1)}$$

This equation shows that the expansion of space (the first term on the left-hand side) depends on the density of matter and energy in the universe (both terms on the right-hand side). Recall that the *density* of anything tells us how much of that stuff there is in a particular volume of space:

$$\text{density} = \frac{\text{amount of something}}{\text{volume it's contained in}} . \quad \text{(eq. 11.2)}$$

This means that *density decreases as volume increases*, because density is inversely proportional to volume. For example, if you have 100 air molecules in a 1 liter box, the density of molecules is 100 molecules per liter. If the number of molecules stays the same, but the volume increases to 100 liters, then the density is 100/100 = 1 molecule per liter. (See Figure 11.1.) That's for something called a *number density*: The *number* of molecules per liter of volume (or number of diseased mosquitoes per cubic meter—whatever). A more commonly used density is the *mass density*, often in kilograms per cubic meter. You can have a density of any sort—a large density of jellyfish on a particular scuba dive, say—but in cosmology, the most useful is usually an *energy density*.

The situation is more complicated when we consider the density of energy, because there are different types of energy and they behave differently. To see this, let's consider the cosmologically important types of energy density, all of which could be quoted in SI units of joules per cubic meter. First, consider matter, both dark and normal. The energy in matter is mc^2, which is constant for some fixed amount of matter. So the energy density of matter (sometimes referred to in cosmology as the "matter density") goes down as the volume goes up, by Equation 11.2. If you take all the matter out of a little box and put it in a big box, then the matter density decreases. More relevant is this relationship: As the universe expands (the box grows), the matter density goes down.

Now consider the energy density of light, which today is much smaller than that of matter. It also diminishes as the universe expands, because of the increasing volume. But remember that the expansion causes it to redshift, too! So not only does the number of photons per unit volume decrease, but the *energy of each photon* decreases as well. The energy density in light therefore decreases *faster* than that for matter, in an expanding universe. This means that if we imagine rewinding the expansion far enough back toward the beginning, then light, or radiation, must have had a *higher* energy density than matter. Therefore, for the first 72,000 years or so, the universe is said to have been **radiation dominated**. But since the energy density in radiation was decreasing faster than the energy density in matter, the universe underwent a transition to being **matter dominated** after about 72,000

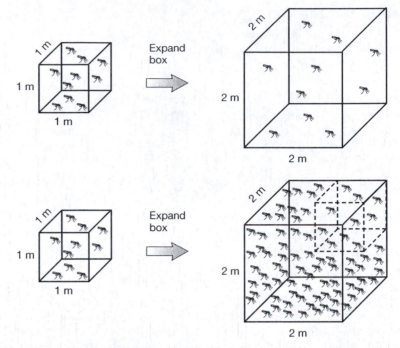

FIGURE 11.1 Volume and Density For familiar objects (air molecules, soccer balls, mosquitoes, etc.), density is inversely proportional to the volume of space the objects fill. For example, as the volume of a box containing 10 mosquitoes is expanded from 1 cubic meter to $2 \times 2 \times 2 = 8$ cubic meters, the number density of mosquitoes changes from 10 mosquitoes/m³ to $10/8 = 1.25$ mosquitoes/m³. Each of the 10 insects now has 8 times more space to move around in, so the number of mosquitoes occupying each cubic meter of space is now only one-eighth of what it was before. Dark energy behaves differently. For instance, the amount of vacuum energy increases as the volume increases, so the vacuum energy density stays constant.

years. (See Figure 11.2) As you'll see when we put together our full cosmic history in the next section, this transition had an important effect on the evolution of the universe.

Dark energy has a different behavior than matter or radiation does. Predicting the evolution of its density is complicated by the fact that we don't yet know the exact nature of dark energy—only that it must act as a repulsive gravitational force. One possibility is a true "cosmological constant" of the sort first proposed by Einstein, like the "vacuum energy" that's associated with the existence of space itself (as discussed in Section 7.4). In this case, the dark energy density remains *constant* as the volume increases: More volume means more space and therefore more vacuum energy, so the energy per unit volume remains unchanged. Both the numerator and denominator in Equation 11.2 increase by the same amount. Figure 11.1 illustrates the concept that vacuum energy density remains constant even as volume increases.

It's also possible that the dark energy is *not* an inherent property of space, but rather is caused by a field associated with some unknown particle—a field similar to the electric field of an electron. If this is the case, then the dark energy density need not remain constant for all time. We'll discuss the possible nature of dark energy further in Section 11.3; for now, we'll assume the simplest possibility: dark energy density that remains constant.

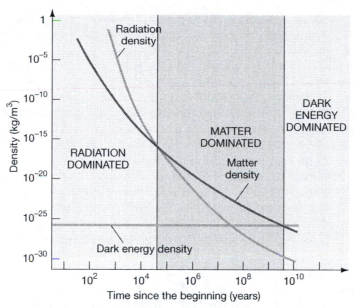

FIGURE 11.2 Eras of Domination This graph displays the energy density of various components of the universe—radiation, matter, and dark energy—on the vertical axis, as they have changed over time (horizontal axis). The energy density of radiation decreases fastest, because the expansion spreads out the photons *and* weakens them via the cosmological redshift. The matter density drops almost as fast, as it spreads out but does not redshift. Dark energy, finally, grows along with the expanding space, so its density never changes. The universe thus started out radiation dominated, then became matter dominated, and is now dark energy dominated. (For this illustration, dark energy is assumed to be a cosmological constant. Other possible forms of dark energy might have energy densities that change with time and thus deviate from the horizontal line. But whatever its true nature, the dark energy density must remain roughly constant close to the present time in order to produce the observed acceleration.)

Since the matter density diminishes as the universe expands, but the dark energy density does not, another transition arises. As of about 4 billion years ago, the universe became neither radiation dominated nor matter dominated any more; it is now **dark energy dominated** and will probably remain that way forever. In the next section, we'll see what effect dark energy domination will have on the future of the cosmos. Figure 11.2 shows graphically how these transitions are brought about by different components of the universe's total energy density evolving differently in the expansion.

Now that we've considered every major form of energy density in our universe—radiation, matter, and dark energy—we can return to Equation 11.1, the Friedmann equation. We can simplify it by incorporating *all* forms of energy density (including vacuum energy and other possible forms of dark energy) into a *single* density term, written as ρ_{total}. We can simplify further by noting that the first term of the equation involves the expansion rate (how fast the scale factor is changing). This rate can be expressed numerically with the famous Hubble constant, H, in order to relate it to the observable rate of galaxy recession. So the Friedmann equation can be written as

Math describing the expansion rate via the value of the Hubble constant H + the curvature of space	$=$	Math involving ρ_{total}, the average density of *everything* (matter, energy, and dark energy) .

(eq. 11.3)

(Notice the use of H rather than H_0: The subscripted "0" is a shorthand notation used by cosmologists to refer to the value of the Hubble constant *today*. Without the "0" subscript, H refers to the value of the Hubble constant (expansion rate) at any time. So if you want to calculate something from the Friedmann equation about the state of the universe when it was 10 years old, you would plug in the values for H, curvature, and density that were correct at age 10.)

If we look at Equation 11.3 as a purely mathematical statement, without thinking about its meaning, then its three terms become very simple looking: $A + B = C$. If C just happens to equal A, then B would have to be zero in order to satisfy the equals sign. In other words, Equation 11.3 reveals that there is a special value of the total density for which the curvature is zero. This value is known as the **critical density**, ρ_{crit}. It is the total energy density, of all forms combined, that must be present in the universe if the universe is *flat* (i.e., if geometry is Euclidean; see Chapter 6). The numerical value of the critical density—the density needed for the universe to be geometrically flat—is evidently connected to the value of the Hubble constant (because $C = A$). For a measured Hubble constant of $H_0 = 71$ (km/s)/Mpc, the critical density of our universe is currently equivalent to having only about 6 hydrogen atoms in each cubic meter of space.

An important measurement in cosmology is how the *actual* average energy density in the universe compares to the calculated *critical* density, because this ratio determines the overall shape of space that we discussed in Section 6.1 (flat, open, or closed). For this reason, cosmologists often express the total average density of matter and energy in the universe in comparison to its critical density by using a **density parameter**, symbolized with a capital Greek "omega":

$$\Omega_{total} = \frac{\rho_{total}}{\rho_{crit}} = \frac{\text{total density}}{\text{critical density}} \ . \qquad \text{(eq. 11.4)}$$

This notation gives you a quick sense of the density of the universe with respect to something meaningful—the critical density needed for geometrical flatness—rather than just some number of joules per cubic meter, say. For example, $\Omega_{total} = 1$ would mean that the actual total density equals the critical density in Equation 11.4. So if astronomers somehow measure Ω_{total} and find that it equals 3, then the density of the universe must be three times greater than the critical density, and therefore space must be highly curved overall.

Similarly, cosmologists often write actual measured densities of different kinds of matter and energy as fractions of the critical density. For any kind of matter or energy "X," its contribution to the total density is given by

$$\Omega_X = \frac{\rho_X}{\rho_{crit}} = \frac{\text{density of X}}{\text{critical density}} \ , \qquad \text{(eq. 11.5)}$$

where "X" might be "b" for baryonic matter, "dm" for dark matter, "m" for all matter, "r" for radiation, "de" for dark energy, etc.). For example, the density parameter for all forms of matter has been measured to be $\Omega_m = 0.27$. If you wanted to calculate the actual density of matter, you'd multiply 0.27 by ρ_{crit}. But normally, we work with density *parameters* rather than actual densities. One reason for this is that the numbers are more manageable and understandable that way. A matter density parameter of 0.27 means that there's enough matter in the universe to account for 27%

of the energy density needed to make space geometrically flat. As you'll learn in the next section, our universe *does* appear to be flat, so 27% of the critical density means 27% of the actual density. Therefore, 27% of the energy in the universe takes the form of matter.

Another reason to use Ω is that it has no units; it's a *fraction* of the critical density, not a density itself. This is nice because when you use the density parameter Ω, you don't have to worry about whether it should be expressed as a mass density (kg/m^3), a number density (number of particles/m^3), or an energy density (J/m^3).

EXERCISE 11.1 Densities of the Past

Test your understanding of this section by answering the following questions about an observed quasar with a redshift of $z = 3$:

(a) How much bigger or smaller was ρ_m when the quasar light we observe today was emitted, compared with ρ_m today?

(b) How about ρ_r?

(c) How about ρ_{de}?

(*Hint:* Imagine a cube-shaped volume of space, and think about how much the length of the imaginary cube's side has changed over time due to the cosmic expansion.)

Global Curvature or Lack Thereof

The density parameter gives us a tidy way to summarize the relationship between average density and the overall curvature of the universe, as given in Table 11.1. From the table, you can see that the curvature of the universe is connected to its average density. This means that *if we know the density, we can figure out the curvature, and vice versa*. That's really just a restatement of what we learned in Chapters 6 and 7. General relativity tells us that matter and energy curve spacetime, so their total density in the universe bends the global fabric of spacetime and determines the overall shape of space. (To be complete, we should mention that there might be more to the "shape" of the universe than just its curvature; see "Pondering Infinity and the Shape of Space" at the end of this chapter for a little more on this topic.)

We've seen that counting up all the matter and energy in the universe is very difficult, especially since we can't see most of the stuff we're trying to count! But the connection between density and the large-scale curvature of space provides a way around this problem. We can look at how light travels through space in order to see the overall curvature, and then we can infer what the density must be to

TABLE 11.1   Total Density and Global Curvature
The average total density of the universe determines the global (overall) curvature of space.

If the density parameter is	then the global curvature of space is	and the universe is
$\Omega_{total} > 1$	Spherical	Closed
$\Omega_{total} = 1$	Flat	Flat
$\Omega_{total} < 1$	Hyperbolic	Open

produce that average curvature. This technique is similar to what we saw already in Chapter 6 for gravitational lensing around clusters of galaxies, except that now we're looking for the bending of light *due to any overall curvature of space*, not just around a few specific objects.

What we really want to know is, "Do parallel light rays remain parallel over very large distances (billions of light-years)?" If they *do* remain parallel, then the universe is flat overall, and thus the total density must be equal to the critical density. If parallel light rays bend together (converge), then space is closed. If parallel rays spread apart (diverge), then space is open. (Recall Figure 6.2.)

So how can we determine whether light travels in straight lines over very large distances? To understand how to approach this question experimentally, we must return to the same concepts of Euclidean geometry we applied in Sections 2.5 (parallax) and 3.3 (standard rulers). The key insight we need is that the farther away an object is located from us, the smaller its angular size will be, as illustrated in Figure 2.2 back in Chapter 2. Specifically,

$$\frac{\text{physical size}}{2\pi \times \text{distance}} = \frac{\text{angular size}}{360°} \, . \tag{eq. 11.6}$$

We don't normally stop to think about it, but our everyday sense of how far away things are is based on the inverse relationship between distance and angular size described by Equation 11.6. And this relationship in turn depends on the assumption that light travels in straight lines. When that assumption fails to hold, we can be fooled into thinking an object is closer or farther than it really is. Even the best of us can be tricked in this way. For example, in the 1991 movie *Robin Hood: Prince of Thieves*, Azeem hands Robin Hood a small telescope to get a better look at approaching attackers. Unfamiliar with telescopes (which work by bending light), Robin draws his sword in panic as he looks through the scope, because it looks like the sheriff's men are right on top of him.

The surprise occurs because our brains assume that light travels in straight lines and "compute" the distance on that basis. But when a telescope, or curved space, or anything else bends the light during its journey to our eyes, the relationship between angular size and distance is modified. (See Figure 11.3.)

With all this in mind, we can identify what is needed in order to measure the curvature of space. If we could find a distant object of *known size* and *known distance* away from us, we could observe its angular size and compare it with what we expect for an object of that size and distance in a flat universe. If the angular size we observe is the same as what we calculate it should be (as in Figure 11.3(a)), then the universe is flat. If the observed angular size is bigger than we calculate it should be (Figure 11.3(b)), then the universe is closed. Finally, if the observed angular size is smaller than we calculate (Figure 11.3(c)), then the universe is open. Stated another way, we are checking whether or not all the angles in each triangle shown in Figure 11.3 add up to 180°, as they should in a flat universe. For example, in frame (b), the triangle formed by the two light rays and the height of the penguin appears bowed outward, making its interior angles larger than those of the Euclidean triangle in frame (a).

It turns out that, thanks to cosmic microwave background observations, we *do* have objects that meet all these criteria for determining the curvature of space: We know their angular size, their distance from us, and their actual physical size. These standard rulers are the most prominent spots in the temperature variation

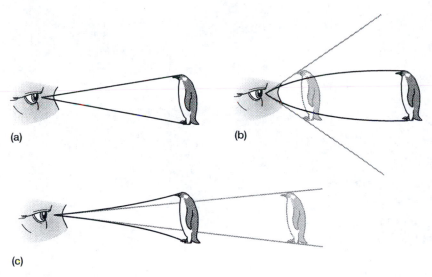

(a)

(b)

(c)

FIGURE 11.3 Bending of Light and Apparent Size (a) When we see the angular size of an object (why not a penguin?) whose physical size we know, we infer its distance by assuming that the light from the head and feet of the penguin has traveled along straight lines to reach our eye, forming the triangle indicated. (b) If the light is bent along its path so that it converges (due to a lens, spherically curved space, or something else), then the object appears closer than it really is (i.e., its angular size is bigger than expected). (c) If the light is bent so that it diverges along the way (such as by a curved, hyperbolic space), then the object appears farther away than it really is (i.e., its angular size is smaller than expected). Solid penguins represent where the penguin actually is, and ghosted penguins indicate where a penguin of the same size appears to be, due to the light bending, which changes the angular size. Notice that the (dark, nonghosted) triangle in each frame, formed by the two light rays and the height of the penguin, is different each time. If the light rays bend, then the interior angles of the triangle no longer add up to 180°, as they do in the normal Euclidean triangle in frame (a); all three interior angles are larger in frame (b), and all three are smaller in frame (c), than they are in frame (a).

of the CMB and have an angular size of about 1° on our sky. It's important that we use a very distant source to measure the overall curvature of space (and the CMB is the most distant source ever observed), because, although space appears flat at a glance, its curvature on very large scales could escape our notice. Therefore, we need a large-scale test, such as a giant triangle whose interior angles we can measure.

The base of the triangle is formed by the 1° hot and cold spots in the CMB, whose physical size can be calculated once we know what caused the spots to form (which we *will* know in Chapter 12). That distance, together with the measured angular size of the spot and the distance to the last scattering surface, is enough to calculate the other two angles. Thus, one can actually check whether they add up to 180°.

The triangle used in this method is shown in Figure 11.4. The results of the measurement are usually expressed with Ω_{total}. Since each Ω is a fraction of the critical density of the universe, any departure in the value of Ω_{total} from 1.0 indicates global curvature. An Ω_{total} less than 1 would be negatively curved; an Ω_{total} more than 1 would be positively curved. Astronomers measure $\Omega_{total} = 1.02 \pm 0.02$, so light does travel in approximately straight lines, and the universe is approximately flat. As we'll see in Chapter 12, there's good reason to believe that the real number should be precisely equal to 1.

FIGURE 11.4 Curvature of the Universe Global curvature is determined by looking for the geometric differences described in Chapter 6: Do the interior angles of a triangle formed by us and opposite sides of a CMB spot really add up to 180°? Alternatively, are distant objects really as far away as they appear to be? The measurements say yes. Compare the straight and arcing triangles in this figure with those in Figure 11.3.

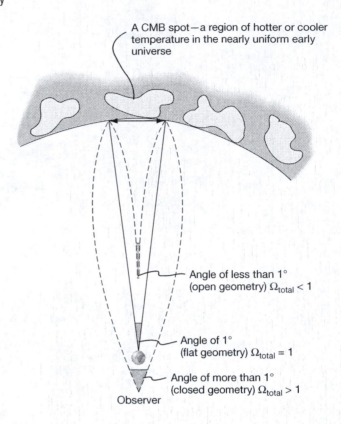

A CMB spot—a region of hotter or cooler temperature in the nearly uniform early universe

Angle of less than 1° (open geometry) $\Omega_{total} < 1$

Angle of 1° (flat geometry) $\Omega_{total} = 1$

Angle of more than 1° (closed geometry) $\Omega_{total} > 1$

Observer

11.2 History of the Universe

Now that you know the significance of the contents of the universe (from the previous section) and the evolution of the universe (from the previous chapter), you can ask specifically about its history. Some of that history you already know, and therefore some of what we cover in this section will not be new. The purpose here is to show the *flow* of events, in chronological order, from the beginning until now. Figure 11.5 traces out the series of events described in this section. Ideally each event in cosmic history will follow as a necessary consequence of previous events, taken together with the expansion and the laws of physics. You should expect the earliest moments to be the least well understood, but as time goes on we reach events that are known with great precision.

Our starting point is not the *actual* beginning. Unfortunately, no one knows the science of the actual beginning. Somehow, an ultra-hot and ultra-dense expanding universe came to be, and it was filled with some form of energy. This is our starting point; we hope you won't mind that it occurs a ten-millionth of a trillionth of a trillionth of a trillionth of a second *after* time zero.

The Universe Cuts Off Its TOE

We'll start at very early times with an incredible density of energy and work our way forward (by expansion and cooling) from there. So that you have something definite in mind, you can imagine the early universe as a sea of different types of

Color Gallery 2

(a)   (b)

COLOR FIGURE 36 HST Images of Gravitational Lenses   (a) Faint, blue spiral galaxies in the distant background are warped, magnified, imaged, and reimaged several times by the lens, which is the massive galaxy cluster Abell 2218 (the collection of mostly yellow elliptical galaxies). Multiple images of the spiral galaxies in the background surround the lens in a pattern of thin, concentric, circular arcs, as indicated by arrows. (b) Many of the blue images scattered around this lensing system are images of the *same* blue galaxy, physically located behind the (yellow) lensing cluster.

COLOR FIGURE 37 Gravitationally Lensed Galaxies and Quasars   The galaxy cluster pictured here is gravitationally lensing some background galaxies; for example, the two similarly yellowish and elongated ones toward the right side are actually images of the same galaxy. Even more impressive, five images of the same quasar (bright spots), physically located behind the cluster, are visible here, with a hint of the galaxy that the quasar lives in visible as a smear of light going through each quasar image.

COLOR FIGURE 38 Spiral Galaxy Viewed Edge-On
Galaxies seen from an edge-on perspective, like this one, help us to study the matter in the universe. By measuring the Doppler shift of the starlight at each distance from the galactic center, astronomers can calculate how much matter the galaxy has, and where that matter is located. From this type of study, we find that most of the matter is invisible, or *dark*, matter.

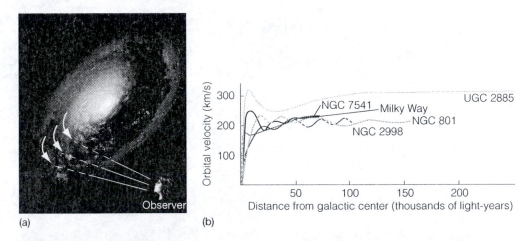

(a)

(b)

COLOR FIGURE 39 Galactic Rotation Curves
We describe the motions of stars in galaxies that we view edge-on by constructing a *rotation curve*. Unlike the velocities of objects in the solar system, the velocities of stars do not decrease as one observes stars that are farther away from the center of their galaxy—the rotation curve does not slope downward. This implies that the mass of the galaxy extends well beyond the stars whose motions we are observing.

COLOR FIGURE 40 The Influence of Dark Matter The long strip of material streaming out to the right from visible galaxy UGC 10214 has been torn loose by something, probably a large mass of dark matter nearby.

COLOR FIGURE 41 Exotic Dark Matter Separated From Normal Matter This image shows a blend of different data from a region in space where two galaxy clusters have passed through one another. Most of the *normal* matter (intra-cluster gas) in the clusters has collided; this is shown by x-ray data (pink) at the center. Most of the *total* matter, however, is shown in blue; it has been located by gravitational lensing, which identifies the presence of matter by its gravity. Because most of the matter (blue) is not in the same place as most of the normal matter (pink), this image suggests the presence of "nonnormal," or "exotic," dark matter in the blue regions to cause the gravity there. Exotic dark matter is not made of protons, neutrons, or electrons the way normal matter is. (Note that the comparable sizes of the blue and pink regions do not indicate that there are comparable amounts of normal matter and exotic dark matter; the false coloring shown here is not calibrated in that way.)

COLOR FIGURE 42 Distant Galaxies in the Expanding Universe Faint red smudges, circled in green, reveal small galaxies from very early times, when the universe was about 5% of its present age. Analyses indicate that these early galaxies should have contained hot, ultraviolet-peaked stars. The red color seen here instead is due to the *cosmological redshift.* This light from far away traveled across the expanding universe for a long time, so its wavelength expanded too, causing the light to redden. These small galaxies are thought to have merged over time into the larger ones we're familiar with today. (Such modern galaxies are also seen in this image.)

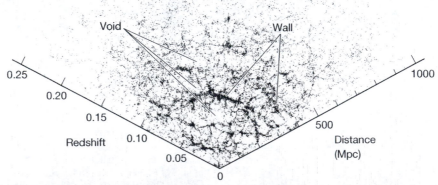

COLOR FIGURE 43 Distribution of Galaxies These data from the Sloan Digital Sky Survey show nearly 67,000 galaxies, each represented with a dot. We are located at the vertex at the bottom. The scale outward is listed both in redshift and in Mpc (a million parsecs, or a little over 3 million light-years). A pattern of walls (rows of galaxies) and voids (regions without many galaxies) can be seen in the data. The thinning density of galaxies far from us is because more distant galaxies are harder to see, not because they're not there.

Temperature: 2.715 K   2.725 K   2.735 K

COLOR FIGURE 44 Isotropic Temperature of the CMB This is what the CMB all over the sky would look like to a detector sensitive to 0.01 K temperature differences. Any temperatures less than 2.72 K or greater than 2.73 K would have shown up as a different shade of gray (according to the temperature scale on the figure), but every direction on the sky is uniform at this sensitivity level. Compare this figure with the temperature map of Earth in Color Figure 45, which shows much wider variations in temperature at different locations.

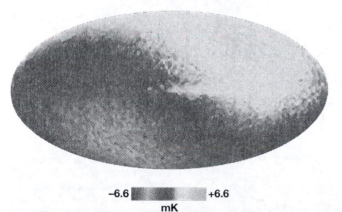

COLOR FIGURE 45 Temperatures on Earth For comparison with the CMB map (Color Figure 44), here is a temperature map of the whole Earth, also flattened into an oval. Temperature variations from 0 °C to about 30 °C are clearly visible as different shades; red is warmer and blue-violet is colder. In both this map and the CMB map, a full sphere of temperature information is shown on a flat oval.

−6.6   +6.6
mK

COLOR FIGURE 46 Temperature Variations in the CMB The shading in the plot shows small temperature variations of a few thousandths of a Kelvin (a few mK, or millikelvins) away from the average of 2.725 K. Because half the sky appears slightly warmer than average, and the other half appears slightly cooler, it is reasonable to suspect that the Earth's motion through space is responsible for the equal and opposite redshift and blueshift. (See Color Figure 47.)

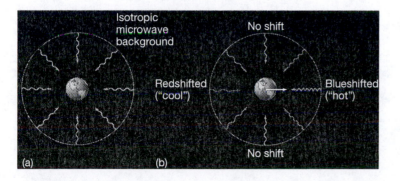

COLOR FIGURE 47 Moving through the CMB Our motion causes an otherwise isotropic distribution (uniform) of background radiation to appear hotter in the direction we are moving toward and cooler in the direction we are moving away from.

(a)

(b)

COLOR FIGURE 48 Cosmic Microwave Background Anisotropy (a) The variations in color indicate variations in temperature, just as in the maps of the sky and the Earth in the preceding Color Figures 44, 45, and 46. But here the variations, also called *anisotropies*, are only some hundred thousandths of a degree. (b) Since CMB radiation is everywhere, it comes at us from every direction, so we can also view it "wrapped" around us (imagine Earth at the center of the sphere).

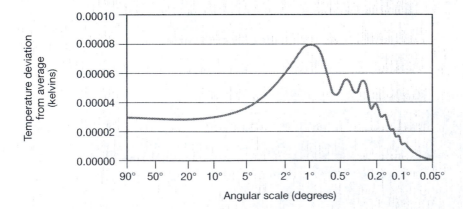

COLOR FIGURE 49 The CMB Power Spectrum The CMB anisotropy is actually far from random, and contains lots of valuable information about the universe. Cosmologists tally up its statistics—temperature information for spots of each size—and graph it as a "power spectrum" (shown here). The meaning of the power spectrum is rich and complicated, and reveals detailed information about the early universe, as discussed in Chapter 12.

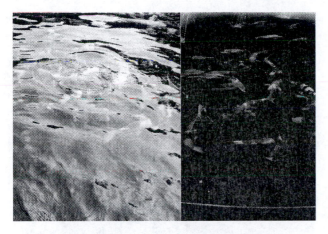

COLOR FIGURE 50 Polarization and Glare   Photographs of fish in a pond without (left) and with (right) a polarizing filter in front of the camera. The filter blocks light that has reflected off the surface of the pond, making it easier to see the light coming from the fish below the surface. Polarization, whose effects are easily seen here, is an important property of light—including CMB light.

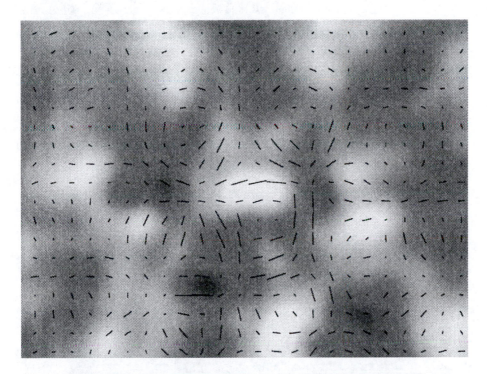

COLOR FIGURE 51 Measured CMB Polarization   Dashes indicate the strength and direction of polarization of light coming from the direction on the microwave sky where each dash is located. The overall pattern is a complex blend of polarization patterns from a variety of distant hot and cold spots. This polarization pattern found in the CMB provides detailed information about the big bang origin of the microwaves, and about the subsequent birth of the first stars.

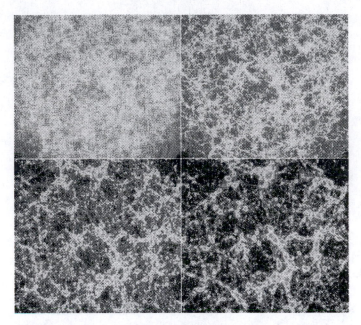

COLOR FIGURE 52 Simulated Time Sequence   In the first frame, we begin with a nearly-smooth universe, like the one the CMB all-sky maps reveal. Over billions of years, structures coalesce gravitationally in a hierarchical fashion, until the observed spongy large scale structure of the universe emerges.

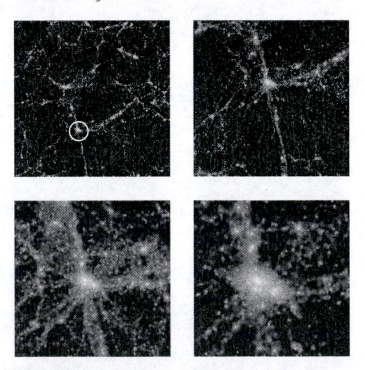

COLOR FIGURE 53 Simulated Structure in the Universe Today
Here we see four successive zooms on the same simulated patch of space. Notice the voids in the upper frames and the rich cluster (circled) forming in the lower frames.

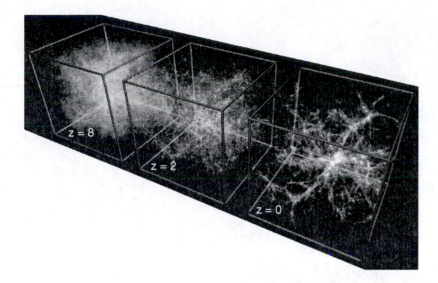

COLOR FIGURE 54 Simulation of Structure Formation This is a 3D simulation of structure formation in the universe. Like the 2D views in Color Figures 52 and 53, this simulation follows matter from a smooth initial state to a clumpy end state, based on the action of gravity in the expanding universe.

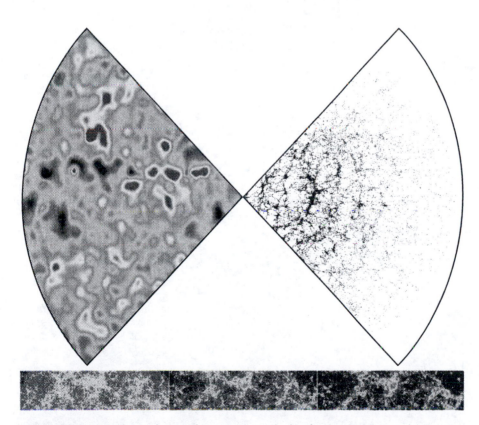

COLOR FIGURE 55 The Evolution of Structure In the first few hundred thousand years, the universe had tiny density fluctuations of only 1 part in 100,000 (left). Today the universe has massive walls and voids 100 million light-years wide (right). Computer simulations show how the universe got from one state to the other (left to right across the bottom).

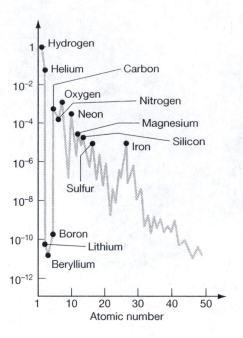

COLOR FIGURE 56 The Composition of the Universe This graph summarizes the measured abundances of various elements in the universe. Hydrogen, the most common element in the universe, is assigned an abundance of 1, and the amounts of all other elements are expressed as a fraction of the amount of hydrogen. Notice that the vertical scale is organized into powers of ten, so these data reveal that hydrogen and helium are *much* more abundant than all the other elements.

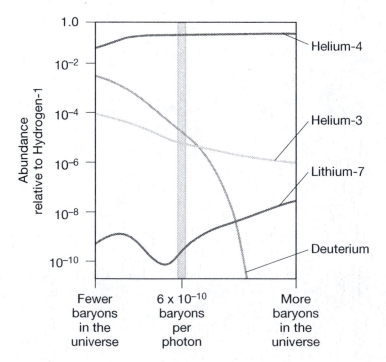

COLOR FIGURE 57 Elemental Abundances Predicted by Big Bang Nucleosynthesis The big bang theory allows researchers to predict how much of each variety (or "isotope") of the lightest elements should exist. Lines on this graph show how the predicted abundance of each isotope varies with the total amount of normal matter (or "baryons"—as contrasted with dark matter) found in the universe. The vertical stripe shows the amount of normal matter that is allowed by observations. Therefore, lithium-7, for example, must be found within a range of abundance specified by just the intersection of the vertical stripe with the red curve. (The abundance for each isotope is given on the vertical axis.) Remarkably, the observed abundances of *all* these isotopes fall within the narrow ranges predicted by the big bang theory.

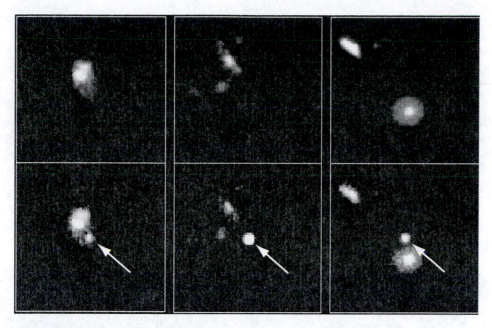

COLOR FIGURE 58 Distant Supernovae
Images of three very distant galaxies before (top) and after (bottom) a type Ia supernova explosion (denoted with arrows) within the galaxy.

The supernova is nearly as bright as its entire host galaxy! Light from a distant supernova can be used to measure how far away the host galaxy is.

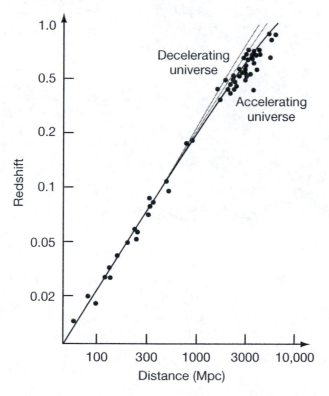

Decelerating universe

Accelerating universe

Redshift

Distance (Mpc)

COLOR FIGURE 59 Evidence for Dark Energy
Data points indicate measured redshifts (caused by the expansion of the universe) and distances (measured with type Ia supernovae) for a collection of galaxies. Solid lines show the predicted relationships for decelerating, constant, and accelerating expansion models. Thus, the data suggest that the expansion of our universe is accelerating, or speeding up! The unknown agent responsible for the accelerated expansion is called *dark energy*.

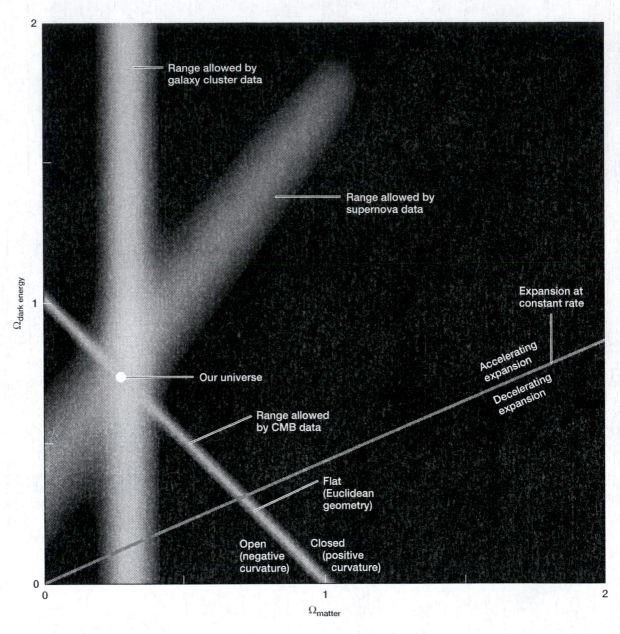

COLOR FIGURE 60 Density and Destiny   In a universe with both matter and dark energy, the relative amounts of each determine the fate of the universe in the distant future. Those relative amounts are shown on this graph of Ω_{de} versus Ω_m. Permissible properties of our universe, according to three major cosmological data sets, are shown as thin strips on the graph: acceleration data from type Ia supernovae, data on the motions of galaxies from studies of galaxy clusters, and cosmological data extracted from the CMB. Importantly, all three data ranges intersect in one region of the graph. That region evidently describes our actual universe: a universe with nearly three times more dark energy than matter. In addition, lines are drawn in to indicate the borderlines for global geometry (open, closed, or flat?), acceleration (or deceleration?), and destiny (expand forever or recollapse?).

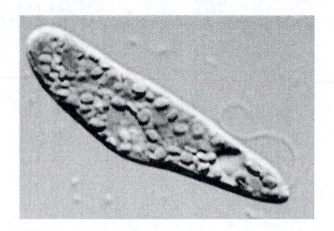

COLOR FIGURE 61 Living Cell, Under the Microscope The laws of the physical universe allow for the emergence of life, at least on one planet in the universe. This is a view of a single-celled organism called Euglena under a microscope.

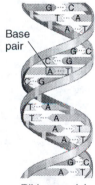

Base pair

Ribbon model   Computer model

COLOR FIGURE 62 The Double Helix Shape of the DNA Molecule "Base pairs" (horizontal links) are mounted on a twisting twin backbone of "deoxyribophosphate." These chemical structures manage to encode the detailed "recipe" for making an organism, such as a person.

5 cm

COLOR FIGURE 63 The Famous Meteorite ALH84001, Which Flew Here from Mars This rock spawned a flurry of research into the possibility of extraterrestrial microbial life. Four separate possible biosignatures, including a possible fossil (right), were found in the meteorite, but it's difficult to know if they really are signs of life.

(a)

(b)

(c)

(d)

COLOR FIGURE 64 Near-Term Frontiers for Humanity
(a) Planned underwater resort ("Poseidon Mystery Island," Fiji),
(b) local space (the International Space Station is shown, from
NASA), (c) a hotel on the Moon (imagined for 2040), and
(d) a small but growing and potentially permanent settlement
on Mars.

COLOR FIGURE 65 Important Outcomes of Cosmic History   Everything in this figure has emerged as part of the history of our universe. The process by which all of them formed includes the expansion of the universe, the formation of galaxies within dark matter halos, the nuclear reactions in stars and supernovae, and universal laws such as the law of gravity, to name just a few examples.

COLOR FIGURE 66 Outcomes of the Universe   This baby seal and our emotional response to it are two remarkable outcomes of the evolution of the universe.

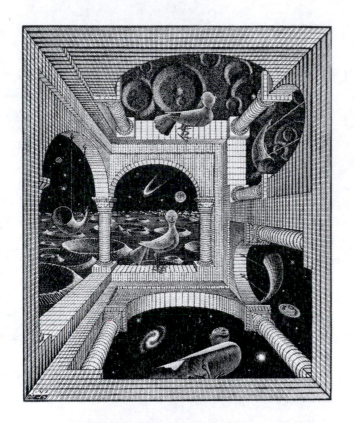

COLOR FIGURE 67 M. C. Escher's "Other World." Our own artistic creations are embedded within the background of everything else the unfolding universe has created.

COLOR FIGURE 68 Your Cosmic Context Each of our lives is set within a much larger story and depends on events that took place in the ultra distant past.

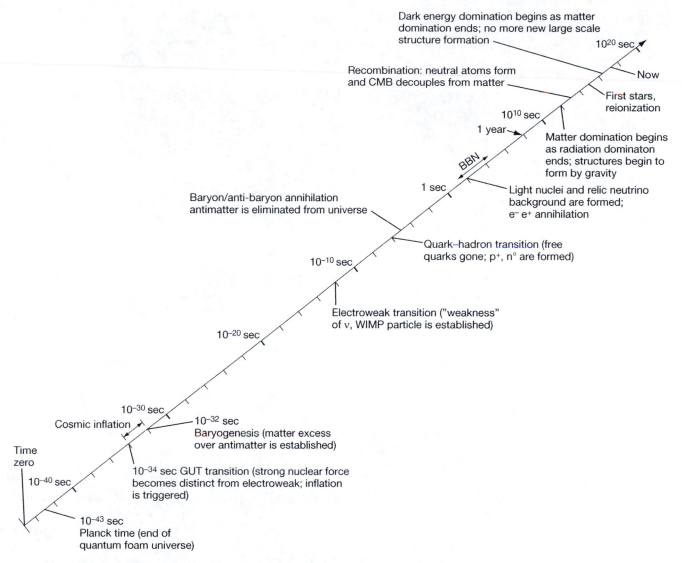

Dark energy domination begins as matter domination ends; no more new large scale structure formation

10^{20} sec

Now

Recombination: neutral atoms form and CMB decouples from matter

First stars, reionization

10^{10} sec

1 year

Matter domination begins as radiation dominaton ends; structures begin to form by gravity

BBN

1 sec

Light nuclei and relic neutrino background are formed; $e^- e^+$ annihilation

Baryon/anti-baryon annihilation antimatter is eliminated from universe

Quark–hadron transition (free quarks gone; p^+, $n^°$ are formed)

10^{-10} sec

Electroweak transition ("weakness" of ν, WIMP particle is established)

10^{-20} sec

10^{-30} sec

Cosmic inflation

10^{-32} sec
Baryogenesis (matter excess over antimatter is established)

Time zero

10^{-40} sec

10^{-34} sec GUT transition (strong nuclear force becomes distinct from electroweak; inflation is triggered)

10^{-43} sec
Planck time (end of quantum foam universe)

FIGURE 11.5 Milestones in the History of the Universe   Notice that the logarithmic scale here means that most of this timeline is taken up by the first minutes of the universe's existence. Refer back to this figure as you read about each event in Section 11.2.

particles running around and colliding, much like the interior of a star. The particles themselves might be completely different from those in the universe today, but that's okay, just as long as their energies decrease with the expansion of spacetime, like a redshifting photon; all known particles do this.

The very early universe is difficult to imagine because the conditions were so unlike what we experience today. Enough energy was packed into each bit of space that many different kinds of particles could freely pop in and out of

FIGURE 11.6 Quantum Foam If you zoom way, way, way, way, way in on the fabric of spacetime, then you'll find its geometry to be warped and complicated by the energy inherent in random particle activity there. The universe probably had to be this way at very early times.

existence and transform from one particle into another. Nothing was static—not even spacetime! We know that the geometry of spacetime is warped by the matter and energy in it. Our starting point for the history of the universe is the time when every little speck of space was so densely crammed with energy, that random collisions and interactions of every sort carried enough energy to sharply distort the space around them. The state of the universe in this condition is sometimes called a **quantum foam**. (See the artistic rendering of quantum foam in Figure 11.6.)

The attempt to understand the physics of this era requires that we push our best theories to their limits. We'll need quantum mechanics for the interactions of particles, and we'll need relativity for the curvature of spacetime. Each great theory carries with it one or two *fundamental constants* of nature—important numbers that characterize the theory. In the present context,

- *Quantum mechanics* governs the world of the very small and therefore employs the very small "Planck constant," $h = 6.63 \times 10^{-34}$ kg m²/sec, which sets the scale at which quantum effects are noticeable. For example, the small size of h plays a key role in determining the size of the region where a particle's energy is likely to be concentrated.

- *Special relativity and general relativity* govern spacetime, motion, and gravity, and therefore employ the speed of light, $c = 3 \times 10^8$ m/s, and the gravitational constant, $G = 6.67 \times 10^{-11}$ m³/(kg s²). The constants c and G together play a key role in determining how much energy or mass is needed to cause significant bending of spacetime.

By combining all three constants together (h, c, and G), physicists determine the length and time scales on which both quantum mechanics and general relativity are important:

$$\bullet \quad \sqrt{\frac{hG}{c^3}} \approx 10^{-35} \text{ meter}$$

$$\bullet \quad \sqrt{\frac{hG}{c^5}} \approx 10^{-43} \text{ second}$$

These quantities, called the **Planck length** and **Planck time**, respectively, are of particular relevance to early-universe cosmology. They can be used in two ways. First, they are the distance and time scales at which random spacetime distortions become large and strongly affect the outcome of any interaction. The universe was a quantum foam prior to the Planck time, when the scale factor of the universe was so small as to localize any particle's energy to within a tiny Planck length.

Second, anything taking place *earlier* than the Planck time, corresponding to scales *smaller* than the Planck length, does so according to physical laws that somehow blend quantum mechanics with relativity. Each theory by itself is inadequate to describe what happens at earlier times than the Planck time or smaller distances than the Planck length. Such a combined theory, generically called **quantum gravity**, does not yet exist. And if it did exist, it would be exceedingly difficult to test: What laboratory instruments could resolve such tiny fractions of a meter and a second? A quantum gravity theory would encompass all of the currently known forces and interactions, and is therefore sometimes referred to as a **TOE**, or **theory of everything**. The physics profession has made some theoretical progress in this area (the area of everything), including the theories of *loop quantum gravity* and *M-theory* (which absorbed *superstring theory*, which absorbed *string theory*, and is still commonly called string theory). But none of this work appears to be particularly close to completion.

What we know of the Planck time is inferred by *extrapolating* theories that have been tested at much lower energies. To address what happened *before* the Planck time would require extrapolating a theory (quantum gravity) that doesn't even exist yet! In other words, all this discussion of ultrasophisticated physics tells us something definitive: Current laws of physics are incomplete and therefore unreliable at scales below the Planck length or time.

Cosmologically, then, we can't realistically talk about things that happened within the first 10^{-43} second. We hope you can live with this restriction. The earliest moment we can talk about is the Planck time; before that, we just don't know anything. (Be wary of popular science articles—and even some textbooks—that make grand claims about how the universe was once a geometric *point*, infinitely smaller than even the Planck length. If that was ever the case, it was earlier than the Planck time, which makes it *unknowable* by any physics available today.)

The Universe Eliminates Its GUT

As the universe expands and cools, the defining characteristic of the Planck era disintegrates: Gravity becomes distinct from the other natural forces (strong nuclear, weak nuclear, and electromagnetic—the ones governed by quantum mechanics). Particle interactions no longer involve enough energy to significantly alter the spacetime around them, and their outcomes can be calculated without the

FIGURE 11.7 An Example of Symmetry Breaking Liquid water is symmetrical in the sense that it is uniform: You can move to a different location and it still looks the same. It is, in fact, homogenous and isotropic, just like the universe on large scales. As water freezes, this symmetry is broken by nonuniformities that crystallize into the ice.

equations for gravity. About 10^{-34} second later, when the universe is about 10^{-34} second old, a similar process ensues: The strong nuclear force disentangles from the weak and the electromagnetic forces. This type of process is called *symmetry breaking*, of which a much more familiar example is displayed in Figure 11.7.

There is no single dominant theory to govern this time in cosmic history, between 10^{-43} and 10^{-34} second, although there's already a name for it: **GUT**, or **grand unified theory**. We realize that this feels like a letdown so soon after the theory of everything: The GUT refers to "everything except gravity." Still, we have good reason to suspect that the transition which ends this GUT era was an important one. It probably triggered a process called *inflation*, which answers some big cosmological questions. But we're going to make you wait until the next chapter before we return to that topic.

So first an era of quantum gravity is ruined as the expansion and redshift of the universe together begin to drain particles' energy, so that they cease to make spacetime "foamy." Then, as the universe cools further, the "GUT superforce" is eliminated, breaking the former symmetry of the strong, weak, and electromagnetic forces. This recurring loss of symmetry transforms the universe, step by step, from a featureless quantum foam governed by one unified theory of everything to the universe we are familiar with—rich in its variety of matter, energy, and physical laws. Pay attention to this theme throughout the rest of this section (11.2).

The Universe Unfairly Favors Matter over Antimatter

One important prediction of grand unified theories is the existence of new types of particle interactions that violate yet another symmetry. As noted back in Exercise 9.2, nuclear reactions normally *conserve baryon number*. That is, normally, there's no way to create protons, for example, without *also* making antiprotons. (If this language isn't connecting for you, we suggest that you bite the bullet and revisit Chapter 9, especially Section 9.5.) The GUT prediction, however, is that the apparent symmetry between baryons and antibaryons is slightly broken. The universe is slightly *biased*; it does not treat matter and antimatter equally. It treats matter just a little bit better.

All of this might sound like a fairly technical little detail that doesn't immediately link back to the rest of the cosmology you've been learning about. In point of fact, however, this little detail is responsible for the existence of matter of any sort! Remember that matter and antimatter *annihilate* when they collide and that the early universe was a rapid series of random particle collisions. If the universe treated matter and antimatter equally and without prejudice, then random collisions would have annihilated *both*, and we'd have no matter (or antimatter) at all. Or, at the very least, some small, isolated pocket of matter would be in continual danger of being destroyed by some equally common pocket of antimatter.

Instead, our universe appears to be made entirely of matter. If there ever was much antimatter (which we certainly think there was), then it must have been destroyed. Now, all of this could be viewed as an overstatement. After all, there is no obvious way to tell matter and antimatter apart from a distance: The light emitted (or absorbed, or reflected) by matter is completely identical to that emitted by antimatter. The only ironclad evidence of "matter, not antimatter" is local: Our spacecraft don't spontaneously explode when traveling through space or landing on (anti)Mars. But neither do we observe any massive energy blasts in deep space from stars colliding with antistars or galaxies with antigalaxies. Various other arguments make it highly unlikely that baryonic antimatter exists in significant quantities anywhere.

Baryogenesis (define it for yourself by checking the root words) may have taken place shortly after the GUT transition, somewhere around 10^{-32} second. At that time, the average particle energy was around a hundred trillion (10^{14}) times greater than the energy $m_p c^2$ needed to "build" a single proton. That's like saying that it costs practically nothing to make a proton—mere pennies in a universe of trillionaires. A random collision between massless photons, for example, would easily suffice:

$$\gamma + \gamma \to p + \bar{p} \ . \qquad\qquad \text{(eq. 11.7)}$$

In this reaction, a proton and antiproton are created. The original two photons apparently respect a symmetry between baryons and antibaryons. What is needed in the early universe is a process that doesn't. It has to violate the symmetry once in a while. It need not be a frequent violation: Once in a billion reactions would do. That's enough to satisfy the conditions for baryogenesis—and the creation of the material world.

This idea is not as contrived as it may appear. Infrequent violations of similar conservation laws are observed in the decay of particles called K^0 mesons. And since the grand unification idea introduces a preference for baryons over antibaryons, we have a plausible, though poorly tested, mechanism for establishing a universe of matter. In reality, reactions like Equation 11.7 are probably not where the extra baryons are produced. More likely, a tiny excess of matter over antimatter particles was produced in the decay of some as-yet undiscovered particle that existed just after the GUT transition. In the next few pages, you'll see that this tiny asymmetry can make a big difference.

The Universe Abandons the Weak

The next early-universe symmetry to break down as the expansion drives down particle energies is that between the electromagnetic force and the weak nuclear force. At low energies, corresponding to everyday temperatures here on Earth, all four natural forces are distinct. That is, we measure gravity, electromagnetism, and the strong and weak nuclear forces as different kinds of forces. But at higher energies the interactions meld together, and the same processes that can be accomplished, for example, electromagnetically become possible by the weak nuclear force as well. This particular case is called the **electroweak unification**, and as you already know, it is not the *only* such unification. (To refresh your memory on the merit of unifying aspects of physics in general, you might want to revisit "Einstein's Legacy of Unification" at the end of Chapter 6.)

The universe continues to expand and cool, and at an age of about 1 trillionth (10^{-12}) of a second, the electroweak force splits in two. Weak interactions require more energy than electromagnetic ones, and as the temperature of the universe forever drops below about 10^{15} K, particles become too energy poor to afford energy-expensive weak reactions. As a result, weak interactions acquire their weakness: Particles that interact via the weak nuclear force, such as neutrinos, will soon and forevermore essentially cease to interact with normal matter. WIMPs, for this reason, go from being "matter" to being "dark matter."

The temperature and density at this stage of the early universe are still far beyond what can be re-created in physics laboratories. But in giant accelerators, particles can be smashed together at very high speeds, transforming most of their kinetic energy into other particles. The laws of physics operating at the end of the electroweak unification and everything that follows, therefore, can be tested by direct experimentation today. Remarkably, that permits us great confidence in our understanding of the early universe, reaching back to the first trillionth of one second!

The Universe Confines Its Quarks

Skipping ahead by a millionth of a second, we reach another transitional point, which has also been simulated in particle physics laboratories. Prior to this time, the universe was too hot for protons and neutrons to coexist "peacefully." To understand this point, first consider electrons. Electrons are believed to be *fundamental* particles, in that they're not constructed by assembling smaller particles. Therefore, if you crank up the temperature around an electron, all you get is a high-energy electron.

Baryons, however, are *composite* particles. Three *other* particles, called **quarks,** make up each baryon. (To the best of our ability to probe such things, quarks themselves appear to be fundamental.) There are six known types of quarks, and the two lightest "flavors" live inside protons and neutrons. Protons are made of two "up" quarks and one "down" quark; neutrons are made of two downs and one up. So when you crank up the temperature around a baryon, it dissolves into its component quarks. This is similar to the disintegration of iron nuclei in the temperatures leading up to type II supernovae (Section 9.4). Prior to about 10^{-6} second, the universe was a *primordial soup* of fundamental particles, including quarks.

A particle physics property called *confinement* prevents quarks from existing on their own today. Instead, all quarks live inside protons and neutrons. Even when physicists deliberately create free quarks in a laboratory, the quarks immediately become confined inside other particles. So, after the first 10^{-6} second, there were baryons, but no free quarks, in the universe. This transition is most commonly called the **quark–hadron transition**, implying that quarks became confined inside *hadron* particles. A hadron is either a baryon (made of three quarks) or a *meson* (made of one quark and one antiquark); but mesons are unstable and decay rapidly, so you can feel free to ignore the "meson/hadron" side of things and think of the transition as the quark–*baryon* transition if you like.

(By the way, in our earlier discussion of baryogenesis, Equation 11.7 was secretly a simplification of the true state of affairs. At that time it was much too hot for protons, meaning that if you made one, it would instantly dissolve into free quarks. The excess of baryons over antibaryons that we claimed had formed at that time was really an excess of quarks over antiquarks, but it makes no difference: The only quarks to survive all wound up inside baryons, and the only antiquarks to survive all wound up inside antibaryons. So, immediately after the quark–hadron transition, the situation we described earlier came about through the operation of Equation 11.7, and a slight excess of baryons prevailed over antibaryons.)

The Universe Annihilates Most of the Material World

Soon after the quark–hadron transition, the universe had expanded and cooled to the point where the average energy carried by any particle became less than the energy $m_p c^2$ needed to make a baryon. Before that, a reaction like Equation 11.7 could run in both directions:

$$\gamma + \gamma \to p + \bar{p} \quad \text{and} \quad p + \bar{p} \to \gamma + \gamma \; . \tag{eq. 11.8}$$

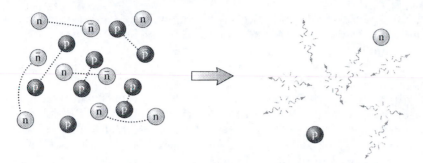

FIGURE 11.8 Mass Annihilation of Baryons   The universe was populated with protons, neutrons, antiprotons, and antineutrons in *almost* equal numbers until the energy density became too low to make new ones whenever a pair, such as a proton and an antiproton, annihilated. But if there was a slight asymmetry of baryons over antibaryons to begin with, then some excess baryons (protons and neutrons) will be left over after the annihilation episode.

After 10^{-5} second or so, however, only the second reaction could occur. You could destroy a baryon and an antibaryon, but there wasn't enough energy in colliding photon pairs to create any more baryon–antibaryon pairs. So, together with the reaction

$$n + \bar{n} \rightarrow \gamma + \gamma \ , \qquad \text{(eq. 11.9)}$$

the universe annihilated all available pairs of baryons and antibaryons, and the vast majority of the normal matter in the universe vanished forever.

Luckily for us, the previous episode of baryogenesis had created a tiny excess of baryons over antibaryons. Without any antiparticles to annihilate with, the excess baryons became the only normal matter in the universe. Thus, despite the reactions in Equations 11.8 and 11.9, which initially provided for equal numbers of baryons and photons, the sweeping annihilation of baryons and antibaryons left only the few excess baryons to populate the material world. (See Figure 11.8.) Measurements of the abundances of the light elements produced by BBN place the remaining baryons in a ratio of almost *two billion* photons to every *one* baryon. The fact that we observe so few baryons relative to photons certainly seems to require that something like "baryogenesis followed by mass annihilation" must have happened, since reactions such as Equation 11.7 would have created far more baryons than just those few. Still, it's enough, and here we are—baryon-based life forms—alive and well to prove it!

We should mention that this annihilation process applies to normal baryonic matter and, a little later on, to electrons, for which there's an asymmetry similar to that for baryons. Weakly interacting species, however, including neutrinos and WIMPs, have far less frequent collisions. Researchers' calculations imply that neutrinos, WIMPs, and their respective antiparticles should have survived the early universe without annihilating and are still out there today.

The Universe Imprisons Baryons but Liberates Neutrinos

During the universe's first second, a series of weak nuclear reactions interchanged protons and neutrons back and forth. (Refresh your memory with Exercise 9.5 and particularly with the discussion of BBN in Section 10.3.) For example, one of these two-way reactions was

$$\nu + n \longleftrightarrow p + e^- \ . \qquad \text{(eq. 11.10)}$$

This type of reaction gets increasingly rare as the universe expands, because all the particles separate more and more, and because the redshift is lowering the temperature.

By the time the universe is 1 second old, proton–neutron interchanging reactions like Equation 11.10 are being brought to an end by the expansion. This not only fixes the ratio of protons to neutrons, but also causes neutrinos, which interact via the weak force, to *freeze out*. That's cosmologist-speak for "Neutrinos cease to interact with anything." In other words, they *decouple* from the rest of the contents of the universe and proceed to travel freely. These neutrinos, as well as their antineutrino counterparts, continue to fly unhindered. All together, they form a "cosmic neutrino background," sometimes called the *relic neutrino background* (RNB), and we know a great deal about it, just as we do about the photon CMB. If astronomers ever manage to measure it, it will be a great achievement for cosmology, giving us observational details from the first second!

While neutrinos retire from the world of normal matter, the normal matter becomes extremely active. Electrons and their antimatter counterparts (positrons) annihilate at this point, just as baryons and antibaryons did earlier. The slight matter-over-antimatter excess leaves as many electrons as there are protons. This annihilation takes place once the temperature of the universe drops below the electron's mass energy, $m_e c^2$, the point below which the universe can no longer afford to make any more electrons or positrons.

Meanwhile, the baryons respond to the new temperature regime. As the expansion cools the universe enough to fuse nuclei like deuterium without their breaking apart again, the period of big bang nucleosynthesis begins. According to details you've already studied in Chapters 9 and 10, the universe becomes populated with hydrogen-1, helium-4, and traces of other light elements. The expansion, cooling, and absence of stable mass-5 and mass-8 elements combine to bring BBN to an end—and then, for the first time in this detailed cosmic history, the universe has a relatively long wait before anything substantially new happens!

The Universe Sees Matter Overthrow Radiation

It's approximately 72,000 *years* before the next major transition. The last transition we discussed took place early in the first hour, and the seven we discussed before that all took place in the first second. At the 72,000 year mark, an event we examined in Section 11.1 takes place: The formerly radiation-dominated universe becomes matter dominated. The type of "domination" we're referring to here is gravitational; gravity is caused by spacetime curvature, and that curvature is caused by *matter and energy*. The transition to matter domination occurs when the expansion has sufficiently weakened the energy in light (radiation) by the cosmological redshift, leaving the matter in control.

The importance of gravitational domination by matter is as follows: Galaxies naturally tend to assemble themselves. That is, galaxies are made of matter—both dark and normal—and this matter holds itself together by gravity, so that its stars don't just fly away, for example. Before matter domination, it was impossible for galaxies to form, because the "gravity of light" was much stronger than that in matter. That is to say, the energy density of light caused bigger warps in the shape of space than the relatively smaller energy density of matter did. Therefore, wherever there was an excess of radiant energy, there would also be a gravitational dip in space, so matter would be pulled there, too. And since light zips around at the speed of—well, light—there's no way to take a "lump" of light and "put it somewhere." Thus, galaxies and other structures can't form until the era of matter domination begins.

The Universe Imprisons Electrons but Liberates Photons

At an age of 379,000 years, expansion had cooled the universe down to the point that it was no longer too hot for neutral hydrogen atoms to exist. Formerly separated from nuclei, the electrons now begin to combine with the nuclei and make normal atoms. The situation is the same for hydrogen-2, hydrogen-3, helium-3, helium-4, and lithium-7. Neutral gas, rather than ionized plasma, comes to occupy the universe.

The transition from free electrons to bound ones makes the universe transparent to light. Photons now travel freely across space for the first time, without any charged particles in their way. As the photons travel, they become more and more redshifted, and eventually one species in the Milky Way Galaxy observes them and names them the cosmic microwave background!

The Universe Builds Objects and Changes Its Mind about Electrons

Over the next few hundred million years, the dark matter halos assemble by gravity. As a collection of dark matter grows more massive, its gravity gets stronger, and it pulls in as much material as it can, while neighboring halos do the same. Baryonic matter is pulled in, too, and clouds of gas accumulate inside these early halos. Within the clouds, massive stars form by the same mechanism: Gravity pulls in as much gas as it can "reach." Figure 11.9 depicts these structure formation processes.

The first stars probably "turned on" only a handful at a time—say, 5 or 10 per halo. These early-universe halos were comparable to the physical size of a globular cluster (Chapter 3), rather than an entire galaxy halo. In any case, the first stars were "metal poor," because there were no elements beyond those few produced by BBN. Models of such stars indicate that they would have been extremely massive, hot, and bright. Their ultraviolet light, in particular, was sufficient to reionize hydrogen atoms, thereby liberating a fleet of electrons and partially undoing some of what the recombination process had accomplished just before the CMB formed. This is the same *reionization* event discussed in Section 10.3 that left its mark in the polarization of the CMB.

As the billions of years rolled onward, structures continued to form. Halos grew. More baryonic matter entered these halos to form galaxies. Within each galaxy, new stars formed, lived, and died, and then more stars repeated the cycle. Supernova explosions enriched the interstellar medium of galaxies with all the elements beyond hydrogen and helium. Meanwhile, galaxies and their halos gravitationally swarmed and merged together, making massive galaxy clusters. Galaxy collisions within clusters transformed some spiral galaxies into ellipticals, and as each cluster gathered itself together, a larger pattern of superclusters, walls, and voids began to emerge.

In other words, the universe organized itself into what we observe today. We'll fill in some more of the details of this process in the next chapter.

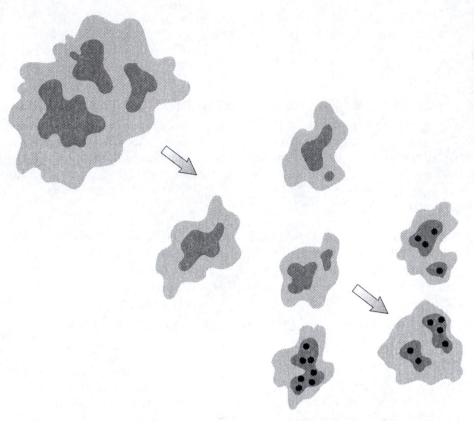

FIGURE 11.9 Structure Formation   Dark matter halos (clumps of dark matter held together by their own gravity) gravitationally pull in clouds of baryonic matter (hydrogen and helium gas). In time, the baryons clump together, too, enabling pockets of particularly compressed gas to form stars. On larger scales, you can imagine networks of these halos and galaxies beginning to assemble into clusters, walls, and voids (more on this in Chapter 12).

The Universe Sees Dark Energy Overthrow Matter

At about 9.5 billion years old, our universe ceased to be matter dominated, and the large-scale structure formation processes just described were brought to an end. The cosmic density of matter decreases over time, since the amount of matter remains the same, but the volume of space continually expands. Dark energy, by contrast, appears to behave differently, such that as the universe grows, the amount of dark energy grows with it. Therefore, it was only a question of time before matter, exotic and baryonic alike, lost its domination to dark energy.

On the basis of our present understanding of the universe and its contents, there is no reason to expect this situation ever to change again. Once dark energy comes to power, it never relinquishes that power. When matter was the dominant gravity source on large scales, the expansion was slowing down accordingly, since the gravity of matter tends to pull things together. Matter therefore battles the expansion. But dark energy's gravity is reversed; it repels, like an "antigravity." Ever since dark energy domination began, the expansion of the universe has been speeding up. In the next section, we'll look at the bleak future this implies. But if you need to have your mind set at ease *right now*, we offer you the following glimmer of an upside: By the time dark energy really messes up the universe you know and love, you'll be much too dead to care.

EXERCISE 11.2   Ancient History

The events described in this section have mostly been distant and ancient, with periods of transition from one type of force to another or one type of gravitational domination to another. To help this history feel more personally meaningful, try to invent an analogous history of yourself. For all 11 of the milestones we just presented as subsections, come up with a corresponding milestone in the history of your existence. Start as early as you like (conception? birth? arrival somewhere?), and make the milestones in your life match up to the universe's milestones in as much detail as possible. For example, most of the milestones in the history of the universe happened very rapidly, very early on. Some milestones are distinct transitions that change the *behavior* of the universe, while others change the *appearance* of the universe. Some changes are permanent, others not. You should select milestones in your life that mimic these types of details.

11.3 Destiny of the Universe

We now know how our universe got to its current state. But what does the future hold? Will space keep expanding forever, or will the expansion slow down, stop, reverse, and recollapse? Before dark energy reasserted itself into the awareness of cosmologists, these questions of *destiny* were thought to only depend on total *density*.

Without dark energy, $\Omega_{total} > 1$ meant not only that the universe was closed geometrically, but also that it would eventually stop expanding and recollapse, just as a ball thrown up into the air turns around and falls back to Earth under the influence of gravity. Similarly, $\Omega_{total} < 1$ meant that the universe would expand forever, like a rocket launched from Earth with enough speed that it overcomes Earth's gravity and never returns. And $\Omega_{total} = 1$ meant that the universe was exactly at the boundary of these two possibilities, expanding just barely fast enough that the expansion would keep slowing down, but never reverse (analogous to a rocket moving precisely at escape velocity from the surface of Earth).

But all that was valid only until the discovery of dark energy. In 1998, two separate teams of astrophysicists independently announced the results of their research. Each team discovered, to the surprise of just about everybody in the field, that light from distant supernovae was redshifted less than expected. If the redshift, which indicates how much the universe has expanded since the supernova went off, is smaller than expected, then the universe must have been expanding less in the past than it is now. We saw what this means in Section 7.4: The expansion of the universe was slower in the past and therefore must be speeding up.

In other words, the supernova research measures the acceleration of the universe, an effect of dark energy. In detail, the researchers can connect the observed rate of acceleration to the amount of dark energy Ω_{de} in the universe. Thus the research also constrains the amount Ω_m of matter in the universe, because matter

*Saul Perlmutter (born 1959) is a researcher at the Lawrence Berkeley National Laboratory in California and head of the "Supernova Cosmology Project," which was the first to announce evidence for an accelerating universe from distant supernovae. Another team, called the "High-z Supernova Search Team" and led by **Brian Schmidt** (born 1967) of the Mount Stromlo and Siding Spring Observatories in Australia, obtained the same results favoring universal acceleration. Schmidt's team was able to confirm that type Ia supernovae could in fact be accurately treated as standard candles (Section 3.2). The team announced its results very quickly after Perlmutter's team did, thus validating the remarkable discovery of both teams.*

opposes dark energy, its attractive nature tending to slow down the expansion. Therefore, the supernova research constrains the values of both Ω parameters. One team reported these constraints in the following way (bracketed comments added for clarity):

> "Depending on the method used to measure all the spectroscopically confirmed SNe Ia [type Ia supernovae] distances, we find Ω_{Λ} [i.e., Ω_{de}] to be inconsistent with zero at confidence levels from 99.7% to more than 99.9%. Current acceleration of the expansion is preferred at the 99.5% to greater than 99.9% confidence level. The ultimate fate of the universe is sealed by a positive cosmological constant [dark energy]. Without a restoring force provided by a surprisingly large mass density (i.e., $\Omega_{m} > 1$) the universe will continue to expand forever."
>
> —Adam Riess et al. (High-z Supernova Search Team), "Observational Evidence from Supernovae for an Accelerating Universe and a Cosmological Constant."

As you can see from this quote, the supernova results help cosmologists narrow down a range of viable Ω values for the universe. But they do more than that: They also help to identify how the connection between the *density* and *destiny* of the universe is changed by the presence of dark energy.

The gravitational effect of matter is to slow down the expansion of the universe, and the gravitational effect of dark energy is to speed up the expansion. So even if $\Omega_{total} > 1$ (so that the universe is geometrically closed), if dark energy is a significant fraction of that density, then the universe will *not* recollapse. In Sections 7.4 and 11.1, we discussed the evidence that dark energy is in fact the largest contributor to Ω_{total}. Current evidence points to a value for Ω_{de} of about 0.73, out of $\Omega_{total} = 1.0$; matter makes up the remaining 0.27.

Figure 11.10 is a graph with Ω_{m} on the *x*-axis and Ω_{de} on the *y*-axis. Since our universe ought to have a single value for each, our universe can be identified by a single point on the graph, whose coordinates *are* those Ω values. The supernova data are displayed as a "strip" running diagonally uphill across the graph; this gives the range of Ω values that the universe can have in order to be consistent with the supernova results. In other words, the supernova data (assuming that they're correct) rule out all combinations of coordinates on the graph (all properties of the universe) outside of that upsloping strip.

Two other, similar data strips are shown as well: one based on observations of galaxy clusters and one based on measurements of the CMB. Notice that all three strips converge in a particular region of the graph, and *only* in that region. This convergence didn't have to happen; if, for example, one of the data sets were totally wrong for some reason, then it wouldn't necessarily cross the other two strips exactly where they cross each other. But they all *do* meet in one place, and this is how we determine Ω_{m} and Ω_{de} in our universe: on the basis of the combination of various types of data that are all mutually consistent with these values.

The nature of the universe—how it expands or accelerates, and its global curvature—depends just on the Ω values. Once those values are known, cosmologists can calculate the behavior of the universe, using the Friedmann equation and a few others. The Ω values for our universe suggest that it will expand eternally, growing colder and less dense at an ever-accelerating rate. Figure 11.10 also shows the relationship between the density of matter, the density of dark energy, and the fate of the universe by indicating boundaries on the graph between universes that acceler-

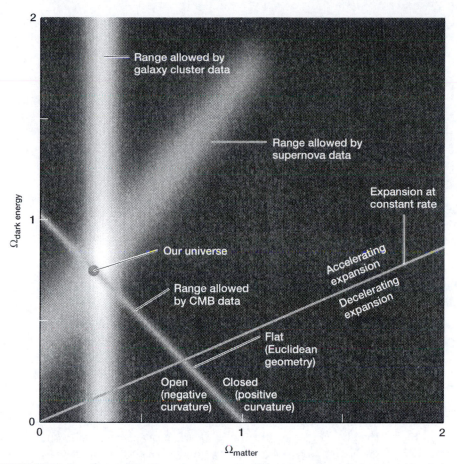

FIGURE 11.10 Density and Destiny In a universe with both matter and dark energy, the relative amounts of each determine our fate in the distant future. Those relative amounts are shown on this graph of Ω_{de} versus Ω_m. Permissible properties of our universe, according to three major cosmological data sets, are shown as thin strips on the graph: acceleration data from type Ia supernovae, data on the motions of galaxies from studies of galaxy clusters, and cosmological data extracted from the CMB. Importantly, all three data ranges intersect in one region of the graph; that region evidently describes our actual universe. In addition, lines are drawn in to indicate the borderlines for global geometry (open, closed, or flat?), acceleration (or deceleration?), and destiny (expand forever or recollapse?) (◆ Also appears in Color Gallery 2, page 28.)

ate vs. decelerate, universes that expand forever vs. recollapse, and universes that are flat, open, and closed. It's definitely worth spending some time studying this figure!

Since dark energy is accelerating the rate of expansion, each collection of galaxies will grow increasingly isolated from all the others. For us, the light from all galaxies beyond our own neighborhood will be redshifted more and more, and the number of galaxies inside our cosmological horizon will decrease, until eventually only nearby galaxies will be visible to us at all. To highlight the contrast with our present experience, think back to the Hubble Ultra Deep Field image (Figure 2.9), with thousands of galaxies visible in a region of the sky much smaller than that occupied by the full moon. Our ultradistant descendants might be lucky to see *one* galaxy in a region that same size. But how far into the future this will happen and whether more extreme events might await our descendants depend on the unknown details of what dark energy actually is. We can speculate on the possibilities; here are a few popular ones in cosmology today:

POSSIBILITY #1: The dark energy is a cosmological constant.

In this case, the dark energy density will remain forever at its current value, because it is an intrinsic property of space itself, like vacuum energy. The space between dense clusters of galaxies will expand at an ever-accelerating rate, because on that scale (greater than about 10 million light-years) the density of matter will continue to fall off as space expands. Thus, the cosmological constant will always dominate the dynamics of the universe outside of clusters.

Inside dense clusters, the matter density is relatively constant and is much higher than the average density throughout the universe. So, within these concentrated regions, matter will always dominate over the cosmological constant, but such clusters will become increasingly cut off from the rest of the universe.

In this scenario, all galaxies outside our local neighborhood will have moved beyond our cosmological horizon by a few trillion (10^{12}) years from now. Within this isolated "universe" of nearby galaxies, and everywhere else, stellar activity will eventually run down. Stars will die out and become white dwarfs, neutron stars, or black holes. With most of the material trapped in stellar embers, no new stars will form after about 10^{14} years from now. Eventually, even fundamental particles and black holes will probably decay, leaving nothing but radiation spread nearly uniformly throughout a vast, cold, and almost empty universe.

POSSIBILITY #2: The dark energy is "quintessence"—energy associated with an unknown particle.

Rather than being energy intrinsic to the vacuum of space itself, the dark energy might be associated with the presence of an unknown particle. This general idea has been given the name "quintessence," after the Greek for "fifth essence." If it turns out that dark energy is quintessence, then its energy density can vary with time, rather than remain constant like vacuum energy. It may remain roughly constant for a certain period, mimicking a cosmological constant and causing the universe to expand at an accelerated rate for a while. But since the strength of the quintessence field can change, its energy density may increase or decrease at some point.

If the dark energy density increases with time (as it might, according to some models), then it will grow to exceed the matter density in smaller and smaller regions, and even regions that are now gravitationally bound (such as our Milky Way or even our solar system) will eventually be ripped apart by the repulsive influence of dark energy. If the increase in dark energy density continues indefinitely, it could eventually even rip atoms apart as the dark energy density grows to dominate smaller and smaller regions, no matter how the material in those regions is held together. (Don't worry about your atoms being ripped apart; you'll be ripped from the Earth long before that ever happens.)

If the dark energy density is decreasing with time, then eventually matter could dominate the density of the universe again (though with a lower density than now, because the expansion is spreading out the matter in the meantime). In addition, it's possible that the dark energy density will eventually become negative. A negative dark energy density is an attractive force (like normal matter) and would halt the expansion at some point in the future. In that case, our universe will recollapse, returning to a hot, dense state like that from which it came.

So if the dark energy is quintessence, the destiny of the universe depends on its detailed behavior. This means that we can't predict with certainty what the fate of our universe will be.

POSSIBILITY #3: The dark energy is part of a recurring cycle in a universe with extra dimensions.

We learned in Section 4.3 that the matter we see makes up only a tiny fraction of the contents of the universe. Similarly, maybe the 13.7 billion-year-old region of space we've been referring to as "our universe" is only a tiny fraction of all that there is. As we discussed in Section 11.2, current theories don't tell us what happened before the Planck time, 10^{-43} second after "time zero." So we know only that the universe was very hot and dense prior to that. It's possible that the universe did not actually begin in a state of infinite density at that time. Rather, we may be in the latest cycle of a universe that repeatedly returns to the conditions of the Planck era.

This idea is the basis for a cyclic model proposed in 2002 by Paul Steinhardt of Princeton University and Neil Turok of Cambridge University. Extrapolated from the physics of string theory (Section 11.2), the idea is that our universe of three spatial dimensions (plus time) is embedded within a higher dimensional space, along with at least one other 3-D universe like ours. Another universe could be right next to us, perhaps much less than a millimeter away, but in a "hidden" direction we don't know how to point toward.

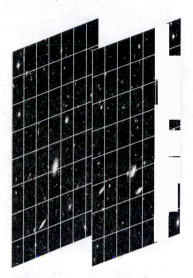

FIGURE 11.11 Cosmology on Your Brane Our universe may be a 3-D version of one of the 2-D branes illustrated here. Our brane may interact with other nearby branes, possibly causing major changes or events inside our brane, such as initiating a big bang!

Like most things related to higher dimensional spaces, this idea is much easier to visualize if you reduce the number of dimensions. So imagine that our universe is a 2-D surface like a sheet, embedded in a 3-D space along with another sheet (essentially another "universe"; see Figure 11.11). In this picture, the universes are reminiscent of membranes. In general, the lower dimensional universes embedded within a higher dimensional overall universe are referred to as branes for short. (We could call them "membranes," but that word is reserved for 2-D surfaces, and in general, branes can have more dimensions. Ours, if we really live on one, has three spatial dimensions.)

Within the context of string theory, researchers can estimate how the branes should interact. Under reasonable assumptions, the result is that they come together and move apart at regular intervals. When they collide, the infusion of energy from their interaction produces the hot, dense early state of a "big bang" inside one or both branes.

This theory is relevant to our discussion here because the dark energy arises naturally in the theory, as a result of an interaction between branes as they get very close together. And it offers the possibility of explaining other mysteries of cosmology that we'll discuss in Chapter 12. Plus you get to keep a straight face when you make statements like "The expansion of space may be accelerating due to the interaction of our brane with another brane that lies several Planck lengths away from us in the direction of a fourth spatial dimension." Try that one the next time you want to strike up a casual conversation with a stranger!

The cyclic model is still quite speculative; even the string theory it emerges from is not experimentally tested, well-established science. So the specifics of the model are susceptible to future details of string theory. Various string-theory-based models (with no "time zero") have been proposed. Each predicts subtle differences in the structure of CMB anisotropies—differences that might be detected by future observations. So we'll have to wait and see.

Pondering Infinity and the Shape of Space

Does the universe have an edge somewhere, or does it extend forever? Or is it finite, yet unbounded (like the surface of a sphere)? And what about time? Did the universe have a beginning, and will it have an end? The cyclic model of the universe renews the possibility that even within the basic framework of the big bang theory, our universe may be much more than 14 billion years old. Ours may be just the latest in a long history of repeated expansions.

Infinity itself is a counterintuitive concept. If you start with an infinite set of things, it's possible to add more things to the set and still be left with something just as infinite as the original. Imagine a number line, infinite in length to the left and right. Now multiply every number on the line by 2. The line has expanded (every segment of the line that you could look at is now a factor of 2 longer), but it's still the same infinite set of points, no bigger than the original. This is one way to see that if the universe is infinite, it can expand without necessarily expanding *into* anything, as we discussed at the end of Chapter 7.

If our universe is indeed flat, as all evidence indicates, does this mean that it is infinite in extent? General relativity allows us to calculate the local curvature of spacetime on the basis of the local distribution of matter and energy. You learned in Chapter 7 that because the universe is homogeneous and isotropic on large enough distance scales, we can also use general relativity to figure out the overall average curvature of space. In this chapter, you saw that observations point to a flat universe overall. This conclusion brings to mind a universe that is infinite in all directions, the 3-D version of an infinite, flat sheet of paper.

But there are other possibilities for the shape of space—or **topology**—that are still consistent with the flat space required by observations and general relativity. We've actually already seen one of the simplest such shapes back in Chapter 1 (Figure 1.3). In this virtual universe of the "Asteroids" video game, space is flat (e.g., the shots you fire from your spaceship travel in straight lines), but not infinite. (If you move off the screen to the right, you reappear at the corresponding point on the left side of the screen.) Analogously, our universe could be flat, but interconnected in such a way that traveling far enough in any direction will take us "off the screen" and back onto the opposite side. The transition would be continuous, just as in the video game.

Or if you're not familiar with this type of video game, then consider the following inspiration for a simple and famous example:

> "Your idea of a donut-shaped universe is intriguing, Homer. I may have to steal it."
>
> —*Dr. Stephen Hawking (cosmologist), appearing as himself on* The Simpsons.

Imagine an ant walking without turning on the surface of a donut. No matter what direction the ant walks, his path will eventually lead him back toward where he started. So, in a donut-like universe, we would never run into an edge or boundary; just a recurring sense of déjà vu.

If we *are* living in a 3-D version of a video game or a donut's surface, it should be possible to find out by looking at the universe on large scales. Looking farther away than the size of the 3-D "screen" (or the size of the donut) is really looking at the same things over again, so we should be able to see patterns repeat. In 2003, hints of this pattern were found in the WMAP all-sky CMB map. Since the CMB map shows light from extremely distant parts of the universe, it gives us a good

FIGURE 11.12 Soccer Ball Universe In our hypothetical "Asteroids" video game universe, in which the topology generates connections between otherwise widely separated regions of space, one could travel without turning and yet continue to pass the Earth many times over.

large-scale view in which we can look for repeating patterns. The actual analysis is much more complicated than that, and more detailed CMB data may be helpful; it should be coming in the next few years.

Some analyses favor a complicated shape to the universe, in some ways similar to a 3-D version of the surface of a soccer ball. This surface is a dodecahedron—a geometric figure constructed from 12 pentagonal sides that loops back on itself like the "Asteroids" video game screen or the donut, so that leaving through one pentagon-shaped face returns you through another. (See Figure 11.12.) If the universe is in fact analogous to a video game, the implications would be quite amazing. First, it would mean that the universe looks bigger than it really is, similar to the way a large mirror hanging on the wall can give the impression that it's a window into an adjacent room that doesn't really exist. Second, it would mean that the universe is *finite*, which could have a profound effect on our understanding of the cosmos and, more personally, of our place in it.

We'll have to wait for more data and more research on this matter. But if space turns out to have this sort of interesting topology, then it's noticeable just on very large scales, ranging from only somewhat smaller than our cosmological horizon to much, much larger. So if we are living in a video game universe, then its screen must be enormous!

Quick Review

The big bang theory provides a framework for understanding the entire history of the universe. In this chapter, you've learned about some of the key stages in the history of the universe, driven by the expansion of space, which causes the density of matter and radiation to decrease with time. The density of dark energy appears to work differently, and that difference could determine the ultimate fate of the universe.

Try the following crossword puzzle to test your knowledge of key terms and concepts from Chapter 11:

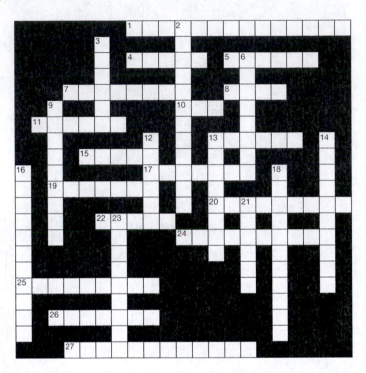

ACROSS

1 The left side of the Friedmann equation includes this number describing the expansion rate.

4 Three of this type of particle make up each baryon.

5 An example of a TOE: _____ theory

7 Just before BBN, the universe became flooded with this type of particle, which should still be out there.

8 Universe in which omega (total) is less than 1

10 Theory that unites the strong and electroweak interactions

11 Size of some 3D part of the universe

13 Universe with positively curved global curvature

15 Density parameter symbol

17 Another term for the future, or fate, of the universe

19 Only in a flat universe can one use ordinary Euclidean geometry to calculate the distance to an object of known physical size by measuring its _____ size.

20 If dark energy takes the form of a cosmological constant, then the amount of dark energy inside some volume _____ as that volume increases.

22 Hypothetical "surface" embedded in a space with more than three spatial dimensions

24 Scientists can't accurately describe the universe before this moment in cosmic history without a theory of "quantum gravity."

25 Equation linking the expansion rate, curvature, and density of the universe.

26 Amount of something per unit volume

27 If dark energy takes the form of _____, then its density could be changing over time, so the future of the universe is uncertain.

DOWN

2 Creation of the ingredients of the material world, based on a slight bias favoring certain matter particles over their antimatter counterparts

3 Type of energy associated with otherwise empty space; behaves as a cosmological constant

6 The study of the overall shape of the universe

9 After about 72,000 years, the universe became _____ by matter.

12 Free quarks ceased to exist after the early universe's quark–_____ transition.

13 Density of a universe that is geometrically flat

14 What the universe is dominated by today

16 Wildly curved spacetime resulting from ultrahigh-energy particles in the very first moments of the universe's existence

18 Transition during which WIMPs and neutrinos ceased to interact significantly with normal matter

21 Class of models of the universe in which the big bang is a recurring process, rather than an absolute beginning

23 The component of the early universe that dominated all others in terms of gravity, or how it affected the shape of space

Further Exploration

(M = mathematical content; I = integrative—builds on information from more than one chapter; P = project idea; R = reflection; D = suitable for class discussion)

1. (R) What is the destiny of the universe? Write a few sentences summarizing your understanding of what will happen to our universe in the distant future.

2. (MI) Suppose there is another universe whose expansion rate right now is the same as ours: 71 (km/s)/Mpc. But in that other universe, $\Omega_{de} = 0$ and $\Omega_m = 25$. Five billion years from now, how will the expansion rate in that universe compare to ours? (Will it be bigger than ours, smaller than ours, or about the same?) Briefly explain your reasoning.

3. (M) Looking out the window of an airplane, you see a soccer field on the ground below. Assuming that the field is 100 meters long, how far from it are you if the length of the field occupies 1/2° (the angular diameter of the full Moon) in your view? Give your answer in meters.

4. (MI) The CMB photons have been redshifted by a factor of almost 1100 since they were freed to travel through mostly transparent space. (a) Calculate how much higher the average density of matter in the universe was at the time the CMB photons were released, compared to the average density of matter now. (b) Do the same for the *number* density of the CMB photons. (c) Do the same for the energy density of CMB radiation. (*Hint:* Imagine a volume of space as a box, and think about how much the length, width, and height have changed. Also, think about how much wavelengths have changed.)

5. (PD) Here's a big project: Research the emerging field of string cosmology by reading "The Myth of the Beginning of Time," by Gabriele Veneziano, *Scientific American*, May 2004. Use the references in the article and any others you find elsewhere (online, etc.) to write a paper or give a presentation on this topic. In addition to covering the main ideas of string cosmology, specifically address the following topics:
   • What mysteries of cosmology does string cosmology help to explain?
   • In what ways would the big bang theory, as presented in this book, be affected? For example, are the BBN and CMB results wrong or misleading in some way? Are galaxy redshifts in trouble?
   • How does string cosmology alter our understanding of what is meant by the "beginning" of our universe?

6. (MR) What does the following equation refer to? What is the significance of the particular redshift z that solves the equation $\rho_{m,0}/(1 + z)^3 = \rho_{r,0}/(1 + z)^4$? (*Hint:* Connect the quantity $(1 + z)$ to the scale factor a, and think about how the scale factor relates to energy density. Refer back to the text, exercises, and figures of Section 11.1.)

7. (P) Create a piece of artwork of some sort—a poster, a computer image or animation, etc.—on the history of our universe outlined in Section 11.2.

8. (RD) In your own words (a few sentences), describe the Planck era.

9. (RD) In your own words (a few sentences), explain how the early universe undergoes *transitions*, such as the GUT, electroweak, quark–hadron, or BBN. Don't go into the details of each transition; instead, what are the common features among them all? What changes trigger them? How are things different before and after one of the transitions?

10. (MRI) How does general relativity imply that the universe, and therefore time itself, had a definite beginning? Be as specific as you can: Think about the Friedmann equation and the scale factor, and how quickly it grew during various periods of cosmic history.

11. (RP) Speculate on how a string/brane approach might soften the "sudden beginning" described in the previous

question. Answer in just a few sentences. (There are many popular science articles on this topic on the Internet in case you want some help with your answer. However, since popular science articles are not always correct, make sure you look at more than one, just to corroborate the information you're getting.)

12. (RD) Use Figure 11.10 to answer the following questions about the properties of our universe:

(a) Find our universe on the graph, and use the three intersecting "data strips" to come up with an approximate range of values allowed by the data for Ω_m and Ω_{de}. What shape would your range of values look like on the graph?

(b) Suppose that someone claims that their new measurement of Ω_m is 0.9, and that Ω_{de} is still around 0.7. What sort of universe would that imply?

(c) Suppose that new evidence suggests that the data from type Ia supernovae (the upsloping data strip) are all wrong for some reason. How would that change the location of our universe on the graph? How would it change the size of the range of values allowed—the "spread" you found in part (a)?

(d) What region on the graph corresponds to a universe that's expanding with no acceleration or deceleration and no curvature? How well do the data match up with this possibility?

13. (RD) Try to explain why each of the three "data strips" shown in Figure 11.10 is sloped the way it is:

(a) The data from galaxy clusters are based on the motions of galaxies within the clusters. Why do these data appear in a vertical strip? (*Hint:* That it occupies a vertical strip means that it spans all possible values of Ω_{de}, with no preference for any particular value.)

(b) The data from the CMB are essentially a reflection of the CMB-based measurement that Ω_{total} is approximately equal to 1. Why does this data strip slope downward between the coordinates (0, 1) and (1, 0)?

(c) The data from type Ia supernovae measure the acceleration of the expansion of the universe. Why does this data strip slope upward?

The Story of Structure

"From so simple a beginning endless forms most beautiful and most wonderful. . . ."
—Charles Darwin, The Origin of Species

The big bang theory provides a context within which to work. It consistently fits together core observations that are very difficult to explain in any other theory yet proposed. But having the framework is a long way from being able to explain everything and understand in detail how it all happened. The situation is similar to realizing that matter is made of atoms, but not knowing how each atom behaves in every instance. The framework is important and a great achievement, but we still have much to learn.

As we discussed in Chapter 1, just because some details come into question does not mean that the basic model is in danger of falling apart. In the practice of science, the controversies and the details often get the most attention, but it's important to keep in mind that those debates usually occur *within* a solid, time-tested framework. A controversy about the specific properties of one fundamental particle (the neutrino, say) would not call into question the very idea that the universe is made up of particles. Similarly, a disagreement about the details of how structures such as clusters of galaxies formed does not call into question the basic idea that they formed in an expanding, cooling universe.

In fact, one of the benefits of having a solid theoretical framework is that it actually *invites* questions that might not have occurred to us without it. The big bang theory raises many new and interesting questions that guide research and investigation: If the universe is expanding now, what will happen in the future? Will it keep expanding forever, or collapse someday? Is the expansion rate speeding up or slowing down? How do structures coalesce out of material in an expanding space? How were the structures we see today (from superclusters, to galaxies, to stars and planets) formed out of the nearly uniform hot early universe? How were the conditions at the beginning of the universe (the strength of the fundamental forces, the kinds of particles that could exist, the rate of expansion, etc.) determined? Why were these initial conditions suitable for life to evolve in the universe billions of years later?

The story of structure formation is our focus for this chapter. As advertised, we'll start with the big bang theory, which holds that the contents of the universe were once crammed together in a hot plasma. The CMB reveals that this plasma was remarkably isotropic and uniform. Everywhere in the universe was hot and dense and very much the same as everywhere else.

The universe today is far from uniform. We see a galaxy of a hundred billion stars in one place, a cluster of a thousand galaxies in another region, and almost nothing in between. Galaxies and clusters constitute **structure**. On larger scales, walls, voids, and superclusters are aptly named **large-scale structure**.

This chapter covers a great triumph in modern cosmology: the story of how the universe went from an almost perfectly uniform beginning to its highly structured form today. The smooth early universe revealed by the CMB:

became the lumpy modern universe revealed by galaxy surveys:

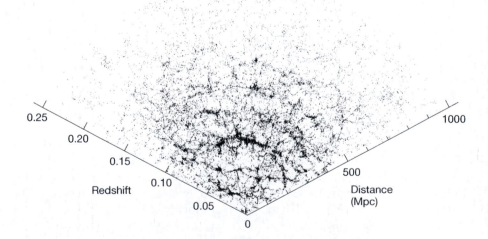

This story covers some beautiful science. If you take the time to carefully absorb the fine details in the story—where the information is hidden in the universe and how our species uncovered it—you'll have no choice but to appreciate its beauty yourself.

In the first two sections of this chapter, we'll investigate the meaning behind all the spots in the CMB and how they spawned the growth of structure in the universe. In the third section, we'll look at how that structure actually emerged. In the last two sections, we'll ask about how the whole process got started, in the earliest moments of our universe.

12.1   Primordial Harmonics

As you saw in the last chapter, the various densities of the universe—of radiation, matter, and dark energy—change over time with the scale factor in a way that you can calculate. At the time of decoupling, when the CMB formed, the universe was matter dominated. We will therefore track the motions of matter in order to understand how the CMB imaged the seeds of structure in the early universe. Just as the light you see tells you about the *objects* that emitted it (e.g., the Moon or the headlights of an approaching car), astronomers use the CMB light to learn about that matter that emitted it. First, we'll try to figure out how to interpret what exactly the CMB is showing us. But you should know right up front that studying the seeds of structure in the CMB is like watching a movie with many little side stories about a lot of different characters: It doesn't all come together until the very end.

CMB anisotropy tells us that the early universe was not perfectly uniform. Rather, it contains spots of various sizes, which we can see when we crank up the contrast in a microwave sky map. For the time being, we will not bother with the actual origin of these disturbances; that will come later in the chapter. For now, we need only acknowledge that the early universe had very weak disturbances.

"Disturbance" is an appropriate description, because changes in one part of the early universe would soon push into other parts. Consider a CMB "hot spot," a blackbody emission whose source region had a slightly higher temperature than its surroundings. Think back to what this means on a particle level: A higher temperature means faster particles. "Blackbody" means that particles collide and thereby redistribute energy. Taking both conditions together, we can conclude with confidence that a hot spot in a particular location is a temporary thing. Regardless of how a region acquires its extra energy, its particles push into surrounding regions and deplete that excess. Then someplace *next to* the original hot spot will become the new hot spot. Then the process repeats, and the energy is transferred to the next region down.

When a particular region heats up, it's because it's being pushed by faster moving particles from neighboring regions. The push *compresses* our region of interest, but because the heating process really means an increase in particle speeds—at the expense of the particle speeds in neighboring regions—there is a rebound. Any region that compresses and heats up subsequently pushes outward again. It *has* to: Its particles are moving fast and they have to go *somewhere*. Our region expands, exchanges energy with surrounding regions, and therefore cools down again. Now the neighboring regions are again compressed; when they reexpand, our region will recompress, and so on.

This sort of recurring compression–expansion–compression process happens all the time here on Earth. When air compresses and expands, your eardrum vibrates and your brain interprets it as *sound*. You can identify which sound you hear (i.e., which musical note) by how fast the compressions and expansions trade off—that is, by the frequency of the *sound wave*. And what causes the air to start vibrating in the first place? A mechanical disturbance, like a vibrating guitar string or your vibrating vocal chords.

In 1970, P. James Peebles and Jer Yu showed that the early universe must have harbored sound waves. Since the nearly uniform plasma of the early universe had no walls or carpets to absorb the sound, the waves would continue for a long time, until the conditions in the universe changed dramatically.

SPECIAL TERMINOLOGY ALERT: Before going any further, let's be absolutely clear on what we mean by "sound"; otherwise the rest of this section will make no sense and your brain could explode. The important points are as follows:

- By "sound," we mean "an oscillating change in the density of a gas" (either air, or the hydrogen–helium mixture that filled the early universe). The density oscillation travels through the gas, away from whatever disturbance caused it. This is how normal sound actually works in your life, but it still might be better to think of it as a repeatedly changing gas density, rather than something audible. Had you been there in the early universe, you wouldn't have heard anything, because the frequencies (timing) at which the gas density changed were much to low (slow) for your ears to register.
- Cosmologists have no way of *hearing* this sound from the early universe. There is no giant space microphone. Rather, they *infer* the existence of sound waves in the early universe from the patterns they notice in the microwave *light* they observe. As you know, the electromagnetic waves that make up the CMB have been traveling relatively unimpeded since very early times; these light waves were affected by the sounds that existed back then.

Figure 12.1 shows how all this works. The most important thing to realize here is that our CMB observations require that the early universe had sound waves and that these waves were stable features of the universe. Today we *see* them in the blackbody emissions of compressed and expanded regions. As we discussed in Section 8.4, hot spots in the CMB come from actual regions of the early universe with slightly higher than average temperatures. When faster particles cause a region to compress, they also cause it to heat up and shine slightly brighter and "bluer" blackbody light. Expanded regions are cooler and shine redder light. In this way, the hot and cold spots in the CMB were caused by the sound waves of the early universe. The exciting part of all this, however, lies in the details.

Your ear and brain together are capable of understanding multiple frequencies at the same time. A musical chord, such as three piano keys played together, makes a distinctive sound that you can appreciate. In a crowded, noisy room, you can pick out a particular voice to eavesdrop on. When a collection of different frequencies all vibrate the same air at the same time, the result is not just a jumbled mess: All the original frequencies are still there and still resolvable. You might imagine that someone with a talent for music might be able to pick out individual notes in a chord just by listening.

The same is done in cosmology with microwave data. Instead of sounds, we measure a sky full of microwaves. And instead of ourselves listening for particular notes, we program a computer to do it, breaking down the observed microwave pattern into its constituent frequencies. From this breakdown, we learn about the strength of the sound waves at each frequency. This gets us a kind of "sound spectrum" of the early universe—a breakdown of exactly what the matter back then was doing—matter that would later go on to form galaxies and halos. But before we reveal the results, think a moment about what sounds you would expect from the primordial plasma. For example, perhaps you're expecting nothing more than noise:

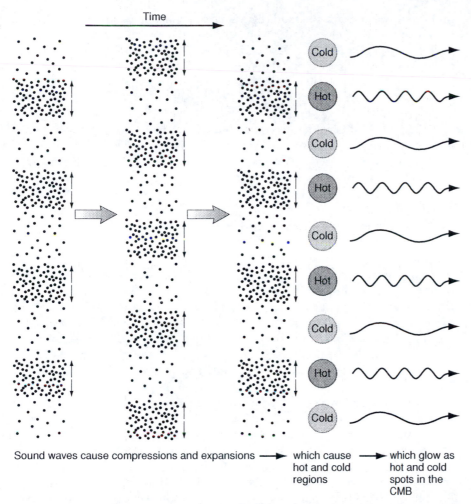

Time

Cold

Hot

Cold

Hot

Cold

Hot

Cold

Hot

Cold

Sound waves cause compressions and expansions ——▶ which cause ——▶ which glow as
hot and cold   hot and cold
regions   spots in the
CMB

FIGURE 12.1 Primordial Sound Waves   This figure shows a time sequence of the alternating compressions and expansions (sound waves) in the same location in space. When the CMB formed, the hotter compressed regions emitted slightly brighter and bluer blackbody light, creating "hot spots" in the CMB. Cooler, expanded regions emitted slightly dimmer and redder light.

a little of every frequency, over some range, with none being particularly special. That expectation turns out to be half right: There does appear to be a blend of every possible frequency over a particular range. But it also appears that the decoupling process which generated the CMB did not capture all of these frequencies equally.

Let's begin to reason everything out by supposing that every possible frequency that could exist did exist, in equal intensities. That is, the universe had no built-in preference for any one type of disturbance over any other; this seems like a reasonable guess. So we'll start with a universe containing minor disturbances of every frequency all over the place, sending sound waves in every direction.

The big bang theory tells us that after a certain amount of time the universe experienced a recombination of hydrogen and a subsequent decoupling of photons from matter (Section 10.3). What's important here is that decoupling was a relatively short-lived *event*: It started, and it ended. Therefore, the photons that were liberated by decoupling, which now make up the CMB, are those which happened to be in flight from various hot and cold regions *when decoupling happened*. In other words, since sound waves cause compressions to come and go, *the CMB should show us only the hot and cold spots that happened to exist at the time of decoupling*, and not all the hot and cold spots that ever existed. The CMB is a snapshot from the early universe, showing us whatever the universe was doing just when that snapshot was taken.

This scenario can be difficult to imagine, so let's try talking about it with an analogy first.

THE ANALOGY: Picture yourself in a room with lamps all around you. But instead of normal lamps, suppose that each one is a special kind of color-changing lamp. It starts out red when you first turn it on, but then cycles smoothly through most of the rainbow: red-orange, orange, orange-yellow, and so on, out to blue. Then it goes back again: blue-green, green, yellow-green, and so on, back to red, and it repeats like this over and over again. The lamps are intended to represent the distant regions of the universe whose light we see as spots in the CMB. The shifting colors represent the way those distant regions cool down and heat up as the primordial sound waves expand and contract them.

Now suppose there are thousands of these lamps around you, but each one has been manufactured with poor quality control, so that the timing is not the same from lamp to lamp. They all rotate through the same colors in the same order, but some do it faster than others. Some take minutes, some take seconds, and some take less than a second to go from red to blue. At any given moment, then, there are a bunch of different colors all around you. The different cycling rates, and therefore the different colors, are intended to represent the primordial sound waves with different frequencies.

You have a camera with you, aimed at a bunch of lamps. You flip the switch on the wall, and all the lamps turn on and begin cycling colors. You put your finger on the shutter button and wait some random amount of time (by humming the full *Star Spangled Banner* and then calculating a 15% tip on a $34.60 restaurant check), and then you press the shutter button. The camera snapshot represents the view of the CMB that we can observe now, which was fixed by what the universe's black-body light looked like right when decoupling took place. The amount of time you wait before taking the picture represents the amount of time between the beginning of the universe and decoupling.

What would your photograph look like? Might it look something like this? (The color version of this figure is in Color Gallery 2, page 22).

As we saw in Chapters 8 and 10, this is the WMAP satellite's map of the cosmic microwave light coming toward us from all around us—an image of the photons that decoupling set free to travel and redshift. The spots, or anisotropies, show us the blackbody light from regions of the early universe that had slightly different temperatures when decoupling happened. And why were there all these regions of slightly different temperature? Because sound waves were alternately compressing and expanding them.

In studying the early universe, it's helpful to ask which types of spots are the reddest and bluest—the maximum deviation from the average. But let's start with the room full of color-cycling lamps. Which lamps will be all the way at the red end of their cycle, or all the way at the blue end, right when you take the picture? The answer depends entirely on the timing, or frequency, of the color cycling. For example, if one lamp started red and just went through the colors to blue right when the camera triggered, then it'll be at the blue extreme. Another lamp that cycles exactly twice as fast would be back to red when the camera triggered. The same is true for three times as fast, and four, and five. The reddest and bluest lamps in the snapshot will be those whose frequency is an *integer multiple* of a particular frequency, but all the frequencies in between produce some in-between color, closer to the average color, like yellow or green.

The lamps are just like the sound waves in the early universe. The average density, or temperature, or color of the universe is false colored around yellow-green in the WMAP image. The most prominent spots, however, are those which deviate sharply from green—the ones that are false colored red and blue. So, *the most prominent spots in the CMB come from regions of maximum and minimum temperature, and not from the regions with temperatures in between.*

Let's take this reasoning a step further. Because hot spots are regions of maximum *compression*, and vice versa, these CMB spots come preferentially from regions where a sound wave achieved maximum compression just at the time of the snapshot (i.e., the time of decoupling). Cold spots similarly come from regions that had just achieved maximum expansion at the moment of decoupling. These regions are the ones where a sound wave of just the right frequency has had exactly enough time to compress the region only once. Or it had exactly enough time to compress and reexpand the region. Or compress, reexpand, and recompress. It's like the situation with the lamps: Only integer multiples of a particular frequency would have maximally compressed or expanded a region right at the time of decoupling, so those are the only regions that show up all the way red or all the way blue on the CMB map. Figure 12.2 shows how the sound-wave frequencies connect to the CMB anisotropies.

Our best measurement of the "snapshot time," from the WMAP mission, is an age of 379,000 years. If a wave has just finished compressing once by then, then the light from that compressed region will stand out in our microwave sky maps, meaning that it will show up as a maximal temperature deviation from the average CMB temperature of 2.725 K. Spots from this first, maximal compression are called the *first peak* in the CMB anisotropy. A wave with double the frequency of the wave that produces the first peak will compress once and reexpand fully in 379,000 years. So the snapshot catches that one at maximal expansion. Such spots constitute the *second peak*. There are subsequent peaks after that, too: The third peak compresses, expands, and recompresses; the fourth expands again after that, and so on. All together, these are called the **acoustic peaks** in the CMB, since they are caused by sound.

The important thing is that if the big bang theory is correct, not only must the CMB exist at all, and as a nearly perfect blackbody with the correct temperature and polarization, but it must also display the correct anisotropies. Sound waves of all

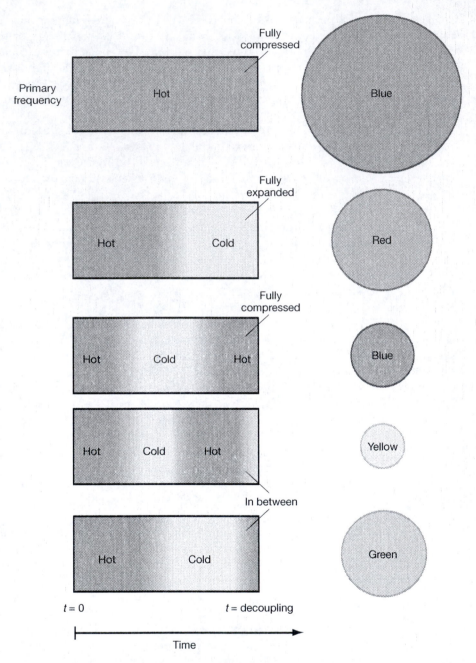

FIGURE 12.2 Sounds and Spots Sound waves of several different frequencies are depicted here, with time running from left (time zero) to right (time of decoupling). Different frequencies compress and expand different numbers of times by the time of decoupling. For the primary frequency, in which there is just one compression and nothing else, the compression leads to one large hot spot. At double that frequency (one compression and then one expansion), the expanded region is relatively cooler. Waves with noninteger multiples of the primary frequency end up somewhere in between compression and expansion, and thus somewhere in between hot and cold. On the WMAP image of the CMB anisotropy, these regions would appear yellow or green. Also shown here, larger frequencies (larger number of cycles before decoupling) have smaller wavelengths and thus generate smaller spots. (Spot sizes are discussed shortly, in connection with Equation 12.1.)

frequencies pervaded the early universe, but only those with the proper frequencies should have generated peaks—or temperature extremes, or spots—on a CMB map.

There's one last connection worth making before we move on. Because the wavelength of a wave is equal to its speed divided by its frequency (Chapter 2), or

$$\lambda = v/f, \qquad\qquad \text{(eq. 12.1)}$$

we can associate every frequency with a physical size: the size of a compressed region, or a CMB spot size. From Equation 12.1, we can expect the peak frequencies (which are integer multiples of one particular frequency) to correspond to wavelengths that are *integer fractions* of one another. For example, the first peak is caused by a wave that compresses, but doesn't yet start to reexpand, by the time of decoupling. The second peak is caused by a sound wave of double the first peak's frequency, and so on. The second peak, then, has half the wavelength of the first peak. To compress *and reexpand* a region—that is, to change it *twice* at a given sound speed v—in the same amount of time (until decoupling) requires that the region be half as big. The pattern continues like this: The third peak changes three times and has a wavelength one-third the size of the first peak's wave.

The importance of the wavelength is that it specifies the size of a CMB spot. Spots caused by the second and third peaks are one-half and one-third the size of first peak spots, respectively, as we saw in Figure 12.2. Since we measure sizes on the sky with angles, this means that spots of the second peak are half as many degrees across as those of the first peak.

All of this is measured in microwave light and analyzed by computer software in great detail. The result is usually shown in the form of a graph, such as Figure 12.3, giving the amount of temperature fluctuation measured for spots of each frequency or spot size. (It doesn't matter which, since they're related by Equation 12.1.) This graph is called the CMB **power spectrum**. Plan to spend some time studying it; it's one of the most important results in modern cosmology, and it's complicated enough

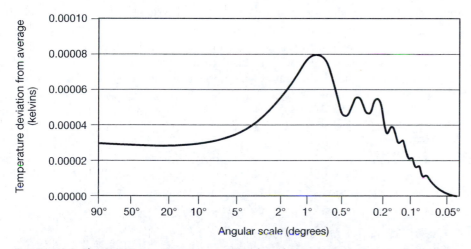

FIGURE 12.3 The CMB Power Spectrum.   This is how the power spectrum is usually plotted, but it's a bit confusing. The plot gives the amount of "temperature deviation from average" in CMB spots (on the *y*-axis) versus spot size (on the *x*-axis). But the *x*-axis is somewhat backwards: Big angular scales (big spots) are on the left, and small ones are on the right. Acoustic peaks are tall points on the curve, with the "first peak" being the tallest, the "second peak" being the next one to its right, and so on. (◆ Also in Color Gallery 2, page 22.)

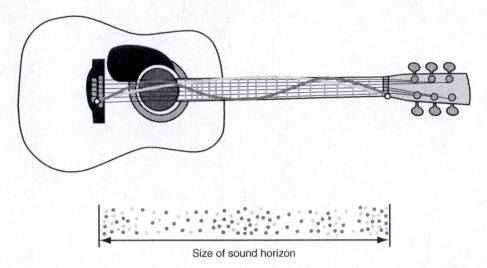

Size of sound horizon

FIGURE 12.4 Harmonics in the Primordial Plasma This figure connects primordial plasma density oscillations with various waves on a guitar string. A harmonic overtone (darker wave and darker spots) is *superimposed* on the primary frequency wave (lighter wave and lighter spots). Notice how cleanly this is handled by varying densities of plasma in various regions, shown here by the number of particles (dots): You can have waves of different sizes in the same region at the same time. In the figure, superimposed sound waves are depicted by having more particles (dots) lined up with a wave's lowest point on the guitar string. At the endpoints where both waves are low, there are lots of both dark and light dots.

that no one just "gets" it with a quick glance. Some of the details will be discussed in the next section, but for now, you should locate the acoustic peaks and notice their angular sizes. Even though this information is plotted logarithmically, you should notice that the first peak occurs at about 1° on the sky; subsequent peaks are indeed one-half, one-third, one-fourth, etc., of that one degree size.

The sound waves of a musical instrument are produced by vibrations, like those of a guitar string. The string can oscillate such that its length is half the wavelength, but it can also oscillate with integer fractions of this half-wavelength, such as $\frac{1}{2}, \frac{1}{3}, \frac{1}{4}, \frac{1}{5}$ of it, etc.

Because the ends of the string are pinned down, which prevents them from moving, the string can resonate only at its primary frequency and at these particular fractional frequencies, which are called **harmonics** (also known as "overtones" or "harmonic overtones"). The "ends" of the *cosmological* "guitar string" are "pinned down" by the fixed distance over which a sound wave could have traveled in 379,000 years—a size called the *sound horizon*. (See Figure 12.4; the sound horizon is similar to the cosmological horizon from Chapter 7, but is based on the speed of sound rather than the speed of light.) Primordial acoustics is all about fitting a few particular wavelengths within the sound horizon. The CMB anisotropy map is literally an image of the harmonic sounds of the early universe.

EXERCISE 12.1   Acoustic Waves

(a) Convince yourself of the compression–expansion explanation of sound by drawing a picture of one person humming a constant note, another person hearing that note, and the air in between. Capture a particular moment of time. You can represent the sound source with vocal chords modeled similarly to guitar strings, and you can represent the person hearing the sound by drawing his or her eardrum responding to local air motions. Most importantly, draw air molecules as dots, and show how they arrange themselves between the two people.

(b) Somewhere on your picture, draw a few arrows on some air molecules to indicate which direction they're moving. Be careful here! The motion is caused by recurring compression and expansion, and not by traveling in a particular direction. Sound doesn't equal wind!

(c) How would your picture differ if the frequency were changed to a lower note? (*Hint:* How is wavelength related to frequency?)

EXERCISE 12.2   The Power Spectrum

Examine Figure 12.3 in order to answer the following questions:

(a) List the *x*- and *y*-coordinates of the first peak (the highest point on the curve). One is an angle, and one is a temperature deviation. State what each one means. For example, a temperature deviation of what and from what?

(b) Estimate (with numbers) the temperature deviation of the first four acoustic peaks

(c) What's going on to the left of the first peak? What spot sizes does that part of the graph refer to? Why are there no peaks there?

12.2 Precision Cosmology

Not only does the analysis from the previous section provide highly specific data in support of the big bang theory, but it also allows us to work in the other direction. If we take it as given that the big bang theory—sound waves and all—gives the correct explanation of the CMB anisotropy pattern, then we can use measurements of its photons to determine other cosmological information. We list several big ones here; for some of them, you'll need to refer back to Figure 12.3.

Power on Small Scales

In the CMB power spectrum, you'll notice that there is a plateau in the curve to the left of the acoustic peaks and a rapid falloff to their right. The plot's *x*-axis displays small angular scales on the right, and large ones are on the left. We'll start on the right, with the small spots.

After the first few acoustic peaks, the overall trend is downhill. You'll notice higher numbered peaks on the way down, but apart from these peaks, the curve itself heads downward to zero. This means that, on very small scales, the CMB displays vanishingly small anisotropies: Temperature deviations from 2.725 K cease to exist in regions of the sky smaller than about 0.05°. This phenomenon is due to an effect called **Silk damping** in cosmology, although it also applies to sound waves in general. Sound waves dissipate whenever the medium they travel through has such low density that the typical distance one of its particles travels between collisions with other particles is larger than the sound's wavelength. Effectively, it means that the sound wave is "trying" to compress a region so small that there are almost no particles in it to compress—or at least, those particles are too few and far between for their interactions to keep up with the sound wave. And if particles don't interact, then there's no way for a compressed region to communicate that compression to neighboring regions—no way for the sound wave to oscillate at the speed its frequency demands. It's why you can't hear sounds in space (except in movies: Hollywood directors are apparently not familiar with Silk damping); the sound dies for lack of air. (Incidentally, the "Silk" in "Silk damping" is named for cosmologist Joseph Silk and has nothing to do with silk fabric.)

As the early universe expanded, distances between particles were constantly growing. With that growth, sound waves began to dissipate, starting with high frequencies and working toward lower ones. The CMB snapshot captures the scale of the transition, with acoustic peaks fading away at small angles. But more than just allowing us to recognize the damping in the power spectrum, this knowledge of Silk damping allows us to calculate various properties of the early universe on the basis of our observations of the damping.

Power on Large Scales

At the left side of the CMB power spectrum are the relics of disturbances that today occupy a few degrees or more on the sky. Since the primary sound tone in the CMB—the first peak—comes from regions sized perfectly for a compression wave to compress just once, any larger regions must not have compressed yet. The first peak is about 1° in size, so as we look at regions 5°, 10°, or 50° in size, we should not expect any significant amount of compression to have occurred in time to appear in the CMB, nor should we expect the associated peaks.

That's fair enough. But then why isn't the temperature variation zero on large scales? In Figure 12.3, it appears to hover around 0.00003 K, which might *sound* small, but *all* of the anisotropy is small, and the first peak is only about 0.00008 K. We're left with sound waves that didn't have time to compress or expand large regions, but still produced anisotropies. This tells us something quite important about the disturbances that caused the sound waves. It tells us that although the sounds are too low (in pitch, or frequency) to show up as peaks in the CMB, *the disturbance itself*—the one that created all the sound—is still evident.

What if the disturbance is not mechanical, but rather, *gravitational*? Figure 12.5 shows this comparison. In frame (a), we have the mechanical case, in which some windlike agent initially disturbs the particles in a region. Later in time, the particles will compress, reexpand, etc. In frame (b), the gravitational case, preexisting disturbances in the shape of spacetime cause the initial compressions and thereby initiate the sound waves. Notice how either way, the resulting acoustics are the same, so the CMB power spectrum would be the same.

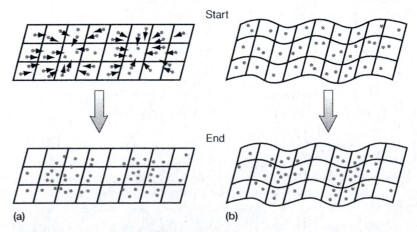

FIGURE 12.5 Early-Universe Gravitational Disturbances (a) A mechanical disturbance has set the particles in motion, generating alternating compressions and expansions: a sound wave. (b) A gravitational disturbance does the same: Due to an initially warped space, the particles gravitate together in a pattern of alternating compressions and expansions. A sound wave can be produced in this way, too.

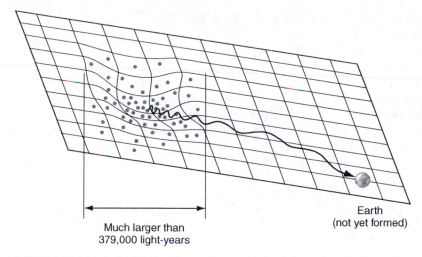

Much larger than
379,000 light-years

Earth
(not yet formed)

FIGURE 12.6 The Sachs-Wolfe Effect   In a gravitational dip too large for its particles to compress by the time of decoupling, the photons will have to expend energy to
climb out of the dip. Therefore, these photons are redshifted, giving the false appearance that they come from a cold region. The effect works in the opposite sense, too,
generating blueshifted light from large gravitational hills.

Now, if a gravitational disturbance, like a dip in Figure 12.5(b), were too large
to produce any appreciable sonic compression in the first 379,000 years, then the
region would not have heated up as a blackbody in that time. However, when photons from that region decouple and fly away, they still must climb out of that gravitational depression. This costs them energy, as you saw in Chapter 6, in the form
of a gravitational redshift. And when the spectrum of a blackbody is redshifted, it's
still the spectrum of a blackbody, only colder.

In other words, gravitational blueshifts and redshifts caused hot and cold spots
(respectively) in the CMB, from gravitational hills and valleys (respectively). This
effect, called the **Sachs–Wolfe effect** and depicted in Figure 12.6, is responsible
for the anisotropy we observe at large angular scales. That anisotropy is the relatively featureless, flat part on the left side of the power spectrum graph, since there
are no acoustics on those scales to spice things up by generating peaks. (The
Sachs–Wolfe effect also operates on smaller scales, but on those scales it is offset
by other phenomena.) We'll note in Section 12.3 that a related effect, called the
"integrated Sachs–Wolfe effect," arises when CMB photons enter and exit gravitational dips after decoupling, rather than starting out in a dip. This aspect of the
effect turns out to be helpful in measuring dark energy.

Now that it has been found in CMB data, the Sachs–Wolfe effect confirms
that the universe was somehow born with bumps and depressions, of all sizes, in
the shape of spacetime. Keep this in mind later in the chapter, when we back up
even further in time, to discuss where those initial wiggles may have come from.

Baryons and Dark Matter

Remember that at the time of recombination and decoupling, the universe was
matter dominated: In any given volume of space, there was more energy in the
form of matter than in the form of radiation (light) or dark energy. We know that
matter and energy cause spacetime to curve. Therefore, because gravitational effects can be manifested in the CMB, as we just saw, we should also investigate the

effects of gravitating matter, and not just the curvature that the universe was born with.

The CMB should be able to tell us about the matter density of the early universe by how much that matter's gravity enhances compressions (hot spots). This is true, but it actually tells us more than that: It distinguishes between types of matter. *For the purposes of this section, when we say "dark matter," we mean exotic dark matter (i.e., nonbaryonic WIMP particles).* (Baryonic dark matter usually refers to MACHOs, which are objects like white dwarfs and neutron stars that formed much later anyway.)

The reasoning goes as follows:

- Both dark matter and baryonic matter gravitate in the same way, and that gravity curves space inward. Because sound waves also compress matter and energy inward, the compressions are enhanced by the gravity from dark matter and baryons alike.

But—

- Baryonic matter interacts with itself and with radiation, so that when it compresses, its particles collide and speed up (heat up), allowing them to rebound and push harder against the surrounding medium. That is, baryonic matter builds up *pressure* and reexpands. However, *exotic dark matter, by definition, doesn't do this!* Dark matter is *weakly interacting* at best, so once dark matter compresses into gravitational dips, it does *not* generate collisions, heat, or pressure; it just stays there.

The difference in the behavior of baryons and dark matter means that both work together to enhance compressions, but they work against one another during reexpansions. The baryon pressure tends toward reexpansion, but the matter's gravity tends to keep the baryons compressed. One of the real gems in the cosmic microwave background is that it records this conflict between baryons and dark matter. What difference should it make in the microwave data? The effect should show up in the acoustic peaks, since that's where the effects of compression and expansion of sound are observed. Peak by peak, then,

THE FIRST PEAK is caused by matter and energy vacating gravity "hills" and entering gravity "valleys." The hills and valleys that cause this first peak are large enough that, in 379,000 years, this happens only once. This creates the first compression on degree scales, but the effect of pressure has not yet caused reexpansion. So we expect the first peak to be enhanced—hotter valleys and colder hills—by the joint gravity of baryons and dark matter. (See Figure 12.7.)

THE SECOND PEAK is caused on scales half as big as those of the first peak. On these smaller scales, matter and energy have compressed into valleys and reexpanded back onto the hills. Only baryons rebound like this, and they are hindered by the gravity of the exotic dark matter, which remains in the valleys. Therefore, the second peak should be weakened, exhibiting milder temperature deviations than the first peak, since baryons and dark matter are no longer working together. (See Figure 12.8.)

THE THIRD PEAK arises from compression, reexpansion, and recompression. On the recompression, baryons and dark matter work together once again, so the third peak should be enhanced.

FIGURE 12.7 Teamwork and the First Peak Baryons and WIMPs join forces. Both gravitate, and since matter and energy tell spacetime how to curve (Chapter 6), both tend to deepen the preexisting valley, enhancing the temperature contrast in the CMB on these (degree-sized) scales.

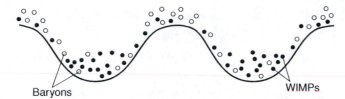

Baryons   WIMPs

FIGURE 12.8 Dissent and the Second Peak   Since only baryons experience a pressure to reexpand out of the valleys while the exotic dark matter remains in them, the baryon reexpansion is hindered by its gravitational attraction to the dark matter. Therefore, the second peak (and each following even-numbered peak) shows subdued temperature contrast.

You see the trend: Odd-numbered peaks are enhanced, and even-numbered peaks are suppressed. But, onward toward smaller scales, Silk damping is increasingly wiping out acoustic waves altogether. Cosmologists are therefore very interested in the second peak, especially in comparison with the first and third. The difference in the second peak is caused by the *difference between normal and dark matter*.

Look back at Figure 12.3. Notice that the first peak is much higher than the second one. You could try to blame this on the way the power spectrum is descending anyway, except that the second peak is roughly level with the third! Clearly, the second peak is suppressed. Cosmologists can calculate how the power spectrum curve should look with various amounts of baryonic and dark matter. Therefore, the CMB provides an independent measure of the total matter density (Ω_m, from the first and third peaks) and of the baryon density (Ω_b, from the second peak). Recall from Chapter 11 that each Ω indicates the amount of that kind of matter or energy, expressed as a fraction of the critical density. Since our current measurements of Ω_{total} give a value of 1.0, the fractional values that follow represent the fraction of the total energy in the universe that's composed of each type of matter. By subtracting the baryons from the total matter, we can extract Ω_{dm} (the percentage of dark matter) from the CMB, too. We measure

$$\Omega_m = 0.27, \text{ or } 27\%,$$
$$\Omega_b = 0.044, \text{ or } 4.4\%, \text{ and therefore,}$$
$$\Omega_{dm} = 0.23, \text{ or } 23\%.$$

It's another indication that our overall understanding of the universe is correct, that dark matter can be measured in many different ways, in comparable quantities. We consistently find evidence for dark matter in the universe today, through rotation curves of galaxies, the dynamics of galaxy clusters, and gravitational lensing. We also find plenty of evidence for it in the early universe, both in the limited abundance of baryons revealed by the baryon-to-photon ratio of BBN (Chapter 10), and in the CMB, as we have just described.

Cosmological Parameters

At this point, it's obvious that the CMB is rich with cosmological information. The last several pages pointed out some of the more important pieces. We'll finish the job in this section by listing some particular numbers that have been measured by means of CMB observations. For some of the measurements, the uncertainties are not equal on the + and − sides; therefore, we list those measured ranges in brackets [ ].

- $\Omega_{total} = 1.02 \pm 0.02$. (Fractional) density of the universe.
- $\Omega_b = 0.044 \pm 0.004$. (Fractional) baryon density.
- $\Omega_m = 0.27 \pm 0.04$. (Fractional) matter density, including both dark and baryonic matter.
- $\Omega_{de} = 0.73 \pm 0.04$. (Fractional) dark energy density.
- $\Omega_r = 0.0000491 \pm 0.0000001$. (Fractional) radiation density (CMB photons).
- $\Omega_v < 0.015$. (Fractional) density of light neutrinos.
- $T_{CMB} = 2.725 \pm 0.002$ K. CMB blackbody temperature today.
- $\eta = 6.1$ [range: 5.9 to 6.4] $\times 10^{-10}$ baryon per photon. Baryon-to-photon ratio.
- $z_{eq} = 3233$ [range: 3023 to 3427]. Redshift at which matter and radiation have equal energy densities.
- $t_{eq} = 72{,}000$ years [range: 64,000 to 82,000 years]. Age of universe at matter–radiation equality.
- $z_{dec} = 1089 \pm 1$. Redshift at which radiation decouples from matter.
- $t_{dec} = 379{,}000$ years [range: 372,000 to 387,000 years]. Age of universe at decoupling.
- $z_{reion} = 20$ [range: 11 to 30]. Redshift of reionization.
- $t_{reion} = 180$ million years [range 100 to 400 million years]. Age of universe at reionization.
- $H_0 = 71$ (km/s)/Mpc [range: 68 to 75 (km/s)/Mpc]. Hubble constant (expansion rate) today.
- $t_0 = 13.7 \pm 0.2$ billion years. Current age of the universe.

12.3 Cold Halos, Galaxies, and the Dark Universe

Based on the cosmological parameters just listed, cosmologists have determined that

- The universe was radiation dominated until $t = 72{,}000$ years after the beginning, and
- The universe was then matter dominated until $t = 9.5$ billion years, and
- The universe has been dark energy dominated ever since.

In our 13.7 billion-year-old universe, the era of dark energy domination is fairly recent. And the era of radiation domination is ancient history. For most of cosmic history, the universe has been matter dominated. Observations at medium and high redshifts clearly demonstrate that very little has changed structurally in the universe since the onset of dark energy domination. The same basic pattern of clumps and voids present in the universe today was also present back in the matter-dominated era. This suggests that it was matter, not dark energy, that drove the gravitational clumping into galaxies and halos, clusters, and walls.

Prior to matter–radiation equality, photons controlled the gravitational structure of spacetime, so not until 72,000 years or so after the beginning could matter begin to clump together. That's because matter didn't have as much gravitational influence as photons early on. So even if there was a clump of matter somewhere, the gravity from the photons all around would tend to smear out the clump, instead of letting the clump's gravity pull more particles in. Still, 72,000 years is very early; the CMB didn't decouple from matter until 379,000 years. Both of these times are important to the formation of structure. Consider the latter for a moment: We often describe decoupling as photons breaking free from matter, but really, there are two important corrections to think about:

IMPORTANT CORRECTION #1: It's not all types of matter that the photons broke free from; it's almost entirely the electrons. Electrons are electrically charged, as are nuclei, so if you push some electrons, the nuclei will follow: Electrons and nuclei are electrically coupled to one another. Most of the mass–energy of the electrons and nuclei is in the nuclei, which are made of baryons. In other words, photons were connected to baryons, with electrons as middlemen. So photons didn't decouple from *all matter* at this time; they decoupled only from *baryonic matter*. Exotic dark matter had been free for more than three hundred thousand years, ever since matter–radiation equality.

IMPORTANT CORRECTION #2: Photons didn't just break free from (baryonic) matter. Photons and baryonic matter decoupled *from each other*. You could just as accurately view things the other way around: that the baryons broke free from the photons.

All of this, taken together, tells us how structure formation worked:

FIRST, at the age of matter–radiation equality, dark matter decoupled from the primordial plasma and started to cluster together into early halo structures.

SECOND, at the age of CMB decoupling, baryonic matter decoupled from the photon field and started to cluster together into clouds of neutral gas.

However, since the universe was matter dominated by then, and because there's more dark matter than baryonic matter, the strongest gravity sources were the already forming dark matter halos. Baryons fell into these existing dark halos, which means that early structure formation was really orchestrated by the dark matter. The baryonic matter—which formed stars, galaxies, and people—followed the dark matter's lead. Dark matter doesn't gravitate any better than baryonic matter, and both types of matter participated in structure formation; it's just that dark matter got a head start. (See Figure 12.9.)

Then, since dark matter barely interacts with itself, it's hard for it to slow down enough to settle into small structures. Without collisions between dark matter particles, those particles have no means of losing energy, so they can't slow down. (Imagine how a swarm of flies would balloon outward if each fly's speed increased by, say, 20 times. If the flies can't slow down, then the swarm can't shrink.) That's why halos are so big. Baryons, on the other hand, do collide with one another and therefore do slow down as they fall into a halo's gravity. That's why they are able to form galaxies that are much smaller than the halos they're in, even though the halos started clumping together earlier. Dark matter started clustering *first*, but baryons are *better* at it.

Simulating Structure Formation in the Universe

To study cosmic structure formation, then, we need to do two things. First, we need to model the interactions of dark matter halos as they move, grow, and reorganize

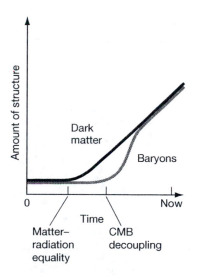

FIGURE 12.9 When Does Structure Construction Begin?   As time goes by, along the horizontal axis of this plot, the expansion liberates exotic dark matter first (at the time of matter-radiation equality) and baryonic matter later (at the time of decoupling). So the dark matter starts clustering first, and when the baryons are freed, they form structures (galaxies) within the foundations (halos) that the dark matter already started.

under their own gravity. Second, we need to model the behavior of baryonic gas as it does the same within the halos. Fortunately, both tasks rely on well-known physics—gravity and gas dynamics—and on the cosmic expansion, which opposes gravity. The problem isn't our knowledge of the necessary physics; the problem is simply tracking an entire observable universe's worth of particles. Fifty years ago, the solution would have required an endless supply of expendable labor (called "grad students"), but today researchers use computers (and grad students).

The computer-based research has been highly successful when programmed under the assumption that the dark matter is made of WIMPs, or at least something like WIMPs. The defining aspect of this category of dark matter models is that the dark matter particles are massive (the "M" in "WIMP"). This means that, with some fixed amount of energy, they wouldn't be accelerated to ultrafast speeds the way low-mass particles would. Speed matters because kinetic energy depends on speed, gravitational binding depends on gravitational *and* kinetic energies (review Section 4.1), and structures form by gravitational binding.

We therefore need software to simulate the behavior of relatively slow-moving dark matter. By analogy with the connection between temperature and particle speeds, this type of WIMP-like dark matter is called *cold* dark matter. (For contrast, neutrinos have very little mass and therefore travel at nearly the speed of light; they are considered *hot* dark matter.) The type of computer program used for studying cosmic structure formation is usually called a **CDM simulation**, where CDM stands for "cold dark matter." Here's how it works:

1. You begin by defining some region of space, and within that space, you tell the computer where all the matter particles start out. Researchers have good knowledge of this initial condition, based on CMB anisotropy maps.

2. Next, you program the computer with the equations describing relevant physical processes, such as gravity and spacetime expansion via the scale factor (including the accelerated expansion caused by dark energy).

3. The computer then calculates what happens in some short time span, much less than the current age of the universe. For every particle, the computer adds up all the gravitational forces acting on that particle from every *other* particle. Under the influence of the calculated cumulative force, the original particle moves; the computer calculates how far, how fast, and in what direction. It does this for each particle in the simulation. (In practice, the software may not need to track particle by particle, because more computationally efficient routines can be used. Still, the basic concept behind the simulation is the same.) The simulation also enlarges distances between particles in accordance with the growth of the scale factor.

4. The software takes all the particles in their new locations, after they have moved, and repeats the same set of calculations all over again. Now that every particle has moved, the computer recomputes the cumulative gravity acting on each, and the scale factor growth, and moves them again. It continues like this until it has simulated all 13.7 billion years of the history of the universe.

5. Meanwhile, the computer also runs computations on the behavior of the baryonic matter (i.e., normal gas). These computations simulate the organization of that matter into galaxies within halos.

When scientists run these simulations, they find strong support for models in which structures form *hierarchically* over time, meaning that small halos form first, and some of them lump together into larger halos later. This happens by two mechanisms:

- *Accretion.* Halos gravitationally attract more matter particles and grow over time. (We saw the word "accretion" earlier when we discussed black holes in Chapter 6; black holes accrete as well, gravitationally pulling matter onto themselves.)
- *Mergers.* Halos merge to form larger halos.

Some halos end up more isolated than others and may not grow much by either mechanism. Others, located among a host of halos, can grow rapidly by both mechanisms. Over time, halos of many sizes populate the universe and collect around one another, forming clusters and walls by emptying out material from spaces in between—leaving voids.

In other words, all the structure in the universe today is effectively the result of initial fluctuations, gravity, and expansion. The panels of Figure 12.10 show a time sequence of structure formation from a CDM simulation. Take your time and look over the growth of structure in the pictures and the caption. Can you convince yourself that expansion outward, and gravity inward, would amplify the tiny structural seeds (seen as CMB anisotropies at a level of 1 part in 10^5) into the grand structures we see today, as the figure shows? When you feel satisfied about that, look at Figure 12.11, which takes a segment from a CDM simulation and provides increasing magnification of its substructure.

The CDM simulations wonderfully reproduce the structures we see in space: walls and voids, clusters and groups, and even individual galaxies like ours. Let's pause to consider this accomplishment. We've shown that it's possible, by using known laws of physics (and the *known* properties of the *unknown* dark matter!), to get from the tiny wiggles in the CMB, over 13 billion years ago, to all the structure

FIGURE 12.10 CDM Simulation Time Sequence In the first frame, we begin with a nearly-smooth universe, like the one the CMB all-sky maps reveal. Over billions of years, structures coalesce gravitationally in a hierarchical fashion, until the current spongy structure emerges. (◆ Also in Color Gallery 2, page 24.)

FIGURE 12.11 Simulated Structure in the Universe Today Here we see four successive zooms on the same simulated patch of space. Notice the voids in the upper frames and the rich cluster (circled) forming in the lower frames. (◆ Also in Color Gallery 2, page 24.)

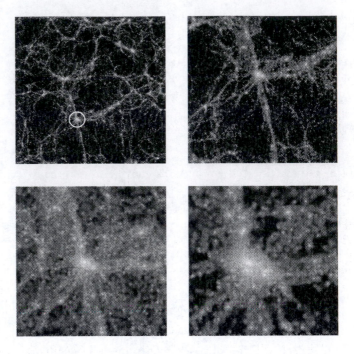

we observe today. Although the details of our understanding are likely to evolve as we learn more, the fact that simulations of the history of structure in the universe match so closely with observations indicates that we're on the right track and provides strong support for the big bang framework.

To summarize the CDM simulation story, we are working with data from the beginning (the CMB) and data from recent times (galaxy surveys), but we fill in the gap in between with simulated history (CDM simulations) because we have no pure snapshot of data along the way. Even so, Figure 12.12 shows a remarkably consistent sequence explaining how we got to this point from the smooth primordial universe.

In order to make CDM simulations match the observed pattern in the universe today, we must include the expansion-enhancing effect of dark energy in the simulations. (See the next subsection.) Today, for example, the accelerated expansion of our dark energy dominated universe has ended the growth of large-scale structures like walls and voids. The large-scale structure that exists now is apparently all that will ever exist. We are perhaps fortunate that we arrived in the universe when we did, so that we can observe that full history documented in the sky.

Evidence for the Dark Components of the Universe

CMB anisotropy data and the galaxy survey data have added much evidence for dark matter and dark energy in the universe. Here, we'll summarize the evidence for each.

EVIDENCE FOR DARK MATTER

- *Galaxy rotation curves* Accounting for the high speeds at which stars travel along their orbits around a galaxy requires a great deal of additional gravity from a great deal of additional mass, beyond what's visible. These speeds are measured in edge-on spiral galaxies via the Doppler effect (See Section 4.3 and Color Figure 35 in Color Gallery 1, page 16, and Color Figures 38 and 39 in Color Gallery 2, page 18.)

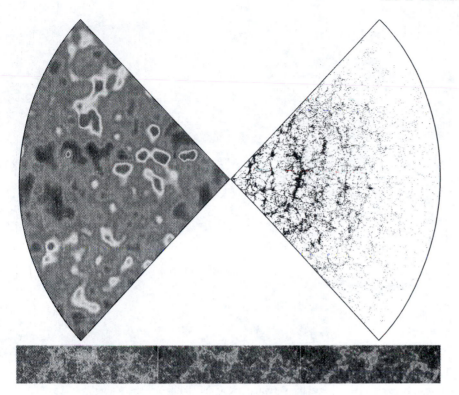

FIGURE 12.12 The Evolution of Structure   In the first few hundred thousand years, the universe had tiny density fluctuations of only 1 part in 100,000 (left). Today the universe has massive walls and voids 100 million light-years wide (right). CDM simulations show how the universe got from one state to the other (left to right across the bottom). (◆ Also in Color Gallery 2, page 25.)

- *Dynamics of galaxy clusters* Accounting for the high speeds at which galaxies travel within galaxy clusters also requires a huge additional mass within the cluster and is also measured via the Doppler effect. Some of the previously invisible mass has been identified as baryonic on the basis of its observed x-ray emission, but most of it remains exotic. (See Sections 4.3 and 10.4, and Color Figure 29 in Color Gallery 1, page 14.)

- *Separation of luminous matter from dark matter* Some observed systems involve dark galaxies, in which there is a galaxy-sized gravity source with no observable starlight, or merging galaxy clusters in which the baryonic matter has been separated from the exotic dark matter. (See Section 4.3, Section 9.5 on WIMPs, and Color Figures 40 and 41 in Color Gallery 2, page 19.) In each case, we find strong evidence for a large body composed solely of exotic dark matter.

- *Gravitational lensing* By studying the mass needed for a gravitational lens (like a galaxy or a galaxy cluster) to deflect the light from a background source (like a galaxy or a quasar), researchers can directly measure the amount of mass in those lens objects. They can even see where that mass is distributed, verifying the existence of dark halos. (See Section 6.4 and Color Figures 36 and 37 in Color Gallery 2, page 17. See also Figure 12.13, which shows a 3-D map of the distribution of dark matter in space, obtained from gravitational lensing.) In addition, some baryonic dark matter is observed by the (much smaller) gravitational lensing effect of a MACHO (such as a small black hole) passing in front of a distant star.

- *CMB anisotropy* The CMB power spectrum is a tally of its spot statistics. Because baryonic matter experiences gravity and pressure, but exotic dark matter experiences gravity only, the density fluctuations of the early universe reveal the distinction between the effects of baryonic and nonbaryonic

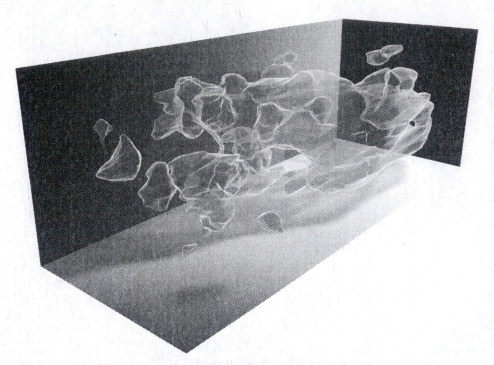

FIGURE 12.13 Map of Dark Matter   Hubble Space Telescope data from studies of gravitational lensing were assembled to reveal this 3-D map of the distribution of dark matter in space. The left side of the image is close to us and therefore shows the current local dark matter distribution; the right side is far away and therefore shows the distant dark matter distribution long ago. Notice how gravity causes the dark matter to get clumpier over time (right to left.)

matter. (See Section 12.2 and Color Figures 48 and 49 in Color Gallery 2, page 22.) These data allows a precise measurement of the amount of each type of matter and confirm that most of it is exotic.

- *Structure formation evidence* The "DM" in "CDM simulation" stands for "dark matter." In order to simulate the formation of structure over cosmic history in a way that matches the observed structure in the universe, cosmologists need to base their simulations on the behavior of exotic dark matter. (See this section and Color Figures 43 and 52–55, in Color Gallery 2, pages 20 and 24–25.) Since the expansion rate is tied to the density of matter and energy in the universe (Sections 7.2 and 11.1), the simulated results can come out right only when the simulations include the right amount of matter (and dark energy).

- *Concordance of the evidence* All of these separate pieces of evidence for dark matter favor the same *amount* of matter: about 27% of the total energy in the universe, with about 23% being exotic dark matter.

EVIDENCE FOR DARK ENERGY

- *Supernova data* Data from distant type Ia supernovae demonstrates that the universe is expanding at an accelerating rate. (See Section 7.4 and Color Figures 58 and 59 in Color Gallery 2, page 27.) In fact, increasingly detailed supernova data have even revealed the transition that occurred from a matter-dominated *de*celerating universe to a dark energy dominated *ac*celerating universe. The transition began about 5 billion years ago—right on schedule, according to the measured densities of matter and dark energy. (See Chapter 11.)

- *CMB data and dynamics of galaxy clusters* CMB data tell us that the geometry of the universe is flat; therefore, the universe must have a density equal to the critical density. (See Section 11.1.) The motion of galaxies within galaxy clusters is based on the amount of (gravitating) matter present, and cluster studies reveal what fraction of that critical density is attributable to matter. Together, these two observations tell us that the remainder of the critical density takes the form of dark energy. (See Section 11.3 and Color Figure 60 in Color Gallery 2, page 28.)

- *Integrated Sachs–Wolfe effect* Measurements of the gravitational redshift and blueshift of the CMB provide another, separate line of evidence for dark energy. (See Section 12.2.) Partly because of the acceleration caused by dark energy, CMB light entering a large gravity source such as a galaxy cluster is gravitationally blueshifted on its way in *by a different amount* than it is gravitationally redshifted on its way out. This observation, which is sensitive to the density of dark energy in the universe, is referred to as the "integrated Sachs–Wolfe effect."

- *Structure formation evidence* Various studies of the manner in which structures like galaxies, galaxy clusters, and walls and voids formed over cosmic history reveal the need for dark energy. (See earlier in this section and Color Figures 52 through 55 in Color Gallery 2, pages 24–25.) CDM simulations tell us when and how many structures formed, an account that matches the observational data only when the right amount of dark energy is included in the simulation. The relevant data include data from galaxy surveys, x-ray observations of the history of the formation of galaxy clusters, and observations on the number and strength of gravitational lensing systems in which the lens object is a galaxy cluster. (See Section 5.2, Section 6.4, and Color Figures 36, 37, and 43, in Color Gallery 2, pages 17 and 20.)

- *Concordance of the evidence* All of these separate pieces of evidence for dark energy favor the same *amount* of dark energy: about 73% of the total energy in the universe.

12.4 Inflation Theory

Having demonstrated the existence of density fluctuations of 1 part in 10^5 in the early universe, and having tracked them all the way to the clusters, walls, and voids of today, we understand how it's possible for tiny seed structures to evolve into astronomically large ones. Now it's time to ask where the seeds came from.

Ideally, we want to explain the origin of structure in terms of some simple, natural phenomenon. This would be more satisfying than explaining it as the result of some earlier event whose origin also needs explaining (as we've done so far, by explaining large-scale structure today in terms of primordial seeds in the CMB, but then needing to ask where the seeds come from). Our most promising explanation, the theory of cosmic **inflation**, is a little of both. Normally, that wouldn't sound too appealing, but inflation sweetens the deal by explaining much more than just the fluctuations: It also addresses several problems that arise within the context of the big bang theory.

The Problems

Here's a list of clues about whatever process imprinted the seeds of structure:

- Whatever happened to make the seeds happened long ago, since they were already there by the time of the CMB decoupling. In fact, they had been there for all 379,000 years previously, too, since they caused sound waves which took that many years to compress. We should expect, therefore, that the crucial event we seek occurred very early—within the first fraction of a second, perhaps.

- Whatever imprinted the spacetime fluctuations did so in the same way everywhere and in every direction, as observed by the *homogeneity and isotropy* of walls and voids today, and of the CMB anisotropy pattern from long ago. As you know, homogeneity and isotropy together define the "cosmological principle," but *why* does our universe obey it?

- The spots in the CMB that are separated by more than about 1° were outside each other's horizon at the time of decoupling, meaning that they should never have been in any kind of communication with one another. Then why do all the spots look the same? How does any one spot "know" that it's "supposed to" deviate from the mean temperature by only 0.003%, maximum? It's like everyone showing up to school one day wearing the same clothes; it wouldn't happen unless they had all phoned each other the night before to plan it. But if they were isolated by their cosmological horizons, then those phone calls couldn't have happened. This is called *the horizon problem*.

To understand the horizon problem, you might want to review the discussion of cosmological horizons in Chapter 7. The basic problem is illustrated in Figure 12.14 in a simplified, *imaginary* universe that has a flat geometry and is *static* (not expanding).

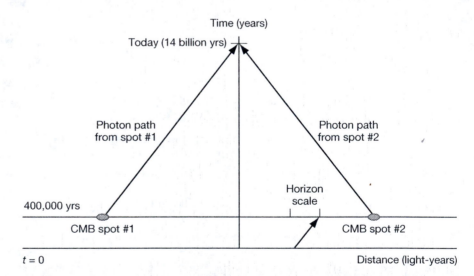

FIGURE 12.14 Horizon Problem We see regions today that were too far apart to have ever communicated 400,000 years after the big bang, when the light first left them, ultimately reaching us today. In this spacetime diagram (with time counting upward from zero on the *y*-axis as usual), two different emitting regions that produce CMB spots were far away from each other when the CMB formed. At that time, the horizon scale (labeled) was small because the universe hadn't existed long enough for light to travel very far. Thus, the two CMB spots could not have had any contact with one another. Today (top of graph), the horizon has grown so that light from both spots has reached us. How can different parts of the universe look the same even though they had never communicated their properties to one another in the entire history of the universe up to then?

Our real universe, of course, is not static, but it's much easier to grasp the concept in the static case, and the logic applies equally well to our expanding universe. If we still assume that cosmic microwave background photons were set free at (roughly) 400,000 years after the beginning of this universe, then Figure 12.14 shows the situation. The crucial point is that we can see regions of the microwave background *now* that were far outside of each others' horizons at the time the light was *emitted*. Observationally, these regions look nearly identical, but they shouldn't if they were unable to contact each other in advance.

- We're used to Euclidean geometry, so it might seem natural that our universe should be that way—that is, flat. But general relativistic cosmology tells us that the universe could have any other curvature, too, so why did it end up perfectly flat? Put another way, since the curvature is tied to Ω_{total}, and since Ω_{total} could have any value, why did it come out to be 1, instead of 0.0000343, or 1.57, or three hundred kajillion? This is called *the flatness problem*.

A Solution

In the 1980s, cosmologists realized that these observations could be explained with a single event, called *inflation*. It works in the following way:

Only about 10^{-34} second into the existence of our universe, the transition from the GUT era to the post-GUT era somehow produced an ultrarapid accelerated expansion of the universe. The source of this expansion is something similar to the form of dark energy called quintessence that we discussed in Section 11.3. Just as dark energy is making the universe accelerate now, some other agent did the same back then. The acceleration-causing dark energy-like agent quickly overcame matter and radiation, the agents of deceleration. In effect, the universe got stretched out.

General-relativity-based cosmology explains the action of inflation with the Friedmann equation, which connects the expansion rate, the curvature, and the various densities of the universe to one another. In the presence of a huge dark energy-like agent in the universe, however, much of this equation essentially goes away because the curvature term and the matter and radiation densities become small enough to ignore by comparison with this huge dark energy density. The equation then boils down to something quite simple:

$$\text{The expansion rate} = \text{a constant that quantifies the dark energy}$$
$$\times \text{ the scale factor ,} \qquad \text{(eq. 12.2)}$$

or

[how fast distances grow] is proportional to [how big they are already] . (eq. 12.3)

This is the mathematical condition for *exponential* growth. Imagine that you were to grow in this way, so that when you're small, your growth is slow, but as you get bigger, your growth speeds up. It's a runaway effect, and pretty soon your head hurts and your home needs a new roof. This is how inflation affects the universe. It grows a little, then faster, then faster, until the growth is explosively fast.

Alan Guth (born 1947) has spent most of his research career studying the interplay between particle physics and cosmology. His most famous achievement in this area was his 1981 publication of inflation theory, an addition to the standard big bang cosmology. With the aid of important contributions from Andrei Linde, Andreas Albrecht, and Paul Steinhardt, Guth put forth a theory that provides a framework for understanding the earliest moments in our universe.

The theory of inflation was proposed by Alan Guth in 1981. Inflation is one of the few major theoretical advances of modern times that was initially accomplished with nothing more than pen and paper. The theory called for a period of exponential growth of the universe's scale factor, brought on by the ultra-high-energy particle physics of the very early universe. In his paper introducing the theory, Guth wrote that it could resolve several issues concerning big bang cosmology:

> "The standard model of hot big-bang cosmology requires initial conditions which are problematic in two ways: (1) The early universe is assumed to be highly homogenous, in spite of the fact that separated regions were causally disconnected (horizon problem); and (2) the initial value of the Hubble constant must be fine tuned to extraordinary accuracy to produce a universe as flat (i.e., near critical mass density) as the one we see today (flatness problem). These problems would disappear if, in its early history, the universe supercooled to temperatures 28 or more orders of magnitude below the critical temperature for some phase transition. A huge expansion factor would then result from a period of exponential growth, and the entropy of the universe would be multiplied by a huge factor when the latent heat is released. Such a scenario is completely natural in the context of grand unified models of elementary-particle interactions."
>
> —Alan H. Guth, "Inflationary Universe: A Possible Solution to the Horizon and Flatness Problems."

In inflation models, this exponential growth is short lived. Whatever condition started it subsided in 10^{-32} of a second, or maybe faster, and inflation was over. Inflation predicts that, during that time, the scale factor grew at least 10^{43} times bigger. So what good did inflation do? Broadly, it took space and blew it up to such an absurdly enormous size that whatever was near you to begin with ended up hopelessly far away. This is shown in Figure 12.15, which you should refer to as you read how inflation fixes all the problems:

- *Inflation creates homogeneity and isotropy.* Whatever uneven, lumpy mess might have existed in some region before inflation was so dramatically stretched out by inflation that the lumps are all incredibly far away now. That means homogeneity if none of the lumps are anywhere within our currently observable universe. And it means isotropy, since even if all the lumps are to the left and none to the right, it doesn't matter: They're all beyond our horizon anyway. So why does the universe obey the cosmological principle? It very well might not! But inflation made sure that the *observable* universe does.

- *Inflation solves the horizon problem*, because regions of the CMB that are outside of each other's horizon *used to be in contact before inflation separated them!* That's why their spots all look the same; that's why they all knew how to "dress" exactly alike.

- *Inflation solves the flatness problem.* No matter how curvy our patch of the universe was before inflation, it got blown up to such a large scale that it looks flat now. Just as the Earth looks flat because it's so big (especially in Nebraska), the universe looks flat because inflation made it so big. If we could zoom out to view the universe, the way observing the Earth from space changes our perspective, we might see that we're really on some wildly curved part of the universe. But we'd have to somehow zoom out enough to see much more than just our observable universe.

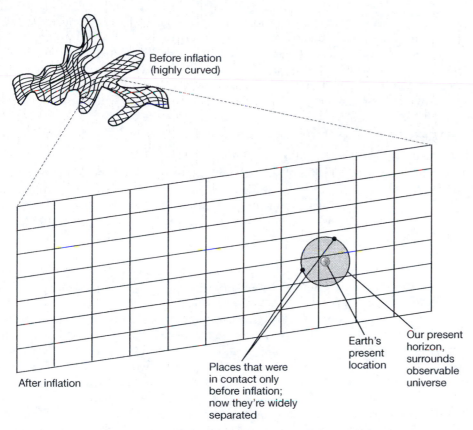

Before inflation
(highly curved)

After inflation

Places that were
in contact only
before inflation;
now they're widely
separated

Earth's
present
location

Our present
horizon,
surrounds
observable
universe

FIGURE 12.15 Cosmological Inflation   (a) An ugly and twisted universe is radically expanded by inflation. (b) Afterward, all the ugly twistedness is still there, but only on scales much too big to notice. Our entire observable universe would be a small patch in a hugely blown-up region. (It's a little like zooming in on a complex impressionistic painting until all you can see is "blue.")

It's worth pointing out just how problematic the flatness problem, in particular, was. We can calculate how parameters of the universe changed over time during the expansion, from the beginning until now. In particular, Ω_{total} determines the curvature. It turns out that it doesn't matter what kind of universe we live in; Ω_{total} always moves away from 1 over time if it wasn't *exactly* 1 to begin with. (See Figure 12.16.) In order for the universe to be anywhere close to its measured flatness today (which is $\Omega_{total} = 1.02 \pm 0.02$, from WMAP), it had to be incredibly flat in the past. Inflation solves this problem by saying that it *was* incredibly flat in the past, just after inflation.

For example, if the universe were slightly positively curved ($\Omega_{total} > 1$) at the time of BBN, when the universe was 1 second old, then it would be *hugely* (and positively) curved now. In fact, the way the Friedmann equation evolves over cosmic time, in order for the universe to be positively curved at BBN and flat now, the curvature at BBN must have been

$$\Omega_{total} \text{ (at BBN, 1 sec)} = 1.000000000000000000XXXXX \ldots , \text{ (eq. 12.4)}$$

where the X's are digits that are permitted to deviate from zero! This kind of incredible precision must have a very specific reason. (Earlier, at the Planck time, the problem was even worse: The accuracy needed to be 1 part in 10^{60}.) The flatness problem is thus a huge problem, but inflation skirts it easily by saying that it doesn't matter if the universe was more curved *before* inflation, because it certainly won't be *after*.

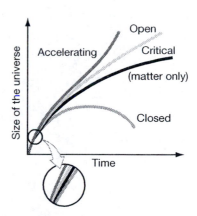

Size of the universe

Time

Accelerating

Open

Critical

(matter only)

Closed

FIGURE 12.16 Deviation from Flatness   As discussed in Chapters 7 and 11, the curvature of spacetime depends on the density of matter and energy. In universes with different compositions of matter and energy, the various energy densities (omegas) change rapidly, so, for the universe to be anywhere near flat now requires that it was almost perfectly flat early on. The graph shows that even very different universes all share this nearly identical beginning.

We generally assume that, in actuality, $\Omega_{total} = 1.0000 \ldots$ today, rather than 1.02. It would be a bit of a coincidence if the universe's deviation from flatness just happened to become noticeable right when humans started conducting microwave experiments to measure it! More likely, that deviation will take place at some arbitrary time in the very distant future. Regardless of the details, however, any measurement of even near-flatness today, as by WMAP, is one line of evidence supporting inflation.

Structure of All Sizes

So far, we've solved a bunch of problems, but we didn't do what we originally set out to do: explain the origin of structure. In fact, at this point, it probably sounds like quite the opposite: Inflation should *erase* structure! In the domain of classical physics, that's exactly correct. But we know that at small scales—which get blown up by inflation—the classical gives way to the quantum. The real glory of inflation theory is that it takes natural, inevitable quantum effects and turns them into the seeds of structure in our universe.

Remember from Section 11.2 that, at the Planck scale, which is a tiny (and therefore unobservable) 10^{-35} meter, quantum-mechanical effects start to randomly distort the shape of spacetime. That is, the random quantum energy on small scales is so great that it has enough gravity to warp spacetime. As inflation starts expanding the warps caused by these **quantum fluctuations**, more quantum fluctuations of energy continuously produce new wiggles in spacetime. (There's no way to turn off quantum physics.) The new wiggles are then expanded at the same inflationary rate, but because they got less of a head start than the ones that expanded in the previous round, they don't end up quite as big. As those new wiggles grow, newer ones emerge, and grow, and so on. (Read Section 12.5 if you're interested in learning more about the meaning and cosmological importance of quantum fluctuations.)

All this activity creates a very simple and specific outcome: Inflation imprints curves *of all sizes* into the fabric of spacetime, and *with the same level of distortion at every size*. This is called *scale invariance*. If some other theory were to compete against inflation, that other theory would have to match this prediction somehow, because the WMAP data—which arrived after inflation theory made the prediction—do indeed reveal scale invariance of the CMB anisotropy.

Note how naturally inflation creates such scale invariance. Quantum fluctuations absolutely had to exist and therefore had to be amplified by inflation. As one scale grew, a smaller one replaced it, generating lasting curves of all sizes in space. These curves were the seeds of structure in the universe and the source of anisotropy in the CMB. When dark matter and baryonic matter started to compress into the curves, the primordial acoustics were set in motion.

Did Inflation Really Happen?

Did inflation really happen? At first glance, it might seem like an awfully contrived way to solve the horizon and flatness problems. So take your healthy skepticism, and like a true scientist, demand that inflation theory make some specific, testable predictions. Inflation theory, in general, makes three:

TESTABLE PREDICTION #1: Inflation causes Ω_{total} to be $1.000 \ldots$, to a high decimal precision. (Actually, it's not entirely clear that this should count as a "testable prediction": Inflation theory was *designed* to explain flatness, so we shouldn't be too impressed when it explains flatness. But even if it's not really a "prediction," it's still a noteworthy result.)

TESTABLE PREDICTION #2: Inflation produces initial variations in density that are equally strong over a wide range of distance scales and that leave an imprint we can observe in the CMB.

TESTABLE PREDICTION #3: Inflation floods the universe with *gravitational waves* (introduced in Section 6.4), which might have a small, but observable, effect on the polarization of the CMB. Future microwave experiments will look for this evidence.

Testable predictions 1 and 2 have passed their tests, according to the high-precision WMAP data set. But to really accept inflation into our hearts and souls, we'd like to know what actually caused it. What particle? What physics? Why did it act like a cosmological constant? Why did it stop acting that way and give rise to normal matter and radiation? Why did it last just as long as it did?

Ideally, we want a GUT model that explains all these things in terms of some natural, inescapable features of particle physics. There are many such models proposed and being studied today (unless today is an important national holiday), but none could really be called compelling yet. (Cosmologists Edward W. Kolb and Michael S. Turner define a "compelling" model as "'compelling' and 'beautiful' to people other than those who proposed it!")[1]

It will be hard, but not impossible, to obtain such a compelling model. It's hard because the model needs to describe the physics that takes place at the high temperatures existing at 10^{-34} second—temperatures around 10^{28} K—which we can't even come close to re-creating in a laboratory. It's not impossible, however, because we do have the observed universe to measure, and some model might become "compelling" if it makes a specific prediction that is validated by future CMB polarization measurements. (WMAP probably isn't sensitive enough for that.) In other words, most of the cosmology community will probably consider the case for inflation and new GUT physics compelling if testable prediction #3 is met.

Even though inflation theory appears solid to most cosmologists, sound scientific practice encourages them to imagine alternatives that could be distinguished by observations or experiments. (In other words, it's always good to ask questions like "What about this other idea?") For example, the cyclic model we discussed in Section 11.3 also offers the hope of explaining the same puzzles addressed by inflation. It's too early to tell how it will pan out, but as more precise CMB observations are collected, it should be possible to evaluate the success and the details of inflation theory more critically. As is true with any theory, then, future observations hold the power to solidify or jeopardize inflation.

EXERCISE 12.3   The Power of Exponential Growth

On your first day of work at a new job, your boss offers you a choice of salaries. One option is a fixed salary of $5000 per month. Or you can be paid on a daily basis throughout the month, starting with only 1 cent the first day, 2 cents the second day, 4 on the third day, etc. Each day you are paid double the amount you got on the previous day, until the end of the month. (Assume that a month has 21 workdays.) Which option should you take? Work out your answer mathematically, then explain how your answer relates to inflation in the early universe.

---

[1]Edward W. Kolb and Michael S. Turner, *The Early Universe*, Westview Press, 1990.

12.5 Supplemental Topic: Quantum Fluctuations in Cosmology

There is something very powerful about the idea that all the structure in the universe is the result of random quantum fluctuations. It's remarkable in its apparent simplicity, but to some it might sound like a cop-out. ("Where did structure come from? I don't know, let's blame quantum fluctuations!") But quantum fluctuations were not dreamed up in order to explain structure. Rather, quantum fluctuations are a necessary consequence of well-established physics. Quantum physics has been validated in many ways (including its application to everyday technologies, such as computers, cell phones, solar cells, etc.). And random fluctuations in vacuum—the same "vacuum energy" effects that might account for dark energy, as discussed in Section 7.4—are documented in laboratory experiments. These fluctuations are always present, so inflation had no choice but to amplify them.

The quantum origin of structure also seems to deal the final Copernican blow to our egos.[2] First, we're not the center of the solar system, or the Galaxy, or the universe. Then, most matter out there isn't even made of baryons as we are, and most of the universe's composition isn't matter anyway. Finally, it seems that we're the ultimate result of a quantum process which is inherently *random*.

Normally, quantum mechanics affects only the ultrasmall. It becomes relevant to cosmology, however, when the ultrasmall affects the ultrabig.

Primordial Black Holes

Stephen Hawking, who holds Isaac Newton's former position at Cambridge University, made a tremendous scientific leap in 1974 when he worked out some complicated mathematics to describe the effects of quantum fluctuations in the highly curved spacetime at the event horizon of a black hole. This research attempts to combine quantum mechanics with general relativity into *quantum gravity*; it's an important first step toward a theory of everything.

Usually, quantum fluctuations are understood in terms of the creation of "virtual" particles: Out of otherwise empty space, particles and antiparticles are spontaneously generated in pairs, and in a very short time they annihilate with one another and are gone. (That's why they're only *virtual*.) Hawking's research showed that the curvature near a black hole tends to separate these particle pairs, because sometimes one falls into the black hole and the other doesn't. If they separate like this, they can't annihilate; if they don't annihilate, then they're no longer virtual: they become *real*. In other words, real particles are being continuously generated at the surface of a black hole. (See Figure 12.17.) Most of the emerging particles are photons, which means that black holes shine like stars—very, very, very dim stars. (Chapter 7 of Hawking's popular book *A Brief History of Time* is called "Black Holes Ain't So Black" for this reason.) Now, if you're worried about conservation of energy (and who isn't?), you can relax. In order to become real, the emerging particles take energy away from the black hole, so it loses mass (mc^2 energy) in the process. The black hole literally radiates its mass away.

---

[2] This way of looking at inflation was introduced by Eric Chaisson and Steve McMillan, in their book *Astronomy Today* 6e (San Francisco: Addison-Wesley, 2008.)

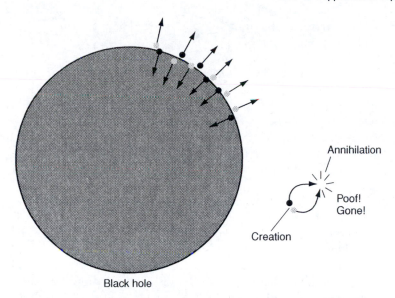

FIGURE 12.17 Hawking Radiation Quantum physics holds that pairs of "virtual particles" are always being created and then annihilated everywhere in the universe. Near the event horizon of a black hole, however, the two particles can be separated by the extreme curvature (gravity), so that one particle falls in. The particle that doesn't fall in is now labeled as real (rather than virtual), because it survives (rather than annihilates). In this way, a gravitational event horizon manufactures particles

The ultrafaint glow from black holes, called **Hawking radiation**, becomes important to cosmology if there are many such radiating holes all over the universe. Today, black holes are formed by exploding stars. These holes are at least several times the Sun's mass and barely radiate at all. Bigger supermassive black holes (SMBHs) radiate even less. But if a black hole could somehow be made microscopically small, it would shine significantly. Hawking pointed out that the high density of the universe when it was less than 1 second old could have produced black holes of a variety of sizes, including microscopic ones.

So far, there is very little evidence for such *primordial black holes* (PBHs). This isn't terribly surprising even if they do exist, since Hawking radiation is quite weak. A certain type of observed gamma-ray burst appears to be consistent with the radiation from a tiny black hole as it radiates the last bits of its mass away, but this interpretation is uncertain. One tantalizing idea is that a fleet of microscopic black holes would behave like WIMPs and could account for what we have classified as exotic dark matter. However, based on how little radiation we see from space that could be Hawking radiation, we conclude that Ω_{pbh} is probably less than 10^{-8}, which is far too little to account for the exotic dark matter.

The Inflationary Event Horizon

The mechanism by which virtual particles become real enough to warp spacetime is just like Hawking radiation from a black hole. At the black hole's event horizon, particle pairs get permanently separated to create the radiation. Inflation actually does the same thing: Due to the acceleration it causes in the universe, a particle pair can be separated by the inflating spacetime! There are three very interesting connections to be made in this regard:

VERY INTERESTING CONNECTION #1: Inflation increases the distance between two particles, so that one particle in a pair quickly ends up outside the cosmological horizon of its partner particle. In other words, an accelerating universe creates a *cosmological event horizon* beyond which all events are unknowable, just like a black hole's event horizon. Unlike a black hole's event horizon, which exists only where there's a black hole, the cosmological event horizon surrounds every point in the inflating space.

VERY INTERESTING CONNECTION #2: By the *equivalence principle* of general relativity (from Section 6.2), there is no measurable difference between the effects of a uniform acceleration and a uniform gravitational field. Therefore, if a gravitational event horizon translates virtual particles into real ones, then so does an acceleration-borne event horizon! Inflation therefore had to make Hawking radiation. Thus, inflation made virtual particles real and amplified their effects on spacetime into the seeds of structure.

VERY INTERESTING CONNECTION #3: The universe is accelerating right now, even without inflation, because of dark energy. The acceleration rate is very slow and the cosmological scale is enormous, so dark energy isn't building important structural seeds from quantum fluctuations right now. But a cosmological event horizon does exist: At a radius of about 16 billion light-years from here, galaxies are being accelerated outside of our cosmological horizon (which is currently 46 billion light-years in radius). Even though we see sources from beyond the cosmological event horizon now, we're seeing light that they emitted toward Earth in the past. Light they emitted after passing beyond our cosmological event horizon will not arrive. (Notice that a cosmological *event* horizon is different from an ordinary cosmological horizon: Light from beyond the cosmological horizon *hasn't arrived yet*, but light from beyond the cosmological event horizon will *never arrive again*.)

Today, a typical virtual particle has nowhere near enough energy to significantly influence the local shape of spacetime. But under the conditions at the onset of inflation, some particles had sufficient energy to slightly affect the curvature of space near them. Once a virtual particle became real, its gravitational influence would "freeze" into the fabric of spacetime. The inflationary expansion of space stretched those curves to larger sizes. When matter, especially baryons, responded to the frozen-in curvature, primordial sound waves were generated.

The Hawking mechanism produces the lightest particles more easily than massive ones. Photons (mass = 0) are produced easily, neutrinos (mass = barely above zero) are produced as well, and gravitational waves, or *gravitons* (mass = 0) are also produced. The photons from inflation can't be observed, since the universe remained opaque to them for 379,000 years. Neutrinos have a similar problem: Even though they're weakly interacting, they didn't break free until about 2 seconds, when the density of the universe dropped enough for them to travel without interacting with something. Only gravitons would have been able to travel freely since inflation. If we could measure them, this would give us an image of the universe when it was *far* less than 1 second old.

These potentially observable inflation-produced gravitational waves have much too long a wavelength and too weak an intensity to be detected by gravitational wave observatories, such as LIGO or LISA (Section 6.4). However, the gravitational push and pull of a passing gravitational wave should have left a measurable trace in the primordial plasma. Thus, we can reasonably hope to measure inflation's gravitational waves by their effect on the early universe, as recorded in the CMB.

In looking for such an effect, we note that gravitational waves propagate differently from photons. While photons can be linearly polarized in a particular direction as they travel, the polarization of a graviton swirls around like a corkscrew. Therefore, in addition to the polarization patterns in the CMB that were caused by electron scattering (Section 8.5), we can look for swirling patterns as well (Figure 12.18). This effect is extremely weak and has not been observed so far, but it could be observed by future, more sensitive CMB experiments.

FIGURE 12.18 CMB Polarization from Gravitational Waves If inflation really happened, then it should have produced a fleet of gravitational waves. Those waves would have created swirling polarization patterns in the CMB, which might be observed by future CMB instruments.

The Very, Very, Very Early Universe

In Section 12.2, we described the first 10^{-43} second as the Planck era, when quantum fluctuations were energetic enough to strongly warp spacetime. We noted that the universe, in this state, is sometimes said to be a *quantum foam*. We claimed that we do not know how to describe the universe as it existed prior to this time because we have no well-tested theory of quantum gravity. Perhaps string theory, or something similar, will fill that void. But for now, let's just make some thoughtful guesses about the Planck universe.

In a radiation-dominated universe, which existed after the inflationary epoch and prior to matter–radiation equality, the scale factor grows in proportion to the square root of age:

$$a(t) \sim \sqrt{t}\,.$$

(eq. 12.5)

The square root isn't particularly important; what *is* important is that if t goes up, a goes up, and vice versa. The equation also means that when t was zero, so was $\sqrt{t}$, and therefore a was zero, too. A zero scale factor means zero separation between any two points in the early universe. This condition demands infinite density along with zero size; it's called the *initial singularity*. Because of this $a = 0$ singularity, many people (and many textbooks) say that, according to the big bang theory, the entire universe emerged from a single *point*. (The center of a black hole is a singularity as well—a singular point in space—but the big bang's *initial* singularity would be equally pointlike in time.)

Apart from the immediate difficulty of trying to imagine a universe that's geometrically flat, possibly infinite, and yet still a point, there are other reasons that we shouldn't be too casual about calling the universe a point. One reason is that we don't really know that the universe was radiation dominated before inflation, so Equation 12.5 might not apply. Another reason is the quantum foam: At 10^{-43} second and earlier, *unknown* quantum effects are *known* to have dominated the structure of spacetime.

You'll recall from Section 8.2 on the Bohr model for quantizing the hydrogen atom that the nature of quantum physics is to make things operate in discrete steps, rather than a smooth continuum. An electron can have *this* energy or *that* energy, but none in between. One popular way to view the Planck epoch is to imagine quantizing spacetime itself in this way: little chunks of space and time, rather than a smooth universe. For the sake of quantum honesty (which would be a great name for a romance novel), we really shouldn't talk about a point universe if the universe might actually be built from little distinct pixels of *fixed* size, like a low-resolution digital camera image (which would make a great cover for the romance novel).

Another obstacle to truly knowing the earliest moments is inflation, which flattens any preexisting curves and imposes scale invariance on any new ones; by its very nature, inflation tends to erase the evidence for what the universe was like before it began! Thus, we find ourselves free to at least imagine any preinflationary condition. We might reason that any sort of quantum gravity condition is likely to render space-time vulnerable to the whims of quantum uncertainty. And by that rationale, we could even imagine that the universe, as it was at 10^{-43} second or sooner, simply popped into existence, a quantum fluctuation *of space and time*, rather than *in* it. We started this section by asking when little quantum effects are relevant to the study of the big universe; well, who knows? Perhaps the universe *is* some sort of quantum fluctuation!

How Mystery Became History

Now that you've read this far, go tell someone that you can confidently describe the universe as it was 13.7 billion years ago, when it was much less than 1 second old. Mention this to the stranger seated next to you on your next flight. Tell your friends on the ski lift. And the next time your family is gathered together for a meal, explain to them how you and others educated in cosmology now have an established description of the very early universe, even though absolutely no light could ever be observed from that era because the universe was opaque at that time. (If you have a family of cosmologists, then this will be old news to them, so instead, say it when dining with someone else's family—especially if you're relatively indifferent about being invited back.)

To anyone who has not studied cosmology, your claim of detailed knowledge of the universe, especially the early universe, sounds absurd. This response is natural skepticism, consistent with the best traditions of science! After all, how much *should* someone accept without seeing the evidence? And how much personal effort is involved in seeing the evidence?

In the latter half of the 20th century, scientific cosmology achieved a major unification. A field that used to be guesswork and speculation has graduated to the level of a high-precision, data-driven science. We now have a working theory of the evolution of the universe that spans all of cosmic time, with an astonishingly reassuring suite of independently measured confirmations. Think of all the lines of evidence for dark matter or dark energy. Think of all the different aspects of cosmology that are simultaneously validated by the WMAP data set alone! Then figure in the abundances of the light elements, the Sloan Digital Sky Survey, the dynamics of clusters, gravitational lenses, computer simulations, and so forth, all agreeing on the same picture of the universe.

Other sciences established similar unifying cornerstones in the last century. The standard model of particle physics, molecular genetics, medicinal chemistry, the complete life cycle of stars—all these and many more have become "hard science," subject to multiple independent experimental and observational tests.

Cosmology, then, is not alone. Its landscape is much bigger than that of other fields, and its "timescape" is vastly older. But in much of modern science, the easy stuff is already done, and new discoveries are possible only in the realms of the very small, the very large, or the very complicated. The evidence for these newer discoveries is not always of the sort that you can readily see for yourself. It takes time and study. But the payoff is tremendous.

A hundred years ago, when the knowledge base was much simpler to catch up on, astronomers were starting to learn about galaxies, except for the 90% of them that's dark matter. We didn't know where any of the elements that we're made of came from. No one knew how the universe began, when the universe began, or even *if* the universe began. These must have seemed like unanswerable questions. What incredibly comprehensive and detailed measurements would have to be made to address them! What a remarkable variety of specific theoretical predictions would all have to be satisfied by observations before we could trust anything we manage to discover! And yet much of what used to be considered "the great mysteries of the universe" we now call the *history* of the universe, reaching back to the first tiny, tiny fraction of a second.

Quick Review

In this chapter, you've learned how the progression from an incredibly smooth early universe to the lumpy universe we see today can be explained within the big bang framework. Tiny quantum fluctuations provided the initial seeds for structure. Amplified by cosmic inflation, these small fluctuations left their mark as anisotropies in the CMB. With the right mix of dark matter, baryonic matter, and expansion, gravity and time turned the small early anisotropies revealed by the CMB into the clumps revealed by galaxy surveys.

Try the following crossword puzzle to test your knowledge of key terms and concepts from Chapter 12.

ACROSS

5 The second acoustic peak is suppressed relative to the first and third because gravity works against the _____ that causes baryonic matter to reexpand.

6 Type of "quantum radiation" from the event horizon of a black hole, named for the scientist who predicted it

7 Problem solved by inflation, regarding why the total density of the universe is so close to the critical density

8 Stands for "cold dark matter"

9 The Sachs–Wolfe effect is caused by _____ blue and red shifts, instead of actual hot and cold regions.

11 Musical notes with one-half, one-third, etc., of some primary wavelength

12 The instantaneous creation of the universe, if we ignore quantum gravity and extrapolate the big bang theory back to time zero, is called the initial _____.

15 Ultrarapid burst of accelerating expansion in the very early universe

18 Waves of alternating compression and expansion of the medium they travel through

19 When two smaller halos combine into one large halo

23 High points on the CMB power spectrum graph

24 It began to cluster into structures when the universe transitioned from radiation to matter domination.

DOWN

1 Type of expansion caused by inflation, where the bigger the scale factor is, the faster it grows

2 Quantum _____ left an imprint of tiny curves in the fabric of the universe, which were amplified by inflation.

3 Problem solved by inflation theory, involving extreme homogeneity among parts of the universe that should never have been in contact with each other

4 Washes out higher numbered acoustic peaks

10 The size of a CMB spot caused by primordial acoustics is based on the _____ of the sound wave.

13 Computer process demonstrating that small initial density variations in the expanding universe have grown into the large-scale structure that's observed today

14 A way to display information about CMB anisotropy based on the temperature deviation for various spot sizes

16 The _____ is caused by baryons and dark matter working together.

17 Mechanism by which halos (and other objects) grow more massive as they pull in more matter by gravity

20 Event that sealed existing anisotropies into the CMB.

21 This type of particle released by inflation may have left a particular observable polarization pattern in the CMB. (*Hint:* See also Section 6.4.)

22 Large-scale examples of objects referred to by this general term include galaxies, clusters, walls, and voids.

Further Exploration

(M = mathematical content; I = integrative—builds on information from more than one chapter; P = project idea; R = reflection; D = suitable for class discussion)

Exercises preceded by "*" are based on material from supplemental Section 12.5.

1. (MI) What percent of the total matter in the universe is composed of neutrinos? What percent of the total nonbaryonic dark matter? What percent of the total dark matter?

2. (R) Describe in detail, with pictures, how a "string telephone" works. You can build one by connecting the bottoms of two paper cups with a taut string; speak into one cup, and someone else can hear you in the other cup.

3. (MRI) How do we know that dark energy will dominate the future of the universe the way it dominates the present? Answer as completely as possible.

4. (MR) Assume that the national economy cycles from expansion to recession repeatedly with a period of 12 years

for a full cycle. (Start at an average level, rise to maximum expansion, fall to maximum recession, and return to the starting point in 12 years.) You may assume (incorrectly) that economic expansions and recessions last equally long and that they follow a rising and falling wavelike pattern in time.

(a) What's the frequency?

(b) What state is observed 50 years after the starting point of the cycle?

(c) Consider five other countries with similar cycles and periods of 2, 4, 6, 8, and 10 years, respectively. Assume that all six nations start together at the beginning of their economic cycles. Which nations will be fully expanded or fully recessed in 15 years? How about 18 years? How about 21 years?

(d) In one paragraph, connect this question to cosmology.

5. (RD) Imagine that, instead of decoupling happening at a particular time, it happens for each sound wave during its first maximum compression, so that different sound wavelengths have different decoupling times.

(a) Sketch the power spectrum for this case. Write captions to explain what's going on on each part of your graph.

(b) Would the Sachs–Wolfe effect reverse the red and blue shifts on large scales, as it does in the real universe? Why or why not?

(c) Would you expect structure formation to be hierarchical in this universe, as in the real one? Why or why not?

6. (RPDI) Make concept maps for key topics in Chapter 12. That is, write down different concepts, such as "homogeneity," all over a page and connect them by a web of lines to some central concept and to other, related concepts. Do one concept map for each of the following central concepts:

(a) Dark matter

(b) Dark energy

(c) Baryonic matter

(d) Inflation

(e) Quantum physics (as it relates to topics in this book only)

7. (RD) What's your opinion of inflation theory? Answer in two or three paragraphs, and substantiate your position with relevant points of science.

8. (R) Address the following issues:

(a) Why does baryonic matter generate pressure?

(b) Why doesn't nonbaryonic dark matter generate pressure?

(c) How is this difference manifested in the first few acoustic peaks of the CMB?

9. (P) Make a poster of the CMB power spectrum. Add labels and pictures to explain the significance of each important part of the graph in terms of what information that part provides us about the universe.

*10. (RDI) Use the equivalence principle (from Chapter 6) to explain *how* we will eventually stop seeing a source in our night sky that is currently within our cosmological horizon, but outside our cosmological *event* horizon. Will it just wink out one night, fade in intensity over a few years, or something else?

11. (RI) Summarize the role of polarization in the CMB. What does it offer to those who study cosmology? Answer in connection to

(a) Small-scale polarization patterns

(b) Large-scale polarization patterns

(c) Any "corkscrew" polarization patterns

12. (RD) Write two or three paragraphs on the following topic: "Was the scale factor ever zero?" In what circumstances would the answer be yes? In what circumstances would it be no? If the scale factor was once zero, what does that mean?

13. (MR) A billion years from now,

(a) will Ω_{de} be bigger or smaller than it is now? Will it be bigger or smaller than 1?

(b) will Ω_m be bigger or smaller than it is now? Will it be bigger or smaller than 1?

(c) will Ω_{de} and Ω_m still add to 1?

Explain any ambiguity in your answers.

14. (RD) Say everything you can about a universe in which

(a) Ω_{total} is less than 1 today.

(b) Ω_{total} is greater than 1 today.

Your answers should include at least five separate points for each part, but try for as many as you can think of.

15. (RDP) Write an essay, make a presentation or video clip, or produce some other form of persuasive communication about what our society should do about the ever-growing body of meaningful scientific knowledge that's becoming harder and harder for the average person to learn about. (Imagine trying to learn the content of this chapter without reading the preceding chapters!) Use the following comment to guide your thinking:

It's part of the legacy of the modern world that the human "database" of knowledge has become enormous and highly specialized. It implies a policy decision to be made: Since more and more knowledge exists, how much should each individual have? Do we choose to become less broadly educated and more deeply specialized? Or do we spend more years in school? Or do we give up on some of the basics—perhaps those things which can be easily referenced or calculated by computer, like vocabulary and arithmetic—to free up more time to cover newer content? The disparity between what each of us *could* know and what each of us *does* know is always on the rise.

13 The Emergence of Complex Life

"This same spectacular transformation continued into the future, carrying these atoms into the form of the galaxies, and then into that of the molecules and cells, and then into the very form of the human and the elephant and the blue spruce and the Mississippi River."

— *Brian Swimme*

Life is strikingly absent from our tale of the universe so far, despite being a pervasive feature of our experience on Earth. Our investigation has revealed remarkable things not found on Earth: huge jets of material, exploding stars, black holes, beautiful patterns of gas and dust thousands of light-years across. But no life. In all the vastness of the universe we have discovered, every single living thing we know about with any certainty exists only here, on one small planet. Why is this so?

Probably it is a consequence of the vastness itself, combined with the subtlety of life. Even if there were another life-bearing planet like the Earth relatively nearby, we would be unable to recognize it with our current technology. From far away, we're restricted to seeing things that involve large amounts of energy, in order that we can still see them even after the inverse square law has taken its toll on the flux of light emitted. But such brightly energetic environments tend to be hostile to life as we know it, which seems to favor the relatively quiet, stable places in the cosmos. Since we observe only what we are *able* to observe, we should consider the possibility that the lifeless universe we see might be a filtered and biased sample of what's actually out there.

Even though life has not been central to our discussion of the universe so far, it has always been present in the background. After all, it is only because of our role as living observers that we know what we know about the universe. Hidden beneath the surface of our investigations lurk deeper questions: What conditions are necessary for the emergence of life? How is it that these conditions are realized in our universe? In this chapter, we explore the nature of life in order to return to ourselves, as the observers who emerged within the cosmos and started wondering about how it all fits together.

How can we integrate the universe we have been studying—the universe of dark matter and baryons, microwaves and neutrinos—together with the emergence of living observers? Let's begin by recognizing the following points:

- *Life exists in the universe.* Therefore, the physical properties of the universe, like the strength of gravity or the mass of an electron, are at least compatible with life.

- *Life might be a natural consequence of the universe.* That is, perhaps any universe like ours would always produce life (and maybe lots of it), just as it produces stars.

- *The existence of life as we know it is really a property of the whole universe, not just an isolated feature of the planet Earth.* For starters, the elements needed as the building blocks for life were formed long ago in other stars and in the big bang.

These points suggest a perspective for thinking about biological life in the context of cosmology. In Chapter 11, we listed important cosmic events chronologically, but we could approach those events differently. What if we organize the history of the universe by increasing *complexity* instead? It started out remarkably *simple*, with a gas of subatomic particles, the same everywhere. Physics at the particle level governed the synthesis of light elements and the decoupling of the CMB. But then it got lumpier. Dark matter, followed by baryons, self-organized into a hierarchy of structures based on relatively isolated galaxies. Within galaxies, gas clouds fragmented into stars, which added complexity by constructing heavier elements. These heavier elements later drifted through space and accumulated in new star systems, where they condensed into yet another generation of stars and planets. There seems to be a general trend in the universe toward assembling structures of increasing complexity out of simpler building blocks. The intricate pattern of a snowflake spontaneously forms from simple water molecules under the right conditions. Spiral arms form in galaxies of stars. On widely different scales, the same organizing trend is evident.

Where does the complexity story go next? Not to large-scale space, which is being accelerated apart by dark energy, but to planetary surfaces, where geological and chemical processes change the scenery over time. It seems very likely that the emergence of life is also a part of this general trend toward increasing complexity. Here on Earth, at least, life has woven itself fully into the dynamics of the planet, altering the composition of air, water, and soil. Biospheres, on Earth and perhaps elsewhere, add tremendous complexity to the universe.

If life is common throughout the universe, then are there other civilizations out there? Might they (or we!) colonize the Galaxy? These are great questions, so far without answers. But imagine how complete an understanding of the universe we might someday construct, one in which we comprehend the most complex topic the universe may have to offer: the emergence of life. By the scientific method, we might hope to someday validate a new *astrobiology* theory by finding the extraterrestrial life it predicts.

We'll start with our only firm reference point: life on Earth. (If you want to go straight to life *beyond* Earth, you can skip ahead to Section 13.2. However, studying how life came about on Earth will grant you a more informed perspective for thinking about the possibility of life forming somewhere else.)

13.1 Supplemental Section: Life on Earth

This section is organized into five subsections: (1) the physical composition of living things on Earth, (2) how living things might have emerged on Earth, (3) what makes them alive, (4) how they've changed over time, and (5) why they behave the way they do. You might want to ready yourself for the relatively nonastronomical nature of what you'll read in this section. It's fascinating in its own right, and it's useful to help you appreciate the scientific possibility of life elsewhere (Section 13.2), but it involves very few supermassive black holes.

The Stuff of Life

Let's begin by considering the living cell. Figure 13.1 shows a view of a cell under a microscope. This is a typical cell in that it has a **membrane**—a wall that separates what's outside from what's inside—and various substructures inside.

To go further, we appeal to chemistry: What are living cells made of? Broadly, the organic molecules of living cells break down into four categories. (Brace yourself: A *lot* of boldface vocabulary is coming! Hopefully, it will just refresh your memory of boldface vocabulary that you've encountered before.)

- **Proteins** form much of the physical structure of cells; that is, major parts of cells are built from different and specialized protein "bricks." In addition, a certain category of protein, called an **enzyme**, is essential for the biochemical functioning of the cell. There are thousands of different enzymes in living cells, and their presence either enables or speeds up many important cell functions, such as cell reproduction. (Similarly, their absence disables these functions or slows them down.) Therefore, proteins are critical to what a cell is *and* what it does.

- **Carbohydrates** include sugar, starch, and cellulose (wood). Carbohydrates are the primary energy storage molecules in cells.

- **Lipids**, also known as *fats* and *oils*, are long molecules that do not dissolve in water. For that reason, they tend to stick to one another and thereby isolate themselves from water. (Think of trying to mix oil and water: The oil forms large blobs and the water stays out.) Lipids tend to be flexible and tend to repel water, which makes them useful material for building cell membranes. They can also

FIGURE 13.1 Living Cell, Under the Microscope   A view of a single-celled organism called *Euglena* under a microscope (◆ Also in Color Gallery 2, page 29.)

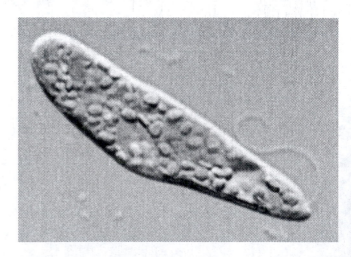

store energy, although many of us over 30 years of age find ourselves in a relentless quest to store our energy in the form of carbohydrates rather than lipids!

- **Nucleic acids** include **RNA** (ribonucleic acid) and **DNA** (deoxyribonucleic acid). These nucleic acids are used for storage (DNA) and processing (RNA) of *genetic* information—information related to characteristics, or traits, passed down from one generation to the next. DNA essentially constitutes the "instruction manual" or "recipe" that tells a cell what to do. Broadly speaking, DNA instructs its cell on how to build the right proteins, and those proteins make the cell work.

All of these molecules are called **polymers**, made by stringing a collection of simpler molecules together, end to end (a process called "polymerization"). Since they're used by living organisms, the ones just listed are **biopolymers**. For example, a polysaccharide is a type of carbohydrate made from a chain of simpler sugar molecules. It is noteworthy that the major components of living cells are just chains of simpler molecules; we are evidently built from *complex* sequences of *simple* ingredients:

- Proteins are polymerized from smaller components called **amino acids**. (Proteins are chains and amino acids are the chain links.)
- Carbohydrates are polymerized from **sugars**.
- Lipids are polymerized from **fatty acids**.
- Nucleic acids are polymerized from *sugars*, *phosphates*, and *nitrogenous bases*, which assemble into **nucleotides**, the individual chain links in nucleic acid chains.

The simple ingredients themselves are made from even simpler ingredients. For example, Figure 13.2 shows a chemical bonding diagram for a nucleotide. The bonding arrangement itself isn't particularly important for our present purpose; what's important is the list of chemical ingredients. *All* these biomolecules are made primarily from the same four chemical elements, which are *by far* the most common elements in living cells. Listed in descending order by the number of atoms of each kind found in cells, they are as follows:

1. Hydrogen (H)
2. Oxygen (O)
3. Carbon (C)
4. Nitrogen (N)

Apart from these four elements, a few amino acids have an atom of sulfur (S), and nucleotides have an atom of phosphorus (P). But the vast majority of the atoms in biopolymers are H, O, C, and N.

As we can note from Figure 5.10, the most abundant elements in the universe (again, by number of atoms) are

1. Hydrogen (made by big bang nucleosynthesis, BBN)
2. Helium (made mostly by BBN)
3. Oxygen (made by stars)
4. Carbon (made by stars)
5. Neon (made by stars)
6. Nitrogen (made by stars)

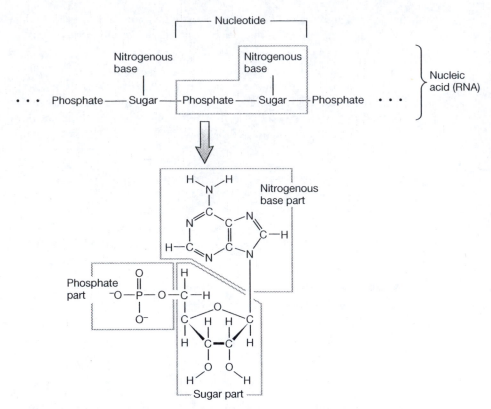

FIGURE 13.2 The Ingredients of Genetic Information   Genetic information is stored in nitrogenous bases, which connect to an alternating "sugar–phosphate–sugar–phosphate . . ." backbone. Notice how it's the same few basic elements that do the whole job here. The same is true for all major biomolecules. (RNA is depicted here.)

The only elements that differ between the living things list and the universe list are helium and neon, both of which are "inert"—chemically unreactive—due to their electron configurations. So even though there are lots of helium and neon atoms around, they don't form chemical bonds and therefore *can't* be a part of the biomolecules we're discussing.

Let's summarize what we have so far: Living organisms on Earth are made of cells. Cells are made mostly out of four types of biopolymers. Those biopolymers are made out of a small list of component molecules (e.g., amino acids), each of which is essentially made from only four elements (plus a little sulfur and a little phosphorus). Apart from chemically inert gases, those four elements are the most common elements in the universe.

We can therefore characterize the material form of life as being hierarchically assembled in successively more complex arrangements of the same four(ish) highly abundant ingredients. Nature evidently does the same thing that we would do to build any complex system (think of a car, perhaps): It manufactures a bunch of common parts (tubes, wires, seats, tires, etc.), using readily available raw materials (metal, plastic, fabric, rubber, etc.), and connects them together into subsystems (engine, electronics, passenger compartment, etc.). The living cell is constructed (*somehow*) by manufacturing a bunch of common parts (biomolecules: amino acids, sugars, fatty acids, nucleotides), using readily available raw materials (H, O, C, N, plus S and P) and connecting them together into subsystems (biopolymers: proteins, carbohydrates, lipids, nucleicacids).

Now we need to think about what's behind the word "somehow" in the previous sentence.

The Origin of Life

Historically, the usual scientific hypothesis for life's origin is the *primordial soup*. (A word of warning here: To a cosmologist, the "primordial soup" usually refers to the universe before the first second—a sea of baryons and electrons, photons and WIMPs, etc. To a biochemist, "primordial soup" refers to the ingredients of amino acids, sugars, etc., drifting around in a body of water on Earth. The difference is about nine billion years.) The primordial soup idea starts with chemicals and ends with life; chemistry turns into biology. Before reading any further, it might be useful to give that concept some thought.

EXERCISE 13.1   Creepy Thought Experiment

Imagine that you had a machine that could disassemble someone, atom by atom, and keep a record of where each atom goes. Then you could put the person back together later with the same machine. Assume that it all happens in about a minute, so that your guinea pig person dies from being disassembled, rather than from hunger or suffocation. Pick someone that you'd like to have out of the way for a while, but you don't necessarily want to kill, and run this thought experiment with him or her in mind. The question is, When you reassemble the person, will he or she be alive again? Answer in four parts:

(a) What's your immediate instinctual answer? And what's the first reason that pops into your head?

(b) Now give the problem a little more thought. Give your answer, and justify it. For example, if you say yes, then explain how the reassembled collection of atoms starts functioning again. If you say no, then explain what prevents the reassembled person from functioning again.

(c) Does it make it any easier to consider something simpler than a human being? Try one or more alternatives, like a cat, or a tree, or a single-celled organism.

(d) Change the question so that you're no longer disassembling the person first. Instead, you're simply assembling a person or a tree from a collection of all the correct atoms, which you put in all the correct places, but these particular atoms have never been in a living thing before.

There is no correct answer to any of this. We don't have the technology to perform this experiment (yet!). Just try to flesh out what you *think* would happen, and try to understand why you think whatever you think.

A good first step in the origin of cellular life, given the apparent simplicity of its molecular construction, seems to be "put all the ingredients in the ocean, let them get stirred around by water currents and maybe heated by sunlight, and poof! Life appears!" Of course, the entire premise that the miraculous "spark of life" boils down to "primordial soup" and "poof" sounds ridiculous. It sounds so ridiculous, in fact, that the idea wasn't even seriously tested with the scientific method until 1952, by a graduate student. The result was a two-page publication in 1953.

Stanley Miller (born 1930) and Harold Urey (1893–1981), two American chemists, conducted the first major origin-of-life experiment, known as the Miller–Urey experiment. Their research showed that biomolecules such as amino acids can form spontaneously from relatively common chemical reactions in a relatively simple environment. They effectively transformed the subject of the origin of life into an experimental science.

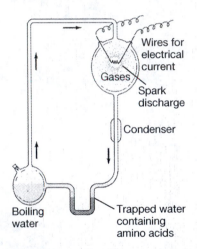

FIGURE 13.3 The Miller–Urey Experiment   This apparatus is designed to simulate the young Earth in all the ways that matter for the spontaneous assembly of simple organic molecules: oceans, atmosphere, and energy sources such as geothermal heat and lightning.

Stanley Miller, aided by his advisor, Professor Harold Urey, attempted to simulate the conditions on the very young Earth, conditions that existed about four billion years ago, roughly when life first appeared, according to various lines of fossil evidence. The apparatus is shown in Figure 13.3. Miller and Urey made an "ocean" with a water reservoir, to host the primordial soup. They heated it, to simulate geothermal warmth, which would have been a major energy source so soon after the Earth formed. They shocked it with electrical arcs, to simulate lightning. Above the simulated ocean, they added a simulated atmosphere, by using published theoretical predictions of what the atmosphere should have been like. Then they left the experiment alone, heating and arcing for a week, after which they analyzed what chemicals had formed. Here's what Miller reported:

> "The water in the flask was boiled, and the discharge was run continuously for a week. During the run the water in the flask became noticeably pink after the first day, and by the end of the week the solution was deeply red and turbid. . . . The red color is due to organic compounds adsorbed in the silica. Also present are yellow organic compounds . . . It is estimated that the total yield of amino acids was in the milligram range. In this apparatus an attempt was made to duplicate a primitive atmosphere of the earth, and not to obtain the optimum conditions for the formation of amino acids."
> —Stanley L. Miller, "A Production of Amino Acids Under Possible Primitive Earth Conditions."

Prior to that week, there was *nothing* organic in the apparatus. After the week was up, however, it contained amino acids and other biomolecules. This was huge. The **Miller–Urey experiment** sparked (no pun intended—well, maybe a little) a whole new field of study: *experimental* origin-of-life science. After all, if amino acids are so easy to form naturally, then maybe the rest of living tissue is, too. Suddenly, the path to understanding the "spark of life" looked like a series of questions that might actually be answerable.

Among the biomolecules formed in the Miller–Urey experiment, and in subsequent versions of the experiment carried out by other scientists, were five particular biomolecules called *nitrogenous bases*. DNA and RNA each use four of these five bases to store and process all the genetic information in a living being. The nitrogenous bases in DNA show up in pairs, cleverly called **base pairs**. It is the *sequence* of these base pairs in DNA that determines all genetic information.

The Miller–Urey experiment demonstrates that all the genetic *components* of DNA, which mastermind just about every aspect of living cells, form *automatically* in a primordial soup! Once they are assembled, they form a "double-helix" shape (Figure 13.4), with nitrogenous base pairs arranged like the rungs of a ladder on a twisting "backbone" of deoxyribose (a sugar) phosphate.

The genetic structure of DNA is usually described in layers. Particular three-letter sequences of base pairs are called **codons.** A single codon instructs the cell to use a particular amino acid. A string of amino acids connected together in the right order, ranging up to tens of thousands of amino acids, makes a protein, which is encoded in DNA by an equally long string of codons. That string of codons is called a **gene**. Finally, the full set of all the genes in an organism is its **genome**. If your genome is like an entire *book* of information about you, then base pairs are the individual *letters* in the book, codons are the *words*, and genes are the *sentences* (or paragraphs, or chapters, depending on the length of the particular gene). Miller and Urey basically found out how to make the letters.

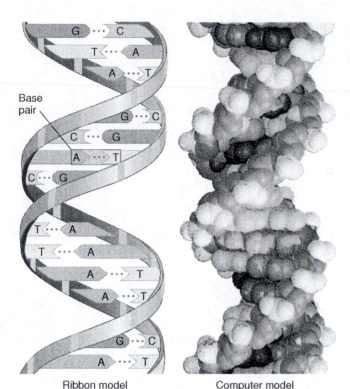

Base pair

Ribbon model   Computer model

FIGURE 13.4 The Double Helix Shape of the DNA Molecule   "Base pairs" (horizontal links) are mounted on a twisting twin backbone of "deoxyribophosphate." (Deoxyribose is a sugar, similar to the one shown in Figure 13.2.) (◆ Also in Color Gallery 2, page 29.)

Others have since conducted variations and improvements upon the original Miller–Urey experiment. Some involved adding energy sources such as ultraviolet light and natural radioactivity, and some involved modifying the composition of the Miller–Urey atmosphere. All of these experiments demonstrated the spontaneous generation of *many different and important* biomolecules, including

- all the amino acids commonly found in living cells
- multiple sugars
- multiple fatty acids
- all the nitrogenous bases in DNA and RNA

To achieve greater quantities of these biomolecules than are produced by Miller–Urey type experiments alone, we might consider another interesting possibility: that some (or all) of life's ingredients did not form on Earth! Since the Miller–Urey result tells us that it doesn't take much to build these ingredients, they might have been assembled on comets, or asteroids, or other planets and then delivered here as meteorites. Early in the history of the solar system, a heavy bombardment of meteors rained down on the Earth and other bodies. (You can see the resulting craters all over the Moon, for example, where there is no natural erosion to "erase" them, as there is on Earth.)

In 1969, in the town of Murchison, Australia, a carbon-rich meteorite fell from space. The *Murchison meteorite* resided in the seemingly sterile main asteroid belt for most of the history of the solar system. But when scientists analyzed its composition, they found most of the same types of organic molecules, including amino acids, as were produced by Miller and Urey. Evidently, we are correct to think that forming organic material from inorganic material doesn't take much if it can happen in a laboratory *and* on an asteroid.

Either way—whether from our oceans, or from meteorites, or from both—we have a solid basis for assuming that amino acids and other simple organic molecules did exist on the young Earth. Eventually, someone will probably figure out which source of these organic molecules is most important to our history. But we really can't wait until they do; we have to move on to the next event needed for the formation of life, which is polymerization. This is where we start to run into trouble.

We'll begin with nucleic acids and proteins. In modern cells, all proteins are built in the following way:

STEP 1   Part of a DNA strand "unzips" from its normal double-helix shape into two single strands, as shown in Figure 13.5. (Remember that the rungs of the ladder in the DNA (Figure 13.4) are formed from nitrogenous base pairs, and that the particular sequence of base pairs constitutes the information stored by the DNA.)

STEP 2   A molecule called mRNA, or *messenger* RNA, is formed by matching up nitrogenous bases with a section of one of the DNA half-strands. This process, called **transcription**, essentially copies the DNA's encoded information onto the mRNA. As the mRNA is assembled, link by link, along the DNA half-strand, completed sections peel away and the DNA rezips behind it.

STEP 3   A molecule called tRNA, or *transfer* RNA, "reads" the mRNA and matches it up with a specific sequence of amino acids. This process is called **translation.** The newly formed sequence of amino acids—the protein—is the result of the whole process.

Why is this three-step process so problematic? Because each step takes place only in the presence of *enzymes*. About *twenty* different enzymes are needed in total. But what is an enzyme? It's a type of protein! In other words,

You need proteins to make proteins.

As for DNA, the odds of randomly assembling a useful DNA sequence—one that codes for a useful protein—can be astronomically small, because proteins can have thousands of amino acids in a specific sequence. In modern cells, this problem is solved by DNA replication, in which existing, correctly sequenced DNA copies itself. But the problem this raises for the origin-of-life theory is this:

You need DNA to make DNA.

The odds of randomly assembling a sequence of amino acids into a biologically useful protein can also be astronomically small. (See Further Exploration Exercise #7 at the end of this chapter.) We might just put off the problem of how the first such protein or DNA ever formed, despite the tremendous improbability, since it's hard to answer. But even if a useful DNA segment or protein just magically popped into existence, we'd *still* be stuck: Without the genetic information found in DNA, proteins can't construct copies of themselves. And although DNA is considered "self-replicating," this, too, requires an enzyme! So,

You need DNA to make proteins, but you need proteins to make DNA!

The situation is like this:

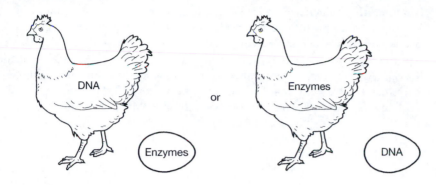

DNA

or

Enzymes

Enzymes

DNA

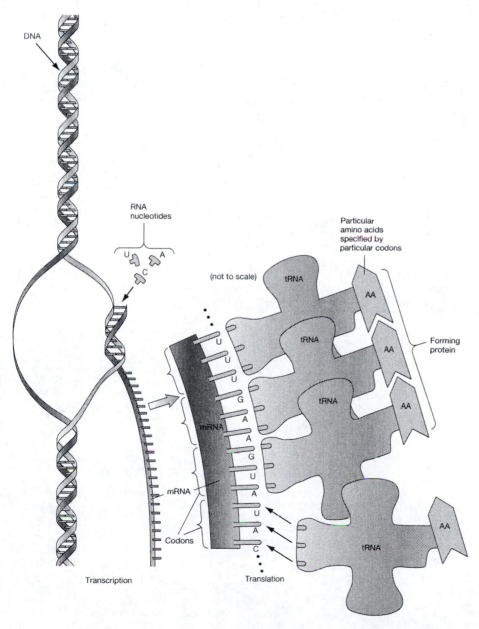

DNA

RNA
nucleotides

U · A
C

(not to scale)

tRNA

Particular
amino acids
specified by
particular codons

AA

tRNA

AA

mRNA

tRNA

AA

mRNA

Codons

AA

tRNA

Transcription

Translation

Forming
protein

FIGURE 13.5 Your DNA, Hard at Work. The double-helix partially unzips and then mates with mRNA, which subsequently mates with tRNA, which snaps together the amino acids on its tails to form a protein. The DNA zips up again when it's done; that way, its friends don't make fun of it. Now, all this is great, except that the enzymes made *by* this process are needed *for* the process, so how did it all happen the first time?

It would be cruel for us to just leave it at that, because the last few decades have seen mounting evidence for a possible solution to this problem. Some RNA has the ability to serve as its own enzyme and has been shown to polymerize on its own under certain conditions. This discovery leads to the extremely promising **RNA world hypothesis**. It proposes that, in the early "RNA world," primitive life needed neither DNA (which can't function without enzymes) nor enzymes (which, unlike DNA, can't store genetic information for replication). Some form of RNA that handled *both* functions was the Earth's only mechanism for handling *either* function. Figure 13.6 maps out this RNA world idea.

If the RNA world really existed, that would answer another issue as well. RNA is much more prone to incurring copying errors than DNA is. So early RNA might have introduced random changes in its own sequencing during replication, until some of those accidents happened to be useful—that is, until they assembled RNA strands whose coding matched that of useful proteins. Then, once life (or something like life, perhaps) was more established, the less error-prone DNA was able to take over as the genetic basis for life. In other words, errors were actually *helpful* at first, because nothing worked at that time anyway, so there was really nothing to lose by randomly modifying the base-pair sequences. But once useful sequences had formed, DNA became more valuable because it rarely messed them up.

In any case, we identify the first hole in the origin-of-life theory:

HOLE #1 IN THE THEORY: Assembling the major biopolymers, including DNA, presents a problem. How do we get past the chicken-and-egg paradox with DNA, RNA, and enzymes? The answer is likely to involve the RNA world hypothesis, but the specifics are far from certain.

You see that we have some leads on the polymerization problem, but nothing complete and convincing just yet. Still, we shouldn't quit there. Maybe you can imagine a more solid resolution. (See Exercise 13.2.) Let's say hypothetically that some new research comes along and solves this issue completely. What else would remain to be explained about the origin of life on Earth? Right now, we have biopolymers (by assumption), but we still need cells.

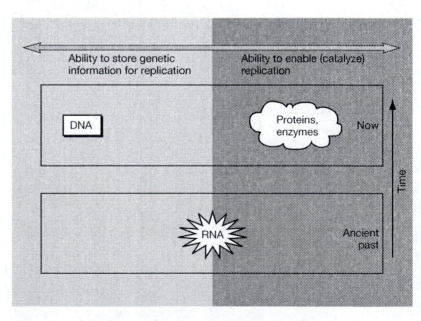

FIGURE 13.6 The RNA World Hypothesis   Time marches forward as you climb up the figure, and the twin roles of early RNA—handling genetics and enabling reactions—are replaced individually by DNA and enzymes, respectively.

EXERCISE 13.2   Plugging Up Hole #1

Invent a hypothetical solution to hole #1. It may be based on an RNA world, if you like that idea, but it must go beyond the RNA world and specify enough details to explain how biopolymers first formed on the early Earth. Don't worry about being scientifically *correct.* You're not expected to develop the actual solution to the problem. Merely present what the solution might *look like.* (One or two paragraphs ought to do it.)

Our purpose is to try to figure out how wide a hole we're dealing with. Is it something that you can imagine being plugged easily (a two minute explanation appears one day on the television news or a sixty minute program on PBS)? Or is it prohibitively difficult to even *imagine* what form a solution could take?

The Nature of Life

To make cells, we need biopolymers to build their own membranes. This process turns out to be quite plausible. Cell membranes are made primarily from lipids, proteins, or both. Under the right conditions, each of these materials can form a simple membrane. For example, lipids in water can spontaneously form into sheets that will wrap around to form hollow spherical membranes called *vesicles* when squeezed through a small opening. Mixtures of biopolymers in water are also observed to spontaneously form hollow spherical droplets, called *coacervates*, made from proteins (Figure 13.7). They automatically form in such a way that the concentration of biopolymers inside the coacervates is higher than that outside.

So cell-like structures are evidently capable of forming themselves if the ingredients are present. A more interesting question is "How close to 'life' does that get you?" We've already come quite far in the journey toward living cells (all the way from the big bang to vesicles and coacervates!). How will we know when we're actually there?

To answer this question, we'll have to think about what it means to be alive. For the most part, we can all recognize life when we see it, but defining it tends to be more difficult. For example, any reasonable definition of life would surely

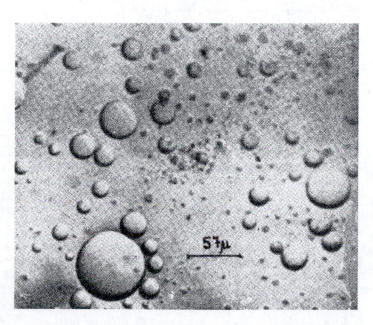

FIGURE 13.7 Coacervates: Precursors to Living Cells? When the major biopolymers (discussed earlier in this chapter) are mixed in water, cell-like droplets called coacervates form automatically, including a protein-based membrane that encloses a higher concentration of organic material inside than out. Lipid membranes called vesicles can spontaneously form under the right conditions, too.

include something like "the ability to reproduce." But mules, which are bred from a donkey and a horse, ordinarily *can't* reproduce, so do you want to change the definition so that it no longer requires reproduction, or do you want to tell yourself that mules are nonliving, like rocks? (To be fair, these "rocks" can carry you down and up the Grand Canyon, metabolizing and releasing waste much of the way!)

Other definitions face other difficulties. "Response to stimuli," for example, is another frequently accepted characteristic of life. But what about an animal that's temporarily dormant, such as a frozen frog that returns to normal when it thaws? And for that matter, what about a computer, which responds to inactivity by activating the screen saver, until someone touches the mouse. That's a response to a stimulus, but the computer is not alive. You may say, "Aha! The computer needs to respond to stimuli *and* reproduce! The trick is that it must meet *several* different requirements!" But what if a computer controls a factory that builds other computers (and thus reproduces) *and* responds to stimuli? For any definition you invent, someone can probably find a counterexample. Perhaps the fact that we can't even define life tells us something deep about its nature.

To proceed, then, let's give up on trying to define life perfectly; we'll tolerate some exceptions and just aim for the main ideas. Our objective is to apply some criteria for life to coacervates or vesicles, to see how close they are to being alive. We don't have to start from scratch; we'll start with biology textbooks, which typically list some or all of the following attributes of life:

- Heritable reproduction ("heritable" means that offspring resemble their parents)
- The ability to undergo evolution by natural selection (we'll be more specific about this shortly)
- Metabolism (using energy, to move or grow, for example)
- Growth and development (like a caterpillar turning into a butterfly)
- Movement (blinking, chewing, swimming, etc.)
- Response to stimuli (like the way flies evade you when you try to swat them)
- Adaptation (like protective shells on turtles)
- Ordered structure (cell structure, arms and legs, branches and blossoms, etc.)

Most biologists would agree on the first two characteristics, and maybe the first three; others are less firm. (Trees don't move much, for example, but are alive.) Even without a clear definition of life, one could argue that vesicles and coacervates do have some of these important attributes.[1]

For example, if a large droplet smashes against a rock and breaks into little pieces, the pieces' membranes are likely to re-form with some of the same DNA inside them that was inside the original large droplet. This sounds a lot like heritable reproduction, from the (larger) parent to the (smaller) offspring. Picture a droplet's membrane letting certain things in and certain things out. That sounds something like metabolism (taking in food and releasing waste). A droplet containing DNA that makes proteins which make the membrane stronger might be considered as possessing a survival advantage that could be passed on to its offspring. The point is, although vesicles and coacervates are not generally considered to be alive, they do appear to be tantalizingly lifelike. So let's designate another gap in our understanding of the origin of life on Earth:

---

[1] This argument was nicely articulated by Frank Shu in his book *The Physical Universe, An Introduction to Astronomy* (Sausalito: University Science Books, 1982).

HOLE #2 IN THE THEORY: Coacervates and lipid vesicle structures exhibit some of the properties mentioned in the definition of life. However, these entities do not match any modern organisms. (Compare Figure 13.7 with Figure 13.1.) To genuinely fill this hole, we would probably need to *build* a single-celled organism in the laboratory.

Life is difficult to nail down, and there may not be an easily defined line that separates it from the inanimate world. It may be more like a spectrum ranging from "more alive" to "less alive." But as before, we'll just suppose that this hole is plugged and that primitive life on Earth has somehow formed (chemically). Now all we need is for that primitive life to diversify into forests (that burn and regrow), and whales (that sing and play), and bugs (that plague our homes, our crops, and our picnics, but are surely good for something, or so we're told), and even a civilization with the desire and the talent to explain all these things in terms of processes that began 13.7 billion years ago!

The Evolution of Life

Skipping over hole #2, we now resume from a world full of primitive, reproducing, membrane-encased cells. At this point, *selection* processes took over the evolution of life on Earth. The process of **natural selection** is based on remarkably simple logic:

- *If* a population exhibits *variations* (like differently shaped beaks within some species of bird),
- *and if* its members *reproduce heritably* (offspring have beaks that resemble their parents' beaks),
- *then* the population will evolve by *survival of the fittest*, with survival-enhancing traits spreading throughout the population. (If a particular shape of beak is better for cracking open seeds in order to eat them, then, over time, all the birds' beaks will evolve toward the better shape.)

The natural selection mechanism thus adapts a population to its environment. Which traits survive in the population depends largely on which individuals survive, since the living are far more likely to reproduce than the dead! Any trait that helps you live long enough to reproduce provides an advantage. Over time, that advantage spreads through the population, as the helpful gene mixes into more and more individuals from one generation to the next. Natural selection can result in changes within a species or even the creation of a new species.

Natural selection operates for single-celled organisms and for multicellular ones, but we still need to explain the transition from single-celled to multicellular life. How this happened is not well known. Fossil evidence indicates that multiple multicellular lines emerged between about 1.8 and 0.9 billion years ago; they were only single-celled before that, as far as we know. Some of these new multicellular organisms may have gone extinct. However, 600 million years ago, multicellular organisms rapidly diversified into a rich variety of plant and animal life. Pieces of this transition are partially understood; for example, a sponge is an animal that contains loose groupings of cells—each of which resembles certain single-celled organisms—rather than real multicellular tissues. As such, a sponge probably represents an early form of multicellular life. However, very little of this history is known in satisfying detail, so let's designate

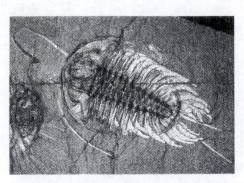
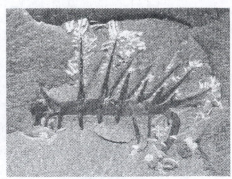

FIGURE 13.8 Fossils from the Cambrian Explosion   The Burgess Shale fossil set, a sample of which is shown here, reveals a rich variety of body structures for early aquatic animals, all from between 500 and 600 million years ago.

HOLE #3 IN THE THEORY: The transition to multicellular life needs further study. Currently, we lack the compelling data we would need to accept any particular proposal for how this shift occurred.

At this point, we'll skip over hole #3, as we have with the others before it, and resume the story of life on Earth with a diverse crowd of multicellular organisms. The "Burgess Shale" fossil set from the *Cambrian explosion*—which saw the rapid proliferation of a great variety of life-forms—shows the assumption of diverse multicellular life to be valid. Figure 13.8 presents some of the fossil evidence from this time. Natural selection can explain the changing forms of the Cambrian explosion and beyond, although selecting purely for survival traits was not the *only* process responsible.

Another important mechanism for biological change over the generations is **sexual selection**, in which the traits that help you to reproduce—in addition to those that help you to survive—are favored in a population. One interesting theory proposes that human brains evolved their powerful intellectual capabilities for just that reason: to attract a mate. Among our distant ancestors, an ultrasophisticated brain (eventually capable of mastering cosmology!) might have been more valuable for impressing a potential mate than for enabling "ordinary" survival advantages, like hunting in groups or cooking with fire. Sexual selection could be responsible for human intelligence if smarter humans were more likely to have sex. (Of course, we all know that this wasn't true in high school, but that's okay, since most reproduction happens after high school.) The evidence for this idea is not particularly compelling at present, so you're welcome to think of the growth of human intelligence as "hole #4." However, *some* sort of *social* benefit—even if it isn't particularly related to sex, like the ability to employ complex language or interpret the facial expressions of others—is likely to explain our great intelligence.

Natural and sexual selection change our biosphere over time in a process called **evolution**: the theory that all life today resulted from accumulated genetic changes that started with common ancestors. Evolution is a well-documented laboratory science today; not only can you look for evolved traits in modern organisms, but you can use a captive population of a fast-reproducing species to literally observe evolution happening. The point is, selection mechanisms, such as natural selection and sexual selection, can produce extraordinary results. Over time, they can transform chemically driven bacteria into intelligent and technological beings. Great complexity can emerge from simple beginnings when selection processes are at work.

The History of Life

At this point in our quest to understand how the universe promotes complex life, we've surveyed the most relevant natural processes as they took place on Earth. Figure 13.9 shows many of the milestones in this history of life on Earth. Among the most important, we have

- 4.6 billion years ago: The Sun and Earth form.
- 3.8 billion years ago: The earliest *biosignatures* (signs of life) are imprinted—

—which show that single-celled life formed less than a billion years after the planet formed, and under harsh conditions. (It was a time of tremendously violent activity, including volcanoes and frequent meteor impacts.) It may stand to reason that simple life therefore forms quite easily, since it happened so "soon" despite the harsh conditions, or "immediately" after the harsh conditions ended.

These biosignatures are found in ancient rocks and face some dispute: Some researchers argue that they may not actually come from living organisms at all.

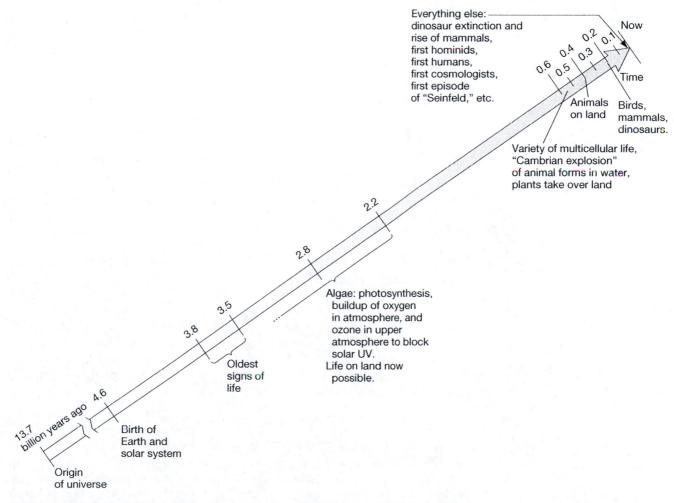

FIGURE 13.9 History of Life on Earth   This timeline shows key milestones in the history of life on Earth approximately to scale, to the extent that the history is actually known.

However, 3.5 billion-year-old *stromatolites*—rocklike structures built by microbes—are widely accepted.

- 2.8 to 2.2 billion years ago: Earth's atmosphere gains oxygen—

—which includes both the molecular form that all animals breathe, O_2 (called "oxygen"), and the molecular form that absorbs most ultraviolet sunlight, O_3 (called "ozone"). It is at this point in history that life *on land* became possible for the first time, because of the protection from ultraviolet sunlight provided by the ozone layer. The first primitive plant life produced all this oxygen as a by-product of photosynthesis (the process by which plants use sunlight to make their own food). It probably took several hundred million years for oxygen to build up in our atmosphere, so photosynthesis had been going on earlier than the dates listed here.

- 1.8 to 0.9 billion years ago: Multicellular life arises.
- 1.4 to 0.8 billion years ago: Bacteria start reproducing sexually—

—which marks a speedup in biological evolution. (There is no particular evidence that bacteria are any happier as a result.) Remember that Darwin's selection mechanism relies on variations within a population. Without sexual reproduction, those variations arise by *mutation*, in which random changes (errors) arise in an organism's DNA. This is relatively rare, which makes evolution by natural selection very slow. But with sex, the major variations are introduced with every generation, since the DNA from *two* parents is *mixed* to make their offspring.

- 0.6 billion years ago: The Cambrian explosion of plants and animals occurs.
- 0.5 billion years ago: Plants become prominent on land.
- 0.4 billion years ago: Animals become prominent on land—

—amphibians and nonflying insects first, followed by reptiles and flying insects.

- 0.2 billion years ago: Birds, dinosaurs, and mammals appear.
- 0.065 billion (65 million) years ago: Dinosaurs go extinct—

—and mammals rise to power. In times of abundant plant and animal food, the larger animals were favored, and dinosaurs ruled over the mammals. But scientists have found evidence that a massive meteor impact changed those conditions and the situation reversed: Little rodentlike mammals could find enough food to survive, but massive dinosaurs, which needed much more food, could not. Mammals were able to inherit—rather than conquer—the Earth.

- 0.002 billion (2 million) years ago: The first *hominids* appear—

—which include the first humanlike species to *walk* on two legs and to have a significant capacity for *language* and *culture*. Interestingly, it has been proposed that one of the most noteworthy distinctions between modern humans and early hominids is the *ability to tell lies*. The ability to deceive others is a significant one, because it means you understand how *they* see things.

- 0.0002 billion (200,000) years ago: The first humans appear—

—whose existence occupies a minuscule fraction of the Earth's history, 0.0002 billion/4.6 billion = 0.004%, and a smaller fraction of the universe's history, 0.001% (how did we compute that one?). We exist at the tip of an interconnected biological, geological, and cosmological history; in each of these time scales, the presence of our species marks only the last few thousandths of a percent or less. Our recorded history is a hundred times shorter (~0.000 01%), and our technological and space-exploring history is shorter still (~0.000 001%).

The Behavior of Life

Cosmological and biological evolutionary history appear poised to explain all the physical forms of life. Might this history help explain the *behavior* of human and other life as well?

Mice can be bred to like alcohol. Some bees carry genes that make them systematically find and remove bacterially infected larvae from their hives. In humans, manic depressiveness (which, statistically, probably affects someone you know) and schizophrenia are both heavily studied genetic conditions. They can be inherited, and they affect emotion and personality. Our behavior is, at least in some ways, chemical.

We all know, even as babies, to hold our breath under water. We all know instinctively that snakes and spiders are dangerous. We all dislike the taste and smell of rotten fruits and vegetables. Most of us intuitively agree on certain aspects of beauty, in landscapes, flowers, and other people. There is an ever-controversial debate over the role of nature (for genetically inherited behaviors) and the role of nurture (for learned behaviors) in determining how we act. The relevant point here is that the two have very different *timing*. Learned behavior becomes a part of you during your life, but evolved behavior *became* a part of you thousands of generations before you were even born. The study of these evolved behaviors from long ago is called **evolutionary psychology**.

We might imagine that thousands of generations ago—say, 50,000 years ago, on the savannas of Africa—the evolved behaviors of the time would have been useful for survival and reproduction in clans and caves. They might have included the discipline and aggression required for successful hunting. They might have included "family values," for successfully raising children. They might have included a subconscious preference for a mate who bears the physical traits important to successful reproduction (angular men and curvy women).

But do all of these qualities continue to provide an advantage in survival or reproduction today? It's fairly easy to imagine how anger, for example, might have once been selected for, possibly providing a benefit when hunting or when defending one's family from intruders. But how often does anger help you survive or reproduce in the modern world? Do you often receive promotions at work after outbursts or anger? Does it lead to tender reproductive episodes? The aggressive response to anger might be helpful in a world of stones and spears, but not so in a world of courts and prisons (and certainly not so in a world of nuclear and biological weapons).

The key to benefiting from the knowledge of your origins may begin with the recognition that your objectives are not necessarily your own! The list of ingredients in your body (and your planet, and your galaxy) was decided by nucleosynthesis, both big bang and stellar. The shape and function of your body and brain were decided by chemistry and evolution. And some of the things you enjoy, or care about, or wish for, were similarly decided in part by a history that long precedes you and your living family. Intelligence and reflection have also evolved from this history. These traits may now allow you to go against some of the behaviors that evolved long ago in a different environment.

What if—through science, study, and soul-searching—we learn to understand the nature of our objectives and our sense of what's important? Can we teach ourselves how to distinguish between our earlier evolved desires and our present ones? Can we *choose* to value cooperation above competition, even in cases where our brains were wired to the contrary? Can we learn to control those emotions that are no longer helpful, and forgive each other for our more primitive behaviors? After all, they're not *our* fault; they're the "fault" of our ultradistant ancestors! Understanding our distant origins may help us to bring about greater peace and prosperity in the present.

13.2   Life on Other Worlds?

If the processes that initiated and evolved all life on Earth are natural ones, stemming from the same general trend toward complexity operating throughout the universe, then it becomes reasonable to ask, "Should we expect this sort of life-bearing process to happen naturally on other worlds, too? And if so, how often?"

Exploring the possibility of life on other worlds is motivated by our curiosity: We just want to know. But perhaps there is a grander purpose driving us to determine whether life is a rare feature in an otherwise lifeless universe, or whether it is the very *nature* of the universe to speckle itself with life. The ultimate answer to this question would represent a capstone for our understanding of cosmology and of who we are.

The answer we seek will occupy a particular location along an axis that runs in two directions:

←───→

All life is   Rare-earth   Technological, intelligent

rare.   hypothesis   life is common.

AT ONE END OF THE AXIS, we are exceedingly rare (or utterly alone) in the universe. This might make us special or unique; or it might make us an accident of no cosmic importance. Either way, we're not part of an interstellar community, and the future evolution of the universe is likely to follow principles with which we're already familiar (e.g., cosmic expansion dominated by dark energy).

AT THE OTHER END OF THE AXIS, the conditions necessary to support life are common in the universe. There might be a large number of habitable planets with the proper chemistry in place, or perhaps life just happens to be easy to make. (On Earth, *extremophiles* support this position. Extremophiles are life-forms that thrive despite living in environments of harsh temperature, radiation, acidity, pressure, dryness, salinity, etc.) It might also be that some fraction of the planets which initiate their own biology routinely develop an intelligent, space-faring civilization for us to encounter someday.

IN THE MIDDLE OF THE AXIS, life is uncommon, but when it does start somewhere, it eventually leads to an intelligent species. Or more likely, according to the *rare Earth hypothesis*, the reverse is true: Simple microbial life might be quite common—perhaps even on other worlds in our own solar system—but the billions of years of continuously mild conditions needed for that life to evolve complexity and intelligence are exceedingly rare. In that case, we got lucky here on Earth.

Fossils from Mars

Our best hope to definitively answer some of our questions about extraterrestrial life would be to find direct evidence for it somewhere. As it turns out, the red planet next door appears to be an especially promising place to look.

For some time, astronomers have known that Mars once supported liquid water. Figure 13.10 shows a satellite photograph of a former Martian water channel; such dried-up rivers are common features on Mars today. Robotic rovers sent to Mars by NASA have discovered several different minerals that are known to form in the presence of liquid water. But evidently, the Martian environment changed. The atmospheric pressure dropped radically as carbon dioxide froze out of the atmosphere and onto the polar ice caps. At low pressure, liquid water can't exist. It can freeze or evaporate, to ice or vapor; it just can't exist as a liquid. (If the idea of cold water vapor sounds unnatural to you, imagine the low air pressure to be like a vacuum, sucking water molecules off the liquid surface and into the air.) Liquid water may exist *underground*, though; Figure 13.11 shows some possible evidence of a recent flow of liquid water from an underground reservoir out onto the surface of Mars.

Since the Martian atmosphere was thicker in the past, it would have included the carbon dioxide that has since frozen onto the polar caps. Carbon dioxide, or CO_2, is a *greenhouse gas*, which warmed the red planet (and contributes to global warming on Earth today). The importance of all this is that Mars used to be both *warmer* and *wetter* in the past, which probably made it a good place to form life. Today, the surface of Mars is cold and dry, so any life that once existed is presumably dead and perhaps buried. Recent evidence for trace amounts of methane in Mars's atmosphere could be produced by bacteria continuing to live underground. But even if there is no life on Mars today, we can still hope to find the *fossilized* remains of living cells in Martian rocks.

Early in the history of the solar system, asteroid impacts on planets were frequent, and violent impacts could eject rocks into space. Geologists have identified about 30 rocks that were once ejected from Mars in this way and have since landed on Earth. (Rocks that come from Mars are identified by air pockets inside them, whose gas concentrations match the Martian atmosphere.) These rocks can

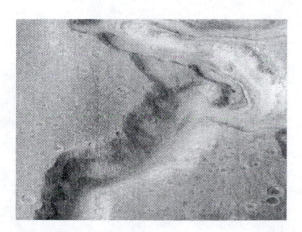

FIGURE 13.10 Dried-up Riverbed on Mars The evidence for liquid water on the surface of Mars is quite compelling, coming from a variety of types of observation. The most visually satisfying, however, are dried-up riverbeds like this one that can be seen from orbit.

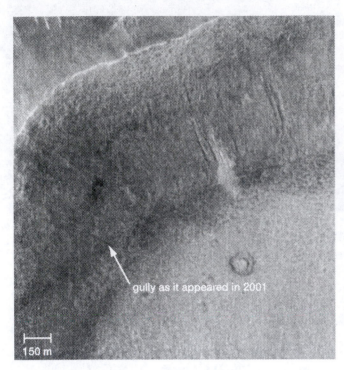
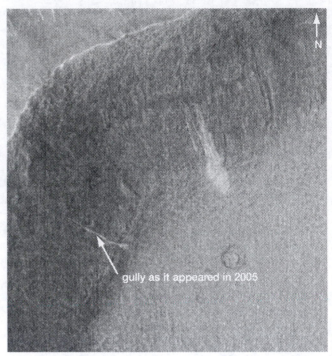

FIGURE 13.11 Martian Gullies   Recent water flows may have emerged from underground reservoirs, partway down the walls of craters like this one. A comparison of the same crater wall in 2001 and again in 2005 shows that the gully channel has changed, possibly because of a new flow of water in recent years. If so, then liquid water probably exists underground on Mars today.

be discovered on the ice fields of Antarctica. The snowy white background makes them easy to spot, and the frigid environment protects them from biological contamination. One such rock has become particularly important.

Figure 13.12 shows the meteorite named "ALH84001"—the first meteorite ("001") discovered in the Allan Hills site ("ALH") in 1984 ("84")—which holds *four independent possible biosignatures*. The possible biosignatures (signs of life) are as follows:

POSSIBLE BIOSIGNATURE #1: *Polycyclic aromatic hydrocarbons.* These are important molecules in organic chemistry. By themselves, they do not necessarily indicate any form of life, since they have also been discovered in interstellar space. But in connection with the other possible biosignatures, their presence might be more naturally interpreted as biological.

POSSIBLE BIOSIGNATURE #2: *Carbonate globules.* Carbonate deposits in the meteorite indicate that liquid (Martian) water once flowed through it. The layering of the deposits is not consistent with simple deposition by water, and on Earth, carbonate globules like those in the meteorite are produced by bacterial organisms. However, recent testing indicates that these globules could also have been produced by a rapid melting process, which probably would have happened in the violent collision that first ejected ALH84001 from Mars anyway.

POSSIBLE BIOSIGNATURE #3: *Magnetite crystals.* The type of magnetite found in ALH84001 is distinctive in size, shape, and crystal structure; on Earth, it is found only when produced by living bacteria. So far, this biosignature has stood up to attempts to challenge its biological origin; the other three are more easily contested. Intensive scientific scrutiny continues on this topic.

FIGURE 13.12 The Famous Meteorite ALH84001, which Flew Here from Mars   This rock spawned a flurry of research into the possibility of extraterrestrial microbial life. Four separate possible biosignatures were found in the meteorite, but are we sure they really signify biology? (See the text!) (◆ Also in Color Gallery 2, page 29.)

POSSIBLE BIOSIGNATURE #4: *Microfossils*. The cutaway view in Figure 13.12 shows the now-famous fossil evidence from Mars. It certainly looks like a fossil (and therefore like definitive evidence of life on Mars!), but is it? Could a nonbiological process make these segmented tubes? Possibly. They're a few hundred nanometers in length (approximately the wavelength of blue light) and less than a hundred nanometers in diameter, which is about a hundred times smaller than most Earth-based bacteria. However, similarly tiny *nanobacteria* or *nanobes* have been found on Earth, too.

There has been enough doubt cast on these biosignatures that probably more than half of the scientific community doubts that they are real signs of life. Still, "more than half" is a very uncommon statistic in science, meaning that things could go either way from here. The robotic space missions we continue to send to Mars are unlikely to resolve this question; we probably need people on Mars who can search for fossil layers and identify, brush, and handle them properly.

The final answer might have to wait until we initiate human exploration of Mars (in 20 years?), although some hope remains for an earlier answer via the magnetite evidence (possible biosignature #3). Until then, you get to enjoy a rare occasion in cutting-edge science, one in which you can form your own opinion and join the debate right along with the experts! Because the scientific evidence is still incomplete, *philosophical* reasoning may prove more enlightening for now. There are two philosophical extremes to consider here:

"OCCAM'S RAZOR": *All things being equal, the simplest explanation tends to be the right one*. The simplest explanation, you might argue, for four different biosignatures in the same small rock is that a single life-form made them all. (See how simple that was?)

CARL SAGAN'S MAXIM: *Extraordinary claims require extraordinary evidence*. If you want to make one of the greatest scientific claims in all of human history—that alien life has been discovered—then you need stunning evidence, better than what you'd need to substantiate lesser scientific claims.

Do you side more with William of Occam (an 11th-century British lord) or with Carl Sagan (a 20th-century American astronomer and writer)? On the subject of Martian fossils, that is the question that will have to serve as the standard by which you judge the evidence for yourself. Meanwhile, as the scientific community tries to catch up with your philosophical inclinations, it can perform other searches in parallel. Beyond Mars, and beyond the rest of our solar system, are there other habitable worlds where some form of life awaits our discovery?

13.3 Extrasolar Planets

The Hubble law, dark matter, the CMB, black holes, and quasars—all of these have been confirmed for decades. Yet the mere *existence* of planets outside our solar system, or **extrasolar planets**, wasn't observationally demonstrated until 1995. Now there are about 250 such extrasolar planets known, with more being discovered all the time. Still, astronomers haven't found many of the kinds of planets we really *want* to find—Earth-like planets—because most of our planetary search techniques really aren't up to the task. Future planet-finding observatories will be able to find other Earths, but current ones are equipped primarily to detect planets with great mass (hundreds of times more massive than Earth, like Jupiter) that orbit close to their parent star (the way Mercury does). There is no such planet in our solar system. Yet, interestingly, we're finding lots of extrasolar planets in this restrictive and unlikely sounding category.

The techniques available for planet finding, most of which connect back to Chapter 2, are as follows:

- *Doppler wobbles.* Far and away the most successful so far, this technique allows astronomers to *infer* the presence of a planet by its gravitational effect on its parent star. As the planet orbits, its gravity pulls on the star, causing the star to move around in a little circle, too. (See Figure 13.13.) The motion of the star is directly observable by its Doppler shift, which will cycle red–blue–red–blue in synchronization with the planet's orbit. Most planets discovered with this method are massive like Jupiter, since a more massive planet has a greater gravitational influence on its star.

One of the more exciting systems discovered by this method centers on the star Gliese 581, about 20 light-years away. Three planets have been discovered orbiting this star (so far). One of these is only about 5 times as massive as Earth and probably only about 50% larger than Earth. Therefore, it ought to be rocky like Earth, rather than gaseous like Jupiter. It makes a complete orbit (a "year") in just 13 days, at a distance 14 times closer to its star than Earth is to the Sun. However, since Gliese 581 is a small red dwarf star, orbiting this closely might not send the temperature through the roof. In fact, one estimate gives a typical temperature range on this planet consistent with the existence of liquid water, meaning that the planet might actually be habitable! Another planet orbiting the same star, about eight times more massive than Earth, also may be habitable.

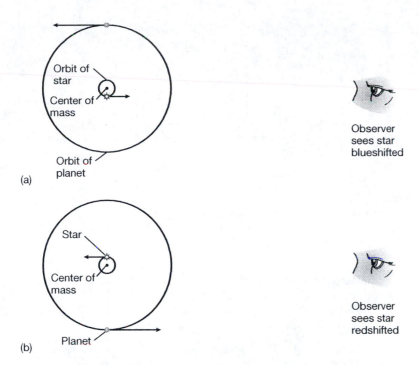

FIGURE 13.13 The Doppler Wobble Technique for Detecting Extrasolar Planets The star's observed motion is affected if the star is pulled on by an orbiting planet. If we observe cycling red and blue Doppler shifts of a star as its planet orbits, thereby catching the star's motions toward and away from Earth, we can partially determine what sort of planet is responsible.

- *Eclipsing systems.* If the planetary system is oriented along our line of sight (Figure 13.14), then we can observe the star dim slightly whenever the planet passes in front of it and blocks a little of the starlight—an event called a *transit*. These systems are detected by their distinctive repeating light curves. The Hubble Space Telescope has already observed several such systems, and NASA's Kepler mission (scheduled for launch in 2008) is expected to find hundreds or more, including Earth-sized planets capable of supporting liquid water.

- *Proper motions.* Face-on planetary systems, unlike eclipsing systems viewed edge on, might produce a detectable proper motion as the planet pulls on the star throughout its orbit. We would see the star move around in a circle on the sky by this technique. (We would not see the planet at all.)

- *Infrared observation.* Planets *reflect* very little visible light compared with their parent stars, but a planet's blackbody *emission* is mostly infrared. (Why?) The Spitzer Space Telescope has succeeded in capturing an extrasolar planet's infrared light already, although we always hope to observe smaller, more Earth-like planets.

- *Gravitational microlensing.* Early in 2006, a planet was discovered orbiting a distant star by the way the presence of that planet altered the gravitational lensing of an even *more distant* star. This method is similar to that used in MACHO searches, described in Section 6.4. It's a tricky business, since it requires

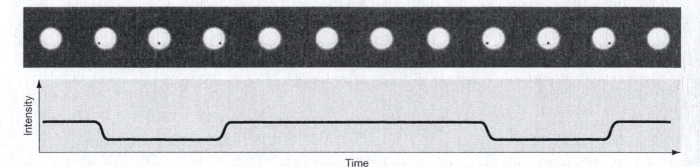

FIGURE 13.14 Eclipsing Planetary System When a planet moves in front of its star, some starlight is blocked from our view. If this pattern repeats on a precise schedule (as seen in this light curve graph), then the fixed orbit of a planet probably caused the pattern.

a helpful alignment of us, the star with the planet, and some other star, in that order. Even so, the first planet discovered by gravitational microlensing is estimated to be only 5.5 times as massive as Earth. Perhaps this method will turn up other Earth-like planets in the coming years.

- *Visual observation.* The Terrestrial Planet Finder (TPF) mission, which was set to launch in 2015, but has been delayed indefinitely for budgetary reasons, is designed to actually see visible starlight reflecting off of extrasolar planets, the way we see planets in our own solar system by the sunlight reflecting off of them. TPF will do so by selectively blocking out light from the parent star, similar to the way you might use your hand to block out the Sun when you look at a bird or airplane. TPF should be able to directly observe light from Earth-like planets, at long last.

It's only a matter of time and money before our techniques are sufficient to find many Earth-like planets, assuming that there are many to be found. Until then, we may note that a solar system like ours, with rocky or "terrestrial" planets close to the Sun and gas giants much farther out, appears to be less common than was once thought, since we've found many extrasolar "hot Jupiters" orbiting close to their star. Perhaps Earth-like planets are rare. But at least we know firmly that planets beyond our solar system exist. We can scour them for clues about how their formation processes compare with those in our solar system; such clues might inform us about how frequently other Earths form. The extrasolar planet story is only just beginning.

EXERCISE 13.3   The Status of Extrasolar Planets

Search online for updates on the status of extrasolar planets. NASA's PlanetQuest site, http://planetquest.jpl.nasa.gov/index.cfm, is a good place to start, and the "new worlds atlas" is a good link to click on when you get there. (The pace at which these planets are being found is rapid enough that if we *tell* you the state of the art in planetary searches, it'll be outdated by the time you read it.) Find out the following:

(a) How many total extrasolar planets are currently known?

(b) How many are less massive than a midsize planet such as Neptune (less than about 20 times the mass of Earth)?

(c) How many have low enough mass to be somewhat Earth-like (less than about 10 times the mass of Earth)?

(d) How many have been detected by *their own* infrared (blackbody) emission, rather than by some optical light observation of the parent star?

(e) Do any of them have much of a chance of harboring life? If so, say which ones and write a few sentences about them. If not, then what's the problem? Are they too hot, too cold, etc.? You may consider only those in the category from part (c) here; we'll just assume, narrow mindedly, that "non-Earth-like" implies "non-life-bearing."

13.4   Intelligence on Other Worlds?

The real jackpot for this chapter is the production of intelligent civilizations in the universe (and by the universe!), and we might as well begin locally with our own Galaxy, since it would take millions of years just to send a message anywhere else. What would it take to generate such a detectable alien civilization? It would take . . .

. . . a star, perhaps like the Sun, stable enough to support evolving life for billions of years.

. . . one or more planets orbiting that star. (Can you think of other locations where life might arise?)

. . . at least one *habitable* planet, which probably means "with a reasonable temperature range" and "with a sufficiently stable climate, supporting the long-term presence of liquid water."

. . . the emergence of life on this habitable planet.

. . . an increase in complexity and diversity, to include at least one intelligent species, which would probably also need to be conscious and self-aware.

. . . a species both willing and able to communicate, using interstellar radio transmissions. Effectively, this means that its members must bother to build radios, and they must produce measurable signals with them. These signals could be either a deliberate attempt at communication, or just TV broadcasts that we happen to intercept. (Of *course* aliens produce and watch TV shows!)

American astronomer Frank Drake summarized the preceding requirements in a single equation, called the **Drake equation**, that is used to compute the *number of intelligent civilizations in our Galaxy with whom we might be able to communicate*. To calculate this number of civilizations, the equation will need to include something about the number of stars in the Galaxy. To arrive at the number of civilizations with the potential for communication, the equation will have to somehow accommodate the fact that stars (and therefore the civilizations that orbit them) differ in age by billions of years, so we seek to count only those which exist and use radio antennas during the same period of time that we exist and use radio antennas.

The Drake equation is

Number of intelligent
civilizations in our
Galaxy with whom we   =
might be able to
communicate

average rate of star formation in the Galaxy[1]
× average number of habitable planets per star[2]
× probability that a habitable planet develops
life[3]
× probability that a planet with life evolves
intelligence[4]
× probability that an intelligent species uses
radio technology[5]
× average lifetime of a radio-active civilization[6]

$$\text{(eq. 13.1)}$$

(The superscripted numbers on the right side are labels, not exponents.) Notice that terms 1 and 6 determine the timing—how often do stars and civilizations pop up, and then, how long before they pop back down again—by giving a "time window" for meeting civilizations. It's like watching fireflies that blink on briefly, at random times, and asking how many you will see if you open your eyes just for an instant. Term 2 translates stars into habitable planets, and its numerical value is open to successively better *measurements* in the coming decades. It is an astronomy question, for which determining the answer is really only a question of money.

Terms 3, 4, and 5, in *principle*, could be measured, too. For example, to measure the probability of developing intelligence (term 4), we just have to count all the planets out there with at least one intelligent species and divide by the total number of planets with life of any kind. Unfortunately, even gobs and gobs of money may not be sufficient to acquire these numbers, since it might turn out that we're the only planet in either category that's close enough to detect some sign of life on (or some sign of intelligence, whatever that sign may be!).

Some *partial* answers may be attainable. For example, there is archeological evidence that many isolated human tribes appear to have developed technologies such as pottery, spears, and jewelry, which might tell us something about term 5. But all of these partial answers must remain partial because they are derived from only one planet and only one civilization (ours). We can't measure the probability of *just any* intelligent species developing technology—unless we find dolphins on other planets making radio transmitters out of clamshells and seaweed.

For the time being, the probability terms reside in the hands of theorists. Perhaps as origin-of-life theory develops, we'll be able to state precisely how life emerged on Earth and then calculate the odds of those same conditions arising on other habitable worlds (term 3). And perhaps computer simulations could reproduce a series of genetic mutations to help constrain the likelihood of emerging intelligence (term 4).

For now, let's just make up some numbers in the Drake equation, to see what types of answers it can generate. You should do this yourself, too, in Exercise 13.1, to guide your own sense of what's possible. The star formation rate (term 1) is about 10 stars per year, but we'll have to invent the other numbers. Here are some imaginable results; see which best matches your own thinking:

OPTIMISTIC RESULT: I expect that most stars have planets; most stars have 1 habitable planet, but a few have none, and a few have 2 or 3. So I'll estimate that the average number of habitable planets per star is 1. I'll suppose that *all* habitable planets develop living inhabitants, 1 in 10 of those gain intelligence, and all intelligent species choose to develop radio technology. Then the probabilities in terms 3, 4, and 5 are 1.0, 0.1, and 1.0, respectively (or equivalently, 100%, 10% and 100%, respectively). I'll also suggest that every civilization lasts until its parent star dies, so maybe 10 billion years is a good estimate for the civilization's lifetime. Then we have

optimistic result = 10 stars form per year × 1 habitable planet per star × 1.0 (any life)
$$× 0.1 \text{ (intelligence)} × 1.0 \text{ (radio)} × 10^{10} \text{ years (lifetime)}$$

$$= 10^{10} \text{ civilizations to communicate with .} \qquad \text{(eq. 13.2)}$$

In that case, the number of civilizations is 10 billion, or roughly 1 per 10 stars. By the interstellar distances determined in Chapter 3, we can estimate that the closest such civilization would be 10 or 20 light-years away, making communication *possible*, but somewhat *boring*, since there's a 20- to 40-year wait for a single round trip of dialogue. ("Hello?"...then 30 years later, "Hi! We're here; are you still there? What's the weather like where you are?") Perhaps traveling that distance might be technically feasible someday, although even at a decent fraction of the speed of light, it would take around a hundred years to get there. (It would take about 200,000 years at our current top spacecraft speeds!)

PESSIMISTIC RESULT: Based on what we learned in Section 13.3, many stars appear to have some sort of planet, but I doubt that many are habitable—let's say 1 planet in 100 star systems. And the probabilities of life, intelligence, and radio technology all seem fairly small to me; let's say 1 in 100 for each. Furthermore, I wouldn't trust an intelligent species to work collaboratively to ensure its survival—just look at us, with our weapons of mass destruction and our tendency toward aggression—not to mention the possibility of a killer asteroid crashing into our planet! So I'll say that the average technological lifetime isn't likely to be much more than ours has been so far, about 100 years. Then

pessimistic result = 10 stars form per year × 0.01 habitable planets per star ×
$$0.01 \text{ (any life)} × 0.01 \text{ (intelligence)} × 0.01 \text{ (radio)} × 100 \text{ years}$$
(lifetime)

$$= 10^{-5} \text{ civilization to communicate with .} \qquad \text{(eq. 13.3)}$$

In other words, with this line of thinking, you conclude that there would be no other civilization in our Galaxy, since the result of the Drake equation is less than 1. If we imagine that all galaxies are similar to one another, then Equation 13.3 implies that, on average, only 1 in 100,000 *galaxies* harbors a single technological civilization. Intergalactic transmission could take a billion years, rendering such communication totally impossible.

In-between possibilities certainly exist. But even then, communication is not really sensible! Even if there were a million civilizations in our Galaxy, the closest would be about a few hundred light-years away. If there were a hundred civilizations in our Galaxy, then the closest would be about 10,000 light-years away. The Drake equation tells us convincingly that interactions with other civilizations like those in *Star Trek* or *Star Wars* are wildly unrealistic, by any foreseeable technology, even if those civilizations are out there.

EXERCISE 13.4   Search Your Feelings, Luke!

Invent all six terms in the Drake equation, using values that you feel are likely to be true or nearly true. Multiply the terms out, and compute the corresponding number of civilizations in our Galaxy. Then invent a rough way to estimate how far away the closest such civilization is, in light-years. (See Chapter 3 for help with this estimate.)

SETI

Even with the most optimistic of assumptions in the Drake equation, the closest civilizations are tens of light-years away. Odds are, your estimate from Exercise 13.4 is much less optimistic. All things considered, even if the emergence of intelligent life elsewhere were probable, we would remain distantly isolated from it.

Popular culture, of course, tells us just the opposite. Sci-fi movies (and alleged UFO sightings) involve alien starships zipping around the Galaxy, encountering one another frequently. This is fun to think about, but is seriously inconsistent with what the Drake equation tells us. Beyond that, the aliens always seem to fly in spacecraft that appear to be a few hundred years beyond our present technology. (*Star Trek: the Next Generation*, for example, takes place in the 24th century.) It's worth noticing what a remarkable coincidence this scenario describes: In a Galaxy whose stars differ in age by *billions* of years, and in which intelligent civilizations evolve from primitive single-celled life over *billions* of years as well, we encounter a civilization whose development just happens to be within a few *hundred* years of our own? In the spirit of "UFO sightings," we might point out that a civilization with the technology to spread through the 100,000-light-year-wide Galaxy would probably be able to hide its presence better than to hover in our sky with a bunch of flashing lights. And why would such an advanced civilization take so much interest in drawing circles in our crops or in mutilating our cattle?[2]

Although these popular versions of alien contact are clearly unrealistic, our fascination with them indicates how profound and exciting the possibility of extraterrestrial life is to us. So, despite the evident rarity or absence of neighboring civilizations, the discovery of one would be of sufficient interest to justify some effort to actually look for them. The Search for ExtraTerrestrial Intelligence, or **SETI**, program has been making radio observations of the stars around us for decades, looking for signs of intelligence (such as alien television broadcasts). Stars emit radio waves normally, so this task really amounts to scanning each star's radio emissions for *patterns*. Based on the radio emission from our solar system, for example, it would be possible for an alien intelligence to distinguish the ordered component, coming from Earth, from the more random component, produced by the Sun. Pattern searches are a matter of signal processing by computers; *your* computer can even help, by running the "seti@home" software available for free from http://setiathome.berkeley.edu. The software is designed to run as a screen saver, processing signals only while your computer isn't busy doing anything else (Figure 13.15).

Do we *want* alien civilizations to know about us? SETI, for example, seeks to *receive* emissions from other star systems, rather than to deliberately *send* messages

---

[2]This argument was beautifully made by Frank Shu in his book *The Physical Universe, An Introduction to Astronomy* (Sausalito: University Science Books, 1982).

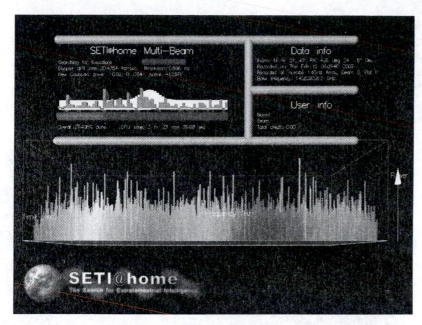

FIGURE 13.15 The seti@home Screen Saver Program   This free downloadable software searches for broadcasts by intelligent aliens by processing radio telescope data from a sample of stars whenever your computer is sitting idle. Visit http://setiathome.berkeley.edu.

to them. (The hypothetical space aliens might still detect our regular radio and TV transmissions, but leading scientists agree that they're likely to lose interest in our culture when they see our "reality TV.")

In fact, "they" might even find one of our exploration spacecraft adrift in interstellar space! NASA's *Pioneer* and *Voyager* spacecraft, launched in the 1970s, are now headed out of our solar system, never to return. Each of these spacecraft has a plaque mounted on it that displays some very revealing information about us. (See Figure 13.16.) The plaque contains information to help establish a universal

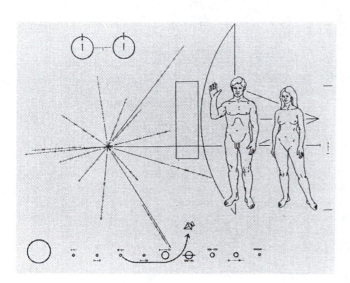

FIGURE 13.16 The Alien-greeting Plaque on the Pioneer 10 Spacecraft   Using symbols described in the text, this plaque communicates information about us, including what we look like and our precise location in the Galaxy, to any enterprising life-form that recovers the spacecraft the plaque is mounted on. That spacecraft has left our solar system and is drifting through local interstellar space.

math-based language with the hypothetical aliens. It also shows a man and a woman standing next to *Pioneer 10*, to show our body size relative to the spacecraft (which the aliens have found if they're reading its plaque). It shows which of our solar system's planets the craft was launched from. And the "explosion" pattern on the left shows our Sun's location among a number of nearby pulsars. Since pulsars are stars that pulse at precise, regular intervals specified on the plaque, these pulsars can be uniquely identified by the aliens. In other words, we've deliberately launched into space a great deal of information about our species and our planet, *including a map of how to find us!* Here's the location of our habitable planet! Here's an indication of our technological capability and our physical form, so that if you're hostile, you're also well informed!

(If you start to panic, hide under a sturdy desk, breathe into a paper bag, and try to remember what you learned from Drake's equation about the likelihood of local technological life-forms. If that doesn't alleviate your fears, try to focus on whatever occupied your attention *before* you realized that if *we* ever found a technologically inferior life form on a nearby world, we'd probably serve it with lemon and butter for $45 a plate.)

The Frontier

Over the first 5 billion years of Earth history, life has shaped the landscape of our planet. Perhaps over the next 5 billion years, life will come to similarly shape the organization of our Galaxy, and perhaps our distant descendants will have a role to play in this future. In this early dark energy dominated era, following the era of structure formation, it's evident that the universe won't soon make any major changes on its own. Perhaps the existence of the universe as a frontier will make it worth the bother of defining what we want the future of our civilization to be like. To that end, we present a series of exercises intended to spark your thoughts—even if you just read them without writing down any answers.

EXERCISE 13.5   Our Future on Earth

Speculate on the future of the Earth's surface, oceans, and skies by describing what they might look like in the future (a few hundred years or a few thousand, as you like). Comment specifically on "resurfacing" (roads, concrete, etc.), climate (global warming, etc.), and population (approaching 7 billion now and growing rapidly). To address these subtopics, cite a few aspects of human nature, as described in Section 13, and how they play into the future you see. A Web search can provide any details you need. One page should suffice.

EXERCISE 13.6   Our Future on Mars

Speculate on the future of human exploration on Mars, in terms of hardware, human colonization, and *terraforming*, in which we deliberately alter the climate of Mars—probably by melting its polar caps to build up an atmosphere that traps heat, and by introducing plants and helpful bacteria—to make it more like that of the Earth. Consider the near and far futures. What would be our purpose there? Will there be families, schools, and so forth? Does our evolved human nature *compel* us to pursue Mars? One page should suffice.

EXERCISE 13.7   Our Future in Space

Look up NASA's "vision for space exploration" on the Internet. (A link to a recent .pdf report on this topic shouldn't be too hard to find.) Write a page summarizing near-term and longer term plans for U.S. space exploration. Don't get bogged down in details; just summarize with a few sentences on topics like "general plans for the Moon," or "human exploration of Mars," or "the search for Earth-like planets." Finally, speculate on a much longer term human future in space: What sort of arrangement of people and spacecraft seems likely for a prolonged presence in space?

EXERCISE 13.8   Artificial Intelligence

According to Hans Moravec, a widely recognized expert on robotics, humanlike robots are likely to arrive as early as the year 2040. These computer-brained, walking machines could be self-aware and even emotional, he claims, and may become more adaptive and more useful than humans. He envisions a future in which robots go to work for us—even at desk jobs—while we stay home!

Starting with his December 1999 *Scientific American* article "Rise of the Robots," evaluate the basis for these claims. Then consider the idea of a future in which robots replace humans as the dominant "species" on the planet. Does this idea seem absurd? or reasonable? or *inevitable?* One page should suffice.

EXERCISE 13.9   Frontier Life

A revolutionary academic paper presented by historian Frederick Jackson Turner in 1893 proposed that the key to the vibrancy of the American mind and spirit was the existence of a *frontier.* This thought is clever, but also worrisome, since the frontier of the American West has been gone for well over a century. Turner offered the following:

"To the frontier the American intellect owes its striking characteristics . . . that practical, inventive turn of mind . . . that masterful grasp of material things . . . that dominant individualism . . . that buoyancy and exuberance that comes from freedom—these are the traits of the frontier, or traits called out elsewhere because of the existence of the frontier."

F. J. Turner, The Frontier in American History. New York: H. Holt & Co., 1920.

Today, the frontier wilderness is all but gone, and Turner's insight may reflect a serious danger. Aerospace engineer and author Robert Zubrin, himself a pioneer in the quest to colonize Mars, asks the following about our fate in the absence of a frontier:

"What happens to America and all it has stood for? Can a free, egalitarian, innovating society survive in the absence of room to grow?

" . . . Currently we see all around us an ever more apparent loss of vigor in our society: increasing fixity of the power structure and bureaucratization at all levels of life; impotence of political institutions to carry off great projects; the proliferation of regulations affecting all aspects of public, private, and commercial life; the spread of irrationalism; the banalization of popular culture; the loss of willingness by individuals to take risks, to fend for themselves or think for themselves; economic stagnation and decline . . . Everywhere you look, the writing is on the wall."

Robert Zubrin, The Case for Mars. New York: Touchstone (Part of Simon & Schuster Inc.), 1996.

Write a two-page essay giving your thoughts on the Turner hypothesis and on Zubrin's claim that its predictions are already coming true, to the detriment of American culture. In your work, comment specifically on what it means if Turner is correct: What is the nature of the human mind if it can't remain free and productive without what Zubrin calls "room to grow?" What explanation could you imagine from evolutionary psychologists for how we ended up this way? Finally, what actions might alleviate our conundrum? Must we endlessly seek a new frontier, conquer it, and move on—as if addicted to a narcotic, but with an ever-growing tolerance to it? Is there any other avenue to follow, apart from eternal expansion or eternal stagnation?

EXERCISE 13.10   Consciousness in the Universe

Write an essay expressing your thoughts about the question "How is it that conscious awareness exists in the universe?" There are at least two possible answers to this question: (1) that conscious awareness was "put there intentionally" and (2) that conscious awareness was an "accidental by-product of other things going on." What do you think about each of those answers, and can you come up with others?

The implications of either answer are profound. It makes a real, operational difference to our perception just to ask the question at all. It forces us to recognize, for example, that the question does have an answer, and the answer we choose has implications for how we live.

If consciousness developed as an intentional "goal" of the universe, this opens up another set of questions. For example, what is that goal and what role do we play in it?

Do you think the universe has a purpose? If so, what purpose(s) can you imagine? Get creative with the possibilities: They don't have to be right; they just have to be *possible*.

On the other hand, if consciousness did *not* develop intentionally, then the universe exists for some other reason (or no reason; it's just doing its universe thing), and we happened as an aside in some way. This possibility also opens up a whole new world to ask about and investigate.

Do you think that we have no significant responsibility to the universe, so that we're free do as we please (within physical constraints) and see what interesting things we can come up with? If you take this "nothing to prove, nothing to lose" perspective, elaborate on how it affects your life.

Law and Order

The *second law of thermodynamics* captures the observation that as time passes, energy tends to spread out and become less accessible (so we can't get our hands on it and use it to do interesting things). For example, the energy in gasoline is concentrated, so you can use it to make your car go. But after you've been driving around for a while, hitting the brakes and knocking aside air molecules that get in your car's way, the energy is spread out all over. The air molecules are moving faster, your brake pads are hotter, etc. It's now harder to access that energy and put it back in your gas tank to power your car again. (You'd have to round up all the little bits of air, the vibrations of atoms in the tires, and so on—good luck!)

No one has ever found a clear exception to the second law, which is often interpreted as a tendency for *disorder* to increase as time goes on. According to the second law, the act of folding your laundry and putting it in drawers creates more disorder than it eliminates. Your motion and your heat output in this process scatter more particles (like air) than they neatly arrange in drawers. (In this sense, doing laundry is very unnatural! Thermodynamically, you'll be more in tune with the overall trend in the universe if you just throw your clothes out the window and forget all about them.)

Interestingly, this overall drive toward evening things out—increasing "disorder"—can create some amazing *ordered* systems in the process. A living cell, for instance, is a spectacularly ordered system, in which every component has to be "just so." The second law requires the process of assembling a cell to create enough disorder *somewhere* to more than account for the cell's intricate order. The universe evidently accepts that trade-off: Sunlight pouring out into space creates a tremendous amount of disorder, but as it flows through the Earth, it triggers the formation of the orderly patterns we see in hurricane clouds, snowflakes, and, most especially, life, to name just a few examples.

It is in the universe's character to order one place at the expense of greater disorder in another place, and it is to this principle that we seem to owe our existence. You can see how it has reordered the surface of the Earth, from molten lava to islands, fields, and forests. And you can see how it has ordered the arrangement of galaxies: fragmentation of matter into halos and stars by gravity, and material enrichment of galaxies by nuclear fusion. Spiral patterns, planetary systems, oceans, and life follow. Meanwhile, dark energy increasingly isolates these galactic systems—islands of order in an otherwise disordered cosmos in which particles fly every which way through the vast spaces in between. At the expense of great regions of disorder, the universe creates small pockets of astonishing complexity and even beauty: waterfalls, wildflowers, coral reefs, Saturn's rings, and so on.

Within an ordered galaxy, we might ask how far the ordering process goes. On Earth, our technological civilization influences the planet's "ordered resurfacing" today, from wilderness to civilization. Now, you might be tempted to view human activity as being distinct from nature and therefore not confined by physical laws like the second law of thermodynamics. But our civilization is part of this universe. For billions of years, the universe slowly evolved without human intervention and even constructed the highly ordered and complex Earth. But that same cosmic evolution process eventually led to us, and when we joined the ordering process on this world, we were able to help accelerate that natural process (with roads, cities, communications networks, etc.).

Perhaps the growth of complexity elsewhere in the Galaxy would also proceed faster if it were aided by an advanced civilization, with greater technology than what we currently have. The Galaxy, if it were transformed in this way, could be *flooded* with permanently staffed space stations, robotic probes, travelers, and electromagnetic transmissions, all of which would be detectable from Earth in any direction we look. The time it would have taken an alien civilization to colonize the Galaxy is millions of years or moderately more. (See Further Exploration Problem 11 to see where this number comes from.) Yet our species evolved over *billions* of years. In other words, our extrasolar neighbors have had *plenty of time* to colonize the whole Galaxy while we were busy evolving! So why haven't they done so yet? They should be everywhere by now, so the burning question (often called the *Fermi paradox*) is *Where are they?*

Perhaps the advanced alien civilization obeys a *Star Trek*–like "prime directive"—a law which demands that its members deliberately hide their existence from "primitive" civilizations like ours. Or perhaps we are the first intelligent species to emerge in the Milky Way. But it could also be that expanding civilizations bump up against a limit before a galaxy *can* be colonized. Massive stars fuse hydrogen into helium, carbon, oxygen, neon, magnesium, silicon, and, finally, iron, but their eventual supernova reduces most of that iron back into protons and neutrons, practically erasing 10 million years of progress. Perhaps civilizations do the same: They arrest themselves with war, dangerous scientific experiments, or the uncontrolled advance of an artificial intelligence that they invent. Or maybe they just lose interest in exploration and define some other role for themselves.

The point is that, as products and agents of the processes that create order in the universe, we are subject to any limitations those processes face. It is perfectly plausible that, because of psychological or technological requirements, *any* attempt at galactic colonization by *any* civilization will *always* yield some outcome *other than* actual colonization. However, if there are no such limitations, and it is indeed the nature of the universe to assemble ever-increasing layers of complexity, then the observation that the Galaxy hasn't been transformed by someone else already might imply that *we* will be the agents of that transformation.

Quick Review

In this chapter, you explored some key aspects of the biochemical basis for life on Earth, to better appreciate what had to occur in order for living organisms to emerge in the universe. Although some of the origin and history of life on Earth is currently unknown, much of it, including the emergence of simple organic materials like amino acids, has been uncovered by scientific investigation. Could the same processes that led to life—even intelligent life—here on Earth also be at work on other worlds? Part of the answer lies in the Drake equation, which addresses the probabilities leading to life in the universe. You saw that some of the numbers appearing in the Drake equation can be found, for example, by studying evidence for possible microbial life on Mars and by studying planets orbiting other stars.

Try the following crossword puzzle to test your knowledge of key terms and concepts from Chapter 13:

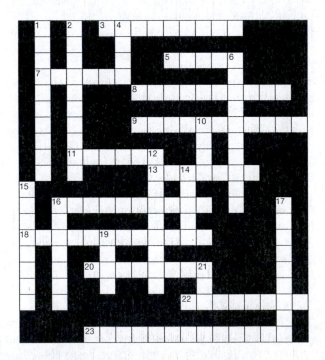

ACROSS

3 Transfer RNA matches up a codon of genetic information with an _____ in the process of making a protein.

5 Equation used to calculate the number of intelligent civilizations in our Galaxy with whom we could possibly communicate

7 Fats and oils

8 Subunit of DNA made from a sugar, a phosphate, and a nitrogenous base

9 Process inside cells of using genetic information stored temporarily in RNA to build a protein

11 Type of protein that enables or speeds up various cellular functions

13 Lipid-based primitive cell membrane

16 The study of evolved behaviors, rather than physical features, is called evolutionary _____.

18 A sign of life, such as a fossil; the oldest one known is 3.8 billion years old.

20 Scientific hypothesis suggesting that DNA and enzymes came about after a period of time when another molecule simultaneously served the purpose of both

22 Smallest unit of genetic information

23 Process inside cells of copying genetic information from DNA to RNA

DOWN

1 Original origin-of-life experiment that simulated the young Earth and produced amino acids and other organic molecules

2 Biologically important molecule, like a protein, that's made from a row of smaller, repeating component molecules

4 Nearby planet with dried-up riverbeds, possible recent water flows, and, possibly, microscopic life in the past

6 Planets orbiting stars other than the Sun

10 The search for radio emissions around nearby stars indicating the presence of a technological civilization

12 Theory that all life on Earth has descended and diversified from distant common ancestors through a series of genetic changes

14 Individual "chain link" in a carbohydrate "chain"

15 Wall surrounding a cell

16 Major ingredient of living cells, constructed from amino acids

17 Darwin proposed two different _____ mechanisms, natural and sexual, to enable life-forms to change over time.

19 A particular sequence of base pairs, found in a particular stretch of DNA, that codes for a particular protein

21 Genetic information storage polymer used by cells as an "instruction manual" or "recipe"

Further Exploration

(M = mathematical content; I = integrative—builds on information from more than one chapter; P = project idea; R = reflection; D = suitable for class discussion)

Exercises preceded by "*" are based on material from supplemental Section 13.1.

1. (PD) Begin with a reading, and end with a two-page paper, covering *one* of the following articles, all of which discuss hot topics in relatively recent science. For these projects, aim for a target audience of your peers, and highlight what you expect them to find most interesting. Each topic deals with either how we got the way we are or how we might choose the way we become:

   - Human exploration of Mars: "The Mars Direct Plan," by Robert Zubrin, *Scientific American*, March 2000. (Or use Zubrin's book *The Case for Mars*.)
   - The fuzzy borderline between life and death: "Are Viruses Alive?" by Luis P. Villarreal, *Scientific American*, December 2004.
   - Biosignatures for the earliest recorded life on Earth and Mars: "Questioning the Oldest Signs of Life," by Sarah Simpson, *Scientific American*, April 2003.
   - A scientific approach to the basis for human morality: "Whose Life Would You Save?" by Carl Zimmer, *Discover* magazine, April 2004.
   - An evolutionary basis for human economics: "How Animals Do Business," by Frans B. M. deWaal, *Scientific American*, April 2005.
   - The near future for the natural resources of planet Earth: "The Bottleneck," by Edward O. Wilson, *Scientific American*, February 2002.

2. (PD) In the 1999 science fiction movie *The Matrix*, Agent Smith (a being of artificial intelligence) tells Morpheus (a human) the following:

   "[A revelation] came to me when I tried to classify your [human] species. I realized that you're not actually mammals. Every mammal on this planet instinctively develops a natural equilibrium with the surrounding environment, but you humans do not. You move to an area and you multiply, and multiply until every natural resource is consumed. The only way you can survive is to spread to another area. There is another organism on this planet that follows the same pattern. Do you know what it is? A virus. Human beings are a disease, a cancer of this planet."

   Agent Smith has a lot to learn about biology. What keeps other mammals from overrunning their environment is not instinct, but *predators* (or other constraints, like geography or disease), which keep the population in check. But because we humans no longer have external predators (and assuming that we don't want to send out a call for advanced space aliens to come to Earth and fill that role for us), we are still faced with the problem Smith points out. We must either find a new place with new natural resources to consume, or adopt some other approach. In a few sentences each, list some ways in which we might respond to this situation, *without* consuming more natural resources from other places. Are there any such efforts in progress already?

*3. (RI) In Section 13.1, you encountered a lot of biochemistry. What is the point of all that detail in your study of the universe? Give it some thought, and answer in one or two paragraphs.

4. (RI) Your body contains mostly carbon, hydrogen, oxygen, and nitrogen. Consider the carbon in your body ($^{12}_{6}C$), and give the *complete* 13.7 billion-year-timeline history of the carbon in your body, with approximate dates, including where it has been in the past, where its nucleons came from, and how it came to inhabit your body.

5. (RI) In the spirit of the essay titled "Law and Order" a few pages back, explain in several sentences apiece by what mechanism the laws of physics could *create* the following ordered structures:
   (a) the elements heavier than hydrogen
   (b) halos and galaxies
   (c) the seeds of structure, seen as CMB anisotropies, that led to walls and voids today
   (d) stars
   (e) rocky planets with an iron core, a silicon–oxygen crust, an ocean, and a gaseous atmosphere (*Hint:* Consider which components are the heaviest and lightest. Where would you expect them to end up?)
   (f) life

*6. (MR) In this problem, you'll demonstrate why codons are *three-letter* "words" instead of some other number of letters. To address this issue, you'll need to know that that DNA's genetic "alphabet" contains only four "letters" (four different nitrogenous bases). If genetic "words" were only one letter long, then no more than 4 amino acids could be specified, because, at best, each letter would be a different amino acid, and there are only four letters. But there are actually 20 biologically useful amino acids commonly used by life on Earth, and therefore we need (at least) 20 distinct "words" (codons) to identify them. Demonstrate mathematically that there are also too few distinct two-letter words to identify the 20 amino acids, but that three-letter words are more than adequate. Therefore, there's no reason to evolve the biochemical apparatus to deal with four-letter words.

(Four-letter words can be hard to deal with using "social apparatus" as well!)

*7. (MR) Typical proteins can be hundreds, thousands, or even tens of thousands of chain links long, where each link is one of 20 commonly used amino acids. Hemoglobin, the protein that allows your blood to carry oxygen to your cells, is a sequence of 574 amino acids, for example. A much simpler human protein, cytochrome c, has only 104 amino acids in a single chain.
   (a) Suppose you want to create cytochrome c in a primordial soup fashion; let's investigate how likely it is to randomly assemble 104 amino acids into the correct sequence. To do this, imagine choosing whatever amino acid happens to float along at each moment to (somehow) attach to your forming protein, and assume that each of the 20 amino acids is equally likely to float along at any given moment. Calculate how many different sequences are possible, by multiplying $20 \times 20 \times 20 \times \ldots$ 104 times, or 20^{104}, to simulate choosing 1 amino acid out of 20, 104 times. (*Hint:* Notice that this number is bigger than 1.0×10^{104}, which means that it might be beyond the capability of your calculator. Working with scientific notation should help you get around this problem.)
   (b) Since life emerged on Earth within a billion years of Earth's formation (probably less), and a year is $365 \times 24 \times 60 \times 60$ seconds, how many "sequencing attempts per second" must have taken place somewhere on Earth during that billion years in order to get it right (to make a single cytochrome c molecule) just once?
   (c) The Earth's volume is that of a sphere, $(4/3) \pi R^3$, where Earth's radius is $R = 6.37 \times 10^8$ cm. Supposing (incorrectly and very conservatively) that the sequencing attempts could occur equally well anywhere on or in the Earth, how many sequencing attempts per second, per cm^3 (cubic centimeters are about the size of dice or a little smaller) must have taken place in the first billion years to make a single cytochrome c molecule?
   (d) On the basis of your results in parts (b) and (c), do you think that proteins could reasonably be expected to form in this purely random way?
   (e) Given the evident reality that proteins *did* form on Earth, comment on what your answer to part (d) implies for their formation mechanism. State any thoughts you have on the matter.
   (f) How might the RNA world hypothesis help here?

8. (MR) Estimate some reasonable numerical values to go into the Drake equation, supposing separately that
   (a) Microbial life has been discovered on Mars, on Europa (one of Jupiter's moons, which has a large liquid-water ocean buried under a surface of water ice), and on Titan (Saturn's largest moon, which has lakes of

liquid methane that could act like water does on Earth, possibly including rainfall).

(b) A thousand years pass by, with no Earth-like extrasolar planets found, no SETI communications received, and no microbes or fossils discovered in our solar system.

In each case, admit it when you can't say anything useful about a particular Drake parameter (e.g., "the radio-using fraction could be anywhere from 0 to 1"). Also, write a paragraph to explain the values you found in (a) and in (b).

*9. (R) In one or two sentences, state the significance of the Miller–Urey results.

*10. (RD) Many important biomolecules have "handedness," which means that they can form equally well in two different shapes, called "left handed" and "right handed." (To see the distinction, consider your hands. There's no obvious difference in the shape of one compared with the other, but to make a photograph of your left hand look *exactly* like your right, you'd need to look at the photograph in a mirror. Now just replace your fingers with C, H, O, and N atoms in your mind's eye.) The proteins used by all life on Earth (as far as we know) are left-handed amino acids exclusively. Use this observation to speculate about

(a) how often living cells might have independently and spontaneously emerged in Earth's past,

(b) whether independent, chemically formed organisms might have competed with one another in the past, and

(c) whether a life-form that uses only left-handed biomolecules would be expected to reproduce in a way that generates offspring with only left-handed biomolecules.

11. (MI) Suppose that an advanced civilization in our Galaxy has invented a space propulsion system capable of traveling at 10% of the speed of light. Some of these aliens fly to some star and spend a few hundred years mining the local planets and asteroids for various raw materials, and constructing new space stations and starships with those raw materials. During that time, their population multiplies, so that there are enough "people" to staff all the new space stations and starships. If they do this each time they move from star to star, come up with a way to estimate how long it would take them to colonize the entire Galaxy. (You'll need to use what you know about galactic distances; show your work and explain your reasoning.) Revisit the concept of the "Fermi paradox" discussed in the essay "Law and Order," a few pages back, and explain why your result might indicate a "paradox."

*12. (R) Summarize the reasoning behind the "RNA world" hypothesis in only a few sentences.

13. (PD) Write a one-page research summary of a proposal by Dr. Geoffrey Miller of the University of New Mexico in which he contends that the reason the human brain evolved such impressive capabilities—wildly beyond what was needed for the success and survival of our distant ancestors on the plains of Africa—was the virtue of sexual selection. That is (Miller suggests), smarter people have better luck attracting a mate with whom to reproduce. In your paper, indicate what role art, music, and comedy play in our lives, according to Miller's hypothesis. (One possible starting point is http://www.edge.org/3rd_culture/miller/.)

14. (MI) For SETI purposes, estimate the number of stars within 50 light-years of us. (That way, we might hope to have a single round-trip radio communication within a single human lifetime.) How many years would it take to check all of these stars, assuming that we can look at one star per half hour for 15 hours a day and with telescope access of 40 days per year? At this rate, is a star-by-star SETI search reasonably useful, or does it just barely scratch the surface?

14

What Does It Mean to You?

> *"We shall not cease from exploration*
> *and the end of our exploring*
> *will be to return to where we started*
> *and know that place for the first time."*
>
> —*T. S. Eliot*

We began this book with the observation that scientific cosmology matters to you personally because it influences your view of yourself in relation to the rest of the universe. With the benefit of the knowledge gained so far, we now return to consider directly what this all might mean to you.

The tone here is more speculative than in previous chapters. We make suggestions regarding how you might look at the information you have learned and find a place for yourself within the universe. But ultimately, this task is up to you. While scientific knowledge enriches and filters your perception, there is plenty of room for creativity and individuality in understanding your own connection to the cosmos.

14.1 What Have You Learned?

Albert Einstein remarked that "education is what remains after one has forgotten everything he learned in school." Long after some of the details of scientific cosmology have faded from your memory, we hope that the following key ideas will remain a part of your education.

The Tools of Science Are Relevant to Your Life

The insights and grand conclusions we've reached in our study of cosmology all emerged from humble beginnings: the simple act of looking up, *observing* the sky, and putting the scientific method to work. By observing your surroundings carefully, asking questions about what you see, proposing theories for how things work, testing your theories by other observations, and refining or discarding theories when they are contradicted by evidence, you can apply the techniques of science directly in your own life. Maybe you'll use the scientific method to help you decide what to believe when you read an advertisement about the health benefits of a new vitamin supplement or when you want to know how seriously to take an environmental concern in the news. You might also continue asking your own "big questions" about the universe and your place within it, using the scientific process to filter information and craft your own cosmic context.

The Universe is Ancient and Immense

Hopefully, we've impressed upon you by now that the universe is huge! As a reminder, you need only look at the Hubble Ultra Deep Field image again (Color Figure 10 in Color Gallery 1, page 5.) and realize that all the galaxies in that image lie within a tiny speck of the night sky. The universe extends at least tens of billions of light-years and probably much farther than that. It appears that it may even be infinite in extent. Our universe might also be only one small part of a system that contains a vast collection of different universes. In any case, the scale of the universe as described in Section 3.3 is almost beyond comprehension, and that realization is worth savoring.

Our known universe is also ancient, though not infinitely old. Fourteen billion years is plenty of time for the emergence of a great variety of remarkable features, but it's still a manageable length of time. In Section 14.3, you'll see some of the key events in the history of the universe on a concrete timeline stretching over 13.7 billion years. Still, we're left wondering what happened before the beginning of our known universe. Was there truly nothing—not even time? Or is our current universe just one blip in an infinitely repeating cycle of universes? Either way, the implications provide a sense of perspective on the events that occur during each of our lifetimes.

The Universe Is Changing

One of the most remarkable insights to be gained from modern cosmology is that the whole universe is *evolving*. As space expands, the density of matter in the universe is decreasing overall. Structures have appeared at different times that had never been seen before in cosmic history: the first atoms, the first stars, the first galaxies, the first planets, and the first life. It also seems that there is a general trend toward greater complexity as time passes.

We can see this kind of evolutionary process reflected in our own lives. Things change over time on widely different scales: Cosmic and biological structures emerge and grow, as do social structures and the patterns of our thoughts and actions. Many people are guided by a desire to improve things: to transform social structures as we develop greater tolerance of differing ideas, to create a society that is more fair to more people, to improve the principles of government, and so on. In our scientific endeavors, we strive to gradually achieve a clearer and clearer understanding of the principles according to which nature seems to work, modifying and refining our old theories to better match observations. As American writer and philosopher Ralph Waldo Emerson pointed out, "[We are] made of the same stuff of which events are made. . . . The mind that is parallel with the laws of Nature will be in the current of events, and strong with their strength." A drawing you make or a sentence you speak emerges in the same flow of history that formed the galaxies and nebulae you saw in Chapter 2.

The Universe is an Interconnected System

In Chapter 13, we gained a glimpse of what's involved in creating the conditions that enabled conscious awareness to emerge on Earth. These conditions are the result of a long series of interrelated cosmic processes. Every part of the universe seems to play some essential role in making this chain of processes possible. For example, conscious life arises from very complex arrangements of matter. But these arrangements require elements that didn't even exist until the first stars emerged after a few hundred million years of cosmic history. The deaths of these stars released new material that was incorporated into future generations of stars and the planets on which conscious life could ultimately form. So the present moment of your existence is not an isolated event: Your existence is embedded within a web of connections upon which it depends. And similar conditions at many other locations in the universe make it likely that life and consciousness are properties of the *universe*, not just Earth.

The vastness of what's around us implies that we should be cautious and humble about our place in the universe. But it's easy to go overboard, to the point that you can feel hopeless about your place in the scheme of things. (See Figure 14.1.) It's worth remembering that physical size and location are not the only indicators of significance. For example, imagine you were a tiny circuit on a little computer chip at a huge microchip fabrication plant. You'd look around and think you were pretty insignificant and irrelevant to this enormous operation. (You're a tiny circuit with eyes and feelings.) But you would be mostly wrong: It takes a big industrial plant and the organized efforts of many resources and workers to produce that tiny circuit.

Similarly, the laws of our universe seem to require a long time for conditions to emerge in which complex life can form. It takes time for matter to clump. Stars must go through many complete life cycles in order to produce the carbon that is a key ingredient for life as we know it. All of this means that the universe must be billions of years old for conscious structures like us to look around and be awed (and needlessly dismayed?) by how big the universe is and how small we are.

Remember from the spacetime diagrams in Section 7.1 that age is intimately linked to size. Our horizon, in a universe that is billions of years old, extends for at least billions of light-years in distance. Thus, it should come as no surprise that, since we *are* here, the observable universe must be big enough to allow us to be here. That is, the universe is not vast in order to humble us or make room for countless alien civilizations; it's vast because it's *old*, and it must be old in order for

FIGURE 14.1 A Grain of Sand   It's easy to be intimidated by the vastness of the universe.

us to be around to notice. Attaining the right conditions for people to exist in the universe takes a long time, and the cosmological horizon has been growing all during that time.

Let's take a deeper look at galaxies as an example of the interconnectedness of the universe. As a resident of one galaxy among uncountably many, you might be inclined to take these cosmic habitats for granted. Galaxies appear to be a dime a dozen in our universe. But in order to support life, a galaxy needs to be about the right size, with the right distribution of stars. The center of our spiral Galaxy, for example, contains a dense bulge of stars. Life probably can't survive there: Too many neighboring stars means too many nearby supernovae that would cause extinctions. And even without dangerous radiation from supernovae and other bright stars, close gravitational encounters with other stars would probably bring frequent and deadly impacts. The edge of our spiral Galaxy, by contrast, has too few stars. That means too few supernovae and therefore too few heavy elements from which to build habitable planets and life-forms. Figure 14.2 shows the crux of this issue. We occupy a place in the Galaxy that's *sufficiently crowded* with stars to have enough heavy elements to make habitable planets, but also *sufficiently deserted* that neighboring stars are generally far away and don't mess with us very often. It's a cosmic version of an American ideal: a better life in the galactic suburbs.

But not only does this mean that certain parts of galaxies are inhospitable to life; it also means that if the laws of physics had populated the universe with some other structure *instead* of galaxies, then life might not be possible *anywhere*. For example, the globular clusters (Color Figure 8 in Color Gallery 1, page 4.) that orbit around galaxies are very old, indicating that they formed very early in the history of the cosmos. We know from CDM simulations that larger structures, such as galaxies, assembled by the merging of smaller pieces, such as the clumps of matter that created globular clusters. If the expansion rate were faster (bigger H_0), or if gravity were weaker (smaller G), or if the matter density were lower (smaller Ω_m), then galaxies might never have assembled. Globular clusters are hostile to life in the same ways that galactic bulges are, so life probably wouldn't have arisen in a universe full of isolated globular clusters.

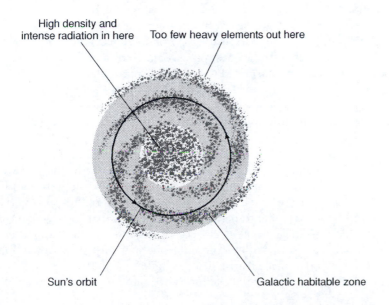

High density and intense radiation in here   Too few heavy elements out here

Sun's orbit   Galactic habitable zone

FIGURE 14.2 The Habitable Part of the Galaxy We seem to occupy the right kind of galaxy, and the right location within it, to make it possible for advanced life to emerge here on Earth.

We'll return to the matter of choosing the optimal values for parameters like H_0 in the next section. Here, we are concerned with the interconnectedness of the processes that lead to habitable galaxies all over the universe. There are a *lot* of different contributing processes throughout cosmic history, which succeed in building galaxies only when all the processes are *combined*.

We can start with the seeds of structure—the first departures from perfect homogeneity, such as the quantum fluctuations discussed in Chapter 12—which were subsequently stretched to useful scales by inflation. The transition from radiation to matter domination set dark matter free early in cosmic history, back when it was sufficiently dense to start clumping together by gravity to make small halos. The rules of atomic physics then determined the timing of recombination, which caused decoupling of photons and set the baryonic matter free to follow its gravitational geodesics into the recently formed halos. The expansion of the universe and the action of gravity allowed these new protogalactic structures to merge and grow, while spreading out into the walls and voids seen on large scales today—an arrangement of a great many habitable galaxies! The history of our Galaxy is a rich, interconnected recipe that required the proper mix of such disparate ingredients as quantum fluctuations, dark matter, and an expanding spacetime that obeys the Friedmann equation.

EXERCISE 14.1   An Interconnected Universe

Identify one or more other examples of the interconnected history that led to the habitable universe we enjoy today. Do so by writing a paragraph similar to the preceding one, but focus on some aspect of our universe other than galaxies in particular. Include several of the following seemingly disparate elements in your story: neutrinos, blackbodies, antimatter, carbon, black holes, and redshift.

14.2 Anthropic Thoughts

What could be the significance of life (including humans) in the cosmos? Does life have a central role? That is, was the universe specifically designed to produce life (Figure 14.3)? Or is life an accidental by-product that could just as easily not have occurred? Or perhaps neither perspective accurately reflects the significance of life.

To shed some light on these questions, we need to think more generally about probability. When are we surprised to see that a particular thing has happened? You are probably surprised when someone rolls a double six while throwing dice, since the odds of that happening are only 1 in 36. But you would be less surprised if the person rolled that double six at some point during a string of a hundred throws of the dice.

How surprising is it that all the circumstances necessary for life came together here on Earth? This question is more complicated than rolling dice. As we saw in Chapter 13, we don't even have a completely unambiguous definition of life, and we don't know precisely what all the circumstances required for life to arise actually are. But for the sake of argument, we can estimate that life is very rare in the naive sense that if we take all the necessary atoms and mix them together, only a very small fraction of all possible combinations will actually produce life. So it seems very unlikely that life will arise from a single "roll of the dice," or even thousands of rolls. But there appear to be trillions of galaxies (each containing hundreds

FIGURE 14.3 A Universe Fit for Life? Perhaps life and consciousness are woven into the fabric of the universe, so that they were an inevitable consequence of the unfolding history. The footprints of life might thus be hidden in many of the clues we uncover as we study the universe. ("Puddle" by M. C. Escher, 1952.)

of billions of stars) within our observable universe. So, even using a pessimistic estimate for the probability of getting life from each roll of the dice, it seems reasonable that out of trillions upon trillions of rolls, a few will lead to life.

The odds we calculate for anything depend on what we consider to be possible outcomes. Life (like other structures in the universe) doesn't come to exist purely by a chance arrangement of its building blocks. There are laws of nature limiting which possibilities can happen. If you fire a cannon at a wall, the laws of Newtonian physics restrict the trajectory of the cannonball and *determine* where on the wall it will hit. We don't yet know all the laws that might determine where, when, and whether life will emerge.

But what if we consider "rolling the dice" to pick the laws of nature themselves? This opens up a whole new realm of possibilities to consider. To get a concrete feel for what this means, you might imagine yourself in charge of a huge panel of knobs that set the laws of nature for the universe, as in Figure 14.4. One knob turns up the strength of gravity, another controls the speed of light, another the amount of electric charge on a proton, etc. Then, in the absence of any other information, we could ask how likely it is, with random settings of the dials, that the settings in our universe will be right to produce life.

Again we are hampered by limited knowledge of the exact requirements for life to arise. Perhaps it can form in structures and under conditions that are vastly different from the structures and conditions in which our own carbon-based life on Earth arose. Still, we can make some reasonable guesses. It's hard to imagine life without a certain level of complexity: Whether in a cell, a computer circuit, or radio waves spread throughout a vast region of space, life needs a way to store information.

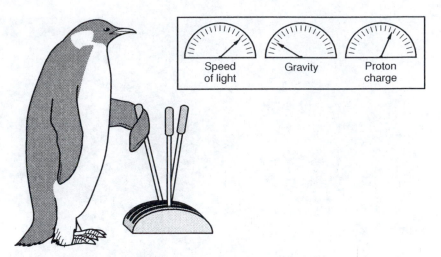

FIGURE 14.4 The "Control Center" for the Laws of the Universe   With a variety of dials to set for different parameters (strength of gravity, speed of light, expansion rate of the universe, proton charge, etc.) and a wide range of possible settings on each dial, how did we happen to be in a universe so well tuned for life?

It turns out that only a very small fraction of the settings on our imaginary "laws of the universe" display panel will produce a universe that supports anything like the complexity that seems necessary for life. Here are a few parameters that appear to be set just right to enable our universe to support life-forms like us:

- *The expansion rate:* Too fast, and matter thins out so quickly that galaxies never get a chance to form; too slow, and the universe stops expanding and recollapses before life ever forms.

- *The strength of gravity:* Too strong, and stars would fuse too fast and die too soon (based on hydrostatic equilibrium); too weak, and fusion would never take place in stars, which means that they wouldn't provide enough luminosity for enough time to support life on orbiting planets.

- *The excess in the strength of the strong nuclear force over the electromagnetic force:* Too much, and rapid fusion would leave no hydrogen left at the end of BBN; too little, and proton–proton repulsion would overcome the strong force, and nuclei heavier than hydrogen could never form. (Either way, there would be no life as we know it.)

When the values of parameters such as these can lie only in a narrow numerical range in order to permit life, they provide an example of what's called *fine-tuning.* Fine-tuning occurs when a theory (or a universe!) works correctly only if you set some particular number to a specific value for no obvious reason. In general, a reliance on fine-tuning is an unattractive feature of any physical theory, and scientists often seek to eradicate it by finding deeper theories that explain what *caused* each parameter to become so finely tuned.

For example, over two thousand years ago, a mathematician and philosopher named Ptolemy created a model of the solar system based on Aristotle's geocentric universe, with Earth at the center. Ptolemy's model included a large array of finely tuned parameters that described a network of nested circles called "epicycles" in which planets looped around in complicated ways as they traveled around us

*Claudius Ptolemaeus (about 85–165 A.D.), better known as **Ptolemy** in English, was of Greek origin, but lived in Egypt. He was an astronomer (and mathematician, and geographer, and philosopher) responsible for inventing the geocentric "Ptolemaic model" of the solar system. Ptolemy's model was replaced when the Polish cleric **Mikołaj Kopernik** (1473–1543), better known as **Nicolaus Copernicus** in Latin, or just plain Copernicus in English, realized that many of the complicated parameters of Ptolemy's model could be eliminated in a heliocentric, or Sun-centered, model with a spinning Earth. Later work by Galileo, Brahe, Kepler, and Newton validated Copernicus's claim.*

(Figure 14.5). Ptolemy needed all of his finely tuned values for each planet in order to preserve Aristotle's philosophical disposition toward the perfection of the circle as a geometric form. Ptolemy described the objects in the solar system in this way:

> "When their motion is viewed with respect to a circle . . ., the center of which coincides with the center of the universe (thus its center can be considered to coincide with our point of view), then we can suppose . . . that they have such a concentric circle, but their uniform motion takes place, not actually on that circle, but on another circle, which is carried by the first circle. . . ."
>
> —Claudius Ptolemaeus, *Almagest.*

Now, ever since Kepler and Newton, we've known that Ptolemy was wrong: Planets orbit the Sun, not the Earth, and in ellipses, not circles with epicycles. Kepler's laws were built upon the heliocentric (Sun-centered) model that was argued for by Nicolaus Copernicus in the 16th century. When Copernicus wrote a book on the heliocentric model, he felt compelled to offer philosophical reasoning (as well as much scientific reasoning) in order to sway readers away from Ptolemy's model, which had held for about 1400 years by then (he also felt compelled not to have the book published until after his death).

> In the center of all rests the sun. For who would place this lamp of a very beautiful temple in another or better place than this wherefrom it can illuminate everything at the same time? As a matter of fact, not unhappily do some call it the lantern; others, the mind and still others, the pilot of the world."
>
> —Nicolaus Copernicus, *De revolutionibus orbium coelestium.*

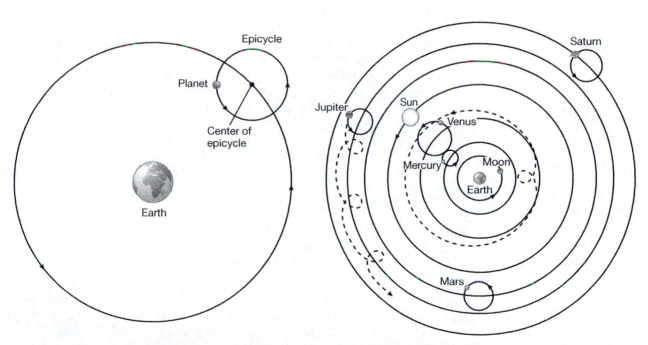

FIGURE 14.5 Ptolemy's Fine-Tuned Epicycles   The epicycles of Ptolemy's Earth-centered model of the solar system are an example of fine-tuning. (a) Epicycles are circular loops that a planet travels around while the loop itself travels around the Earth. (b) If all the parameters of Ptolemy's system are set just right, then the complicated loopy orbits of the planets can be made to match our observations of planetary motion in the night sky. Newtonian gravity eventually provided the reason behind the apparent fine-tuning.

However, it is worth noting that Ptolemy's model, despite being incorrect, was a tremendous success! It accounted for the motions of all the planets known at that time with fairly good accuracy. To do so, it employed a great many numerical parameters to describe the planets' motions around their epicycles and the epicycles' motions around the Earth—numbers that had to be set just right (i.e., fine-tuned). In the heliocentric theory that replaced it, most of that fine-tuning became unnecessary, because if you know a planet's position and motion at any one time, then you can predict it at any other time. That is to say, the newer theory (Kepler's laws explained with Newton's theory of gravity) supplied the underlying reason for the *appearance* of fine-tuning.

Are there similar explanations for the fine-tuning evident in the universe as we now know it, just waiting to be discovered and verified by experiment? This question remains open today. In the absence of a scientific answer, there may be a philosophical one: The **anthropic principle** (from the Greek root word "anthropos," meaning man, as in "anthropology"), which comes in two versions, may offer a helpful perspective.

Why is the universe tuned just right so that human life is possible? One option is that the drive toward conscious life is built into the fabric of the universe in some way, so that the universe was destined to support life all along. This idea, that the universe *must* have originated with the properties that would enable conscious life to emerge, is known as the *strong anthropic principle*. According to its adherents, we should be impressed that our universe is finely tuned to support our existence; it means that *our* presence in *this* universe is no mere coincidence.

A more general point of view is known as the *weak anthropic principle*. This is merely the observation that since *we* are alive, we are biased in the kind of universe we can find ourselves in. Even if the universe never particularly "intended" to support us, we could emerge only in a universe that happens to allow *conscious* life as a possibility. This perspective invites us to wonder whether there might be many universes, each with different properties. Perhaps only a tiny fraction of those support life, but even if just one of them does, one is enough, since that one must be the one we inhabit. In other words, we should *not* be impressed that our universe is finely tuned to support our existence: We *do* exist, so how could the universe be otherwise?

One Universe or Many?

The anthropic principle invites us to consider whether conditions in the universe could be very different elsewhere. So far, we have focused our attention on the observable universe. Already this is a vast stretch of space—a region nearly a hundred billion light-years across containing trillions of galaxies. But the entire universe is probably much, much bigger than the observable part. In fact, no compelling evidence contradicts the idea that the universe is infinite—extending without bound in every direction. And the universe is homogeneous on very large scales: The contents of an imaginary box a few hundred million light-years across are about the same no matter where in our observable universe we place that box.

Pause a moment to ponder what this means. Recall what it felt like the last time you stood and looked out at the ocean or out toward the horizon in an open field. Hold in mind how big the Earth seems from these vantage points, and then look back over the "scale of the universe" tour in Section 3.3, going from the size of Earth up to the scale of the observable universe. Now realize that there is probably a scale on which our entire observable universe is a tiny dot, and maybe another on which billions of those dots again look like a tiny dot, and so on. In such a vast universe, perhaps everything that *can* happen (i.e., everything that's not prohibited by the laws of physics) actually *does* happen somewhere in space and time.

It's analogous to an infinite universe of Lego®s, in which a limited set of different kinds of blocks is repeated over and over again throughout that universe. If the blocks are everywhere assembled randomly into structures, then every possible configuration is bound to occur somewhere. In fact, in an infinite universe, every combination occurs an infinity of times.

The fundamental particles that are the building blocks for structures in our real universe are in this sense just like the Lego®s. *You* are one combination of fundamental particles. So somewhere in an infinite universe, there should be another "you" identical to your familiar self, staying up late finishing your cosmology reading before tomorrow's class. Cosmologist Max Tegmark estimates that the nearest other "you" is some $10^{100,000,000,000,000,000,000,000,000}$ light-years away, far beyond the approximately 4×10^{10} light-years to the most distant visible objects. So there's no need to worry that you might run into this twin copy of yourself anytime soon, or any of the infinite other copies out there even farther away!

The possibility of an infinite universe like this only scratches the surface of the vastness of which we might be a part. We might be part of a reality containing multiple universes, or a *multiverse*. Perhaps inflation created uncountable distinct universes that separated themselves from the background sea of reality. Maybe our 13.7 billion-year-old universe is only one tiny universe among many, like a single bubble forming and growing in a frothy ocean. Or perhaps our universe is but one *brane* among many in a higher dimensional universe, as it would be in the cyclic model we discussed in Section 11.3.

None of the existing multiverse scenarios is well-tested science. Some are motivated by sound reasoning, such as the realization that the universe is remarkably uniform in *both* the distant past (as seen in the CMB) and the distant future (because dark energy will spread things out so much). One could speculate that the future's uniformity somehow sets the stage for the uniform beginnings of a new universe. String theory provides a context in which this idea makes sense. But perhaps the best reason to imagine a multiverse is to "help out" the anthropic principle: Why is our universe just right for our existence? How about blind chance! In a *multiverse*, in which each of its universes has its own settings on the "universe control panel," maybe a few of them were bound to come out just right. And according to the weak anthropic principle, we shouldn't be surprised to find ourselves in one of the good ones.

EXERCISE 14.2   An Identity Crisis?

Write a short essay describing your reaction to the possibility that somewhere far, far away there might be a physical copy of yourself, identical to you in every way (and an infinity of nearly identical copies, expressing every slight variation on you and your life). How does this affect your sense of identity? Does your distant twin experience the same feelings that you do? Does your twin believe that he or she is the "real you"? What makes you unique in that case? Does this scenario take away your responsibility to do anything, since no matter which choice you make, an infinity of "yous" made the same choice and all the other choices available at that instant? Write down your thoughts about these questions and similar ones that may occur to you.

Can Life Survive?

Regardless of how we came to exist in this universe, we wonder about our future prospects. Is it possible that, with unlimited technology, our distant descendants could manage to survive indefinitely, billions, trillions, even 10^{100} years from now?

Physicist Freeman Dyson opened the door for serious consideration of the future of life in the universe in a 1979 paper titled "Time Without End: Physics and Biology in an Open Universe." In the open universe he envisaged, its expansion slowed down with time. (This was before the recent evidence for dark energy causing the expansion to accelerate.) His conclusion was optimistic: Given unlimited technology, but restricted by the laws of physics, an advanced civilization could continue endlessly into the future.

More recently, cosmologists Lawrence Krauss and Glenn Starkman have reconsidered this question in light of our current understanding of the universe with accelerating expansion. Their conclusion is more pessimistic: An accelerating expansion means that as time passes, less and less matter and energy are available within any observer's cosmological horizon. Under these circumstances, as the universe grows colder and more dilute, it's difficult to see how life could persist. But the question is still openly debated, and our knowledge of the universe is still limited, so the fate of life in the distant future remains unknown. In any case, we still have many billions of years left to ponder this question.

EXERCISE 14.3   Eternal Life

The quest for immortality is a powerful theme throughout human history. Rulers have sought to extend their lives by preserving their bodies, or to be remembered through grand structures built during their reign. Performers seek to "live forever" through fame. Writers hope to pen a classic that will be read for generations.

Write a short essay describing your feelings about immortality. Is it comforting to know that human life may last forever, even if your individual life is short? Does it matter that we ever existed if at some point in the future all traces of our existence are gone? Does it concern you that you won't last forever as an individual, even if you are a part of a bigger context that does live on? Or is there something meaningful in transience, so that you don't see eternity as being particularly important?

14.3 This Is *Your* History

Now that we have stretched our thinking to the farthest imaginable reaches of space and time, we return once more to our starting point: you. Your activities in your tiny corner of the universe are the same as they were before: eating, sleeping, reading, going to work or class, talking with friends, trying to convince them to appreciate your taste in music, etc. But with the benefit of new information about your context, you can now see your daily life from a perspective that includes concrete connections to a system that is both ancient and immense.

As you sit there breathing, eating, or reading, you are not isolated and detached from the rest of the universe. You're part of a vast cosmic web of intertwined history and structures. For example, you are located on the surface of a rock nearly 13,000 km in diameter that is spinning more than 1600 km/hour at the equator and traveling around the Sun at 29 kilometers per second. Your activities on this spinning rock are ultimately sustained by the energy of sunlight. This energy in turn originates from nuclear fusion in the Sun and pours out as electromagnetic radiation that reaches you eight minutes after leaving the surface of our star. That solar energy is captured by plants and converted into a form that provides you with nourishment. The substance of your body is composed of atoms that were produced in other stars billions of years ago. All of this context is embedded in a space that formed nearly 14 billion years ago and is expanding with time.

Amidst this network of structures and processes, things have arisen that we take for granted, but that are among the most amazing and wondrous occurrences in the universe. Life. Feelings. Thoughts and conscious self-awareness. Choices. Morality. Laughter. Microwave popcorn. (See Figure 14.6.) These things have emerged as part of the evolutionary development of the cosmos, as surely as other wonders such as galaxies, black holes, planets, and rainbows. All are a consequence of cosmic processes. Our survey of the universe through the lens of science (so far) doesn't tell us what breathes life into some systems or what gives them the experience of conscious awareness and feelings that dominate our own lives. Yet they all are part of the framework we've been describing.

The context we've assembled over the last 13 chapters spells out the interconnected history of events so that we can look at them and ponder what they mean.

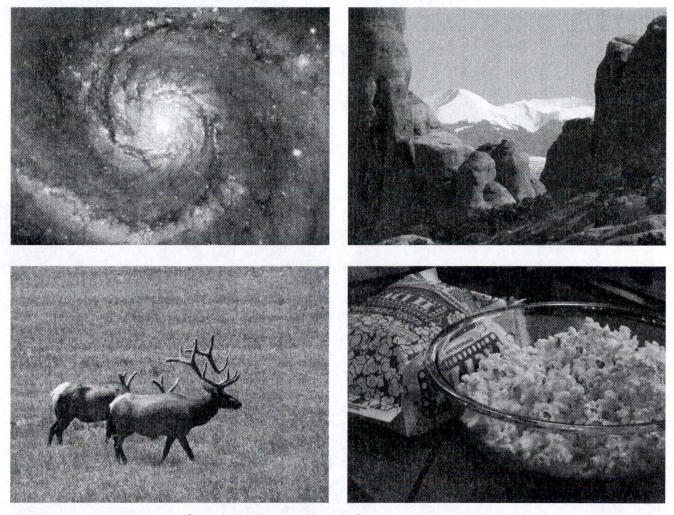

FIGURE 14.6 Important Outcomes of Cosmic History   Everything in this figure has emerged as part of the history of our universe. The process by which all of them formed includes the expansion of the universe, the formation of galaxies within dark matter halos, the nuclear reactions in stars and supernovae, and universal laws such as the law of gravity, to name just a few examples. (◆ Also in Color Gallery 2, page 31.)

FIGURE 14.7 Outcomes of the Universe   This baby seal and our emotional response to it are two remarkable outcomes of the evolution of the universe. (◆ Also in Color Gallery 2, page 31.)

Imagine watching the complete "history of the universe" DVD set, noting the exact moment when the first star formed, when the first carbon atom was produced in a star, and when the first supernova exploded. Watching the whole DVD set, you would see a hot, dense soup of fundamental particles evolve over nearly 14 billion years into such things as the baby seal in Figure 14.7 and, perhaps even more incredible, into the emotional response some people experience at the sight of one.

Think of it: There was an instant in the universe when the first life-form came into existence, another instant when consciousness first emerged as a property of some life-form, and yet another instant when the first written language developed. As intelligence developed, we became increasingly able to appreciate beauty, feel sadness, tell lies, understand humor, and sense how others feel about us (not to mention all the other little mental subtleties that have been ruining our self-image since junior high!). Perhaps these things occurred first on Earth, or perhaps on a distant planet billions of light-years from here.

EXERCISE 14.4   Passing the Torch

Think back and identify the oldest person you can remember talking to as a young child. If you can, find out that person's age, as well as your age when you last spoke with him or her. Otherwise make your best guess of the age difference between you and that person. Now imagine a conversation in which you are 10 years old and this person from a previous generation passes on to you a sense of what the world was like when he or she was 10. For example, if you were 10 when your great-grandmother was 90, then go back 80 years from the date when you turned 10. Now continue the chain backward, imagining successive 80-year intervals and a great-grandparent talking to a great-grandchild in this way. Imagine handing off a "torch of knowledge" representing the experience passed from one person to the next.

How many intervals like this would you have to go back to get to the time of the first written language? The first humans? The first dinosaurs? The first life on Earth? The birth of the solar system? The birth of the universe?

To really *experience* the current moment as part of this cosmic history, it's helpful to make a timeline of selected events. We'll make the history more concrete by counting in units of "(maximum) human lifetimes" (abbreviated *hlt*) of about 100 years. Visualize this scenario as one person (just before he or she dies at an age of about 100) handing off the torch of knowledge to a young child just beginning to learn about the world. Thus, 1 hlt represents the longest time interval from which we can build a continuous chain of people stretching back into history. (See Table 14.1 for a summary of the narrative that follows.)

We begin at the present, in the early 21st century. Within a single human lifetime ago (1 hlt), humans have seen the Earth from space for the first time and realized that the universe is expanding. All of recorded history happened less than about 60 hlt ago since the first known written language began about 5500 years ago. Most of human history passed before written language came about; at least 14,000 hlt (1.4 million years) have passed since humans domesticated fire. Try to imagine all the generations of humans that came and went before written language, compared with the very few that have known life with written records as we do. (Thus, for most of human history, students had to *memorize* entire lectures, since they couldn't take notes in class.)

Expanding the time scale even more, at least 60 million years (600,000 hlt) have passed since mammals first emerged on Earth. Dwarfing even that time span, 320 million years are required to take us back to the beginning of photosynthesis, which emerged relatively soon after the first life on Earth.

Stretching beyond the era of life on Earth, we reach the beginning of our planet and solar system itself, 4.6 billion years ago. Still, we are only about one-third of the way back to the beginning of the known universe. We must go back more than 13 billion years (130 million hlt, in case you're curious) to reach the formation of the first stars and galaxies. At this point, it's easier to talk in terms of time *after* the beginning of the known universe, since we are only a few hundred million years from the big bang.

The first neutral atoms (hydrogen, helium, and lithium) formed about 379,000 years after the big bang, when the universe had cooled enough to allow the nuclei to hold onto their electrons by electrical attraction. This is less than 4000 hlt after the beginning. From this point forward, the universe became mostly transparent, so light was free to stream through space. This is the radiation we detect today as the cosmic microwave background, a remnant from the big bang. When the CMB decoupled, it freed up baryonic matter to cluster in preexisting halos. Those halos began forming as soon as the universe went from being radiation to matter dominated, when it had been around for only 720 hlt.

Further back than this, significant events happened very quickly. The universe was hot and dense, so particles were moving around rapidly and space was also expanding quickly. Nuclei of the lightest elements (hydrogen, helium, and traces of lithium) formed during the first minutes after the beginning, far less than 1 hlt. Early in the first second, most of the matter and antimatter particles in the universe annihilated each other to produce photons. A slight asymmetry of matter over antimatter left us with a cosmos made up of matter, from which our familiar everyday objects formed. It also left the universe with about a billion photons for every particle of matter. By this time, the temperature had fallen to about 10 billion kelvins, and the density of the universe was about 100,000 times that of the Earth. One teaspoon of matter would have weighed about 1.5 tons. Finally, we reach back to only 10^{-43} second after the beginning: the Planck era, at which point our physical understanding comes to an end.

TABLE 14.1 Cosmic Timeline In working backward from the early 21st century, use the following scale: hlt = human lifetime, about 100 years (maximum).[1]

Time After Birth of Universe	Years Ago	Human Lifetimes Ago	Event
13.7 billion years	40	< 1	Humans in space see the whole Earth for the first time.
13.7 billion	40	< 1	First awareness of the cosmic microwave background (light from the early universe).
13.7 billion	80	< 1	Evidence that the universe is expanding, based on redshift–distance relation for galaxies.
13.7 billion	550	6	Printing press invented, changes our ability to communicate and share ideas widely.
13.7 billion	1,700	17	Mayan civilization.
13.7 billion	3,500–4,000	35	Megalithic structures are built in Europe (Stonehenge, etc.).
13.7 billion	5,500	55	Cuneiform writing invented by the Sumerian civilization in Mesopotamia. The first known written language.
13.7 billion	11,000–9,000	100	The practice of agriculture emerges.
13.7 billion	30,000	300	Humans create the first cave paintings.
13.7 billion	50,000–500,000	500–5,000	Symbolic language emerges.
13.7 billion	1.4 million	14,000	Humans domesticate fire.
13.7 billion	2.5 million	25,000	Early humans use stone tools.
13.7 billion	61 million	610,000	Rapid expansion of mammals onto the world scene.
10.8 billion	3.2 billion	32 million	Photosynthesis emerges in blue-green cyanobacteria.
10.5 billion	3.8 billion	38 million	The first life emerges, possibly at great depth within Earth's crust or at hydrothermal fissures in the floor of the deep ocean.
9.6 billion	4.1 billion	41 million	End of the period of heavy bombardment by debris from space; rain falls upon a cooling Earth for the first time.
9.5 billion	4.2 billion	42 million	Dark energy domination begins. Large structures formed during the preceding matter-dominated era are increasingly separated by an accelerated expansion.
9.1 billion	4.6 billion	46 million	Earth and other planets coalesce out of the disk of debris orbiting the newborn sun.
9 billion	4.7 billion	47 million	Birth of our solar system from a cloud of gas and dust (containing the remains of previous generations of stars) coalescing under its own gravity and perhaps triggered by a shock wave from a supernova.
<1 billion	13+ billion	130+ million	Galaxies begin to form.
<1 billion	13+ billion	130+ million	Protogalactic clouds of hydrogen form; the universe differentiates into vast clumps of gaseous matter.

(continued)

Time After Birth of Universe	Years Ago	Human Lifetimes Ago	Event
Few hundred million to present	13+ billion to present	130 million to < 1 (stars are still forming today, as you read)	Birth of stars. These forge heavy elements (atoms heavier than lithium—i.e., carbon, nitrogen, oxygen, calcium, iron, etc.) that are crucial to the future processes of the universe. The elements that are created enrich the galaxies when the stars explode as supernovae or give off some of their material less violently as red giants, etc.
379,000	13.7 billion	137 million	First atoms. The universe cools to the point that electrons can combine with hydrogen and helium nuclei, producing the first neutral atoms, which are now free to start forming galaxies. This also makes the universe mostly transparent, and the light is free to stream through the universe. We detect this radiation today as the CMB.
72,000	13.7 billion	137 million	Matter–radiation equality. The universe ceases to be radiation dominated and becomes matter dominated, enabling (dark) matter structures (halos) to begin forming.
1 minute	13.7 billion	137 million	Production of light elements. Protons and neutrons emerge and form the nuclei of the simplest chemical elements: heavy hydrogen (deuterium), helium, and traces of lithium.
10^{-5} sec	13.7 billion	137 million	Particle–antiparticle annihilation. All the antiparticles in the universe annihilate almost all the particles, creating a cosmos made up of matter (because there was about 1 extra particle out of every billion particle–antiparticle pairs that annihilated into photons) and photons. By this time, the temperature had fallen to about 10 billion degrees Kelvin, and the density of the universe was about 100,000 times that of the Earth. One teaspoon of matter would have weighed about 1.5 tons.
From 10^{-32} to 10^{-10} sec	13.7 billion	137 million	Emergence of the four fundamental forces: gravity, electromagnetism, strong nuclear force, weak nuclear force.
10^{-35} to 10^{-32} sec	13.7 billion	137 million	Inflation.
10^{-43} sec	13.7 billion	137 million	Planck era; current physics can't describe earlier times.
0	13.7 billion	137 million	The beginning of present universe.
≤ 0	13.7 billion	137 million	The "great mystery": It's not even clear what it means to talk about time "before" the beginning of time!

[1]Thanks to Connie Barlow and Michael Dowd for the collection of timeline ideas (http://thegreatstory.org/great_story_beads.html) from which this timeline is adapted.

EXERCISE 14.5   Personalize the Timeline

The main purpose of the cosmic timeline is to help you see yourself within the context of history. Using Table 14.1 as a starting point, write your own personal timeline as a table of events in the history of the universe that are significant to you. Include both "objective" phenomena and personal events in your life that are important, and connect them as much as possible. (For example, what events in ancient history made the calcium in the ice cream you ate on your first birthday?)

The important lesson to remember is this: *You* are literally a product of everything we've been talking about in cosmology. Every bit of material that is a part of you has a history and a story to tell. It can tell a story of its journey through the unimaginable heat and pressure of the big bang, of the cooling as the universe expanded, of the first light, of the first time when the universe was transparent, when it became part of a star and a galaxy, and when it passed through other generations of stars, into a cloud that formed the Sun and Earth, and now you. The universal trend toward increasing complexity is expressed in everything from the formation of matter to the formation of a sentence that you speak to a friend (or the profanity that you speak to a complete stranger in traffic). The story continues far beyond you: What future events will take place in the little bit of space that you can hold between your hands, a region of space that was once hotter and denser than the core of our Sun?

14.4 So What? Finding Your Cosmic Context

In Chapter 1 (Exercise 1.6), you considered the question "What image do I hold of myself in relation to the universe?" As a student of scientific cosmology, your intellectual understanding of the universe has changed, perhaps dramatically. But what difference does this new knowledge make for how you live? It will make a difference only if you find an individual connection for yourself to this newly understood universe. This kind of connection is more than just listing the key facts about the cosmos and its history. It requires a separate step of actively bringing the information into your personal philosophy.

There is much room for interpretation and your own way of seeing your personal significance within the cosmos. As philosopher William James points out, the day-to-day image you hold of yourself

> is not a technical matter; it is our . . . sense of what life honestly and deeply means. It is only partly got from books; it is our individual way of just seeing and feeling the total push and pressure of the cosmos.

In this final section, we invite you to consider how best to feel the "push and pressure of the cosmos"—your cosmic context. To help spark your own ideas, we provide a few examples of how different people choose to incorporate this information.

EXERCISE 14.6   Describe Your Universe Again

"Describe your universe" (as in Exercise 1.7) again, taking into account what you have learned about cosmology. Compare your answer here with the essay you wrote in Chapter 1, noting how new information has affected your perspective on the universe.

Solace in the Face of Difficulty

For many people, the universe expands dramatically during their study of cosmology. The solar system you studied in grade school was almost too big to comprehend. To absorb the idea that the solar system itself is but a speck in the cosmos can be a perspective-changing experience. Students sometimes react to this newfound vastness by noticing that their own problems feel smaller and more manageable in comparison. Sure, being stuck in traffic for hours, losing your job, or failing a test are tragedies of profound personal significance. But your troubles are small in comparison to all the events happening right now, throughout the cosmos: Countless new stars and planets are being born and also dying in supernovae. New elements are forming in stellar thermonuclear furnaces. Funny-looking green creatures with one eye and three arms may be stuck in traffic just like you, but a billion light-years away in hover cars that use acorns for fuel.

This perspective is relatively new in human history. As recently as the early 1900s, even professional astronomers were unaware that the cosmos extends beyond our own Milky Way. The knowledge gathered in the late 20th and early 21st centuries provides each of us a new way to see beyond some of the difficulties in our lives. There are bigger things going on out there, so you might be able to recalibrate your thoughts toward designing your place in the world, rather than fretting about whether or not you spent enough money on someone's birthday present. (And even then, you may feel dimly aware that if the *universe* could consciously judge you, it would find your concerns to be petty!)

> All the while I'm working as if everything depends on what I do, I remember that nothing depends on what I do.
>
> —Ram Dass (American psychologist and spiritual leader)

Cosmic Artists

We've seen that change is a dominant theme in the cosmos. Tiny quantum fluctuations within the hot, dense early universe evolved into clusters of galaxies now drifting amid mostly empty space. Stars have life cycles whose duration depends on their masses, ending their lives peacefully in some cases, but with dramatic explosions in others. In all cases, stars release new building blocks into their surroundings—the raw material for new stars and planets. Wonders continue to emerge as time passes, condensing out of the backdrop of energy within the framework of space and time.

From this knowledge comes a perspective that views humanity as part of a cosmic creative process. Perhaps each person's life is a work of art or an experiment within the vast universal collection. From that perspective, maybe what matters isn't the specific image of yourself you choose, but rather just that you *do* choose some specific way of seeing yourself in the context of the universe and live wholeheartedly according to that. That way, we each contribute something to the universe just by how we try to live. Every human creation (such as the M. C. Escher print in Figure 14.8) reflects the creative universe we are part of.

This perspective is well expressed in the words of 20th-century anthropologist Theodosius Dobzhansky:

> The creation...is not an event which happened in the remote past but is rather a living reality of the present. Creation is a process of evolution of which [humanity] is not merely a witness but a participant and a partner as well.
>
> —from *The Biological Basis of Human Freedom.*

FIGURE 14.8 M. C. Escher's "Other World" (1947)   Our own artistic creations are embedded within the background of everything else the unfolding universe has created. (◆ Also in Color Gallery 2, page 32.)

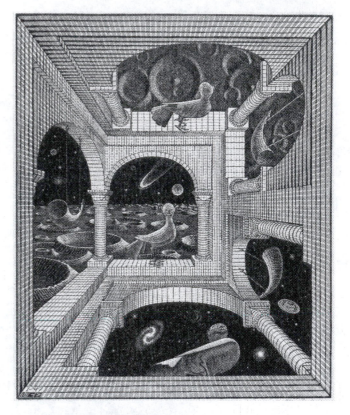

Contributors to a Cosmic Plan

Each religious tradition gives its own perspective on humanity's relationship to the cosmos, along with guidance for how we should live. Within the context of a particular religion, the scientific investigation of the cosmos can be seen as a way to enrich and deepen our appreciation of the cosmic order. For example, the Dalai Lama describes how insights from the physics of relativity and quantum mechanics have enriched his Buddhist perspective on the relativity of time and the dynamic, transient nature of substance in the universe. Here he expresses a key point in any view that uses scientific cosmology to enrich a religious cosmology:

> One can take science seriously and accept the validity of its empirical findings without subscribing to scientific materialism. I have argued for the need for and possibility of a worldview grounded in science, yet one that does not deny the richness of human nature and the validity of modes of knowing other than the scientific. I say this because I believe strongly that there is an intimate connection between one's conceptual understanding of the world, one's vision of human existence and its potential, and the ethical values that guide one's behavior. How we view ourselves and the world around us cannot help but affect our attitudes and our relations with our fellow beings and the world we live in.
>
> —from *The Universe in a Single Atom: The Convergence of Science and Spirituality*

Random Specks

The perspectives described so far all serve to counter in some way the sense that we are merely random specks in a vast, uncaring universe. But suppose that we really are insignificant in a cosmic sense. Physicist Steven Weinberg summed up this perspective in one sentence:

> The more the universe seems comprehensible, the more it also seems pointless.
>
> —from *The First Three Minutes*

From this point of view, must we despair for the absence of a meaningful life? Perhaps not. Even in that case, the reality of our experience still is quite tangible. Writer Olaf Stapledon emphasized this point of view with these words:

> And if after all there is no Star Maker, if the great company of galaxies leapt into being of their own accord, and even if this little nasty world of ours is the only habitation of the spirit anywhere among the stars, and this world doomed, even so, even so, I must praise. But if there is no Star Maker, what can it be that I praise? I do not know. I will call it only the sharp tang and savour of existence. But to call it this is to say little.
>
> —from *Star Maker*

It may be that "random specks" is the bleakest of cosmological viewpoints. Yet it can also be the most liberating (in that we are individually free to define our own value systems, since the universe hasn't done it for us) and, at the same time, the most demanding upon us (in that we are obligated to do so).

> There is then a simple answer to the question, "What is the purpose of our individual lives?" They have whatever purpose we succeed in putting into them.
>
> —A. J. Ayer, 20th-century British philosopher

EXERCISE 14.7 Life in the Absence of External Meaning

Suppose you somehow learned with certainty that your actions and thoughts hold no significance within the external universe—nothing outside of the meaning you create for yourself. What would you do in that case? How would you choose to live? You would still have the feeling that some things matter more than others; you would still make choices; and you would still be faced with daily life just as you are now. What might you choose as the things that matter? Write 300 words (about one double-spaced page) describing how you might define values that are purely your own.

These few examples barely begin to explore the possible ways to see yourself within the universe (or even just within the universe as we currently understand it). It's likely that deeper and more complete perspectives lie in directions that have yet to be discovered. So keep your perspective open to new ideas! As Shakespeare's Hamlet advises,

> There are more things in heaven and earth, Horatio, than are dreamt of in your philosophy.

An Open Future

Until recently, cosmologists thought that our universe was geometrically open ("saddle shaped") because $\Omega_{matter} < 1$. Observations of the cosmic microwave background now provide solid evidence that $\Omega_{total} = 1$, so the universe is actually *flat*, or at least very nearly so, in the technical, geometric sense we discussed in Chapter 11. Still, our universe is probably "open ended," in the sense that it will continue expanding forever. Despite the scientific progress we've outlined in this book, many questions in cosmology remain open as well.

While the ability to keep learning and modifying any theory is a key strength of the scientific method, it's also disconcerting when you're trying to understand how you fit into the big scheme of things. How do you build a meaningful role for yourself within a cosmos that might turn out to be different from what you thought it was? Perhaps there is a lesson in this very fact, that our lot as conscious creatures is to live with uncertainty. The "final truth" may forever be a mystery to us, so we are forever driven to explore and learn, our knowledge expanding into the distant future, just like the universe.

In this spirit, we end with three open questions you'll likely hear more about in the coming years.

Open question #1: What's the matter in the universe made of?

The ordinary, baryonic matter we're made of is only a small fraction of the total matter in the universe (about one-sixth, the evidence indicates). So the basic nature of most of the stuff in the cosmos is a big unknown. We know that this matter must be some exotic form, not like the stuff we find in our familiar surroundings on Earth. That is, we know what it is *not*. But we may be far from knowing what it actually *is*. Weakly interacting massive particles (WIMPs) are a strong candidate, but so far these are theoretical constructs invented to outline the properties this exotic matter must have in order to match with observations. In the absence of a direct WIMP detection, we must remain open to the possibility of surprises ahead.

Open question #2: What is the dark energy?

It's puzzling enough that ordinary baryonic matter is only a small fraction of all the matter in the universe. But the mystery deepens: Evidence we discussed in Chapter 7 indicates that most of the energy in the universe isn't matter or radiation of *any* kind. Rather, about 73% is in the form of dark energy responsible for the accelerating expansion of the universe. It's sobering to realize that the universe is composed mostly of something we know almost nothing about.

Open question #3: What is the ultimate fate of the universe?

We've said already that the universe will probably continue expanding indefinitely. But *probably* is a key word. Without knowing the specific nature of dark energy, we cannot know the ultimate destiny of our universe with certainty. Will the universe expand forever, growing colder and dimmer into the distant future? Or will it someday recollapse, returning to the state of intense heat and pressure of the big bang from which it was born?

The answers to these questions lie in the future, an extension of the timeline that includes you and everything else in the universe. What will you create within this unfolding cosmic context?

Quick Review

In this final chapter, you've looked at how all the material in the previous chapters ties together into *your* cosmic context. In this light, you read about the universe's size, behavior, and interconnectedness. You saw that some properties of the universe appear to be fine-tuned to allow the emergence of human life. The strong and weak anthropic principles are two differing ways to interpret the significance of this fine-tuning. Human existence (so far) is only the briefest of blips at the very tail end of an enormous timescape that preceded it. So are you just a random speck? And if so, does that license you to worry less, for example? Or do you directly contribute to a grand cosmic plan? Such questions can affect your daily decisions, and yet they can be meaningfully informed by the science of the entire universe.

Try the following crossword puzzle to test your knowledge of key terms and concepts from Chapter 14:

ACROSS

1 The _____ part of a spiral galaxy is neither too close to the center nor too close to the outer edge.

4 Extra circular loop proposed by Ptolemy to describe the motions of planets

8 If certain properties of the universe, such as the strength of gravity, were changed slightly, then _____ could never have come to exist.

9 Setting one or more numerical parameters very precisely for no obvious physical reason

10 Version of the anthropic principle which proposes that the universe's properties are the way they are in order to accommodate human life

11 Principle that constrains theories of the universe by insisting that they allow for the possibility of human existence

DOWN

2 The likelihood of life enduring forever is diminished because the _____ of the universe causes less and less energy to be available within any cosmological horizon as time goes by.

3 Hypothetical collection of universes in which ours is only one among many

5 It's possible that many separated "bubble universes" exist because they were "blown up" by the process of _____.

6 Changing over time, the way the universe has been

7 Responsible for initiating the historical shift from a geocentric cosmology to a heliocentric one

Further Exploration

(M = mathematical content; I = integrative—builds on information from more than one chapter; P = project idea; R = reflection; D = suitable for class discussion)

1. **(MI)** Express the age of the universe in different units of time. For example, in this chapter we found that the universe was 137 million "(maximum) human lifetimes." Now compute the age of the universe in terms of the approximate time span for each of the following instead:
   (a) lifetime of a very short lived star (i.e., how many such stars could live and die, in sequence, between time zero and now?)
   (b) lifetime of an average star, like the Sun
   (c) lifetime of a very long lived star
   (d) some sort of average time between mass extinctions on Earth
   (e) time since humans first appeared on Earth
   (f) time since life first formed on Earth
   (g) time since humans started exploring space

2. **(MI)** For each of the following cosmic periods, compute what percent of the present age of the universe that period occupied. Show enough decimal places to see the precision of the number (e.g., 99.999994%).
   (a) helium nuclei exist (i.e., for what percent of the present age of the universe have helium nuclei existed?)
   (b) neutral atoms exist
   (c) protons exist
   (d) radiation domination
   (e) matter domination
   (f) dark energy domination
   (g) the CMB exists (time since last scattering)
   (h) the "dark ages"
   (i) galaxies exist
   (j) our solar system exists
   (k) life span of a Sun-like star
   (l) life on Earth exists
   (m) mammals exist
   (n) humans exist
   (o) the United States exists
   (p) nuclear weapons exist
   (q) you exist

3. **(IRD)** Describe something in cosmology you would like to know about that we did not cover in this book, and explain how you can learn about it on your own. You might choose a topic we did not cover in detail, something you have heard about that we didn't discuss, or a question you have that you would like to see researchers answer. What are some resources that will aid you in finding out what you don't know? What experiments or observations could you make yourself?

4. **(IRP)** Choose one idea from cosmology that is significant to you, and express your knowledge in whatever artistic way you think best captures the idea: through a drawing, painting, dance, song, short story, poem, etc.

5. **(IRD)** What do you perceive as the top five key features or properties of the universe that science tells us we live in? Write a short essay describing those properties and explaining why they are important to you.

6. **(IRD)** What qualities or properties do you think the universe must have in order for you to feel that you have a place in it and that your life is meaningful? (In other words, what are the key features or properties of your "ideal universe?"). Make a list of at least three such properties, and write a few sentences explaining why each is important to you.

7. **(IRD)** Identify one way in which you live differently because you know something specific about the world that was learned through science. Write a few sentences describing how you make different choices or act differently because you know this piece of scientific information.

8. **(IRDP)** If you subscribe to the "random speck" theme from Section 14.4, then it may seem that the more science you learn, the more empty your life becomes! (This is especially true if it means that you're constantly finding little technical flaws in space movies, such as hearing exploding ships, which couldn't really happen without air to support the sound waves.) The purpose of this question is to turn things around: Try to identify ways in which you can use scientific knowledge deliberately to make yourself happier.
   (a) List one or two aspects of cosmology that you have learned from this book and that could affect your thinking in a positive way (perhaps to make you feel better on a daily basis).
   (b) Expand the question beyond cosmology: What sort of scientific knowledge could you directly *use* in order to improve your life? Feel free to google your way to happiness here. Find a study or a theory, and explain how you can behave differently in your life to capitalize on that scientific knowledge. For example, does a particular type of exercise schedule regulate chemicals in your brain that affect your emotional state (giving you a "runner's high")? Or can you capitalize on the short life span of taste bud cells (a few weeks) in order to train yourself to enjoy the taste of healthier foods? Find another example like these that could make a difference for you.
   (c) Find a popular scientific article, a chapter from a book, etc., to read about a recent study on how the mind works. A good source for such articles is the magazine *Scientific American Mind*. Pick one article, read it, and use it to create a recipe for improving your happiness. For example, if you read that rap

music helps regenerate your creative ability when you're tired (we just made that up), then you might create a list of four or five occasions to listen to rap, like "right before I brainstorm with my homies."

9. (MRI) Imagine that you are immortal and that you have been traveling around the universe ever since it formed (in some sort of faster-than-light, protective spaceship). Break your existence down into approximately five different periods so that you can describe how the universe and the "cool stuff" to see and do was different from one period to the next. For each period, state how many years it lasted, what was going on in the universe, and roughly how far you had to travel to see whatever there was to see.

10. (RD)
   (a) Write down, in your own words, the meaning of the strong and weak anthropic principles.
   (b) For each version of the anthropic principle, what is your opinion on its validity *and* usefulness?
   (c) How do the multiverse concept and the anthropic principle connect to one another?

11. (IRD) What remains to be solved in scientific cosmology? List off as many as yet unanswered questions about the universe as you can. You should be able to come up with at least 5, and a good answer would probably exceed 10. For each question, give an estimate, with a brief reason for that estimate, of when you think that question will be answered, if ever.

12. (MRD) Revisit the cosmic timeline of Table 14.1 and replace the data in the "human lifetime" column with the life spans of some other organisms, to imagine what history might look like from their perspective. For example, how do events look to a tree that lives for 10,000 years or to an insect that lives only a few days? (You don't need to get carried away; just quote the changes to the table that seem interesting. Compared with the life of a person, a tree, or an insect, for instance, 13.7 billion years is a very long time ago, so that's not really worth changing the table over. But people, trees, and insects would each have a very different sense of the time span of recorded human history.)

Bibliography and Credits

BOOKS

Abbott, Edwin. *Flatland: A Romance of Many Dimensions*. Boston: Little, Brown, and Company, 1884.

Barrow, John D.; Tipler, Frank J.; Wheeler, John A. *The Anthropic Cosmological Principle*. Oxford, UK: Oxford University Press, 1988.

Barrow, John D. *The Constants of Nature*. New York: Random House (Vintage Books), 2002.

Bartusiak, Marcia. *Archives of the Universe*. New York: Pantheon Books, 2004.

Becker, Wayne M.; Kleinsmith, Lewis J.; and Hardin, Jeff. *The World of the Cell*, 5th edition. San Francisco: Benjamin Cummings, 2003.

Bennett, Jeffrey; Donahue, Megan; Schneider, Nicholas; and Voit, Mark. *The Cosmic Perspective*, 3rd edition. San Francisco: Addison Wesley, 2004.

Bennett, Jeffrey; Shostak, Seth; and Jakosky, Bruce. *Life in the Universe*. San Francisco: Addison Wesley, 2003.

Binney, James and Merrifield, Michael. *Galactic Astronomy*. Princeton: Princeton University Press, 1998.

Binney, James and Tremaine, Scott. *Galactic Dynamics*. Princeton: Princeton University Press, 1987.

Campbell, Neil A.; Reece, Jane B.; Taylor, Martha R.; and Simon, Eric J. *Biology: Concepts and Connections*, 5th edition. San Francisco: Benjamin Cummings, 2006.

Carroll, Bradley W. and Ostlie, Dale A. *An Introduction to Modern Astrophysics*. Reading, MA: Addison Wesley, 1996.

Chaisson, Eric and McMillan, Steve. *Astronomy Today*, 5th edition. Upper Saddle River: Prentice Hall, 2005.

Chaisson, Eric. *Cosmic Evolution: The Rise of Complexity*. Boston: Harvard University Press, 2002.

Coble, Kim and Duncan, Todd L. *The Big Bang Happened Here*. Unfinished manuscript, 2003; Coble, Kim; Duncan, Todd L., and Tyler, Craig. *The Big Bang Happened Here*. Unfinished manuscript, 2004.

Cramer, John. *How Alien Would Aliens Be?* Writers Club Press, 2001.

Drake, Stillman (translator): *Discoveries and Opinions of Galileo*. New York: Doubleday Anchor Books, 1957.

Duncan, Todd. *An Ordinary World: The Role of Science in Your Search for Personal Meaning*. Hillsboro, OR: The Science Integration Institute, 2001.

Einstein, Albert; *Mein Weltbild*. Seelig, Carl (editor); and Bargmann, Sonja (translator). *Albert Einstein: Ideas and Opinions*. New York: The Modern Library, 1994.

Ember, Carol R.; Ember, Melvin; and Peregrine, Peter N. *Anthropology*, 10th ed. Upper Saddle River: Prentice Hall, 2002.

Ferris, Timothy. *The Whole Shebang*. New York: Simon and Schuster, 1997.

Feynman, Richard P. *What Do You Care What Other People Think?* New York: W. W. Norton & Company, 1988.

Filkin, David. *Stephen Hawking's Universe*. New York: Harper Collins, 1997.

Galilei, Galileo, translated by Drake, Stillman. *Dialogue Concerning the Two Chief World Systems*. Berkeley: University of California Press, 1953.

Gilmour, Iain and Sephton, Mark A. *An Introduction to Astrobiology*. Cambridge, UK: Cambridge University Press, 2003.

Gleick, James. *Isaac Newton*. New York: Vintage, 2004.

Goldsmith, Donald. *The Hunt for Life on Mars*. New York: Penguin Group (Dutton), 1997.

Gott, J. Richard III. *Time Travel in Einstein's Universe: The Physical Possibilities of Travel through Time*. Boston: Houghton Mifflin, 2001.

Greene, Brian. *The Elegant Universe*. New York: W. W. Norton & Company, 1999.

Gribbin, John. *In Search of Schrödinger's Cat*. New York: Bantam Books, 1984.

Guth, Alan. *The Inflationary Universe*. Cambridge, MA: Perseus Books, 1997.

Harpaz, Amos. *Relativity Theory: Concepts and Basic Principles*. Wellesley, MA: A.K. Peters, 1993.

Harrison, Edward. *Darkness at Night: A Riddle of the Universe*. Cambridge, MA: Harvard University Press, 1987.

Harrison, Edward. *Cosmology: The Science of the Universe*, 2nd ed. Cambridge, UK: Cambridge University Press, 2000.

Harrison, Edward. *Masks of the Universe: Changing Ideas on the Nature of the Cosmos*. Cambridge, UK: Cambridge University Press, 2003.

Hartle, James, *Gravity: An Introduction to Einstein's General Relativity*. San Francisco: Addison Wesley, 2002.

Hawking, Stephen. *[The Illustrated] A Brief History of Time*. New York: Bantam Books, 1988, 1996.

Hawking, Stephen. *The Universe in a Nutshell*. New York: Bantam Books, 2001.

Hetherington, Norriss S. (ed.). *Cosmology: Historical, Literary, Philosophical, Religious, and Scientific Perspectives*. Hamden, CT: Garland Publishing, 1993.

Hobson, Art. *Physics: Concepts and Connections*, 3rd edition. Upper Saddle River: Prentice Hall, 2003.

Hoyle, Fred; Burbidge, Geoffrey; Narlikar, Jayant V. *A Different Approach to Cosmology*. Cambridge, UK: Cambridge University Press, 2000.

Jones, Mark H. and Lambourne, Robert J. A., editors. *An Introduction to Galaxies and Cosmology*. Cambridge, UK: Cambridge University Press, 2003.

Klug, William S. and Cummings, Michael R. *Concepts of Genetics* 6th ed. Upper Saddle River: Prentice Hall, 2000. *The Early Universe*. Reading, MA: Addison Wesley, 1990.

Krauss, Lawrence M. *Atom: An Odyssey from the Big Bang to Life on Earth . . . and Beyond*. Boston: Little, Brown & Company, 2001.

Levin, Janna. *How the Universe Got Its Spots: Diary of a Finite Time in a Finite Space*. New York: Random House (Anchor Books), 2002.

Liddle, Andrew. *An Introduction to Modern Cosmology*, 2nd ed. West Sussex: Wiley, 2003.

Lightman, Alan. *Dance for Two*. New York: Random House (Pantheon Books), 1996.

Lightman, Alan. *The Discoveries*. New York: Random House (Pantheon Books), 2005.

Lunine, Jonathan I. *Astrobiology: A Multidisciplinary Approach*. San Francisco: Addison Wesley, 2005.

Nash, Roderick Frazier. *Wilderness & the American Mind.* New Haven: Yale University Press, 2001.

Nelson, David L. and Cox, Michael M. *Principles of Biochemistry,* 4th edition. New York: W. H. Freeman & Company, 2005.

Peacock, John A. *Cosmological Physics.* Cambridge, UK: Cambridge University Press, 1999.

Peebles, P. James E. *Principles of Physical Cosmology.* Princeton: Princeton University Press, 1993.

Pinker, Steven. *How the Mind Works.* New York: W. W. Norton & Company, 1997.

Pinker, Steven. *The Blank Slate.* New York: The Penguin Group (Viking), 2002.

Plaxco, Kevin W. and Gross, Michael. *Astrobiology: A Brief Introduction.* Baltimore: Johns Hopkins University Press, 2006.

Postlethwait, John H.; Hopson, Janet L.; and Veres, Ruth C. *Biology! Bringing Science to Life.* New York: McGraw Hill, 1991.

Primack, Joel R. and Abrams, Nancy Ellen. *The View from the Center of the Universe.* New York: Riverhead Books, 2006.

Rees, Martin J. *Before the Beginning: Our Universe and Others.* San Francisco: Addison Wesley, 1997.

Rees, Martin J. *Our Cosmic Habitat.* Princeton: Princeton University Press, 2003.

Ross, Hugh. *The Creator and the Cosmos.* Colorado Springs: NavPress, 1993.

Rybicki, George B. and Lightman, Alan P. *Radiative Processes in Astrophysics.* New York: Wiley, 1979.

Ryden, Barbara. *Introduction to Cosmology.* San Francisco: Addison Wesley, 2003.

Shu, Frank. *The Physical Universe.* Mill Valley, CA: University Science Books, 1982.

Stapp, Henry. *Mind, Matter, and Quantum Mechanics.* 2nd edition. Springer, 2004.

Swerdlow, N.M. *The Babylonian Theory of the Planets.* Princeton: Princeton University Press, 1998.

Swimme, Brian and Berry, Thomas. *The Universe Story.* San Francisco: HarperCollins, 1992.

Taylor, Edwin F.; Wheeler, John Archibald. *Spacetime Physics.* W.H. Freeman, 1966.

Tegmark, Max. *Parallel Universes* in *Science and Ultimate Reality.* Edited by John D. Barrow, Paul C. W. Davies, Charles L. Harper, Jr. Cambridge UK: Cambridge University Press, 2004.

Thorne, Kip S.; Misner, Charles W.; Wheeler, John Archibald. *Gravitation.* W.H. Freeman, 1973.

Turner, F. J. "The Frontier in American History." New York: H. Holt & Co., 1920.

Von Baeyer, Hans Christian. *Warmth Disperses and Time Passes: The History of Heat.* New York: Random House, 1999.

Ward, Peter D. and Brownlee, Donald. *Rare Earth.* New York: Springer-Verlag (Copernicus), 2000.

Weiner, Jonathan. *The Beak of the Finch.* New York: Random House (Vintage Books), 1994.

Wilson, Edward O. *The Future of Life.* New York: Knopf, 2002.

Wright, Robert. *The Moral Animal.* New York: Random House (Vintage Books), 1994.

Young, Louise B. *The Unfinished Universe.* Oxford, UK: Oxford University Press, 1993.

Zimmer, Carl. *Evolution: the Triumph of an Idea.* New York: HarperCollins, 2001.

Zubrin, Robert. *The Case for Mars.* New York: Touchstone, 1996.

PAPERS AND ARTICLES

Ahmad, Q. R. et al. "Direct Evidence for Neutrino Flavor Transformation from Neutral-Current Interactions in the Sudbury Neutrino Observatory." *Physical Review Letters* 89 (2002): 011301.

Alcock, C. et al. (the MACHO collaboration). "MACHO LMC Microlensing Results." *Astrophysical Journal,* 542 (2000): 281.

Allègre, Claude J. and Schneider, Stephen H. "The Evolution of the Earth." *Scientific American.* (1994)

Asimov, Isaac. "The Relativity of Wrong." *The Skeptical Inquirer,* Volume 145, Number 1. (Fall 1989): 35–44.

Atreya, Sushil K. "The Mystery of Methane on Mars & Titan." *Scientific American* (May 2007).

Beckman, John E. and Casuso, Emilio. "The Evolution of the Light Elements, Be and B (also Li), in the Galaxy." *Chinese Journal of Astronomy and Astrophysics,* Volume 3 (2003): 105–114.

Bennett, C. L., et al. "First Year Wilkinson Microwave Anisotropy Probe (WMAP) Observations: Preliminary Maps and Basic Results." *Astrophysical Journal* Suppl. 148 (2003): 1.

Bennett, Charles L.; Hinshaw, Gary F.; Page, Lyman. "A Cosmic Cartographer." *Scientific American* (January 2001).

Bennett, Charles H.; Li, Ming; and Ma, Bin. "Chain Letters and Evolutionary Histories," *Scientific American* (June 2003).

Bernstein, Max P.; Sandford, Scott A.; and Allamandola, Louis J. "Life's Far-Flung Raw Materials." *Scientific American* (July 1999).

Borgani, Stefano. "Cosmology with clusters of galaxies."

Bucher, Martin A.; Spergel, David N. "Inflation in a Low-Density Universe." *Scientific American* (January 1999).

Caldwell, Robert R; Kamionkowski, Marc. "Echoes from the Big Bang." *Scientific American* (January 2001).

Cayrel, R.; Hill, V.; Beers, T.C.; Barbuy, B.; Spite. M.; Spite, F.; Plez,B.; Andersen, J.; Bonifacio, P.; François, P.; Molaro, P.; Nordström, B.; and Primas, F. "Measurement of Stellar Age from Uranium Decay." *Nature,* Volume 409 (8 February 2001): 691.

Clack, Jennifer A. "Getting a Leg Up on Land." *Scientific American* (December 2005).

Cline, D. B., Matthey, C., and Otwinowski, S. "Evidence for a Galactic Origin of Very Short Gamma Ray Bursts and Primordial Black Hole Sources." *Astroparticle Physics* 18 (2003): 531–538.

Cline, David B. "The Search for Dark Matter." *Scientific American* (March 2003).

Combes, Françoise. "Ripples in a Galactic Pond." *Scientific American* (October 2005).

Conselice, Christopher J. "The Universe's Invisible Hand." *Scientific American* (February 2007).

Coulson, David; Crittenden, Robert G.; and Turok, Neil. "Polarization and Anisotropy of the Microwave Sky." *Physical Review Letters* 73 (1994): 2390–2393.

Cowan, John J.; Pfeiffer, B.; Kratz, K.-L.; Thielemann, F.-K.; Sneden, Christopher; Burles, Scott; Tytler, David; Beers, Timothy C. "r-Process Abundances and Chronometers in Metal-poor Stars." *Astrophysical Journal* Volume 521 (10 August 1999): 194.

DeWaal, Frans B. M. "How Animals Do Business." *Scientific American* (April 2005).

Dobb, Edwin. "Without Earth There Is no Heaven: The Cosmos Is not a Physics Equation." *Harper's* Volume 289, Issue 1737 (February 1995): 33–41.

Dvali, Georgi. "Out of the Darkness." *Scientific American* (February 2004).

Dyson, Freeman J. "Time without End: Physics and Biology in an Open Universe." *Reviews of Modern Physics* Volume 51, No. 3 (July 1979): 447–60.

Eisenstein, Daniel J., et al. "Detection of the Baryon Acoustic Peak in the Large-Scale Correlation Function of SDSS Luminous Red Galaxies." *Astrophysical Journal* 633 (2005): 560–574.

Fang, T. et al. "Chandra Detection of O VIII Lyman-alpha Absorption from an Overdense Region of the Intergalactic Medium." *Astrophysical Journal Letters*, 572 (2002): L127.

Ferrarese, Laura; Merritt, David. "Supermassive Black Holes." *Physics World* (June 2002).

Friedmann, A. "On the Curvature of Space." *General Relativity and Gravitation* Volume 31, No. 12, (1999): 1991.

Gonzales, Guillermo; Brownlee, David; and Ward, Peter D. "Refuges for Life in a Hostile Universe." *Scientific American* (October 2001).

Gould, Stephen Jay. "The Evolution of Life on the Earth." *Scientific American* (October 1994).

Green, Anne M. "Viability of primordial black holes as short period gamma-ray bursts." *Physical Review* D 65 (2002): 027301.

Hansen, Brad M. S.; Brewer, James; Fahlman, Greg G.; Gibson, Brad K.; Ibata, Rodrigo; Limongi, Marco; Rich, R. Michael; Richer; Harvey B.; Shara, Michael M.; Stetson, Peter B. "The White Dwarf Cooling Sequence of the Globular Cluster Messier 4." *Astrophysical Journal Letters*, Volume 574 (1 August 2002): L155.

Hansen, Brad; Richer, Harvey; Fahlman, Greg; Stetson, Peter; Brewer, James; Currie, Thayne; Gibson, Brad; Ibata, Rodrigo; Rich, R. Michael; Shara, Michael. "HST Observations of the White Dwarf Cooling Sequence of M4." Submitted to *Astrophysical Journal* (2004).

Hedman, Matthew. "Polarization of the Cosmic Microwave Background." *American Scientist Magazine* Volume 93 (May-June 2005).

Hillebrandt, Wolfgang; Janka, Hans-Thomas; and Müller, Ewald. "How to Blow Up a Star." *Scientific American* (October 2006).

Hinshaw, G., et al. "Three-Year Wilkinson Microwave Anisotropy Probe (WMAP) Observations: Temperature Analysis." submitted to *Astrophysical Journal* (2007).

Hogan, Craig J.; Kirshner, Robert P.; Suntzeff, Nicholas B. "Surveying Space-time with Supernovae." *Scientific American* (January 1999).

Hu, Wayne; White, Martin. "The Cosmic Symphony." *Scientific American* (February 2004).

Hu, Wayne. "CMB Temperature and Polarization Anisotropy Fundamentals." *Astrophysical Journal* 170 (2007).

Jackson, Randal (curator). "PlanetQuest." NASA Jet Propulsion Laboratory website, <http://planetquest.jpl.nasa.gov/index.com>

Johnson, Steven. "Fear." *Discover* (March 2003).

Johnson, Steven. "Laughter." *Discover* (April 2003).

Johnson, Steven. "Love." *Discover* (May 2003).

Kane, Gordon. "The Mysteries of Mass." *Scientific American* (July 2005).

Kaplinghat, Manoj and Turner, Michael S. "How Cold Dark Matter Theory Explains Milgrom's Law." *Astrophysical Journal* 569 (2002): L19.

Kearns, Edward; Kajita, Takaaki; and Totsuka, Yoji. "Detecting Massive Neutrinos." *Scientific American* (August 1999).

Kirkman, David, et al., "The Cosmological Baryon Density from the Deuterium to Hydrogen Ratio towards QSO Absorption Systems: D/H towards Q1243+3047." *Astrophysical Journal* Suppl. 149 (2003): 1.

Knop, R. A. et al (The Supernova Cosmology Project). "New Constraints on Omega_M, Omega_Lambda, and w from and Independent Set of Eleven High-Redshift Supernovae Observed with HST." *Astrophysical Journal* 598 (2003): 102.

Kovac, J. M., et al. "Detection of Polarization in the Cosmic Microwave Background Using DASI." *Nature*, Volume 420 (19/26 December 2002): 772.

Krauss, Lawrence M. "Cosmological Antigravity." *Scientific American* (January 1999).

Krauss, Lawrence M., and Starkman, Glenn D. "The Fate of Life in the Universe." *Scientific American* special edition, "The Once and Future Cosmos" (December 2002): 50–57.

Lahav, Ofer and Liddle, Andrew R. "The Cosmological Parameters 2006," article for "The Review of Particle Physics 2006." *Journal of Physics* G33 (2006): 1.

Larson, Richard B. and Bromm, Volker. "The First Stars in the Universe." *Scientific American* (December 2001).

Lasota, Jean-Pierre. "Unmasking Black Holes." *Scientific American*, (May 1999).

Lemonick, Michael D. "Before the Big Bang." *Discover* (February 2004).

Lineweaver, Charles H. and Davis, Tamara M. "Misconceptions about the Big Bang." *Scientific American* (March 2005).

Loeb, Abraham. "The Dark Ages of the Universe." *Scientific American* (November 2006).

Luminet, Jean-Pierre; Starkman, Glenn D.; and Weeks, Jeffrey R. "Is Space Finite?" *Scientific American* (April 1999).

Luminet, Jean-Pierre; Starkman, Glenn D.; and Weeks, Jeffrey R. "Is Space Finite?" *Scientific American* special edition, "The Once and Future Cosmos" (December 2002).

Luminet, Jean-Pierre. "The Shape of Space After WMAP Data." *Brazilian Journal of Physics* 36 (2006): 107–114.

Luminet, Jean-Pierre; Weeks, Jeffrey R.; Riazuelo, Alain; Lehoucq, Roland; and Uzan, Jean-Philippe. "Dodecahedral Space Topology as an Explanation for Weak Wide-Angle Temperature Correlations in the Cosmic Microwave Background." *Nature*, Volume 425 (9 October 2003): 593.

Magueijo, Joao. "Plan B for the Cosmos." *Scientific American* (January 2001).

McDonald, Arthur B.; Klein, Joshua R.; and Wark, David L. "Solving the Solar Neutrino Problem." *Scientific American* (April 2003).

Meléndez, Jorge and Rámez, Ivan. "Reappraising the Spite Lithium Plateau: Extremely Thin and Marginally Consistent with WMAP." *Astrophysical Journal*. 615 (2004): L33.

Milgrom, Mordehai. "Does Dark Matter Really Exist?" *Scientific American* (August 2002).

Miller, Stanley L. "A Production of Amino Acids Under Possible Primitive Earth Conditions." *Science*, Volume 117 (15 May 1953): 528–529.

Moravec, Hans. "Rise of the Robots." *Scientific American* (December 1999).

O'Connor, J. J. and Robertson, E. F., "Biography of Aleksandr Aleksandrovich Friedmann." *School of Mathematics and Statistics at the University of St. Andrews, Scotland*, <http://www-history.mcs.st-andrews.ac.uk/Biographics/friedmann.html>

Orgel, Leslie E. "The Origin of Life on Earth." *Scientific American* (October 1994).

Ostriker, Jeremiah P. and Steinhardt, Paul J. "The Quintessential Universe." *Scientific American* (January 2001).

Page, L., et al. "Three Year Wilkinson Microwave Anisotropy Probe (WMAP) Observations: Polarization Analysis." *Astrohpysical Journal* (2007).

Panagia, Nino. "High Redshift Supernovae: Cosmological Implications." Nuovo Cimento B120 (2005): 667.

Parker, Eugene N. "Shielding Space Travelers." *Scientific American* (March 2006).

Peebles, P. James E. "Making Sense of Modern Cosmology." *Scientific American* (January 2001).

Percival, Will J. "Cosmological constraints from galaxy clustering."

Perryman, Michael. "Hipparcos: The Stars in Three Dimensions." *Sky and Telescope* (June 1999): 40–49.

Phillips, N. G. and Kogut, A. "Constraints on the Topology of the Universe from the WMAP First-Year Sky Maps." *Astrophysical Journal* (2006): 820–825.

Reid, I. N. "Low-mass Binaries in the Hyades - A Scarcity of Brown Dwarfs." *Astronomical Journal*, Volume 114 (December 1997): 161.

Riess, Adam G. et al. "Identification of Type Ia Supernova at Redshift 1.3 and Beyond with the Advanced Camera for Surveys on HST." *Astrophysical Journal* 600 (2004): L163.

Riess, Adam G. et al. "Type Ia Supernova Discoveries at $z > 1$ From the Hubble Space Telescope: Evidence for Past Deceleration and Constraints on Dark Energy Evolution." *Astrophysical Journal* 607 (2004): 665.

Riess, Adam G.; Turner, Michael S. "From Slowdown to Speedup." *Scientific American* (February 2004).

Sanders, R. H. "The Published Extended Rotation Curves of Spiral Galaxies: Confrontation with Modified Dynamics." *Astrophysical Journal*, Volume 473 (1996): 117.

Schwarzschild, Bertram. "Gravitational Microlensing Reveals the Lightest Exoplanet Yet Found." *Physics Today* (April 2006).

Schwarzschild, Bertram. "High-redshift supernovae indicate that dark energy has been around for 10 billion years." *Physics Today* (January 2007).

Schwarzchild, Bertram. "X-ray Absorption Lines Probe the Missing Half of the Cosmic Baryon Population." *Physics Today* (March 2005).

Scientific American Special Edition. "The Once and Future Cosmos." December 2002.

Scott, Douglas and Smoot, George F. "Cosmic Background Radiation Mini-Review." Excerpt from "The Review of Particle Physics." edited by Eidelman, S., et al., *Physics Letters* (2004): B.592, 1.

Serpico, P. D., et al. "Nuclear Reaction Network for Primordial Nucleosynthesis: a Detailed Analysis of Rates, Uncertainties and Light Nuclei Yields." *Journal of Cosmology and Astroparticle Physics* 0412 (2004): 010.

Simpson, Sarah. "Questioning the Oldest Signs of Life." *Scientific American* (April 2003).

Spergel, D. N. "First Year Wilkinson Microwave Anisotropy Probe (WMAP) Observations: Determination of Cosmological Parameters." *Astrophysical Journal* Suppl. 148 (2003): 175.

Starkman, Glenn D. and Schwarz, Dominik J. "Is the Universe Out of Tune?" *Scientific American* (August 2005).

Steinhardt, Paul J. and Turok, Neil. "The Cyclic Model Simplified," from a talk given at Dark Matter 2004, Santa Monica, CA (18–20 February 2004).

Strauss, Michael A. "Reading the Blueprints of Creation." *Scientific American* (February 2004).

Tegmark, Max. "Parallel Universes." *Scientific American* (May 2003).

Tegmark, Max, et al. "Cosmological Parameters from SDSS and WMAP." *Physical Review* (2004).

Tozzi, Paolo. "Cosmological parameters from Galaxy Clusters: an Introduction."

Turner, Michael S. "Dark Matter and Dark Energy: The Critical Questions." To appear in "Hubble's Science Legacy: Future Optical-Ultraviolet Astronomy from Space", eds. K.R. Sembach, J.C. Blades, G.D. Illingworth, & R.C. Kennicutt, ASP Conference Series.

Tyler, C.; Janus, B.; and Santos-Noble, D. "The Race to Build Supermassive Black Holes." *Bulletin of the American Astronomical Society* 36.2 (2004): 42.08.

Udry, S., et al. "The HARPS Search for Southern Extrasolar Planets, XI. Super-Earths in a 3-planet System." *Astronomy and Astrophysics* (2007).

Uwins, Philippa J. R.; Webb, Richard I.; and Taylor, Anthony P. "Novel Nano-Organisms from Australian Sandstones." *American Mineralogist*, Volume 83 (1998): 1541–1550.

Veneziano, Gabrielle. "The Myth of the Beginning of Time." *Scientific American* (May 2004): 54–65.

Villarreal, Luis P. "Are Viruses Alive?" *Scientific American* (December 2004).

von Bloh, W., Bounama, C., Cuntz, M., and Franck, S. "The Habitability of Super-Earths in Gliese 581." *Astronomy and Astrophysics* (2007).

Wamsganns, Joachim. "Gravity's Kaleidoscope." *Scientific American* (November 2001).

Wilson, Edward O. "The Bottleneck." *Scientific American* (February 2002).

Zubrin, Robert. "The Mars Direct Plan." *Scientific American*, March 2000.

Zimmer, Carl. "What Came Before DNA?" *Discover* (June 2004).

Zimmer, Carl. "Whose Life Would You Save?" *Discover* (April 2004).

Zimmerman, Robert. "Seeking Other Earths." *Astronomy Magazine* (August 2004).

*Many of the articles referenced above can be found at xxx.lanl.gov; go to subject search: Physics, and search by author name or title.

Photo Credits

t = top m = middle b = bottom
l = left c = center r = right

CHAPTER 1

1.2 "Ascending and Descending" by M.C. Escher, 1960 © Cordon Art-Baarn-Holland. **1.4** Non Sequitur © Wiley Miller Reprinted with permission of Universal Press Syndicate. All rights reserved. **1.6** Alan Hills, The British Museum/Dorling Kindersley Media Library.

CHAPTER 2

2.1 John Sanford/Astrostock-Sanford. **2.3** Angela Lowman **2.4** Alex Mellinger. **2.5** NASA. **2.6** C.R. O'Dell and S.K. Wong, Rice U./NASA. **2.7** The Hubble Heritage Team/AURA/STScI/NASA. **2.8** Tony and Daphne Hallas. **2.9** S. Beckwith (STScI) and the HUDF Team/NASA, ESA. **2.10** NASA. **2.11** Robert Gendler. **2.12** David Malin, Anglo-Australian Observatory. **2.13** Robert Gendler. **2.14** Royal Observatory, Edinburgh/Science Photo Library/Photo Researchers, Inc. **2.16** Benjamin Cummings. **2.17** NASA. **2.19** Atlas Image obtained as part of the Two Micron All Sky Survey (2MASS), a joint project. J. Carpenter, T. H. Jarrett, & R. Hurt. **2.24** Nigel Sharp, NOAO/NSO/Kitt Peak FTS/AURA/NSF. **2.33 (l)** Harvard College Observatory. **2.33 (r)** Harvard College Observatory.

CHAPTER 3

3.2 (a) Craig Tyler. **3.2 (b)** NASA. **3.2 (c)** NASA. **3.13** Harvard College Observatory. **3.15** Adam G. Riess (STScI) et al., NASA. **3.17** NASA. **3.24 (a)** NASA. **3.25 (a)** Atlas Image. **3.25 (b)** Atlas Image obtained as part of the Two Micron All Sky Survey (2MASS), a joint project. J. Carpenter, T. H. Jarrett, & R. Hurt. **3.27 (c)** Richard Powell/www.atlasoftheuniverse.com/virgo.html. **3.28 (a)** Jim Misti, Misti Mountain Observatory.

CHAPTER 4

4.10 T Rector, Gemini Obs, U. Alaska Anchorage/AURA. **4.16** Jean Charles Cuillandre (CFHT)/Hawaiian Starlight. **4.22 (a)** NASA. **4.24** Neil Trentham, Simon Hodgkin and the INT Wide Field Survey. The Cavendish Laboratory, University of Cambridge, Cambridge, England. **4.27** NASA.

CHAPTER 5

5.5 GMS, GOES-8, Meteosat, SSEC, NCDC, U. Wisc., NOAA. **5.7** NASA.

CHAPTER 6

6.9 Richard Megna/Fundamental Photographs. **6.10** Andrew Fruchter (STScI) et al./WFPC2, HST, NASA. **6.12** Harvard-Smithsonian Center for Astrophysics. **6.13** European Southern Observatory. **6.14** M. P. Muno, UCLA, et al./CXC, NASA. **6.18** NOAO. **6.19** Halton Arp, Max Planck Institute for Astrophysics, Garching, Germany. **6.20** NASA Headquarters.

CHAPTER 8

8.9 NASA. **8.11** WMAP Science Team/NASA. **Page 240 (l)** WMAP Science Team/NASA. **Page 240 (r)** Terraserver. **8.12** www.tackle-tour.com. **8.16** DASI Collaboration/U. of Chicago, 2002.

CHAPTER 9

9.5 NASA/ESA. **9.9 (t)** NASA, ESA, HEIC, and The Hubble Heritage Team (STScI/AURA). **9.9 (b)** Andrew Fruchter (STScI) et al., WFPC2, HST, NASA **9.12** NASA - X-ray: CXC, J. Hester (ASU) et al.; Optical: ESA, J. Hester and A. Loll (ASU); Infrared: JPL-Caltech, R.Gehrz (U. Minn). **9.16** ICRR Institute for Cosmic Ray Research. **9.17** NASA.

CHAPTER 10

10.4 Max-Planck-Institut fur Extraterrestrische Physik (MPE).

CHAPTER 11

11.6 The Elegant Universe: Superstrings, Hidden Dimensions, and the Quest for the Ultimate Theory (New York: Norton & Co., 1999). **11.7** Gary Bartholomew/MIRA. **11.11** S. Beckwith (STScI) and the HUDF Team/NASA, ESA. **11.12** Jeffrey Weeks.

CHAPTER 12

Page 352 WMAP Science Team/NASA. **12.10** Michael S. Warren. **12.11** Joerg Colberg/The VIRGO Collaboration, 1996. **12.12** WMAP Science Team/NASA. **12.13** R. Massey, California Institute of Technology/NASA, ESA.

CHAPTER 13

13.1 Michael Abbey/Visuals Unlimited. **13.7** A.I Oparin/Russian Academy of Sciences. **13.8** Smithsonian National Museum of Natural History. **13.10** Mars Project/JPL/NASA. **13.11** Mars Global Surveyor/ NASA. **13.12** NASA Headquarters. **13.15** Benjamin Cummings.

CHAPTER 14

14.1 Calvin and Hobbes © Watterson. Reprinted with permission of Universal Press Syndicate. All rights reserved. **14.3** "Puddle" by M.C. Escher, 1952. © Cordon Art-Baarn-Holland, 2001. All rights reserved. **14.6 (a)** N.Scoville (Caltech), T Rector (U. Alaska, NOAO) et. Al. Hubble Heritage Team, NASA. **14.6 (b–d)** Craig Tyler. **14.7** Tom Murphy/ National Geographic Image Collection. **14.8** "Other World" by M.C. Escher, 1947. © Cordon Art-Baarn-Holland, 2001. All rights reserved.

COLOR GALLERY 1

Color Figure 1 CICLOPS, JPL, ESA, NASA. **Color Figure 2 (a)** Craig Tyler. **Color Figure 2 (b)** NASA. **Color Figure 2 (c)** NASA. **Color Figure 3** NASA. **Color Figure 4 (t)** NASA. **Color Figure 4 (b)** NASA. **Color Figure 5** John Sanford/Astrostock-Sanford. **Color Figure 6** NASA. **Color Figure 7** T. A. Rector/NOAO/AURA/NSF and Hubble Heritage Team

(STScI/AURA/NASA). **Color Figure 8** C.R. O'Dell and S.K. Wong, Rice U./NASA. **Color Figure 9** Tony and Daphne Hallas. **Color Figure 10** S. Beckwith (STScI) and the HUDF Team/NASA, ESA. **Color Figure 11** Francesco Ferraro, Bologna Observatory/ESA. **Color Figure 12** Stefan Seip. **Color Figure 14 (l)** ROSAT. **Color Figure 14 (l) (c)** MSX. **Color Figure 14 (c)** Akiri Fujii/JPL. **Color Figure 14 (r) (c)** IRAS. **Color Figure 14 (r)** NRAO. **Color Figure 18** Alexander Kalina/Shutterstock. **Color Figure 19** National Optical Astronomy Observatories. **Color Figure 20** T. Rector, Gemini Obs, U. Alaska Anchorage/AURA. **Color Figure 21 (t)** NASA. **Color Figure 21 (b)** NASA. **Color Figure 22 (l)** NASA, ESA, HEIC, and The Hubble Heritage Team (STScI/AURA). **Color Figure 22 (l) (c)** NASA. **Color Figure 22 (r) (c)** Andrew Fruchter (STScI) et al., WFPC2, HST, NASA. **Color Figure 22 (r)** NASA. **Color Figure 23** O. Krause, Steward Observatory, et. al./SSC, JPL, Caltech, NASA. **Color Figure 24** NASA - X-ray: CXC, J. Hester (ASU) et al.; Optical: ESA, J. Hester and A. Loll (ASU); Infrared: JPL-Caltech, R. Gehrz (U. Minn). **Color Figure 25** NASA. **Color Figure 26 (l)** ICRR Institute for Cosmic Ray Research. **Color Figure 26 (r)** SNO. **Color Figure 27 (a)** NASA. **Color Figure 27 (b)** AURA; NASA. **Color Figure 27 (c)** J. Barnes. **Color Figure 28** Jean Charles Cuillandre (CFHT)/Hawaiian Starlight. **Color Figure 29** Jim Misti, Misti Mountain Observatory. **Color Figure 30** D. Wang, U. Massachusetts, et al./CXC, NASA. **Color Figure 31** ESO. **Color Figure 31** X-Ray: NASA / CXC / D. Hudson, T. Reiprich et al. (AIfA); Radio: NRAO / VLA/ NRL. **Color Figure 32** NASA / CXC / D.Hudson, T.Reiprich et al. (AIfA).**Color Figure 33 (t)** NASA. **Color Figure 33 (b)** NASA. **Color Figure 34 (a)** A. Hobart, Chandra X-ray Observatory/CXC. **Color Figure 34 (b)** NASA.

COLOR GALLERY 2

Color Figure 36 (a) Andrew Fruchter (STScI) et al./WFPC2, HST, NASA. **Color Figure 36 (b)** W. N. Colley, U. Virgina, E. Turner, Princeton, and J.A. Tyson, UC Davis/HST, NASA. **Color Figure 37** K. Sharon, Tel Aviv U. and E. Ofek, Caltech/ESA, NASA. **Color Figure 38** Robert Gendler. **Color Figure 39 (a)** NASA. **Color Figure 40** The Cavendish Laboratory. **Color Figure 41** X-ray: NASA/CXC/CfA/ M. Markevitch et al.; Lensing Map: NASA/STScI; ESO WFI; Magellan/U. Arizona/ D. Clowe et al. Optical: NASA/STScI; Magellan/U. Arizona/ D. Clowe et al. **Color Figure 42** R. Windhorst, Arizona State U., H. Yan, SSC, Caltech, et al./ESA, NASA. **Color Figure 45** GMS, GOES-8, Meteosat, SSEC, NCDC, U. Wisc., NOAA. **Color Figure 46** NASA. **Color Figure 48 (a)** WMAP Science Team/NASA. **Color Figure 48 (b)** NASA. **Color Figure 48 (c)** Terraserver. **Color Figure 50** www.tackletour.com. **Color Figure 51** DASI Collaboration/U. of Chicago, 2002. **Color Figure 52** Michael S. Warren. **Color Figure 53** Joerg Colberg/The VIRGO Collaboration, 1996. **Color Figure 54** Max-Planck-Institut fur Extraterrestrische Physik (MPE). **Color Figure 55** WMAP Science Team/NASA. **Color Figure 58** Adam G. Riess (STScI) et al., NASA. **Color Figure 61 + Color Figure 63** NASA. **Color Figure 64 (a)** Peter Bollinger/Shannon Associates. **Color Figure 64 (b)** NASA. **Color Figure 64 (c)** NASA. **Color Figure 64 (d)** NASA. **Color Figure 65 (t) (l)** N. Scoville (Caltech), T Rector (U. Alaska, NOAO) et al. Hubble Heritage Team, NASA. **Color Figure 65 (t) (r)** Craig Tyler. **Color Figure 65 (b) (l)** Craig Tyler. **Color Figure 65 (b) (r)** Craig Tyler. **Color Figure 66** Tom Murphy/National Geographic Image Collection. **Color Figure 67** "Other World" by M.C. Escher, 1947. © Cordon Art-Baarn-Holland, 2001. All rights reserved.

Illustration Credits

The following figures were adapted from Chaisson/McMillan, *Astronomy Today*, 6th ed. (San Francisco, CA: Pearson Addison Wesley, 2008): 2.32, 6.5, 3.6, 3.12, 3.13, 3.28, 4.1, 4.7, 4.17, 4.22(a), 5.02(b), 5.9, 5.10, 6.13, 6.20, 7.7, 9.11, 9.13, 11.2, 12.15, 13.6

The following figures were adapted from Hobson, *Physics: Concepts and Connections*, 4th ed . (Upper Saddle River, NJ: Pearson Prentice Hall, 2007): 11.04, 13.6

The following figures were adapted from Bennett, Donahue, Schneider, and Voit, *The Cosmic Perspective*, 4th ed. (San Francisco, CA: Pearson Addison Wesley, 2007): 4.22(b), 5.2

The following figures were adapted from Campbell, Reece, Taylor, and Simon, *Biology: Concepts and Connections*, 5th ed. (San Francisco, CA: Pearson Benjamin Cummings, 2006): 13.04

Text Credits

Page 10: Aristotle, *De Caelo*. Translated by J.L. Stocks. Lawrence: digireads.com, 2006.

Page 21: Feynman, R. P. *What Do You Care What Other People Think?: Further Adventures of a Curious Character*. New York: W.W. Norton and Company, 2001.

Page 23: Patchett, James M. and Gerould S. Wilhelm. "Designing Sustainable Systems: Fact or Fancy". *Wild Ones* (1997).

Page 31: *Discoveries and Opinions of Galileo* by Galileo Galilei, translated by Stillman Drake, copyright © 1957 by Stillman Drake. Used by permission of Doubleday, a division of Random House, Inc.

Pages 34–35: Galileo. *Discoveries and Opinions of Galileo*. Translated by Stillman Drake. New York: Random House Inc., 1957.

Page 43: Fraunhofer, Joseph. *Edinburgh Journal of Science* Vol. 8 (1828).

Page 64: Wilson, Edward O. *Consilience: The Unity of Knowledge*. New York: Vintage Books, 1998.

Page 66: Gribbin, John. *In Search of Schrödinger's Cat: Quantum Physics and Reality*. New York: Bantam Books, 1984.

Page 82: Leavitt, Henrietta. "Periods of 25 Variable Stars in the Small Magellanic Cloud." *Harvard College Observatory Circular* No.173 (1912).

Page 98: Lightman, Alan. *Great Ideas in Physics*. New York: McGraw-Hill, 2000.

Page 100: Kepler, Johannes. *Epitome of Copernican Astronomy and Harmonies of the World*. Translated by Charles Glenn Wallis. New York: Prometheus Books, 1995.

Page 102: Newton, Isaac. *The Principia: Mathematical Principles of Natural Philosophy*. Translated by I. Bernard Cohen and Anne Whitman. Los Angeles: University of California Press, 1999.

Page 123: Rubin, Vera C., W. Kent Ford, and Norbert Thonnard. "Extended Rotation Curves of High-Luminosity Spiral Galaxies, IV. Systematic Dynamical Properties, Sa → Sc." *Astrophysical Journal*, Volume 225 (1978).

Page 138: Hubble, Edwin. "A Relation Between Distance and Radial Velocity Among Extra-Galactic Nebulae." *Proceedings of the National Academy of Sciences*, Vol. 15 (1929).

Page 141: Lapparent, Valerie de, Margaret J. Geller, and John P. Huchra. "A Slice of the Universe." *Astrophysical Journal*, Vol. 302 (1986).

Page 143: Penzias, Arno A., and Robert W. Wilson. "A Measurement of Excess Antenna Temperature at 4080 Mc/s." *Astrophysical Journal*, Vol. 142 (1965).

Page 166: Einstein, Albert. "On the Relativity Principle and the Conclusions Drawn from It." *Jahrbuch der Radioaktivität und Electronik* 2 (1907).

Page 174: Dyson, Frank W., Arthur S. Eddington, and Charles Davidson. "A Determination of the Deflection of Light by the Sun's Gravitational Field, from Observations Made at the Total Eclipse of May 29, 1919." *Philosophical Transactions of the Royal Society of London*, Series A, Vol. 220 (1920).

Page 194: Lederman, Leon, and David Schramm. *From Quarks to the Cosmos: Tools of Discovery* (Scientific American Library Series, Vol. 28). New York: W.H. Freeman and Co., 1995.

Page 203: Friedmann, A. "On the Curvature of Space." *General Relativity and Gravitation* 31, no. 12 (1999). (First published in *Zeitschrift für Physik*, 1922).

Page 204: Hubble, Edwin. "A Relation Between Distance and Radial Velocity Among Extra-Galactic Nebulae." *Proceedings of the National Academy of Sciences* 15 (1929).

Page 219: Barry, Dave. "Emergency: California Could Use a Jump." *The Milwaukee Journal Sentinel*, February 4, 2001.

Page 246: Kovac, J.M., E.M. Leitch, C., C. Pyrke, J.E. Carlstrom, N.W. Halverson, and W.L. Holzapfel. "Detection of Polarization in the Cosmic Microwave Background Using DASI." *Nature* 420 (2002).

Page 282: Alpher, Ralph A., Hans Bethe, and George Gamow. "The Origin of Chemical Elements." *Physical Review* 73 (1948).

Page 312: Bondi, Hermann, and Thomas Gold. "The Steady State Theory of the Expanding Universe." *Monthly Notices of the Royal Astronomical Society* 108 (1948).

Page 338: Riess, Adam, et. al. "Observational Evidence from Supernovae for an Accelerating Universe and a Cosmological Constant." *Astronomical Journal* 116 (1998).

Page 372: Guth, Alan H. "Inflationary Universe: A Possible Solution to the Horizon and Flatness Problems." *Physical Review* D, Vol. 23 (1981).

Page 390: Miller, Stanley L. "A Production of Amino Acids Under Possible Primitive Earth Conditions." *Science* Vol.117 (1953).

Page 415 (top): Turner, F. J. *The Frontier in American History*. New York: H. Holt and Co.: 1920.

Page 415 (bottom): Zubrin, Robert. *The Case for Mars*. New York: Touchstone, Simon Schuster, 1996.

Page 429 (top): Ptolemy. *Almagest*. Translated by G. J. Toomer. New Jersey: Princeton University Press, 1998.

Page 429 (bottom): Copernicus, Nicolaus. "De revolutionibus orbium coelestium (On the Revolutions of the Heavenly Spheres)". Translated by Charles Glenn Wallis and quoted in Bartusiak, Marcia. *Archives of the Universe*. New York: Pantheon Books, 2004.

Page 438: James, William. *Pragmatism: A New Name for Some Old Ways of Thinking*. West Valley City, Utah: The Editorium, LLC: 2006.

Page 439: Dobzhansky, Theodosius. *The Biological Basis of Human Freedom*. New York: Columbia University Press, 1960.

Page 440 (top): Dalai Lama, *The Universe in a Single Atom: The Convergence of Science and Spirituality*. New York: Broadway Books, 2006.

Page 440 (bottom): Weinberg, Steven. *The First Three Minutes: A Modern View of the Origin of the Universe*. New York: Basic Books, 1977.

Page 441: Stapledon, Olaf. *Star Maker*. Edited by Patrick McCarthy. Middletown: Wesleyan University Press, 2004.

Index

A

Abell 1185 galaxy cluster, 116
absolute zero temperature, 68–69
absorption lines, 74, 136
  CMB (cosmic microwave background) and, 236–237
  from Sun, 229
absorption spectrum, 44
abundances of elements, 272, 280–285
  baseline abundances, 281
  and big bang nucleosynthesis (BBN), 307
  big bang theory and, 302–308
  light elements, 291
  and living cells, 387–388
  quasars and measuring, 308
acceleration, 165–166
  dark energy and, 210–212, 214, 337–338, 378
  elevator thought experiment, 166–167
  and gravity, 171
  inflation and, 378
accretion, 187
  disks, 187–188
  of halos, 365
acoustic peaks, 357
  in CMB, 353, 355
  and matter, 360–361
acoustic waves, 356–357
aesthetic criteria, 10
age. See also astronomical ages
  size and, 424–425
  of universe, 423
Albrecht, Andreas, 372
Alexander the Great, 10
Allan Hills site, 404–405
alphabet paper, 282
Alpher, Ralph, 282
amino acids, 387
  in Miller-Urey experiment, 390
  in Murchison meteorite, 391–392
Andromeda Galaxy, 31, 83, 92
  as blueshifting, 137
  distance to, 92
anger, 401
  angles
  meaning of, 27
triangulation, 50–56
angstroms, 38
angular size, 27–28
animals, appearance of, 400
animist cosmology, 17
amino acids, 392

anisotropies, 238–240. See also CMB anisotropies
  big bang theory and, 301–302
  polarization anisotropies, 247
annihilations, 435
  of antimatter, 327
  of material world, 332–333
  of particles, 275
  WIMP-anti-WIMP annihilations, 280
Antennae Galaxies, 116
anthropic principle, 426–432
antibaryons, 332
  mass annihilation of, 333
antimatter, 275, 435
  annihilation, 327
  bias of universe and, 330–331
  Sun and, 275
antineutrinos, 275
  lepton number for, 275
  and Sun, 276
anti-WIMPS, 279–280
Apollo missions, 49
apparent brightness, 75. See also inverse-square law
  flux, measurement as, 76
apparent size, 87
arc minutes, 55–56
arc seconds, 55–56
Aristotle, 10, 100, 428–429
artificial categories, 39–40
artificial intelligence, 415
artificial satellites, 35
asteroids
  binding energy and, 107–108
  geodesics of, 169–170
  and planets, 403–404
Asteroids video game, 10–11, 342–343
astronomical ages, 291
  big bang theory and, 308–309
  clusters, 154
  CMB measuring, 362
  comparing, 153–155
  elements and, 151–153
  radiometric dating and, 154
  white dwarfs, 154
astronomical distance ladder, 87–88
astronomical units (AUs), 55, 96
atmosphere
  of Mars, 403
  of Sun, 229
atomic absorption, 229
atomic number, 147, 256
  even atomic numbers, 269

atoms, 21, 64. See also bound electrons; nuclei
  absolute zero temperature, 68–69
  Bohr atom, 226–227, 379
  electrical charge of, 66
  formation of, 435
  ionization, 65
  line radiation, 223
  and molecules, 248
  quantum atom, 226–230
  strong nuclear force, 257
  types of particles in, 65
attractive gravity, 103–105
average particle speed, 68
Avogadro's number, 259
Ayer, A. J., 441

B

Babylonian cosmology, 17–18
bacteria, 400
  on Mars, 403
  nanobacteria, 405
Barlow, Connie, 436–437
Barnard's star, 58–59
barred spiral galaxies, 31, 33
Barry, Dave, 219
baryogenesis, 331, 332
  annihilation, 327
baryonic dark matter, 278–279, 311, 360
baryonic gas, 364–369
baryonic matter, 435
  decoupling of, 363
  freeing of, 426
baryonic visible matter, 311
baryons, 260, 359–361
  acoustic peaks and, 360–361
  annihilation, 327
  baryon-to-photon ratio, 306
  and beta decay, 274
  clustering of, 363
  CMB and density, 361, 362
  conserve baryon number, 263, 330
  in dark matter halos, 363
  in early universe, 282–284
  hadrons, 332
  imprisonment of, 333–334
  in iron nucleus, 270
  mass annihilation of, 332–333
  quarks and, 332
  temperature regime, response to, 334
baryon-to-photon ratio, 306
baselines
  abundances, 281
  parallax, effect of, 52

baselines, *continued*
  stars, measurements of, 54
  for triangulation, 51
base pairs, 390, 391
behavior of life, 401–402
benchmark numbers, 96
bending light, 173–174, 324–325
Bennett, Jeffrey, 121
beryllium, 151–153, 281
  interstellar collisions and, 285
  production of, 283
Bessel, Friedrich, 56
beta decay, 273–276
Bethe, Hans, 282
bias, 9–10
big bang nucleosynthesis (BBN), 303
  abundances of elements and, 307
  and baryon-to-photon ratio, 306
  full calculation of, 306
  neutrons in, 304
big bang theory, 282, 290–317, 435. *See
    also* big bang nucleosynthesis
    (BBN)
  and abundances of elements, 302–308
  alternatives to, 310–311
  astronomical ages and, 308–309
  CMB (cosmic microwave background)
    and, 299–302, 309–310, 311
  dissension from, 312–313
  evaluating, 309–313
  expansions and, 294–297
  explosions and, 294–297
  flexibility of, 310–311
  galaxy survey data and, 298–299
  horizon problem, 370–371
  inflation theory and, 372
  observations, explanation of, 297–309
  Olbers's paradox and, 309, 310
  overview of, 291–294
  questions raised by, 347
  redshifts and, 297, 311
  three pillars of, 309–313
binding energy, 106–111
  of iron, 270
  nuclear binding energy, 257–258
binoculars, 30
biomolecules, 387
  in Miller-Urey experiment, 390
biopolymers, 387, 388
  need for, 395–397
biosignatures, 399
  in ALH84001 meteorite, 404–405
biospheres, 385
birth of stars, 264–265
blackbody radiation, 71–72, 220–223.
    See also CMB (cosmic microwave
    background)
  anisotropies showing, 353–354
  conditions for, 220–221
  opacity of object, 221

Sun emitting, 22
  temperature and, 220–221
  Wien law, 222
blackbody spectrum, 72
  and CMB, 146, 300–301
black holes, 128, 177–186, 271. *See also*
    event horizon; SMBHs
    (supermassive black holes)
  accretion process, 187
  defined, 178, 184
  detecting, 179–180
  gas near, 187
  Hawking radiation, 377
  light-capture radius, 181–183
  mass, radiation of, 376–377
  and quasars, 186–191
  relativistic behaviors of, 181–183
  singularity of, 379
  spacetime and, 182–183
blue light, 220
blueshifts, 46, 136
  Sachs-Wolfe effect, 359
  spacetime bending and, 183
blue supergiant stars, 265–266
Bohr, Niels, 226–227
Bohr atom, 226–227, 379
Bondi, Hermann, 312
bootstrapping, 85–86
boron, 151–153, 281
  interstellar collisions and, 285
  production of, 283
bound and unbound systems, 108–110
bound electrons, 223–230
  energy and, 228
  orbitals of, 225
  quantized energy, 225
  transparency of, 232
Brahe, Tycho, 100
brain, evolution of, 398
branes, 341, 431
A Brief History of Time (Hawking), 376
brightness, 28–29. *See also* apparent
    brightness
  comparing, 42
  pulsation of star and, 82–83
  of spectrum images, 41
  of supernova standard candles,
    84–85
brown dwarfs, 128
B-type stars, 74
Buddhist perspective, 440
bulges
  life, hostility to, 425
  Milky Way, bulge of, 112–113
Burbridge, Geoffrey, 312–313
Burgess Shale fossil, 398

C

calcium, 270
Cambrian explosion, 398, 400

Cannon, Annie, 74
carbohydrates, 386
carbon
  in core of high-mass stars, 269
  interstellar collisions and, 285
  in living cells, 387
  production of, 272
carbon-12, 284
carbon-14, 153
  decay, 273–274
carbonate globules, 404
carbon dioxide, 153
  on Mars, 403
Carl Sagan's maxim, 405–406
Carlstrom, John, 245–246
CDM simulation, 364–366, 425
  structure formation evidence from,
    368
  time sequence, 365
cell phone microwaves, 143
cells, 386–389
cellulose, 386
Celsius temperature scale, 68–70
Centaurus, 90
center of universe, 296
Cepheid variable stars, 81–83, 86
  light curve for, 83
Cepheus, Delta Cephei in, 82–83
chain links, 387
chains, 387
Chaisson, Eric, 376
Chandra x-ray telescope, 179
  and Sunyaev-Zel'dovich (SZ) effect,
    234
chemical bonds, 248
chemical reactions, 149
chemical-reorganization, 148–149
chromium, 270
circular orbit velocity, 110–111
cloud rotation and birth of star, 264
clumping, 424, 426
  of galaxies, 204–205
  of matter, 362–363
CMB anisotropies, 238–240, 353
  acoustic peaks in, 353
  and dark matter, 366–369
  and exotic dark matter, 367–368
  first peak in, 353
  harmonic sounds and, 356
  homogeneity/isotropy of, 370
  WMAP image of, 354
CMB (cosmic microwave background),
    143–147, 219, 234–241, 290, 435.
    See also CMB anisotropies;
    inflation theory
  acoustic peaks, 357
  and big bang theory, 299–302,
    309–310, 311

and blackbody spectrum, 146, 300–301
curvature of space and, 324–325
dark energy, evidence for, 369
destiny of universe and, 338
hot and cold spots, 238–240, 245–246, 325–326, 349–350
  decoupling and, 352
    horizon problem, 370–371
    inflation theory and, 370
    most prominent spots, 353
isotropic anisotropies, 240
isotropic temperature of, 145
isotropy of, 238–240
matter density, 360–362
numbers measured by observations, 361–362
in past universe, 292
polarization, 245–247, 379
quasar light and, 236–237
redshifts of, 302, 362
Sachs-Wolfe effect, 359
snapshot, 353, 358
Sunyaev-Zel'dovich (SZ) effect, 233–234
temperatures of, 143, 145–147, 237, 349–350
transparency and, 235–236
up-scattering of photons, 233
CMB polarization, 245–247, 379
CMB power spectrum, 355–356
  and gravitational disturbances, 358–359
  on large scales, 358–359
  on small scales, 357–358
coacervates, 395–397, 397
codons, 390
coldness, 68
colors
  and blackbody radiation, 221–222
  of elements, 43
  Fraunhofer lines, 42–43
  and gas in disk galaxies, 113
  of light, 39
  spectroscopy, 40–46
  of stars, 26, 265–266
  temperature and, 223
compact objects, 128
compression-expansion-compression process, 349–350
compressions. *See also* sound waves
  baryons and, 360
  compression-expansion-compression process, 349–350
  dark matter and, 360
Comte, Auguste, 223
confidence in measurements, 12
confinement property, 332
consciousness, 3–4, 430
  in universe, 416

conservation laws, 263
  baryon number, conservation of, 263
  energy, law of conservation of, 254–255, 263
  lepton number, conservation of, 275–276
conservation of energy, law of, 254–255, 263
conserve baryon number, 330
constellations, 29
continuous spectra, 44
Copernicus, 428, 429
core of stars, 266–267
  collapse of, 270–271
  of high-mass stars, 268
cosmic inflation, 177
  annihilation, 327
cosmic microwave background (CMB). *See* CMB (cosmic microwave background)
cosmic rays, 153
cosmic timeline, 436–437
cosmological constant, 202, 320
  dark energy as, 340
  vacuum energy and, 212
cosmological horizon, 194–196, 378
cosmological principle, 199–200
cosmology, 16–21
Crab Nebula, 271, 272
Cretaceous era, 4
critical density, 322
  omegas and, 323
culture, 400
cuneiform records, 17
curvature
  by black hole, 376
  density and, 323–326
  Einstein equation and, 199
  in event horizon, 184
  global curvature, 323–326
  negative curvature, 164
  positive curvature, 163–164
curvature pulse, 177
curved space, 163–165, 324–325. *See also* black holes
  dark matter and, 174–175
  and general relativity, 168–172
  gravitational lensing, 172–176
  three-dimensional curved space, 164
  Tollman-Oppenheimer-Volkoff (TOV) equation, 178
cyclic universe, 341, 342

D
Dalai Lama, 440
dark ages, 301
dark energy, 202, 209–212, 312, 417, 431
  acceleration and, 210–212, 214, 338–339, 378

CMB measuring density, 362
as cosmological constant, 340
deceleration parameter and, 210–212
defined, 211–212
density of, 320, 321
discovery of, 203
evidence for, 368–369
and expanding universe, 210–212
Friedmann equation and, 201–202
and matter, 336–337
mystery of, 442
as quintessence, 340–341
as recurring cycle, 341
significance of, 337
as vacuum energy, 212
dark energy domination, 321, 327, 362
dark matter, 123–126, 359–361. *See also* dark matter halos; exotic dark matter
  baryonic dark matter, 278–279, 311, 360
  baryons in dark matter holes, 363
  and big bang theory, 310–311
  and brown dwarfs, 128
  CDM simulation, 364–366
  clustering of, 363
  CMB anisotropy data and, 366–369
  components of, 128–131
  freeing of, 426
  galaxy survey data and, 366–369
  gravitational lensing and, 174–175, 367
  hot dark matter, 364
  luminous matter, separation of, 367
  map of, 368
  microlensing and, 175–176
  MOND (MOdified Newtonian Dynamics) and, 130–131
  normal dark matter, 278–279
  as particles, 129–130
  SMBHs (supermassive black holes) and, 181
  and WIMPs (weakly interacting massive particles), 364
dark matter halos, 125–126, 278, 363–369
  formation of, 335, 336
dark sky, reason for, 155–157
Darwin, Charles, 347, 400
daughter stars, 272
Davidson, Charles, 174
death phase of stars, 270–271
deceleration parameter, 209–212
decoupling, 300, 327, 363, 426, 435
  of neutrinos, 334
  reionization and, 302
  as short-lived event, 351–352
deflection of starlight, 173–174
Degree Angular Scale Interferometer (DASI) experiment, 245–246

degrees, 27
de Lapparent, Valérie, 141–142
Delta Cephei, 82–83
density. *See also* energy density; radiation
    CMB measuring, 362
    Einstein equation and, 200
    and electroweak unification, 331–332
    Friedmann equation and, 201–202
    and global curvature, 323–326
    inflation and, 375
    omegas and, 319–323
    in past universe, 292
    of universe, 318–326
    volume and, 319–320
density parameters, 322–323
    and destiny of universe, 338
density waves, 118–119
    and gas clouds, 264
deoxyribophosphate, 391
de Sitter, Willem, 203
destiny of universe, 337–341
deuterium, 261
    in big bang nucleosynthesis (BBN), 303, 307
    reactions, 282
    temperature and, 304
diameter
    of Earth, 88
    of galaxy, 91–95
    of Moon, 88
Dicke, Robert, 143
differential rotation, 117
dim stars. *See* low-mass stars
dinosaurs, 399, 400
disk galaxies, 112–116
    density wave theory, 119
    spiral arms, 116–120
disorder, increase of, 416–417
distances, 63
    astronomical distance ladder, 87–88
    calibration and, 86–87
    extragalactic distances, 76
    and flux, 78
    between galaxies, 293
    interstellar distances, 90
    methods for measuring, 86–88
    parallax, effect of, 52
    relative-distance indicators, 87–88
    and standard candles, 75
    for triangulation, 51
    zero distance between galaxies, 293
disturbances
    on CMB power spectrum, 358
    in early universe, 349
    gravitational disturbances, 358–359
DNA (deoxyribonucleic acid), 2, 387
    double-helix shape of, 390–391, 393
    genetic structure of, 390
    mutation, 400

nitrogenous bases and, 390
proteins and, 392–394
replication, 392
RNA world hypothesis and, 394–395
Dobzhansky, Theodosius, 439
dodecahedron, 343
dome of sky, 27–28
Donahue, Megan, 121
Doppler, Christian, 46
Doppler effect, 46–48
    and binding energy, 109–111
    formula or, 46–48
    radial velocity, measuring, 58
    spiral arms, development of, 116–120
Doppler shift, 136
    big bang theory and, 297
    planet-finding and, 406–407
    recession velocity and, 208
Doppler wobbles, 406–407
double-helix shape, 390–391, 393
Dowd, Michael, 436–437
down-scatters, 233
Drake, Frank, 409
Drake equation, 409–412
Draper, Henry, 74
dwarf galaxies, 92
    MOND (MOdified Newtonian Dynamics) and, 131
    orbital speeds of, 122
Dyson, Frank W., 174
Dyson, Freeman, 432

E

early universe. *See also* CMB (cosmic microwave background)
    disturbances in, 349
    sound waves in, 350
Earth
    appearance of, 2–3
    building blocks for, 2
    and expanding universe, 197
    future on, 414
    life and complexity, 385
    parallax using, 54
    size of, 88
    spherical-Earth theory, 10–11
    temperatures on, 145
eccentricities of ellipses, 100
eclipsing systems, 407, 408
Eddington, Arthur S., 174, 259
Eiffel Tower, 27
Einstein, Albert, 1, 5
    cosmological constant, 320
    on education, 422
    elevator thought experiment, 166–167
    and expanding universe, 198–204
    on falling light, 172
    on light particles, 66–67
    mass as energy, 258

and Newton's theory of gravity, 102
and scientific unification, 190
Einstein Cross, 189
Einstein equation, 198–204
electrical charge, law of conservation of, 263
electrical repulsion, 258–259, 262
electric field, 241. *See also* polarization
electromagnetic radiation, 40
electromagnetic spectrum, 39–40
electromagnetic waves, 37–38
    electric field, 241
    frequency of, 39
    gravitational waves compared, 177
    life, parameters for, 428
electron capture reaction, 271
electron cloud, 65, 227
electrons, 38, 65, 147. *See also* bound electrons; free electrons; scattering
    annihilation process and, 333
    and beta decay, 273–274
    binding of, 149
    energy of, 65, 227–228
    exciting, 228
    as fundamental particles, 332
    imprisonment of, 335
    interaction with photons, 248
    ionization, 65
    lepton number for, 275
    in mass number, 150
    orbitals, 225, 227
    photoelectric effect, 67
    planets, analogy between, 224–225
    Sunyaev-Zel'dovich (SZ) effect, 233–234
electrostatic binding energy, 106
electroweak transition, 327
electroweak unification, 331–332
elements, 64, 147. *See also* abundances of elements; heavy elements
    and age of stars, 151–153
    alphabet paper on, 282
    baseline abundances, 281
    beta decay, 273–276
    of body, 148
    formation of, 151
    Fraunhofer lines and, 43
    in living cells, 387
    radiometric dating of, 153–154
    supernova debris and, 271
    table of elements, 256
elevator thought experiment, 166–167
Eliot, T. S., 318, 422
ellipses, 100–102
elliptical galaxies, 31, 33
    formation of, 114–115
    gas in, 113
    swarming systems, 113

Emerson, Ralph Waldo, 423–424
empty space, 168
energy, 253–256. *See also* dark energy; gravitational energy; kinetic energy; nuclear energy; thermal energy
  binding energy, 106–111
  conservation of energy, law of, 254–255, 263
  of electrons, 65, 227–228
  first law of thermodynamics, 254–255
  forms of, 253–254
  frequency and, 38
  Friedmann equation and, 202
  and general relativity, 168–172
  gravitational energy, 254
  law of conservation of, 263
  in light, 67
  mass as, 258
  particle behavior and, 67–70
  second law of thermodynamics, 416–417
  stars, requirements for, 255–256
  thermal energy, 68
  vacuum energy, 212
  of virtual particles, 378
energy density, 319
  critical density, 322
  eras of domination, 321
  of light, 319–320
  omegas, 322–323
  of past, 323
enzymes, 386, 392
equivalence principle, 167
  gravitational event horizon and, 378
eras of domination, 321
escape velocity, 110–111
Escher, M. C., 6, 427, 439, 440
eternal life, 432
Euclidean geometry, 163
  and curvature, 324
Euglena, 386
European Space Agency, Laser Interferometric Space Antenna (LISA), 177, 379
event horizon, 183–186
  geodesics of, 184–185
  inflation and, 377–379
  Schwarzschild radius, 185
events, coordinates of, 194–195
evidence, 12
  in history, 1–2
evolution, 398
  of life, 397–398
  of universe, 423–424
evolutionary psychology, 401
excited states, 226
exotic dark matter, 278–279, 311
  CMB anisotropies and, 367–368

decoupling of, 363
and pressure, 360
expanding universe, 196, 197–207, 426. *See also* big bang theory
  age and relationship to, 293
  consequences of, 213–216
  dark energy and, 202, 210–212
  deceleration parameter, 209–212
  Einstein, Albert and, 198–204
  explosions v. expansions, 294–297
  Friedmann equation, 200–203
  Hubble law and, 208–209
  life, parameters for, 428
  light in, 213–214
  nuclear reactions and, 285
  outer universe, expansion into, 215–216
  recession and, 204–209
  redshifts and, 292
  scale factor, 198
  structure and, 365
experimental origin-of-life science, 390
experiments, 7–8, 9
explosions. *See also* big bang theory
  expansion contrasted with, 294–297
exponential growth, 375
extragalactic distances, 76
extrasolar planets, 406–409
extraterrestrial life, 409–417
extremophiles, 402

F
Faber, Sandra, 83
Faber-Jackson relation, 83, 86
Fahrenheit temperature scale, 69–70
fate of universe, 442
fats, 386
fatty acids, 387
femtometers, 256
Fermi, Enrico, 274
Fermi paradox, 417
Feynman, Richard, 21
fine-tuning, 428–430
finite age, 196
  of universe, 292–293
finite universe, 215
first law of thermodynamics, 254–255
Fisher, Richard, 83
flat space, 163
flat universe, 441–442
  Friedmann equation and, 202
  inflation theory and, 372–373
flow of heat, 68–69
flux, 76. *See also* inverse-square law
  blackbody radiation and, 222
  distance and, 78
  light curves measuring, 82–83
  rate of energy for, 255
focus of ellipse, 100

force. *See also* strong nuclear force; weak nuclear force
  gravitational force, 171
  between particles, 178
  Tollman-Oppenheimer-Volkoff (TOV) equation, 178
Ford, W. Kent, 123
fossils, 397
  from Cambrian explosion, 398
  from Mars, 403–406
fourth dimension. *See* spacetime
francium, 147
Fraunhofer, Joseph von, 42–43
Fraunhofer lines, 42–43
free electrons, 224
  CMB (cosmic microwave background) photons and, 235
  and light, 232
  opacity of, 232
  photons and, 230–234
  and polarization, 242
  recombination, 300
  scattering, 231–232
free fall, 170, 171
freeze out of neutrinos, 334
frequency
  blackbody radiation and, 71–72
  Doppler effect and, 46
  of electromagnetic waves, 39
  energy and, 38
  of light waves, 36–37
  musical frequencies, 46, 350
  in past universe, 292
  pitch and, 46
Friedmann, Aleksandr, 203
Friedmann equation, 200–203
  and density, 319
  energy density in, 321–322
  inflation action and, 371
  omega values and, 338–339
  and recession, 207
frontiers, 415–416
fusion, 258–263
future of life, 414–415

G
GAIA telescope, 55
Galactic Neighborhood, 94
galactic rotation curves, 123–124, 366
galactic winding, 117
galaxies, 29, 31–33, 88–95. *See also* expanding universe; redshifts
  clumps of, 204–205
  dark matter, 123–126
  diameter of, 91–95
  distribution of, 140–142
  dwarf galaxies, 92
  gravitational domination, 334
  habitable part of, 425–426

galaxies, *continued*
    motion of, 111–120
    peculiar velocity, 207
    recession, 204–209
    recession velocity, 138
    rotation curves, 123–124, 366
    spectrum of, 136
    superclusters, 94–95
    thickness of, 90–91
    types of, 31–32
    voids, 141–142
    weighing, 120–122
galaxy clusters
    dark energy, evidence for, 369
    destiny of universe and, 338, 339
    dynamics of, 367
    escape and circular orbital speeds and,
        110–111
    formation of, 115–116
    intracluster medium (ICM), 126–127
    mass of, 122–126
    and Sunyaev-Zel'dovich (SZ) effect,
        234
    walls, 141–142
galaxy survey, 140–142
    big bang theory and, 298–299
    and dark matter, 366–369
    Sloan Digital Sky Survey (SDSS) of,
        140–142
Galilean satellites, 35
Galileo Galilei, 25, 31
    Jupiter, description of, 34–35
Gamow, George, 282
gas. *See also* gas clouds
    as dark matter, 128
    particles in, 67–68
    quasars and, 187
gas clouds, 264
    and birth of stars, 264
    collisions between, 116
    dark matter and, 127–128
    and density wave theory, 119
    in disk galaxies, 113–114
    planetary nebula, 267
Geller, Margaret, 141–142
general relativity, 168–172, 328. *See also*
        black holes
    and big bang theory, 292
    and event horizon, 184
    and expanding universe, 197
    gravitational lensing, 172–176
    gravitational waves, 176–177
    and infinite universe, 342
    and quasars, 186–191
    and recession velocity, 208
    testing, 172–177
    Tollman-Oppenheimer-Volkoff
        (TOV) equation, 178
genes, 390–391
genetic information, 387, 388

genetics, 401
    stellar genetics, 271–273
genomes, 390
geodesics, 164–165
    of asteroids, 169–170
    and event horizon, 184–185
geological history, 4
geometry. *See also* Euclidean geometry
    spatial geometries, 164
glare, 241
Gliese, 406
global curvature, 323–326
globular clusters, 92–95, 425
    age of, 425
    early-universe halos as, 335
    gas in, 113
    mass-to-light ratio of, 121
    and orbital speeds, 111, 122
    swarming of, 112
gluons, 248
gold, 271
    type II supernovae and, 285–286
Gold, Thomas, 312
graphs, reading, 15–16
gravitational disturbances, 358–359
gravitational domination, 334–335
gravitational energy, 254
    kinetic energy compared, 106–107
    thermonuclear fusion and, 259
gravitational lensing, 172–176. *See also*
        microlensing
    and dark matter, 367
    and quasars, 189
    time delays, 87, 188–190
gravitational mass, 167, 171
gravitational microlensing. *See*
        microlensing
gravitational time dilation, 182–183
    and quasars, 189
gravitational waves, 176–177
    CMB polarization from, 379
    and Hawking mechanism, 378–379
    inflation and, 375
gravitons, 177, 248. *See also* gravitational
        waves
gravity, 101–102. *See also* black holes;
        Newtonian gravity
    acceleration and, 165–167, 171
    attractive gravity, 103–105
    as binding energy, 106–111
    and birth of star, 264
    of compact objects, 128
    Doppler wobbles, 406–407
    equivalence principle, 167
    on large scales, 130–131
    life, parameters for, 428
    quantum gravity, 190, 376
    structure and, 365
    Tollman-Oppenheimer-Volkoff
        (TOV) equation, 178

as warping of space, 168–172
greenhouse gas, 403
Gribbin, John, 66
ground state, 226
G-type spectral lines, 75
guitar galaxies, 116
GUT (grand unified theory), 330–331,
        371
    annihilation, 327
    and inflation theory, 375
Guth, Alan, 372

H
hadrons and quarks, 327, 332
half-life, 153
halos, 426. *See also* dark matter halos
    accretion of, 365
    early-universe halos, 335
    formation of, 435
    mergers of, 365
    of Milky Way, 112–113
halo stars, 112
Halverson, N. W., 246
harmonics, 356–357
Hawking, Stephen, 342, 376
Hawking radiation, 377
    inflation and, 378
heat. *See* temperature; thermal energy
heavy elements, 263–273
    ebb and flow of, 273
    and habitable planets, 425
heavy metals and dark matter, 128
heliocentric model of solar system, 428,
        429
helium, 64
    abundance of, 281
    atom, 66
    atomic number of, 256–257
    big bang nucleosynthesis (BBN) and,
        304, 307
    formation of, 435
    fusion of, 262–263
    isotopes of, 150
    and living cells, 388
    nuclei of, 262–263, 284
    origin of, 271–272
    and PP chain, 261
    production of, 272, 282–285
    spectrum of, 43
    in Sun, 149
heritable reproduction, 396, 397
high-mass stars, 267, 269
    death phase of, 270–271
    evolution of, 268
    life-cycle of, 272
    as manufacturers of material world,
        285
    in old age, 269
High-z Supernova Search Team, 337
Hipparcos telescope, 55

history
evidence in, 1–2
geological history, 4
human lifetimes units, 435
interconnectedness and, 432–438
of life, 3–4
milestones in, 327
timeline of cosmos, 436–437
of universe, 3, 326–337
Holzapfel, W. L., 246
hominids, 399, 400–401
homogeneity, 142
departures from, 426
Einstein equation and, 199–200
and infinite universe, 342
inflation creating, 372
of walls and voids, 370
horizon problem, 370–371
inflation theory and, 371–375
horizontally polarized light, 242
hot and cold spots. See CMB (cosmic
microwave background)
hot dark matter, 364
Hoyle, Fred, 312–313
Hubble, Edwin, 31, 92, 137–138, 297
galaxy classification scheme, 34
on redshifts, 204
Hubble constant, 138
and age of universe, 293
CMB measuring redshift of, 362
critical density and, 322
and recession velocity, 205
recession velocity and, 138–139
Hubble diagram, 138–139
for recession of galaxies, 206–207
Hubble law, 87, 138, 290
and expanding universe, 208–209
galaxy survey and, 140–142
and quasars, 186
recession and, 205–207
Hubble Space Telescope (HST), 31
dark matter, map, 368
eclipsing systems, detection of, 407,
408
Hubble Ultra Deep Field, 32, 339–340,
423
Huchra, John, 141–142
Huggins, William, 43
human lifetimes units, 435
human nature, 63–64
humans, 399
hydrogen, 3, 64, 147
absorption lines, 136
abundance of, 281
big bang nucleosynthesis (BBN) and,
304
Bohr atom for, 228
formation of, 435
isotopes of, 150
in living cells, 387

as nuclear fuel, 260
nuclear power in Sun and, 258–260
origin of, 271–272
production of, 282–285
shell of hydrogen fusion, 267
spectrum of, 43
in Sun, 149
in universe, 150
hydrogen-2, 261
hydrostatic equilibrium, 264
hyperbolic orbits, 108
hyperbolic space, 164

I
ICM (intracluster medium), 126–127
Sunyaev-Zel'dovich (SZ) effect,
233
WIMPs (weakly interacting massive
particles) and, 129–130
immortality, 432
inertia, 167, 168
inertial mass, 167, 171
inertial motion, 168
infinite universe, 215, 431
infinity, 6, 342
inflation theory, 330, 369–375, 426. See
also cosmic inflation
earliest moments, 380
and event horizon, 377–379
and multiverse theory, 431
quantum fluctuations and, 374
testable predictions, 374–375
infrared light, 39, 265
planet detection and, 407
initial singularity, 379
instinct, 401
integer fractions, 355
integrated Sachs-Wolfe effect, 359
dark energy, evidence for, 369
intelligent civilizations, 409–415
Drake equation, 409–412
SETI (Search for Extraterrestrial
Intelligence) program, 412–414
intensity, 76
interconnection of universe, 424–426
intermediate-mass black holes (IMBHs),
180–181
interstellar collisions, 285
interstellar distances, 90
intracluster medium (ICM). See ICM
(intracluster medium)
inverse beta decay, 274–275
inverse-square law, 76–81
Newton's law of gravity, 102–103
ionization, 65
ions, 66
in Sun's atmosphere, 229
iron, 64, 270, 286
abundance of, 272
stability of, 270

irregular galaxies, 31, 33
isotopes, 150
identification of, 256–257
in older stars, 150
for radiometric dating, 153–154
isotropic anisotropies, 240
isotropic radiation, 143
isotropy, 142. See also anisotropies
of CMB (cosmic microwave
background), 143, 238–240
Einstein equation and, 199–200
and infinite universe, 342
inflation creating, 372
of walls and voids, 370

J
James, William, 438
joules, 106, 253
energy density in, 319
watts and, 255–256
Jupiter, 406
Galileo, description of, 34–35
moons of, 36
Jurassic era, 4

K
Kelvin temperature scale, 69–70
Kepler, Johannes, 100–102
Kepler mission, 407
Kepler's laws, 100–101, 429, 430
first law, 100
second law, 100–101
kinetic energy, 68, 254
binding energy into, 258
gravitational energy compared,
106–107
Kolb, Edward W., 375
Kopernik, Miklaj (Nicolaus Copernicus),
428, 429
Kovac, J. M., 246
Krauss, Lawrence, 432

L
language, 400
Large Magellanic Cloud (LMC), 92
large-scale structures, 290, 348
Laser Interferometer Gravitational-
Wave Observatory (LIGO), 177,
379
Laser Interferometric Space Antenna
(LISA), 177, 379
last scattering, 235
laws, scientific, 8
lead
neutrinos and, 276
radiometric dating of, 154
Leavitt, Henrietta, 82, 138
Lederman, Leon, 194
Leitch, E. M., 246
Lemaitre, Georges, 290

lepton numbers, 275
  conservation of, 275–276
Lewis, C. S., 135
lies, ability to tell, 400
life, 3–4, 384–421
  anthropic principle, 426–432
  attributes of, 396
  behavior of, 401–402
  cells, 386–389
  definition of, 395–396
  Drake equation, 409–412
  evolution of, 397–398
  history of, 399–401
  nature of, 395–397
  origin of, 389–395
  on other worlds, 402–406
  probability of, 426–427
  survival of, 432
  as emergent property of the universe, 433
light, 30–31. *See also* black holes; CMB (cosmic microwave background); dark matter; luminosity; photons; polarization; quasars; speed of light
  bending light, 173–174, 324–325
  blackbody radiation, 71–72
  blue light, 220
  from cosmological event horizon, 378
  Doppler effect, 46–48
  electromagnetic spectrum, 39–40
  as electromagnetic wave, 38
  energy density of, 319–320
  energy in, 67
  in expanding universe, 213–214
  falling light, 172–173
  Fraunhofer lines, 42–43
  free electrons and, 232
  gravitational waves, 176–177
  heat and, 70–73
  as information source, 149–150
  particles of, 66–67
  properties of, 26–27
  spacetime diagram for, 195
  as time machine, 49–50
light-capture radius, 181–183
light curves, 82–83
  of supernovae, 85
Lightman, Alan, 98
light-minutes, 90
light-seconds, 49, 90
light waves. *See* electromagnetic waves
light-years, 49, 195
Linde, Andrei, 372
line radiation, 223
lines of sight, 50–51
  Olbers's paradox and, 156–157
line spectra, 44, 73–74
  of galaxies, 136
  of Sun, 75

lipids, 386–387
  vesicle structures, 397
liquid water, 403
lithium
  baseline amounts of, 281
  big bang nucleosynthesis (BBN) and, 304, 307
  formation of, 435
  interstellar collisions and, 285
  nucleus size and, 284
  origin of, 271–272
  production of, 283–285
lithium-7, 151–153
  isotopes of, 150
Local Group, 92
locations, coordinates of, 194–195
long axis of ellipse, 100
loop quantum gravity, 329
low-mass stars, 267
  and dark matter, 128–129
  evolution of, 268
  life-cycle of, 272
  as manufacturers of material world, 285
luminosity, 75
  blackbody radiation and, 222
  Faber-Jackson relation and, 83
  in inverse-square law, 76
  mass-to-light ratio, 121
  measurement of, 75–76
  as power, 255
  Stefan-Boltzmann law and, 222
  of Sun, 76, 121, 255–256
  Tully-Fisher relation, 83
luminous matter, 125

M

MACHOs (massive astrophysical compact halo objects), 128, 277, 360. *See also* SMBHs (supermassive black holes)
  microlensing and, 175–176
Magellan, Ferdinand, 92
magnesium, 269
magnetic energy, 187
magnetic field, 241
magnetism, 37–38
  and birth of star, 264
magnetite crystals, 404, 405
main-sequence fitting, 76, 86
main-sequence stars, 76, 264–265
  gravitational pull of, 265
  relative sizes of, 266
major axis of ellipse, 100
mammals, appearance of, 400
manic depression, 401
Mars
  fossils from, 403–406
  future on, 414
mass

and binding energy, 106
black holes and, 376–377
dark matter, 123–126
enclosed by two orbits, 121
as energy, 258
gravitational mass, 167, 171
inertial mass, 167, 171
and Newton's law of gravity, 102–103
oxygen and, 149
of Sun, 261–262
weighing galaxies, 120–122
massive astrophysical compact halo objects (MACHOs). *See* MACHOs (massive astrophysical compact halo objects)
mass number, 150
mass-to-light ratio, 121
material world, manufacturers of, 285
Mathematical Principles of Natural Philosophy (The Principia) (Newton), 102
The Matrix, 11
matter. *See also* dark matter; mass
  acoustic peaks and, 360–361
  bias of universe and, 330–331
  black holes, matter near, 179
  breakdown of, 311
  clumping of, 362–363
  dark energy and, 336–337
  energy density of, 321
  Friedmann equation and, 202
  mystery of, 442
  spacetime diagram for, 195
matter density, 319
  CMB and, 360–362
matter domination, 319–320, 327, 334–335, 359–360, 362, 426, 435
matter particles, 64–66
  star composition and, 70
matter-radiation equality, 363
McMillan, Steve, 376
measurements
  Moon, distance to, 49
  of motion, 57–59
  speed of light and, 48–49
  triangulation, 50–56
  uncertainty in, 11–12
  units of, 13–14
mechanical waves, 36
mechanistic cosmology, 18–20
medium, 37
membranes, 386
Mercury, 406
mercury, spectrum of, 43
mergers of halos, 365
mesons, 332
meteorites, 149–150
  ALH84001, 404–405
  and biomolecules, 391
  Murchison meteorite, 391–392

radiometric dating of, 153–154
methane, 403
microfossils, 405
microlensing, 175–176
  planet detection by, 407–408
microwaves, 142–147. *See also*
     CMB (cosmic microwave
     background)
  listening for, 350–351
middle age of stars, 265–266
Milky Way, 29, 30, 92
  edge-on view of, 93
  Galileo Galilei and, 31
  halo stars, 112
  infrared light, 39
  MACHOs (massive astrophysical
    compact halo objects) in, 128
  motions of, 112–113
  supermassive black hole (SMBH) in,
    180
Miller, Stanley, 390
Miller-Urey experiment, 390–391
minutes of arc, 55–56
molecules, 64–65
  atoms and, 248
MOND (MOdified Newtonian
    Dynamics), 130–131
Moon
  angular size of, 29
  distance to, 49
  and general relativity, 168–169
  movement of, 30
  parallax for measuring distance to, 55
  size of, 29, 88
Moravec, Hans, 415
motion. *See also* kinetic energy; proper
    motion
  of galaxies, 111–120
  measuring, 57–59
  of Milky Way, 112–113
  of stars, 111–120
mRNA (messenger RNA), 392
M-theory, 329
multicellular organisms, 397
multiverse theory, 431
Murchison meteorite, 391–392
musical frequencies, 46, 350
mutation, 400

N
nanobacteria, 405
nanobes, 405
nanometers, 38
Narlikar, Jayant V., 312–313
NASA
  Chandra X-ray Observatory, 127
  Kepler mission, 407
  Laser Interferometric Space Antenna
    (LISA), 177, 379
  Mars exploration, 403

Pioneer and *Voyager* spacecraft, 413–414
  PlanetQuest site, 408
natural satellites, 35
natural selection, 397–398
  mutation and, 400
nature, 63–64
  v. nurture controversy, 401
nebulae, 29, 30
  Crab Nebula, 271, 272
  planetary nebula, 267, 269
negative curvature, 164
neon, 387
  and living cells, 388
  spectrum of, 43
Neptune, 90
neutral atoms, 66
neutralinos, 280
neutrino glow, 277
neutrinos, 130, 248, 274, 286, 327
  annihilation process and, 333
  CMB measuring density, 362
  in dark matter, 278
  freeze out of, 334
  and Hawking mechanism, 378
  and lead, 276
  lepton number for, 275
  liberation of, 333–334
  and Sun, 276–277, 278
  weak nuclear force, 276–277
neutrion telescopes, 277–278
neutrons, 65, 147
  in big bang nucleosynthesis (BBN),
    304
  in early universe, 282
  in mass number, 150
  mass of, 304
  quarks in, 332
  radiative capture, 282
  temperature and, 304
neutron stars, 128, 271
Newton, Isaac, 101–102, 190
Newtonian gravity, 102–106, 430
  and falling light, 172–173
  MOND (MOdified Newtonian
    Dynamics), 130–131
  and motion of galaxies, 111–112
Newtonian physics, 427
Newton's constant, 102
NGC 2467 region, 110
night sky, 26–28
  Olbers's paradox, 155–157
nitrogen, 270
  interstellar collisions and, 285
  in living cells, 387
  production of, 272
  in universe, 150
nitrogenous bases, 390
  nucleic acids and, 387
nonstatic universe, 203
normal dark matter, 278–279

nuclear energy, 106
  binding energy, 257–258
  in Sun, 259–260
nuclear fission, 263
nuclear fusion, 257
  baryons and, 260
  birth of star and, 264
  in early universe, 282
  gravitational analogue of, 258–259
  iron and, 270–271
  and life of star, 265
  and low-mass stars, 267
  mass and, 258
  PP chain, 260–261
nuclear process, 149
nuclear reactions, 256–259
  effects of, 150
  rules for, 263
nuclear reorganization, 148–149
nuclei
  of atom, 65
  baryons and, 260
  binding energy in, 258
  binding of, 149
  formation of, 435
  of helium, 262–263
  ionization and, 65
  size of, 149, 256
nucleic acids, 387, 392
nucleosynthesis. *See* big bang
    nucleosynthesis (BBN)
nucleotides, 387
  bonding arrangement for, 387–388
number density, 319

O
observable universe, 196, 213, 430
  big bang theory and, 309
  dark energy and, 215
observations, 7, 9
  CMB observations, numbers
    measured by, 361–362
  of extrasolar planets, 408
  Gallileo Galilei and, 25
  of night sky, 27
Occam's razor, 405
oils, 386
Olbers, Heinrich Wilhelm, 156–157
Olbers's paradox, 155–157, 291
  big bang theory and, 309, 310
  and cosmological horizon, 196
old age of stars, 266–270
omegas, 319–323
  and curvature of universe, 325–326
  supernova research and, 337–338
omens, 17–18
onion-skin model, 269
opacity
  and blackbody radiation, 221
  of free electrons, 232

open universe, 432, 442
orbitals, 225, 227
orbits, 99–106. *See also* spiral arms
  bound and unbound orbits, 108
  cause of, 106
  circular orbit velocity, 110
  escape velocity, 110–111
  globular clusters, orbital speeds of, 111, 122
  gravity and, 105
  velocity of objects in, 120
ordered resurfacing, 417
order of magnitude, 15, 95
origin-of-life theory, 410
Orion the Hunter, 26, 29
  HST image of, 30
outcomes of universe, 433–434
oxygen, 3, 64, 147
  interstellar collisions and, 285
  life and, 400
  in living cells, 387
  and mass, 149
  production of, 272
  in Sun, 149
  in universe, 150
ozone, 400

P

parabolic orbits, 108
parallax, 50–56, 86
  defined, 51
  GAIA telescope measuring, 55
  Hipparcos telescope measuring, 55
  spectroscopic parallax, 76
  of stars, 54–56
  uses for, 73
parent stars, 272
parsecs, 54, 55
particles. *See also* matter particles; quarks; WIMPs (weakly interacting massive particles)
  average particle speed, 68
  baryons, 260
  and beta decay, 273–276
  dark matter as, 129–130
  and energy, 67–70
  forces between, 178
  in infinite universe, 431
  inflation and, 378
  of light, 66–67
  photons, interactions with, 248
  scattering, 231–233
  of Sun, 178
  virtual particles, 212, 376
Pauli, Wolfgang, 273–274
peak frequency, 222
peculiar velocity, 207
Peebles, P. J. E., 290, 350
peer review, 9
Penzias, Arno, 142

perception, 11
periodic tables, 147–148
period-luminosity graphs, 82–83
Perlmutter, Saul, 337
Permian era, 4
personal cosmology, 16–17
  science and, 21
perspective and mechanistic cosmology, 19
philosophical beliefs, 9–10
  life, theories of, 405–406
phosphates, 387
phosphorus, 270, 387
photodisintegration, 271
photoelectric effect, 67
photons, 67, 435. *See also* black holes; CMB (cosmic microwave background); decoupling; electrons; inverse-square law; scattering
  atomic absorption, 229
  baryon-to-photon ratio, 306
  blackbody radiation, 71–72
  and bound electrons, 223–230
  charged particles and, 248
  and clumping of matter, 362–363
  creation of, 70
  energy of, 228
  flux, 76
  and free electrons, 230–234
  gravitational waves and, 379
  and Hawking mechanism, 378
  interactions among, 70
  liberation of, 335
  in past universe, 292
  star composition and, 70
  Sunyaev-Zel'dovich (SZ) effect, 233–234
photosynthesis, 400
physical size, 87
Pioneer spacecraft, 413–414
pitch, 46
Planck era, 341, 379, 435
Planck length, 329
Planck time, 327, 329
  flatness problem and, 373
planetary nebula, 267, 269
PlanetQuest site, NASA, 408
planets, 30, 33–35. *See also* orbits
  electrons, analogy between, 224–225
  extrasolar planets, 406–409
  orbital speeds of, 111
plants, appearance of, 400
plasma, 66
platinum, 271
Plato, 10
polarization, 241–247, 300
  big bang theory and, 300–301
  CMB polarization, 245–247, 379
  free electrons and, 242

pattern, 247
  reionization and, 302
  scattering and, 243–244
polycyclic aromatic hydrocarbons, 404
polymerization, 387
  problem, 394
polymers, 387
positive curvature, 163–164
positrons, 275
  annihilation of, 334
  lepton number for, 275
potassium, radiometric dating of, 154
power, 255. *See also* CMB power spectrum
  CMB power spectrum, 355–356
  for stars, 257
  of Sun, 255–256
PP chain, 260–261
  inverse beta decay and, 274–275
present day Earth, 44
pressure
  baryonic matter building, 360
  and birth of star, 264
primordial black holes (PBHs), 377
primordial harmonics, 349–357
primordial plasma density, 356
primordial soup, 332
  and life, 389
principles, 9–10
prisms, 41
probability, 426–427
proof, 10
proper motion
  for planet detection, 407
  transverse velocity, measuring, 58–59
proteins, 386
  building of, 392
  and DNA (deoxyribonucleic acid), 392–394
  polymerization of, 387
protons, 65, 147
  atomic number, 256
  in early universe, 282
  in mass number, 150
  mass of, 304
  PP chain, 260–261
  quarks in, 332
protostar, 264–265
Proxima Centauri, 75, 90
Ptolemaeus, Claudius (Ptolemy), 428–430
Ptolemaic model, 428–430
pulsars, *Pioneer 10* plaque showing, 413–414
pulsation of stars, 82
Pyrke, C., 246

Q

quantized energy, 225

quantum energy states, 227–228
quantum fluctuations, 374, 376–380, 426
quantum foam, 328, 379
quantum gravity, 190, 329, 376
quantum mechanics, 225, 328
quantum physics
  Bohr model and, 379
  Hawking radiation, 377
  vacuum energy and, 212
  validation of, 376
quark-hadron transition, 327, 332
quarks, 248, 332
  hadron transition, 327, 332
  strange stars and, 178
quasars, 186–191
  absorption of light from, 236
  abundances, measurement of, 308
  CMB (cosmic microwave background) light and, 236–237
  gravitational lensing and, 189
  pulses of light from, 213
  SMBHs (supermassive black holes) and, 187–188
quasi-steady-state theory, 312–313
quasi-stellar objects (QSOs), 186
quintessence, 371
  dark energy as, 340–341

R

radar, 49, 86
radial marathon image, 297
radial velocity, 57–59
  Doppler effect, measuring with, 58
radiant energy, 255
radiation. See also blackbody radiation
  black holes and, 376–377
  CMB measuring density, 362
  electromagnetic radiation, 40
  energy density of, 321
  Hawing radiation, 377
  isotropic radiation, 143
  line radiation, 223
  natural radiation, 153
radiation domination, 319–320, 327, 362, 426, 435
  scale factor and, 379
radiative capture, 282
radioactivity
  beta decay and, 273–274
  and Miller-Urey experiment, 391
radio astronomy, 142
radiocarbon dating, 153
radio light, 39
radiometric dating, 153–154
rainbows, 41
  spectra of, 44
Ram Dass, 439
randomness, 6
rare Earth hypothesis, 402–403

recession, 204–209, 290. See also recession velocity
recession velocity, 138, 205–207
  measurement of, 207–209
recombination, 300, 327
  timing of, 426
red dwarfs, 265
  as dark matter, 128–129
red giants, 266
redshifts, 46, 136–140. See also Hubble law
  and big bang theory, 297, 311
  and CMB (cosmic microwave background), 302, 362
  destiny of universe and, 337–339
  and expanding universe, 292
  measuring, 210–212
  nuclear reactions and, 285
  of quasar light, 236–238
  recession and, 204–209
  recession velocity and, 207–209
  Sachs-Wolfe effect, 359
  spacetime bending and, 183
  spectral lines and, 230
redshift surveys, 140–142
  big bang theory and, 298–299
  quasars in, 188
red supergiant, 267
reionization, 301, 302, 327
  CMB measuring redshift of, 362
relative-distance indicators, 87–88
relativity. See also general relativity; theory of relativity
  special relativity, 258, 328
relevance of science, 422–423
relic neutrino background (RNB), 334
religious tradition, 440
reproduction
  behaviors and, 401
  and life, 396, 397
repulsion, 258–259, 262
Riess, Adam, 338
"Rise of the Robots" (Moravec), 415
RNA (ribonucleic acid), 387
  depiction of, 388
  nitrogenous bases and, 390
  world hypothesis, 394–395
robotics, 415
rocks, radiometric dating of, 153–154
rope waves, 36–37
rotation. See also rotation curves
  differential rotation, 117
rotation curves, 122–126
  galactic rotation curves, 123–124, 366
rubber-sheet analogy, 169–170
rubidium, radiometric dating of, 154
Rubin, Vera, 123
rules of universe, 5

S

Sachs-Wolfe effect, 359
saddle-shaped universe, 441–442
Sagan, Carl, 252
  maxim of, 405–406
satellites, 35
scale factor, 198, 213
  and curvature of spacetime, 199
  Friedmann equation and, 201
  in radiation-dominated universe, 379
scale invariance, 374
scattering, 231–233
  last scattering, 235
  polarization and, 243–244
  Sunyaev-Zel'dovich (SZ) effect, 233–234
  unpolarized light and, 243
schizophrenia, 401
Schmidt, Brian, 337
Schneider, Nicholas, 121
Schramm, David, 194
Schrödinger, Erwin, 1
Schwarzschild, Karl, 185
Schwarzschild radius, 185
science
  defined, 7–9
  key elements of, 9–10
  laws of, 8
  relevance of, 422–423
  testing theories, 10–12
  thinking and, 12–13
Scientific American "Rise of the Robots" article, 415
scientific notation, 14–15
scientific unification, 190
second law of thermodynamics, 416–417
seconds of arc, 55–56
self-image, 17
semimajor axis of ellipse, 101
senses and reality, 231
sequences, 147–148
SETI (Search for Extraterrestrial Intelligence) program, 412–414
sexual selection, 398
Shakespeare, William, 441
shape
  double-helix shape, 390–391, 393
  saddle-shape, 441–442
  of universe, 341–342
Shapley, Harlow, 92–95
shell of hydrogen fusion, 267
shock waves and gas clouds, 264
Shu, Frank, 396, 412
sight, 30
silicon, 270
Silk, Joseph, 357
silk damping, 357
single-celled organisms, 397, 399
singularity, 184

SI units, 13–14. *See also* joules
watts, 255–256
61 Cygni, 57
size, 63
   and age, 424–425
   apparent size, 87
   of compact objects, 128
   physical size, 87
   of skies, 95–97
sky. *See* night sky
Slipher, Vesto, 137–138
Sloan Digital Sky Survey (SDSS),
   140–142, 380
Small Magellanic Cloud (SMC), 92
SMBHs (supermassive black holes),
   180–181
   and quasars, 187–188
   radiation of, 377
   Schwarzschild radius of, 185
soccer ball universe, 343
sodium, 270
   absorption spectrum of, 45
   emission spectrum of, 45
   excited sodium atoms, 229
   spectrum of, 43
sodium lamps, 229
solar constant, 255
solar eclipse of 1919, 173–174
solar fusion, 262
solar mass, 121
   and core, 271
solar neighborhood, 91
solar panels, 67
solar system
   elliptical orbits of, 101
   heliocentric model of, 428, 429
   Ptolemaic model, 428–430
   rotation curve, 123
sound, 349
   defined, 350
sound horizon, 356–357
sound waves, 349–350
   of different frequencies, 354
   Doppler effect and, 46
   harmonics, 356–357
   on large scales, 358–359
   primordial sound waves, 351
   silk damping, 357
space. *See also* curved space
   empty space, 168
   general relativity and, 171
   hyperbolic space, 164
   notions of, 162
   and relativity, 168–172
   three-dimensional curved space, 164
   warping of, 168–172
spacetime, 194–196
   big bang theory and, 299
   and black holes, 182–183
   curvature of, 199

diagrams, 195
   fluctuations in, 370
   Friedmann equation, 200–203
   inflation theory and, 374
   nuclear reactions and, 285
special relativity, 258, 328
spectral classes, 74
spectral lines, 229–230
spectroscopes, 41–42, 48
spectroscopic parallax, 76
spectroscopy, 40–46, 229–230
spectrum, 41, 135. *See also* line spectra
   absorption spectrum, 44–45
   of blackbody radiation, 220
   blackbody spectrum, 72, 146, 300–301
   causes of, 45
   CMB spectrum, 144
   common stellar spectra, 74
   emission spectrum, 44–45
   of familiar elements, 43
   of galaxy, 136
   Sun, light from, 42, 44
speed of light, 38, 48
   global curvature and, 324
   redshifts and, 137
   theory of special relativity and, 258
spherical-Earth theory, 10–11
spiral arms, 112–113
   density wave theory, 118–119
   development of, 116–120
spiral galaxies, 31, 32–33
   differential rotation, 117
   disks of, 113–114
   formation of, 114–115
   standard rulers for, 87
Spitzer Space Telescope, 407
sponges, 397
squarks, 280
standard candles, 73–85, 75
   Cepheid variable stars, 81–83
   Faber-Jackson relation, 83
   spectroscopic parallax, 76
   supernovae as, 84–85
   Tully-Fisher relation, 83
standard particle physics, 280
standard rulers, 87–88
Stapledon, Olaf, 63, 441
starch, 386
star clusters, 30
   ages of, 154
Starkman, Glenn, 432
The Starry Messenger (Galileo), 31,
   34–35
stars, 26–28
   birth of, 264–265
   blackbody radiation and, 71–72
   first impressions of, 28–30
   life cycles of, 439
   middle age of, 265–266
   motion of, 111–120

old age of, 266–270
   parallax of, 54–56
   patterns among, 29
   velocities, components of, 57
static universe, 196, 370–371
   cosmological constant and, 202
steady-state theory, 312
Stefan-Boltzmann constant, 222
Stefan-Boltzmann law, 222
Steinhardt, Paul, 341, 372
stellar evolution, 266–270
stellar genetics, 271–273
stellar mass black holes, 180
stellar nucleosynthesis, 282
strange stars, 178
string theory, 329, 341, 379
stromatolites, 400
strong anthropic principle, 430
strong nuclear force, 257
   baryons and, 260
   life, parameters for, 428
structure, 348. *See also* inflation
   theory
   CDM simulation, 364–366
   dark energy, evidence for, 369
   evolution of, 367
   formation, 348
   increasing complexity of, 385
   inflation theory and, 374
   for life, 427
   quantum origin of, 376
   seeds of, 426
   simulating formation of, 363–369
sugars, 386
   nucleic acids and, 387
sulfur, 387
Sun. *See also* orbits
   antineutrinos and, 276
   atmosphere of, 229
   bent light and, 173
   blackbody radiation from, 229
   distance to, 88–89
   elements of, 149
   energy of, 70–71
   Fraunhofer lines, 42–43
   gases on surface of, 67–68
   line spectrum of, 75
   luminosity of, 76, 121, 255–256
   mass, loss of, 261–262
   neutrinos and, 275–278
   nuclear power in, 258–260
   particles of, 178
   photon life in, 70–71
   positrons and, 275
   power of, 255–256
   size of, 89
   spectrum of, 42, 44
   thermal energy and, 68
   thermonuclear fusion in, 259–263
Sunyaev, Rashid, 233–234

Sunyaev-Zel'dovich (SZ) effect, 87, 233–234
superclusters, 94–95
    as large-scale structure, 348
Super Kamiokande neutrino observatory, 278
supermassive black holes. *See* SMBHs (supermassive black holes)
supernovae, 86, 270–271. *See also* type Ia supernovae; type II supernovae
    dark energy and, 210–211, 337–338, 368
    light curves, 85
    neutrinos, production of, 277
    as standard candles, 84–85
superstring theory, 329
supersymmetry theory, 280
survival
    behaviors and, 401
    of the fittest, 397
    of life, 432
swarming stellar orbits, 112
Swimme, Brian, 384
symmetry breaking, 330

T

table of elements, 256
technological civilization, 402, 417
Tegmark, Max, 431
telescopes, 30. *See also* Chandra x-ray telescope; Hubble Space Telescope (HST)
    GAIA telescope, 55
    Hipparcos telescope, 55
    neutrion telescopes, 277–278
    Spitzer Space Telescope, 407
temperature, 68
    absolute zero temperature, 68–69
    in big bang nucleosynthesis (BBN), 303
    and blackbody radiation, 220–221
    blackbody spectrum and, 72
    of CMB (cosmic microwave background), 143, 145–147, 237, 349–350
    and color, 223
    on Earth, 145
    and electroweak unification, 331–332
    flow of heat, 68–69
    in history, 435
    light and heat, 70–73
    and neutrons, 304
    in past universe, 292
    scales, 68–70
Terrestrial Planet Finder (TPF) mission, 408
theories, 7–8
    exceptions to, 11
    testing, 10–12

theory of relativity, 166. *See also* general relativity
    event horizon and, 184–186
    special relativity, 258, 328
thermal energy, 68, 254
    and birth of star, 264
    blackbody radiation, 71–72
thermodynamics
    first law of, 254–255
    second law of, 416–417
thermonuclear fusion, 259–263
    natural occurrence of, 302–303
Thonnard, Norbert, 123
three-dimensional curved space, 164
3-D universe, 341
time
    gravitational time dilation, 182–183
    notions of, 162
    and relativity, 168–172
    and universe, 341
timeline of cosmos, 436–437
time zero, 313, 341
titanium, 270
TOE (theory of everything), 329
Tollman-Oppenheimer-Volkoff (TOV) equation, 178
topology of space, 342
total energy, 107
tracers, 126
transit events, 407
translation process, 392
transparency
    of bound electrons, 232
    CMB (cosmic microwave background) and, 235–236
transverse velocity, 57–59
    proper motion, measuring by, 58–59
triangulation, 50–56
    in astronomy, 53–56
Triassic era, 4
tRNA (transfer RNA), 392
Tully, Brent, 83
Tully-Fisher relation, 83, 86
Turner, Frederick Jackson, 415
Turner, Michael S., 375
Turok, Neil, 341
two-dimensional surfaces, 164
type Ia supernovae, 84, 267
    dark energy, evidence for, 368
    elements produced by, 272
    as manufacturers of material world, 285
    and Sunyaev-Zel'dovich (SZ) effect, 234
type II supernovae, 84, 271
    and gold, 285–286
    as manufacturers of material world, 285
Type Sa galaxies, 32
Type Sb galaxies, 32

U

UGC 10214 Galaxy, 125
ultraviolet light, 39, 266
    CMB polarization and, 301
    life and, 400
    and Miller-Urey experiment, 391
uncertainty, 8
    in measurements, 11–12
units of measurement, 13–14
universe. *See also* early universe; expanding universe; observable universe
    age of, 403
    appearance of, 2–3
    center of, 296
    composition of, 2, 150
    consciousness in, 416
    cyclic universe, 341
    density of, 318–326
    destiny of, 337–341
    evolution of, 423–424
    finite age of, 196
    future of, 4–5
    history of, 3, 326–337
    immensity of, 423
    as interconnected system, 424–426
    nonstatic universe, 203
    rules of, 5
    scale of, 430–431
    shape of, 341–342
unpolarized light, 243
up-scatters, 233
uranium, radiometric dating of, 154
Urey, Harold, 390

V

vacuum energy, 212, 320, 376
variable stars, 81–83
    light curve for, 83
Vega, spectrum of, 74
velocity, 36, 57–59, 207. *See also* recession velocity
    circular orbit velocity, 110–111
    escape velocity, 110–111
    explosions and, 295
    orbit, object in, 120
    radial velocity, 57–59
    transverse velocity, 57–58
Venus, 55
vertically polarized light, 242
vesicles, 395–397
Virgo Supercluster, 94
virtual particles, 212, 376
    energy of, 378
visible light, 39
voids, 141–142, 426
    big bang theory and, 299
    homogeneity/isotropy of, 370
    as large-scale structure, 348
    single, gigantic void, 142

Voit, Mark, 121
volume, density and, 319–320
Voyager spacecraft, 413–414

W

walking, 400
walls, 426
    of galaxy clusters, 141–142
    homogeneity/isotropy of, 370
    as large-scale structure, 348
warm-hot intergalactic medium
        (WHIM), 127–128
warping, 181–183
    gravitational lensing and, 174
water on Mars, 403
watts, 255
    joules and, 255–256
wavelengths, 37, 48
    blueshift, 46
    measurement of, 38
    recession velocity and, 208
    redshift, 46
waves. *See also* electromagnetic waves
    blackbody spectra and, 72
    gravitational waves, 176–177
    and particles of light, 66–67
wave speed, 36

W bosons, 248
weak anthropic principle, 430
weakly interacting massive particles
        (WIMPs). *See* WIMPs (weakly
        interacting massive particles)
weak nuclear force
    interaction, 303
    and neutrinos, 276–277
weightlessness, 170
Weinberg, Steven, 441
Wheeler, John A., 162
WHIM (warm-hot intergalactic
        medium), 127–128
white dwarfs, 128, 267, 269
    ages of, 154
    supporting forces, 178
Wien law, 222
Wilkinson, David Todd, 245
William of Occam, 405, 406
Wilson, E. O., 64
Wilson, Robert, 142
WIMPs (weakly interacting massive
        particles), 129–130, 212, 276–277,
        286, 311
    acoustic peaks and, 360–361
    annihilation process and, 333
    anti-WIMPS, 279–280

dark matter and, 277–280
detection of, 279–280
WMAP (Wilkinson Microwave Isotropy
        Probe), 238–240, 245
    astronomical ages and, 308
    of CMB anisotropies, 354
    map of, 353
worldlines, 195–196

X

x-rays, 39
    intracluster medium (ICM),
        126–127
    and lead, 276

Y

Yu, Jer, 350

Z

Z bosons, 248
Zel'dovich, Yakov, 143, 233–234
zero curvature, 164
zero degrees Celsius, 69
zero distance between galaxies, 293
zero scale factor, 379
Zubrin, Robert, 415–416